Neuroimmunology of Sleep

S. R. Pandi-Perumal
Daniel P. Cardinali
George P. Chrousos

Neuroimmunology of Sleep

S. R. Pandi-Perumal
Mount Sinai School of Medicine
New York, NY
USA

Daniel P. Cardinali
University of Buenow Aires
Buenos Aires
Argentina

George P. Chrousos
Athens University Medical School
Athens
Greece

Library of Congress Control Number: 2006940757

ISBN-10: 0-387-69144-8   e-ISBN-10: 0-387-69146-4
ISBN-13: 978-0-387-69144-2   e-ISBN-13: 978-0-387-69146-6

Printed on acid-free paper.

9 8 7 6 5 4 3 2 1

springer.com

Contents

Preface

Both sleep and wakefulness, when immoderate, are detrimental

– Hippocrates (489–360 BC)

The volume *Neuroimmunology of Sleep* covers the topic of sleep and immunity, an area of intense medical and scientific interest. Its subject matter is very complex and touches all facets of our health and well-being. The field started with an astute observation by Hippocrates, that sick patients felt sleepy and withdrawn while sick. This knowledge was passed down many generations of physicians for over 2500 years; however, it was only recently that a strong bidirectional interaction between the brain and the immune system was revealed. Thus, a substance traditionally connected with the immune system, such as eicosanoids and cytokines, was discovered to play major roles in regulating several brain functions, including sleep, whiles the brain itself, including the sleep centers, was shown to influence that activity of the immune system.

The chapters of *Neuroimmunology of Sleep* have been written by experts in the field who have performed primary research in the areas they covered and this is a major plus of the book.

In its first section, the book contains three introductory chapters that cover the complex interactions between the brain and the immune system, the biological neuroendocrine rhythms, and the circadian organization of the immune response.

In the second section of the book, the basic science of the field is briefly reviewed in six chapters. These chapters cover the basics of the biological clock and its interactions with products of inflammation, the roles of VIP and prolactin as immune mediators with effects on sleep regulation, the immune pathways in the brain that influence sleep, the dynamic relations between ageing, sleep and immunity, the neuroimmune correlates of sleep, and the still mysterious interaction between REM sleep and immune function.

The third and largest section of the book focuses on clinical science. Eleven chapters cover the panorama of studies on human sleep and sleep disturbances as they relate to neuroendocrine and neuroimmune functions. The first chapter provides a brief overview on the translational research in the field. The second covers the area on the effects of sleep deprivation on the immune response, the third reviews the existing data in pregnancy, while the fourth chapter reviews that data of an ingenious human experimental model and how it has expanded our understanding of the effects of immune products on sleep. The fifth chapter provides an overview of inflammation and sleep, while the remaining six chapters are disease-specific, covering sleep apnea in adults and children, traumatic brain injuries, depression, chronic fatigue syndrome, and narcolepsy and cataplexy.

The volume *Neuroimmunology of Sleep* is not exhaustive or complete. It is, however, current and will hopefully be useful to sleep researchers, both basic and clinical, as well as to neuropharmacologists, psychiatrists, and physicians who evaluate and treat sleep disorders, including internists, pediatricians, and general practitioners. In addition, this book may be useful to immunologists, pharmacologists, pharmacists, medical students, and clinicians of various disciplines who want to get an overall grasp of the field.

It is our hope that we have succeeded in producing a useful book. As usual, we welcome communications from our readers concerning our volume.

S.R. Pandi-Perumal
D.P. Cardinali
George Chrousos

Acknowledgments

Many individuals played instrumental roles in the development and completion of this new volume entitled *Neuroimmunology of Sleep*. This book provides an introduction to the ever-expanding interface between sleep and immunity. An enterprise of this sort is challenging, and the editors received a lot of help from several people. We are delighted to acknowledge some of these here.

We were fortunate to experience a warm, professional, and highly enthusiastic support from Kathleen P. Lyons, Editorial Director, Biomedicine, Springer, USA. Her commitment to excellence was a strong guiding force throughout the development of this volume. The wonderfully talented people at Springer group made this project a pleasurable one. In particular, we wish to acknowledge the help of Claire B. Wynperle, Editorial Assistant, Springer, USA.

A very special appreciation is also owed to several reviewers who made numerous helpful suggestions. Their candid comments and insights were invaluable.

Finally, we express our gratitude to our families for their patience and support.

S.R. Pandi-Perumal
D.P. Cardinali
George Chrousos

Contributors

M. Bentivoglio
Department of Morphological and
Biomedical Sciences, University of
Verona Medical Faculty, Strada Le
Grazie 8, 37134 Verona, Italy

J.E. Blalock
University of Alabama at Birmingham,
Physiology and Biophysics University,
1918 University Blvd., MCLM 898,
Birmingham, AL 35294-0005, USA

Daniel P. Cardinali
Departmento de Fisiologia, Facultad de
Medicina, Universidad de Buenos Aires,
1121 Buenos Aires, Argentina

O.S. Cajulis
Dental Group of Sherman Oaks, 4910
Van Nuys Blvd., # 210 Sherman Oaks,
CA 91403, USA

Neil S. Cherniack
UMDNJ-New Jersey Medical School,
Newark, NJ, USA
Pain & Fatigue Study Center,
1618 ADMC, 30 Bergen St., Newark,
NJ 07101, USA

Francesco Chiappelli
UCLA School of Dentistry, Los Angeles,
CA 90095-1668, USA

G.P. Chrousos
First Department of Pediatrics,
Athens University Medical School,
Aghia Sophia Children's Hospital,
115 27 Athens, Greece

M.E. Coussons-Read
University of Colorado at Denver and
Health Sciences Center, CB 173, POB
173364, Denver, CO 80217-3364, USA

Y. Dauvilliers
Centre de Référence National sur la
Narcolepsie, Service de Neurologie,
Hôpital Gui-de-Chauliac, INSERM
E0361, 80 avenue Augustin Fliche,
34295 Montpellier cedex 5, France

J.E. Dimsdale
Department of Psychiatry, Behavioral
Medicine Program, University of
California, San Diego, UCSD Medical
Center, 200 West Arbor Drive, San
Diego, CA 92103-0804, USA

R. Drucker-Colin
Instituto de Fisiologia Celular,
Universidad Nacional Autonoma de
Mexico, A.P. 70250, C.P. 04510,
Mexico City, Mexico

E. Esqueda-León
Depto. Biología de la Reproducción,
U.A.M.Iztapalapa, San Rafael Atlixco
186, Col. Leyes de Reforma, Del.
Iztapalapa, CP 09340, Mexico DF

A.I. Esquifino
Departmento de Bioquimica y Biologia
Molecular III, Facultad de Medicina,
Universidad Complutense, 28040
Madrid, Spain

F. Garcia-Garcia
Instituto de la Salud, Universidad
Veracruzana, Av. Luis Castelazo-Ayala
s/n Industrial-Animas, Xalapa,
Ver. 91190, Mexico

R.P. Gaykema
Program in Sensory and Systems
Neuroscience, Department of
Psychology, University of Virginia,
Charlottesville, VA 22904, USA

L.E. Goehler
Laboratory of Neuroimmunology &
Behavior, Program in Sensory and
Systems Neuroscience, Department of
Psychology & Neuroscience Graduate
Program, University of Virginia, 102
Gilmer Hall, P.O. Box 400400,
Charlottesville, VA 22904-4400, USA

Laura-Yvette Gorczynski
Departments of Surgery & Immunology,
University Health Network, MaRS
Centre, Toronto Medical Discovery
Tower, 101 College Street, 2nd Floor,
Rm. 805, Toronto, Ontario, Canada
M5G1L7
Toronto Hospital, Toronto, Ontario,
Canada

R. Gorczynski
Departments of Surgery & Immunology,
University Health Network, MaRS
Centre, Toronto Medical Discovery,
Tower,101 College Street, 2nd Floor,
Rm. 805, Toronto, Ontario, Canada
M5G1L7
Toronto Hospital, Toronto, Ontario,
Canada

M. Haack
Department of Neurology, Beth Israel
Deaconess Medical Center and Harvard
Medical School, East Campus, Dana 779,
Boston, MA 02215, USA

M. Irwin
Cousins Center for
Psychoneuroimmunology, Semel
Institute for Neuroscience and Human
Behavior - David Geffen School of
Medicine at UCLA, Mail Code 707624,
300 Medical Plaza, Suite 3109,
Los Angeles, CA 90095-7076, USA

K. Jhaveri
Department of Pharmacology, Southern
Illinois University School of Medicine,
801 North Rutledge Street, Box 19616,
Springfield, IL 62794-9616, USA

A. Jiménez-Anguiano
Depto. Biología de la Reproducción,
U.A.M.Iztapalapa, San Rafael Atlixco
186, Col. Leyes de Reforma, Del.
Iztapalapa CP 09340, Mexico DF

Krister Kristensson
Division of Neurodegenerative Disease
Research, Department of Neuroscience,
Retzius vag 8, B2:5, Karolinska
Institutet, SE-171 77 Stockholm, Sweden

Tetsuya Kushikata
Department of Anesthesiology,
University of Hirosaki School of
Medicine, Zaifu 5, Hirosaki 036-8563,
Japan

Georges J.M. Maestroni
Center for Experimental Pathology,
Cantonal Institute of Pathology, Locarno,
Switzerland

P.J. Mills
Department of Psychiatry, Behavioral
Medicine Program, University of
California, San Diego, UCSD Medical
Center, 200 West Arbor Drive,
San Diego, CA 92103-0804, USA

D.R. Moradi
UCLA School of Dentistry, Los Angeles,
CA 90095-1668, USA

S.J. Motivala
Cousins Center for Psychoimmunology,
UCLA Samel Institute for Neuroscience
and Human Behavior, Los Angeles, CA,
USA

Janet Mullington
Department of Neurology, Beth Israel
Deaconess Medical Center, 330
Brookline Ave., Boston, MA 02115,
USA

Benjamin H. Natelson
UMDNJ-New Jersey Medical School,
Newark, NJ, USA Pain &
Fatigue Study Center; 1618 ADMC,
30 Bergen St., Newark, NJ 07101,
USA

A.M. Navarro
UCLA School of Dentistry, Los Angeles,
CA 90095-1668, USA

Luigi Nespoli
Pediatric Clinic Insubria University, Via
F. del Ponte 2 21100 Varese, Italy

Luana Nosetti
Pediatric Clinic Insubria University, Via
F. del Ponte 2 21100 Varese, Italy

Mikael Nygård
Division of Neurodegenerative Disease
Research, Department of Neuroscience,
Retzius vag 8, B2:5, Karolinska
Institutet, SE-171 77 Stockholm, Sweden

M.L. Okun
Western Psychiatric Institute and Clinic,
University of Pittsburgh School of
Medicine, Department of Psychiatry,
TDH E1122 Pittsburgh, USA

T. Olivares-Banuelos
Instituto de Fisiologia Celular,
Universidad Nacional Autonoma de
Mexico, A.P. 70250, C.P. 04510, Mexico
City, Mexico

Beatriz Duarte Palma
Department of Psychobiology-
Universidade Federal de São Paulo,
Escola Paulista de Medicina
(UNIFESP/EPM), Rua Napoleão de
Barros, 925, Vila Clementino-SP-04024-
002, São Paulo, Brazil

Maria Palomba
Department Morphological and
Biomedical Sciences, University of
Verona Medical Faculty,
Strada Le Grazie 8, 37134 Verona, Italy

S.R. Pandi-Perumal
Comprehensive Center for Sleep
Medicine, Department of Pulmonary,
Critical Care, and Sleep Medicine,
Mount Sinai School of Medicine,
1176 - 5th Avenue, 6th Floor, New York,
NY 10029, USA

T. Pollmächer
Zentrum für Psychiatrie und
Psychotherapie Klinikum Ingolstadt,
Krumenauerstrasse 25, 85021 Ingolstadt,
Denmark

P. Prolo
Division of Oral Biology and Medicine,
Psychoneuroimmunology, UCLA School
of Dentistry,
CHS 63-090, Los Angeles, CA 90095-
1668, USA

A. Quintanar Stephano
Dept. Fisiología y Farmacología, Centro
de Ciencias Básicas, Universidad
Autónoma de Aguascalientes,
Aguascalientes, México

Contributors

J.A. Rojas-Zamorano
Depto. Biología de la Reproducción,
U.A.M.Iztapalapa, San Rafael
Atlixco 186, Col. Leyes de Reforma,
Del. Iztapalapa CP 09340, Mexico DF

Andreas Schuld
Zentrum für psychische Gesundheit,
Klinikum Ingolstadt Atlixco 186,
Col. Leyes de Krumenauerstrasse 25,
85021 Ingolstadt, Denmark

Deborah Suchecki
Department of Psychobiology-
Universidade Federal de São Paulo,
Escola Paulista de Medicina
(UNIFESP/EPM), Rua Napoleão de
Barros, 925, Vila Clementino-SP-04024-
002, São Paulo, Brazil

Judith Szelenyi
Institute of Experimental Medicine of
the Hungarian Academy of Sciences,
H-1083 Budapest, Szigonz str. 43,
Budapest, Hungary

A.N. Taylor
Department of Neurobiology, David
Geffen School of Medicine, University
of California at Los Angeles, 10833 Le
Conte Avenue, Los Angeles, CA 90095,
USA

Ender Terzioglu
Departments of Surgery & Immunology,
University Health Network, MaRS
Centre, Toronto Medical Discovery
Tower, 101 College Street, 2nd Floor,
Rm. 805, Toronto, Ontario, Canada
M5G1L7
Toronto Hospital, Toronto, Ontario,
Canada

Fumiharu Togo
Department of Neurosciences, UMDNJ-
New Jersey Medical School, Newark,
NJ, USA

Linda Toth
Department of Pharmacology, Southern
Illinois University School of Medicine,
801 North Rutledge Street, Box 19616,
Springfield, IL 62794-9616, USA

R. Trammell
Department of Pharmacology, Southern
Illinois University School of Medicine,
801 North Rutledge Street, Box 19616,
Springfield, IL 62794-9616, USA

S. Tufik
Department of Psychobiology-
Universidade Federal de São Paulo,
Escola Paulista de Medicina
(UNIFESP/EPM), Rua Napoleão de
Barros, 925, Vila Clementino-SP-04024-
002, São Paulo, Brazil

J. Velazquez-Moctezuma
Depto. Biología de la Reproducción,
U.A.M.Iztapalapa, San Rafael Atlixco
186, Col. Leyes de Reforma, Del.
Iztapalapa CP 09340, Mexico DF

Sylvester E. Vizi
Institute of Experimental Medicine of
the Hungarian Academy of Sciences,
H-1083 Budapest, Szigonz str. 43,
Budapest, Hungary

Thierry Waelli
Departments of Surgery & Immunology,
University Health Network, MaRS
Centre, Toronto Medical Discovery
Tower, 101 College Street, 2nd Floor,
Rm. 805, Toronto, Ontario, Canada M5G
1L7
Toronto Hospital, Toronto, Ontario,
Canada

D.A. Weigent
University of Alabama at Birmingham,
Physiology and Biophysics, 1918
University Blvd., MCLM 894,
Birmingham, AL 35294-0005, USA

T. Yasuda
Department of Anesthesiology,
University of Hirosaki School of
Medicine, Zaifu 5, Hirosaki 036-8563,
Japan

Hitoshi Yoshida
Department of Anesthesiology,
University of Hirosaki School of
Medicine, Zaifu 5, Hirosaki 036-8563,
Japan

M.G. Ziegler
University of California, San Diego
Medical Center, Department of
Medicine, 200 West Arbor Drive, San
Diego, CA 92103-8341, USA

Part I
General Concepts

1 Bidirectional Communication Between the Brain and the Immune System

Douglas A. Weigent and J. Edwin Blalock

1.1 Introduction

The idea that a pathway of interaction existed between the brain and the immune system has been suggested for many years. However, only the evidence gathered over the past 25 years has provided important details about the mechanisms. The clues for bidirectional communication early on related to the effects of stress on immune function and that psychological factors could alter the onset and course of autoimmune disease (Ader 1996). A list of selected findings contributing to our understanding is shown in Table 1.1. The classical (Pavlovian) conditioning of host defense mechanisms and antigen-specific immune responses was first suggested and studied in the 1920s (Metal'nikov and Chorine 1926) and later in the 1970s (Ader and Cohen 1975).

A number of years ago, stress was shown to increase the susceptibility of mice to virus infection (Rasmussen, Marsh, and Brill 1957) and lesioning studies of the anterior hypothalamus were shown to be associated with diminished immune reactivity (Cross, Markesbery, Brooks, and Roszman 1980; Stein, Schiavi, and Camerino 1976). Reconstitution of antibody production in hormonally deficient mice was observed after treatment of animals with somatotrophic hormone, thyrotropic hormone, and thyroxin (Pierpaoli, Baroni, Fabris, and Sorkin 1969).

Studies identifying neural receptors on cells of the immune system, and identifying nerve fibers in compartments of lymphoid organs all supported the idea of a dynamic interaction between the immune and nervous system (Besedovsky, delRey, Sorkin, DaPrada, and Keller 1979; Hadden, Hadden, and Middleton 1970; Hadden, Hadden, Middleton, and Good 1971). Later, it was shown that nerve fibers are localized in precise compartments of both primary and secondary lymphoid organs in close contact with T lymphocytes and macrophages (Felten et al. 1987; Williams, Peterson, Shea, Schmedtje, Bauer, and Felten 1981).

In addition, it was shown that the immune system as a consequence of antigenic challenge altered the firing rate of hypothalamic neurons suggesting the bidirectional nature of communication between these two systems (Besedovsky, Sorkin, Keller, and Miller 1977). Supernatant fluids from activated leukocytes could mimic this phenomenon (Besedovsky, delRey, and Sorkin 1981), and it is now clear that a wide range of lymphocyte products influence the synthesis and secretion or release of neuroendocrine hormones and neurotransmitters (Blalock and Smith 1980; Smith and Blalock 1981; Weigent and Blalock 1995).

TABLE 1.1. Major selected discoveries in bidirectional communication.

Principal finding	References
Classical conditioning of the immune response	Metal'nikov and Chorine 1926
Stress increases the susceptibility of mice to virus infection	Rasmussen, Jr. et al. 1957
Lymphocytes have adrenergic receptors that influence immunity	Hadden et al. 1970; Hadden et al. 1971
Reconstitution of antibody production in hormonally deficient mice by somatotropic hormone, thyrotropic hormone, and thyroxin	Pierpaoli et al. 1969
The immune system is subject to classical conditioning	Ader and Cohen 1975
Nervous system is capable of responding to a signal emitted during an immune response	Besedovsky et al. 1977
Lesioning of the anterior hypothalamus is associated with diminished immune reactivity	Stein et al. 1976; Cross et al. 1980
Products of activated cells of the immune system can affect endocrine responses under CNS control	Besedovsky et al. 1981
Lymphocytes were discovered to be a source of pituitary hormones	Blalock 1994; Blalock and Smith 1980; Smith and Blalock 1981
The lymphokine interleukin-1 was demonstrated to induce ACTH production suggesting communication between the CNS and immune system was bidirectional	Woloski 1985
Nerve fibers are localized in precise compartments of both primary and secondary lymphoid organs in close contact with T lymphocytes and macrophages	Williams et al. 1981; Felten et al. 1987
A variety of psychosocial events interpreted as being stressful are capable of influencing immunity	Glaser et al. 1987
Alteration of sleep by infection	Toth and Krueger 1988

Our studies reviewed, in part, below, initially showed that cells of the immune system could be a source of pituitary hormones and neurotransmitters and that immune-derived cytokines could function as hormones and hypothalamic-releasing factors (Weigent and Blalock 1995; Woloski, Smith, Meyer, Fuller, and Blalock 1985). Thus, it now seems clear that the proposition put forward over 20 years ago that the immune system serves as a sensory organ acting as a "sixth" sense (Blalock

1984) is essentially true. What seems clear is that two pathways link the immune system with the brain. One is the autonomic nervous system activity and the other neuroendocrine outflow from the pituitary. The molecules released are recognized by the immune system via specific cell-surface receptors. In addition, it is clear that activation of the immune system is accompanied by changes in autonomic, hypothalamic, and endocrine processes as well as by changes in behavior (Glaser et al. 1987).

Cytokines influence the activation of the hypothalamic-pituitary-adrenal axis and are influenced by glucocorticoids, depending on their concentration (Berkenbosch, Van Oers, Del Rey, Tilders, and Besedovsky 1987). Thus, the exchange of information between the immune system and the brain is bidirectional (Fig. 1.1). The immune system contributes to the brain's function by serving a sensory role while the brain adopts an immune function by participating in and/or coordinating the immune response.

Collectively, this interaction adds another level of complexity to understanding the influences of behavior including sleep (Toth and Krueger 1988) on immunity and vice versa. This chapter will review studies that describe the pathways for bidirectional communication between the brain and the immune system (Fig. 1.1).

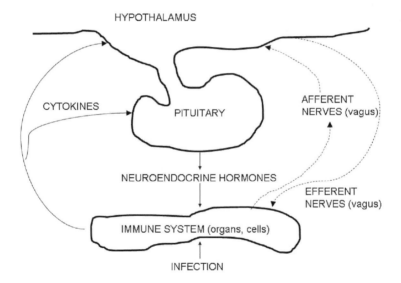

FIGURE 1.1. Neural and hormonal pathways of bidirectional communication.

1.2 Nervous System Communication to the Immune System

The endocrine and autonomic systems exert their effects by secreting molecules that interact with lymphocytes and macrophages (Wrona 2006). The biologically active molecules secreted by the nervous system interact at specific receptors on cells of the immune system (Table 1.2; see also Blalock and Costa (1990), Carr (1991), Clarke and Bost (1989), DeSouza (1993), Gala and Shevach (1993), Harbour, Leon, Keating, and Hughes (1990), Harbour and Hughes (1991), Johnson, Downs, and Pontzer (1992), O'Neal, Schwarz, and Yu-Lee (1991), and Roupas and Herington (1989)).

The interaction may affect a wide range of immune cell activities, including cytokine and antibody production, cell proliferation, and lytic activity and migration (for recent review see Wrona (2006)). Norepinephrine (NE)-induced inhibition of natural killer (NK) cytotoxicity has been shown to occur at several levels, including an influence on NK cell receptor ligation to target cells, NK cell cytokine secretion, and inhibition of the cytotoxic mechanisms in NK cells (Gan, Zhang, Solomon, and Bonavida 2002). Catecholamines have been shown to enhance (Dhabhar and McEwen 1999) or suppress (Dobbs, Vasquez, Glaser, and Sheridan 1993) immune cell function. Thus, catecholamines can enhance the expression of cell-surface differentiation antigens (Singh 1985) and inhibit macrophage-mediated lysis of tumor cells and complement activation (Koff and Dunnegan 1986). Recently, NE and dopamine (DA) were shown to increase lymphocyte activation with augmented T-helper 1 (Th1) and T-helper 2 (Th2)-type cytokine production while the action of NE with dexamethasone resulted in immunosuppression (Torres, Antonelli, Souza, Teixeira, Dutra, and Gollob 2005). Neuropeptides have also been shown to be present in nerve terminals in primary and secondary lymphoid organs and influence immune cell function (Bellinger, Lorton, Romano, Olschowka, Felten, and Felten 1990). For example, in activated macrophages, vasoactive intestinal peptide (VIP) and pituitary adenylate cyclase activating polypeptide (PACAP) inhibit the expression of proinflammatory cytokines through effects on the nuclear translocation of transcription-factors-like nuclear factor kappa beta (NF-κB) (Ganea, Rodriquez, and Delgado 2003).

They also promote the survival of Th2 effectors (Delgado, Gonzalez-Rey, and Ganea 2004). Substance P (SP) facilitates lymphocyte migration, proliferation, production of immunoglobulin-A (IgA), and phagocytosis (Feistritzer et al. 2003). Recently, it has been shown that the expression and function of full-length and truncated receptors for SP in acute monocytic leukemia cells (THP-1) demonstrate unique signaling pathways between undifferentiated and differentiated cells and ligand interaction (Lai et al. 2006).

There are also data that support a role for the parasympathetic nervous system in influencing the immune system. Both muscarinic and nicotinic acetylcholine (Ach) receptors have been detected on T lymphocytes and macrophages (Tracey 2002). The cholinergic anti-inflammatory pathway, via the vagus nerve, may control the inflammatory response by inhibiting cytokine release from macrophages bearing nicotinic Ach receptors (Pavlov, Wang, Czura, Friedman, and Tracey 2003). Thus, since lymphoid tissues are innervated with sympathetic and parasympathetic fibers and contain specific receptors on their cell surfaces, it seems reasonable to perceive a mechanism by which the nervous system may communicate or regulate specific aspects of the immune response.

TABLE 1.2. Receptors for neuropeptides hormones on cells of the immune system.

Receptor	Binding cell type	References
Adrenocorticotropin	Rat spleen T and B cells	Johnson et al. 1992
β-Endorphin	Spleen	Carr 1991
Thyrotropin	Neutrophils, monocytes, B cells	Harbour 1990
Prolactin	Human and mouse T and B cells	O'Neal et al. 1991; Gala and Shevach 1993
Growth hormone	Rat, mouse, human PBL, spleen, thymus	Suzuki et al. 1990; Roupas and Herington 1989
Growth hormone releasing hormone	Rat spleen, human PBL	Guarcello et al. 1991
Corticotropin releasing factor	Human PBL	De Souza 1993
Thyrotropin releasing hormone	T cell line	Harbour and Hughes 1991
Luteinizing hormone releasing hormone	Rat thymocytes	Blalock and Costa 1990
Arginine vasopressin		Johnson et al. 1992
Somatostatin	Human PBL	Hiruma et al. 1990
Substance P	Mouse T and B cells	Pascual et al. 1991
Calcitonin gene-related peptide	Rat T and B cells	McGillis et al. 1991
Vasoactive intestinal peptide	T cells	Ottaway and Greenberg 1984

1.3 Neuroendocrine Hormone Influence on the Immune System

The immune system, in addition to autonomic nervous activity, is influenced by hormones released by the neuroendocrine system (Table 1.3; see also Brooks (1990), Carr (1992), Clevenger, Sillman, and Prystowsky (1990), Clevenger, Sillman, Hanley-Hyde, and Prystowsky (1992), Foster, Mandak, Kromer, and Rot (1992),

Jain et al. (1991), Johnson, Farrar, and Torres (1982a), Johnson, Smith, Torres, and Blalock (1982b), Johnson, Torres, Smith, Dion, and Blalock (1984), Kelley (1989), Kruger, Smith, Harbour, and Blalock (1989), Matera, Cesano, Bellone, and Oberholtzer (1992), Mathew, Cook, Blum, Metwali, Felman, and Weinstock (1992), McGillis, Humphreys, and Reid (1991), Ottaway, Bernaerts, Chang, and Greenberg (1983), Pascual, Xu-Amano, Kiyono, McGhee, and Bost (1991), Provinciali, Di Stefano, and Fabris (1992), Rouabhia, Chakir, and Deschaux (1991), Smith, Hughes, Hashemi, and Stefano (1992), Stanisz, Befus, and Bienenstock (1986), and Zelazowski, Dohler, Stepien, and Pawlikowski (1989)). A great deal of evidence exists to support both the presence of receptors for neuroendocrine hormones on cells of the immune system as well as the ability of these hormones to modulate specific functions of the various immune cell types (Tables 1.2 and 1.3; see also Weigent and Blalock (1995)).

TABLE 1.3. Modulation of immune responses by neuropeptides.

Hormone	Modulation	References
Corticotropin	Antibody synthesis	Johnson et al. 1982a, 1982b
	IFN-γ production	Johnson et al. 1984
	B-lymphocyte growth	Brooks 1990
Endorphins	Antibody synthesis	Johnson et al. 1982a, 1982b
	Mitogenesis	Carr 1992
	NK activity	
Thyrotropin	Antibody synthesis	Johnson et al. 1992; Kruger et al. 1989
	Comitogenic with ConA	Provinciali et al. 1992
GH	Cytotoxic T cells	Kelley 1989
	Mitogenesis	
LH and FSH	Proliferation	Provinciali et al. 1992
	Cytokine	Rouabhia et al. 1991
PRL	Comitogenic with ConA	Clevenger et al. 1990, 1992;
	Induces IL2 receptors	Matera et al. 1992
CRF	IL1 production	Jain et al. 1991;
	Enhance NK	Smith et al. 1992
	Immunosuppressive	
TRH	Antibody synthesis	Harbour et al. 1990; Kruger et al. 1989
GHRH	Stimulate chemotaxis	Guarcello et al. 1991;
	Inhibit NK activity	Zelazowski et al. 1989
	Inhibit chemotactic response	
Substance P	Stimulate chemotaxis	Stanisz et al. 1986;
	Stimulate proliferation	Pascual et al. 1991
	Modulate cytokine levels	
VIP	Inhibit proliferation	Ottaway et al. 1983; Mathew et al. 1992
AVP	T cell helper	Johnson et al. 1982a, 1982b
	Functions for IFN-γ production	
SOM	Inhibit proliferation	Stanisz et al. 1986
	Reduces IFN-γ production	
CGRP	B cell differentiation	Foster et al. 1992
	T cell chemotaxis	

The receptors for corticotropin (ACTH) and endorphins have been identified on cells of the immune system as well as the ability of these hormones to modulate many aspects of immune reactivity. The binding of ACTH initiates a signal transduction pathway that involves both cAMP and mobilization of Ca^{2+} (Clarke and Bost 1989). Analysis of the effects of ACTH by patch-clamp methods suggests that this hormone can modulate macrophage functions through the activation of Ca^{2+}-dependent K^+ channels (Fukushima, Ichinose, Shingai, and Sawada 2001). ACTH has been shown to suppress major histocompatibility complex (MHC) class II expression, stimulate NK cell activity, suppress interferon-gamma (IFN) production, modulate IL-2 production, and function as a late-acting B cell growth factor (Brooks 1990; Johnson et al. 1982a, 1982b). The production of opioid peptides in immune cells (Smith 2003), and lymphocyte receptors for the opioid peptides share many of the features, including size, sequence, immunogenicity, and intracellular signaling as those described on neuronal tissue. Many aspects of immunity are modulated by the opiate peptides including (1) enhancement of the natural cytotoxicity of lymphocytes and macrophages toward tumor cells, (2) enhancement or inhibition of T cell mitogenesis, (3) enhancement of T cell rosetting, (4) stimulation of human peripheral blood mononuclear cells, and (5) inhibition of major histocompatibility class II antigen expression (Carr 1992; Johnson et al. 1982b).

The anti-inflammatory influences of α-melanocyte-stimulating hormone (α-MSH) and other melanocortins are primarily exerted through inhibition of inflammatory mediator production and cell migration (Luger, Scholzen, Brzoska, and Bohm 2003). These effects occur through binding of melanocortins to melanocortin receptors on cells of the immune system (Lipton and Catania 2003). The in vitro and in vivo inhibitory effects of α-MSH influence adhesion, production of cytokines and other mediators of inflammation, including IL-1, IL-6, IL-10, tumor necrosis factor (TNF)-α, chemokines, and nitric oxide (Luger et al. 2003). The effects of α-MSH, on inflammatory mediator production is thought to occur through the participation of G-proteins, the JAK kinase signal transducer activator of transcription (JAK/STAT) pathway, and inhibition of the activation of the nuclear factor NF-κB (Lipton and Catania 2003).

It has also been shown that cells of the immune system contain receptors for growth hormone (GH) and prolactin (PRL) and that these hormones are potent modulators of the immune response (Gala 1991). A systematic survey of PRL receptor expression by flow cytometry showed that PRL receptors are universally expressed in normal hematopoietic tissues with some differences in density, which could be increased by concanavalin (Con)A treatment. GH receptors from a number of species have been sequenced and GH binding and cellular processing of the GH receptor have been studied in a cell line of immune origin. In the IM-9 cell line, it has been shown that GH stimulates proliferation and that the GH receptor can be down-regulated by phorbol esters (Suzuki, Suzuki, Saito, Ikebuchi, and Terao 1990). A role for GH in immunoregulation has been demonstrated in vitro for numerous immune functions, including stimulation of deoxyribonucleic acid (DNA) and ribonucleic acid (RNA) synthesis in the spleen and thymus. GH also affects hematopoiesis by stimulating neutrophil differentiation, augmenting erythropoiesis, increasing proliferation of bone marrow cells, and influences thymic development (Gala 1991). GH affects the functional activity of cytolytic cells, including T lymphocytes and NK cells (Kelley 1989). GH has also been shown to stimulate the production of superoxide anion formation from macrophages (Edwards, Ghiasuddin, Schepper, Yunger, and Kelley 1988). It is not clear whether GH directly influences intrathymic or extrathymic development or acts indirectly by augmenting the synthesis of thymulin or insulin-like growth factor-1 (IGF-I). There is a vast literature on the ability of IGF-1 to exert immunomodulatory effects (Kooijman, Hooghe-Peters, and Hooghe 1996). Although most reports suggest IGF-1 is proinflammatory (Renier, Clement, Desfaits, and Lambert 1996), it also may exert anti-inflammatory actions via stimulation of IL-10 production in activated T cells (Kooijman and Coppens 2004). The observations suggest that GH may stimulate local production of IGF-I, which acts to promote tissue growth and action in a paracrine fashion. Likewise, PRL can have modulating effects on the immune system (Gala 1991). PRL is involved in the activation of many immunological responses, including stimulating the activity of Th1 and Th2 lymphocytes that alter antitumor cytotoxicity and autoimmunity (DeBellis, Bizzaro, Pivonello, Lombardi, and Bellastella 2005). PRL stimulated the release of IL-1 by macrophages and abolished the stress-induced inhibition of proliferation of peripheral blood lymphocytes (Fomicheva, Nemirovich-Danchenko, and Korneva 2004). Data show that suppression of PRL secretion in mice with bromocriptine increases the lethality of a *Listeria* challenge and abrogates T cell-dependent activation of macrophages (Bernton, Bryant, Holaday, and Dave 1991). Antibodies to the PRL receptor have been shown to abolish PRL-induced proliferation of Nb2 cells (Clevenger et al. 1992). Other studies suggest that PRL may promote survival of certain lymphocyte subsets, modulate the naive B cell repertoire, and promote antigen-presenting functions (Matera, Mori, and Galetto 2001).

1.4 Actions of Hypothalamic-Releasing Hormones

In addition to pituitary hormones, hypothalamic-releasing hormone receptors and their effects have been documented on cells of the immune system (Tables 1.2 and 1.3). Corticotropin-releasing hormone (CRH) inhibits lymphocyte proliferation and NK cell activity (Jain et al. 1991; Smith et al. 1992). The GH-releasing hormone (GHRH) receptor has also been identified on cells of the immune system. The GHRH receptor binding sites are saturable and are found on both thymocytes and splenic lymphocytes (Guarcello, Weigent, and Blalock 1991). Other in vitro findings suggest GHRH may inhibit NK cell activity and chemotaxis and increase IFN-γ secretion (Guarcello et al. 1991; Zelazowski et al. 1989). Recently, GHRH was shown to modulate IL-6 secretion from human peripheral blood mononuclear cells without any significant effect on IL-8 secretion (Siejka, Stepien, Lawnicka, Krupinski, Komorowski, and Stepien 2005). In addition, leukocytes have been shown to respond to thyrotropin-releasing hormone (TRH) treatment by producing thyrotropin (TSH) mRNA and protein (Kruger et al. 1989). Recent work has shown the presence of two receptor types for TRH on T cells. One of these appears to be the classical TRH receptor and is involved in the release of IFN from T cells (Harbour et al. 1990). TRH at very low concentrations enhances the in vitro plaque-forming cell response via production of TSH (Kruger et al. 1989). In this instance, T cells were shown to produce TSH while other studies suggest dendritic cells and monocytes may also produce biologically active TSH. Interestingly, T lymphocytes cultured with T3 and T4, but not TSH nor TRH, showed enhanced apoptosis with reduced expression of Bcl-2 protein (Wang and Klein 2001). The existence of distinct subsets of somatostatin (SOM) receptors on the Jurkat line of human leukemic T cells and U266 IgG-producing human myeloma cells has also been described (Hiruma et al. 1990). Two subsets of receptors may account for the biphasic concentration-dependent nature of the effects of SOM in some systems. Although GH and PRL have immunoenhancing capabilities, SOM has potent inhibitory effects on immune responses. SOM has been shown to significantly inhibit Molt-4 lymphoblast proliferation and phytohemagglutinin (PHA) stimulation of human T lymphocytes and nanomolar concentrations are able to inhibit the proliferation of both spleen-derived and Peyer's patch-derived lymphocytes (Stanisz et al. 1986). Other immune responses, such as superantigen-stimulated IFN secretion, endotoxin-induced leukocytosin, and colony-stimulating activity release, are also inhibited by SOM (Stanisz et al. 1986).

1.5 Immune System Communication with the Nervous System Hormones

There is now convincing evidence to show that the immune system can communicate with the central nervous system (CNS) (Wrona 2006). The invasion of the body by microorganisms activates cells of the immune system, which then releases a complex variety of soluble mediators called cytokines, which include interleukins, interferons, and tumor necrosis factors. These substances modulate the immune response but also influence the brain. This immune-to-brain pathway of communication triggers what we know as the "sickness response." It is this effect which is discussed below and which constitutes the second arm of bidirectional communication.

Cytokines released after infection produce many effects on behavior, including decreased feeding, sexual activity, and pleasure seeking behaviors. In addition, sleep is increased and cognitive function is impaired (Banks, Farr, and Morley 2002). Several mechanisms have been proposed by which blood-borne cytokines convey information to the CNS (Dinarello 2004). These include cytokine acting at the circumventricular organ (CVO), acting at afferent nerves including the vagal, or altering the permeability of the blood–brain barrier (BBB) to immune cells or another substance. In addition to these indirect mechanisms of effects on cytokines on the brain, cytokine transport across the BBB has also been described. The list of cytokine transporters includes IL-1, IL-6, tumor necrosis factor-α, and the IL-1 receptor antagonist (Banks et al. 2002). The cytokine transporters are usually distinguishable from cytokine receptors and interestingly show differences in regional uptake in the brain (Banks, Moinuddin, and Morley 2006). It appears that in the case of fever, exogenous pyrogens (bacterial products and foreign antigens), and endogenous pyrogens (cytokines) from the circulation bind to Toll-like receptors (TLR) and cytokine receptors in the CVO, respectively. Activation of TLR and/or cytokine receptors induces cyclooxygenase-2, which results in the synthesis of prostaglandin-E2 (PGE2) on the brain side of the CVO (Dinarello 2004). The increases in PGE2 stimulate the release of cAMP and other neurotransmitters triggering neurons in the thermoregulatory center as well as initiating behavioral responses. Interestingly, fever in autoimmune diseases appears to be mostly cytokine-mediated whereas fever during infection may be both cytokine- and TLR-mediated (Dinarello 2004).

1.6 Cytokine Influences on the Nervous System

The effects of cytokines and the presence of their receptors in the neuroendocrine system is currently the topic of much research. Of particular interest have been IL-1, IL-2, IL-6, and TNF-α, with particular emphasis on IL-1. A common pattern after immune activation is an increase in ACTH secretion and suppression of TSH release while the pattern for GH, PRL, and LH is less consistent (McCann et al. 1994).

The earliest report with IL-1 showed that it was able to stimulate the release of ACTH (Woloski et al. 1985). Several pathways appear to mediate the influence of IL-1 on neuroendocrine neurons and depend on the route of administration and on whether the cytokine acts at the level of the release and/or the biosynthesis of CRF (Rivest 1995). The increase in CRF mRNA appears to be dependent upon IL-1 in the CNS because central administration of the IL-1 receptor antagonist completely blocked the expression of CRF transcripts in the periventricular nucleus (Kakucska, Qi, Clark, and Lechan 1993). It has also been reported that IL-1 receptors can be demonstrated on the pituitary gland and that CRH can up-regulate IL-1 receptors on AtT-20 (Webster, Tracey, and DeSouza 1991). A finding by us has revealed that low levels of exogenous or endogenous CRH can sensitize the pituitary gland to the direct releasing activity of IL-1 (Payne, Weigent, and Blalock 1994). Therefore, one can conclude that IL-1 functions as a neuromodulator in the hypothalamus to enhance CRH release into the hypophyseal portal blood and that both IL-1 and CRH can sensitize the corticotroph, thus facilitating the release of ACTH.

The evidence suggests that both IL-1 and CRH activate corticotrophs, but they elicit different patterns in the regulation of POMC (Ruzicka and Huda 1995). Thus, IL-1 evoked an early release of β-lipotropin and an intermediate release of β-endorphin while CRH caused an early β-endorphin secretory response. Such a distinct pattern allows the pituitary to be specifically activated and therefore determine the interaction with the immune system. In addition to its effects on the hypothalamic-thyroid axis and the hypothalamic-gonadal axis. Thus, IL-1 inhibits the ovarian steroid-induced LH surge and release of hypothalamic luteinizing hormone releasing hormone (LHRH) in rats (Kalra, Sahu, and Kalra 1990). It also decreases plasma thyroid hormone and TSH levels in rats, probably by suppressing hypothalamic TRH secretion (Dubuis, Dayer, Siegrist-Kaiser, and Burger 1988).

IL-2 is the most potent regulator of pituitary ACTH secretion and is more active than the classical hypothalamic regulator, CRH (Karanth and McCann 1991). In rat pituitary cell cultures at low concentrations, IL-2 elevated ACTH, PRL, and TSH release and inhibited the release of follicle-stimulating hormone (FSH), luteinizing hormone (LH), and GH from hemipituitaries in vitro (Karanth and McCann 1991). It appears that both IL-2 and IL-6, in addition to their effects on hormone secretion, may participate in anterior pituitary cell growth regulation. Both cytokines were found to stimulate the growth of the GH_3 cell line and inhibit the proliferation of normal rat anterior pituitary cells (Arzt, Stelzer, Renner, Lange, Muller, and Stalla 1992). It seems clear that some of the effects of IL-2 may occur directly at the level of the hypothalamus. Thus, IL-2 stimulates the release of GHRH from medial hypothalamic fragments and can stimulate the release of SOM (Karanth, Aguila, and McCann 1993). IL-2 given centrally induces a somewhat different pattern of response than the other cytokines because it stimulated instead of inhibited TSH release and also inhibited GH release, which was stimulated by low dosages of IL-1 and cachectin. As in the case of IL-1, IL-2 inhibited LH release but it also inhibited FSH release. Thus, this cytokine has powerful actions at the hypothalamic level to alter pituitary hormone release as well as act directly on the pituitary. The intracerebral ventricular injection of IL-6 results in an increase in plasma ACTH along with elevated temperature and a decrease in TSH (Spangelo and MacLeod 1990). Another group reported that IL-6 could stimulate the in vitro release of PRL, GH, and LH by dispersed pituitary cells (Bernton, Bryant, and Holaday 1987). The stimulation of PRL release by TRH could be enhanced by IL-6, indicating these hormones may have different intracellular mechanisms to stimulate hormone release (Spangelo, Judd, Isakson, and MacLeod 1989). The in vivo effect of IL-6 could be blocked by the prior administration of antibody against CRH, thus demonstrating that IL-6 stimulates ACTH secretion through the production of CRH (Naitoh et al. 1988).

IFN-γ also has been shown to influence the secretory activity of anterior pituitary cells in culture. In vivo injection of IFN-γ stimulated ACTH, with no effect on PRL and a delayed inhibition of GH and TSH release (Gonzalez, Riedel, Rettori, Yu, and McCann 1990). IFN-γ at physiological concentrations has been shown to inhibit stimulated secretion of ACTH, PRL, and GH of pituitary cells cultured in vitro and stimulated with hypothalamic factors (Vankelecom et al. 1990). These results indicate that IFN-γ may modulate GH secretion from the pituitary gland by both a direct suppressive effect at the level of the pituitary and indirect hypothalamic suppression involving stimulation of SOM release (Gonzalez, Aguila, and McCann 1991).

TNF-α is a cytokine produced by activated macrophages or monocytes that is important in the hormonal response to shock (Kelley, Arkins, and Li 1992). It has been shown that after only 1 h of incubation, TNF was capable of stimulating the release of ACTH, GH, TSH, and PRL from either overnight-cultured dispersed pituitary cells or hemipituitaries; however, the dose for these actions of TNF was 100-fold greater with dispersed cells than with hemipituitaries (Milenkovic, Rettori, Snyder, Reutler, and McCann 1989). In vivo, TNF-α has been reported to stimulate the release of ACTH, PRL, and GH similar to what has been observed with IL-1 (Bernardini et al. 1990). The stimulatory effect on ACTH release was completely inhibited by previous injection of CRH antiserum, suggesting that endogenous CRH serves as a mediator of the response.

In addition to these effects, TNF-α has been suggested to inhibit the hypothalamic-pituitary-thyroid axis at multiple levels (Dubuis et al. 1988) and the hypothalamic-pituitary-gonadal axis (Gaillard, Turnill, Sappino, and Muller 1990). Several recent studies also indicate that cytokines may synergize in the CNS. A form of motivation known as social investigation was used to demonstrate synergy between centrally injected (i.c.v.) IL-1 and TNF-α (Laye et al. 1995). IL-6 and its soluble receptor, when injected i.c.v., have been shown to interact in a way that potentiates fever and anorexia (Schobitz et al. 1995). Thus, the biological activity of cytokines may be dependent on the presence (or absence) of soluble receptors, which may exhibit either agonistic or antagonistic activity.

1.7 Neuroendocrine Hormone Release by Cells
of the Immune System

There is now substantial evidence that cells of the immune system produce neuroendocrine hormones. This was first established for ACTH and subsequently for TSH, GH, PRL, LH, FSH and the hypothalamic hormones SOM, CRH, GHRH, and LHRH (Weigent and Blalock 1995). The evidence supports the idea that neuroendocrine peptides and neurotransmitters, endogenous to the immune system, are used for both intraimmune system regulation, as well as for bidirectional communication between the immune and neuroendocrine systems. Although the structures of these immune-cell-derived peptides are, for the most part, identical to those identified in the neuroendocrine system, both similarities and differences exist in the mechanism of synthesis to the patterns previously described in the neuroendocrine system.

At least two possibilities exist concerning the potential function of these peptide hormones produced by the immune system. First they act on their classic neuroendocrine target tissues. Second, they may serve as endogenous regulators of the immune system. A number of investigators have now been able to demonstrate that such regulation is endogenous to the immune system. Specifically, TSH is a pituitary hormone that can be produced by lymphocytes in response to TRH and, like TSH, TRH enhanced the in vitro antibody response (Kruger et al. 1989). This was the first demonstration that a neuroendocrine hormone (TSH) can function as an endogenous regulator within the immune system. A large number of human hematopoietic cell lines and tumors synthesize and release PRL (Montgomery 2001). The synthesis and secretion of 23 kDa PRL from cells of the immune system is well established, although size heterogeneity is evident (11 to 60 kDa) (Kooijman and Gerlo 2002). Most forms appear to exhibit some biological activity in proliferative assays (Montgomery, Shen, Ulrich, Steiner, Parrish, and Zukoski 1992). The evidence suggests a low constitutive level of PRL expression inducible by IL-2 and inhibited by dexamethasone. In T cells, PRL is translocated to the nucleus in an IL-2-dependent P-13 kinase pathway inhibited by rapamycin (Clevenger et al. 1990). Immune-cell-derived PRL most likely plays a role in hematopietic cell differentiation and proliferation. In another study, antibody to PRL was shown to inhibit mitogenesis through neutralization of the lymphocyte-associated PRL (Clevenger et al. 1992). Furthermore, coordinate gene expression of LHRH and the LHRH receptor has been shown in the Nb2 T-cell line after PRL stimulation (Wilson, Yu-Lee, and Kelley 1995).

Two approaches have been utilized to obtain convincing evidence that endogenous neuroendocrine peptides have autocrine or paracrine immunoregulatory functions. First, an opiate antagonist was shown to indirectly block CRH enhancement of NK cell activity by inhibiting the action of immunocyte-derived opioid peptides (Carr, DeCosta, Jacobson, Rice, and Blalock 1990). Second, we have used an antisense oligodeoxynucleotide (ODN) to specifically inhibit leukocyte production of GH which resulted in reduced basal rates of DNA synthesis (Weigent, Blalock, and LeBoeuf 1991). Another group examining SOM found that antisense oligodeoxynucleotides to SOM dramatically increased lymphocyte proliferation in culture (Aguila, Rodriguez, Aguila-Mansila, and Lee 1996). Additionally, LHRH agonists were found to diminish NK cell activity, stimulate T-cell proliferation, and increase IL-2-receptor expression suggesting an important role for LHRH in the regulation of the immune response (Batticane, Morale, Galio, Farinella, and Marchetti 1991).

Another major function of GH produced by cells of the immune system is the induction of the synthesis of IGF-1, which, in turn, may inhibit the synthesis of both lymphocyte GH mRNA and protein. The results in T cell lines supported a role for locally generated IGF-1 in the mediation of GH action on T-lymphocytes and indicated the effect was mediated via the type 1 IGF receptor (Geffner, Bersch, Lippe, Rosenfeld, Hintz, and Golde 1990). We could detect IGF-1 in primary rat spleen cells by direct immunofluorescence with specific IGF-1 antibodies, immunoaffinity purification, HPLC, and a fibroblast proliferation bioassay. The data showed that IGF-1 was de novo synthesized and similar to serum IGF-1 in molecular weight, antigenicity, and bioactivity (Baxter, Blalock, and Weigent 1991). The major work examining the expression of the IGF-1 mRNA in the mouse lymphohemopoietic system has been done by Kelley and colleagues in macrophages (Arkins, Rebeiz, Biragyn, Reese, and Kelley 1993). Their results establish that murine macrophages express abundant insulin-like growth factor-1 class 1 Ea and Eb transcripts. Further, their data suggest myeloid rather than lymphoid cells are the major source of IGF-1 that is associated with differentiation of bone marrow macrophages (Arkins et al. 1993). Thus, in macrophages, initiation of transcription is primarily within exon 1 that is typical of extrahepatic tissues with a higher percentage of Eb transcripts that is typical of hepatic tissues. The significance of different leader peptides and E terminal domains on IGF-1 is unknown but may influence targeting, processing or stability. Kelley and coworkers have also shown that mature adherent myeloid cells synthesize and secrete a substantial amount of IGF-binding proteins (BPs), whereas less differentiated or nonadherent myeloid cells produce fewer IGF-BPs. Premyeloid cells, mature T cells, and primary murine thymocytes did not synthesize detectable IGF-BPs (Li et al. 1996). Additional gel-shift, Northern blotting, and sequencing analysis showed that the IGF-BP secreted by mature adherent macrophages was IGF-BP4. Taken together, the presence of IGF-1, the IGF-1 receptor and IGF-BPs, particularly in myeloid cells, strongly supports the suggestion that the IGF system is important in hematopoiesis and inflammation. The findings support the existence of a complete regulatory loop within cells of the immune system, and they provide a molecular basis whereby GHRH, GH, IGF-1 and their binding proteins may be intimately involved in regulating each other's synthesis (Weigent, Baxter, and Blalock 1992). Our most recent findings in a T cell-line show that endogenous GH promotes nitric oxide production, up-regulation of IGF-1 receptors and Bcl2 protein along with an inhibition of superoxide formation, clearly establishing a role for lymphocyte GH in apoptosis (Arnold and Weigent 2003).

1.8 Neurotransmitter Release by Cells of the Immune System

Lymphocytes have also been suggested to be important sites of synthesis and action of acetylcholine and catecholamines since they contain both the enzymes necessary for biosynthesis of epinephrine and acetylcholine as well as the relevant receptor system (Bergquist, Tarkowski, Ekman, and Ewing. 1994; Tayebati, El-Assouad,

Ricci, and Amenta 2002; Warthan et al. 2002). The synthesis of catecholamines was shown to increase after mitogen treatment of rat lymphocytes obtained from spleen, thymus, and mesenteric lymph nodes (Qiu, Peng, Jiang, and Wang 2004). The ability to express tyrosine hydroxylase was greatest in mesenteric lymph node cells compared to the spleen and thymus. Although more work needs to be done, the available data suggest that endogenous catecholamines may suppress IL-2 production and thereby modulate T cell expansion (Tsao, Lin, and Cheng 1998). It has been suggested that in the dual regulation of immune function by endogenous and exogenous catecholamines, endogenous catecholamines may be more important since the lymphocytes are responding to antigen and performing an immune response (Qiu et al. 2004). Similar to catecholamines, stimulating lymphocytes with mitogens enhances the synthesis and release of acetylcholine (Kawashima and Fujii 2003). Both muscarinic and/or nicotinic acetylcholine receptor agonists influence intracellular calcium levels, c-fos gene expression, nitric oxide synthesis, and IL-2 production (Fujii and Kawashima 2000; Fujino, Kitamura, Yada, Uehara, and Nomura 1997; Kawashima and Fujii 2003). Further, abnormalities in the lymphocytic cholinergic system have been detected in animal models of immune disorders (i.e., spontaneously hypertensive rat) (Fujimoto, Matsui, Fujii, and Kawashima 2001). Thus, it seems clear that immune function is, in part, under the control of nonneuronal catecholamine and cholinergic systems. The effect, if any, of lymphocyte-derived catecholamines and acetylcholine on neural activity remains to be determined. Recent work has identified a neural mechanism involving the vagus nerve and release of acetylcholine that inhibits macrophage activation termed the "cholinergic anti-inflammatory pathway" (Tracey 2002). The sensory afferent vagus pathway may be activated by low doses of endotoxin, IL-1, or products from damaged tissues. The signal is relayed to the brain where activation of the efferent vagus nerve releases acetylcholine. Acetylcholine acts to inhibit macrophage release of the proinflammatory cytokines tumor necrosis factor (TNF), IL-1 and IL-18, but not the anti-inflammatory cytokine IL-10. Thus, cholinergic neuron participation in the inhibition of acute inflammation constitutes a "hardwire" neural mechanism of modulation of the immune response. Finally, calcitonin gene-related peptide (CGRP) has also been shown to be produced and secreted by human lymphocytes and may be involved in inhibition of T-lymphocyte proliferation (Wang, Xing, Li, Hou, Guo, and Wang 2002). In another more recent study, substance P, the potent mediator of neuroimmune regulation, was shown to be up-regulated in lymphocytes by HIV infection implying it may be involved in immunopathogenesis of HIV infection and AIDS (Ho, Lai, Li, and Douglas 2002). Neuropeptides, by direct interaction with T cells, induce cytokine secretion and break the commitment to a distinct T helper phenotype (Levite 1998). Thus, neurons are not the exclusive source of neurotransmitters and, therefore, provide another instance of shared molecular signals and their receptors between the nervous and immune system.

1.9 Bidirectional Communication and Sleep

The general pattern in both animals and humans is that sleep is increased during infection (Toth and Krueger 1988). In general, animals appear to show an increase in stage 3 and 4 slow-wave sleep (SWS) of nonrapid eye movement (NREM) and decreased rapid eye movement (REM) sleep (Bryant, Trinder, and Curtis 2004). Infection with influenza virus in human causes reduced sleep during the incubation period but increased sleep during the symptomatic period (Smith 1992). Exposure to microbial components such as lipopolysaccharide and muramyl peptides mimics the sleep effect (Mullington et al. 2000). Finally, increases in the intracerebral or plasma levels of TNF or IL1-β result in an increase in SWS duration, whereas antagonizing these cytokines with specific antibodies inhibit SWS (Opp and Krueger 1994). It appears that most proinflammatory cytokines are somnogenic, whereas most anti-inflammatory cytokines are not.

Neuroendocrine hormones produced by the hypothalamus influence sleep. Thus, CRH is a potent inducer of waking while GHRH promotes SWS (Obal, Fang, Payne, and Krueger 1995; Opp and Imeri 2001). These same hormones are produced by cells of the immune system (Weigent and Blalock 1995) and thus the immune and nervous system share regulatory molecules which supports their ability to interact. There are some data to suggest that chronic sleep loss might be detrimental to the immune system. Thus, sleep deprivation in nonimmune mice appeared to impede the clearance of influenza virus along the respiratory tract (Renegar, Crouse, Floyd, and Krueger 2000).

Also, sleep-deprived humans immunized against infection with influenza virus had lower virus-specific antibody titers compared to non-sleep-deprived individuals (Spiegel, Sheridan, and Van Cauter 2002). Despite the evidence that sleep loss has effects on immune function and secretion of cytokines, the significance of these changes on the immune response is not known (Irwin 2002). Finally, illnesses such as chronic fatigue, fibromyalgia and depression show NREM sleep disruption (Moldofsky 1993). Primary sleep disorders are also associated with alterations in immune competence (Sakami et al. 2002). Thus, decreases in the numbers of $CD3^+$, $CD4^+$, and $CD8^+$ T cells and reduced NK cell responses have been associated with chronic insomnia (Savard, Laroche, Simard, Ivers, and Morin 2003).

There is a strong association of HLA molecules and impaired transmission of hypocretin with narcolepsy (Lin, Hungs, and Mignot 2001). IFN-α has been shown to inhibit the production of hypocretin, supporting the hypothesis of a connection between sleep and the immune system. Taken together, the growing amount of data show that there is a reciprocal relationship or bidirectional communication between sleep and immunity.

1.10 Summary and Conclusions

It is now well established that the nervous system and immune system speak a common biochemical language and communicate via a complete bidirectional circuit involving shared ligands such as neurotransmitters, neuroendocrine hormones, cytokines, and the respective receptors. Thus, neurotransmitters, neuropeptides, and cytokines represent the signaling molecules relaying chemical information and depending on the stimulus either neurons or immune cells can be the initial source. The chemical information in turn can be received by both neurons and immune cells since they share receptor repertoires. This complete biochemical information circuit between neurons and immune cells allows the immune system to function as a sensory organ (Blalock 1994).

A sixth sense, if you will, that completes our ability to be cognizant not only of the universe of things we can see, hear, taste, touch, and smell but also the other universe of things we cannot. These would include bacteria, viruses, antigens, tumor cells, and other agents that are too small to see or touch, make no noise, have no taste or odor. Recognition of such "noncognitive stimuli" by the immune system would result in transmission of information to the CNS via the aforementioned shared signal molecules to cause a physiological response that is ultimately beneficial to the host and detrimental to the infectious agent.

References

Ader, R. (1996) Historical perspectives on psychoneuroimmunology. In: H. Friedman, T.W. Klein, and A.L. Friedman (Eds.), *Psychoneuroimmunology, Stress, and Infection*. CRC Press, Boca Raton, pp. 1–24.

Ader, R., and Cohen, N. (1975) Behaviorally conditioned immunosuppression. *Psychosom Med* 37, 333–340.

Aguila, M.C., Rodriguez, A.M., Aguila-Mansila, H.N., and Lee, W.T. (1996) Somatostatin antisense oligodeoxynucleotide-mediated stimulation of lymphocyte proliferation in culture. *Endocrinology* 137, 1585–1590.

Arkins, S., Rebeiz, N., Biragyn, A., Reese, D.L., and Kelley, K.W. (1993) Murine macrophages express abundant insulin-like growth factor-I class I Ea and Eb transcripts. *Endocrinology* 133, 2334–2343.

Arnold, R.E., and Weigent, D.A. (2003) The production of nitric oxide in EL4 lymphoma cells overexpressing growth hormone. *J Neuroimmunol* 134, 82–94.

Arzt, E., Stelzer, G., Renner, U., Lange, M., Muller, A., and Stalla, G.K. (1992) Interleukin-2 and interleukin-2 receptor in human corticotrophic adenoma and murine pituitary cell cultures. *J Clin Invest* 90, 1944–1951.

Banks, W.A., Farr, S.A., and Morley, J.E. (2002) Entry of blood-borne cytokines into the central nervous system: Effects on cognitive processes. *Neuroimmunomodulation* 10, 319–327.

Banks, W.A., Moinuddin, A., and Morley, J.E. (2006) Regional transport of TNF-alpha across the blood-brain barrier in young ICR and young and aged SAMP8 mice. *Neurobiol Aging* 22, 671–676.

Batticane, N., Morale, M., Galio, F., Farinella, Z., and Marchetti, B. (1991) Luteinizing hormone-releasing hormone signalling at the lymphocyte involves stimulation of interleukin-2 receptor expression. *Endocrinology* 129, 277–286.

Baxter, J.B., Blalock, J.E., and Weigent, D.A. (1991) Characterization of immunoreactive insulin-like growth factor-I from leukocytes and its regulation by growth hormone. *Endocrinology* 129, 1727–1734.

Bellinger, D.L., Lorton, D., Romano, T., Olschowka, J.A., Felten, S.Y., and Felten, D.L. (1990) Neuropeptide innervation of lymphoid organs. *Ann N Y Acad Sci* 594, 17–33.

Bergquist, J., Tarkowski, A., Ekman, R., and Ewing, A. (1994) Discovery of endogenous catecholamines in lymphocytes and evidence for catecholamine regulation of lymphocyte function via an autocrine loop. *Proc Natl Acad Sci USA* 91, 12912–12916.

Berkenbosch, F., Van Oers, J., Del Rey, A., Tilders, F., and Besedovsky, H. (1987) Corticotropin-releasing factor-producing neurons in the rat activated by interleukin-1. *Science* 238, 524–526.

Bernardini, R., Kammilaris, T.C., Calogero, A.E., Johnson, E.O., Gomez, M.T., Gold, P.W., and Chrousos, G.P. (1990) Interactions between tumor necrosis factor-α, hypothalamic corticotropin-releasing hormone, and adrenocorticotropin secretion in the rat. *Endocrinology* 126, 2876–2881.

Bernton, E., Bryant, H., Holaday, J., and Dave, J. (1987) Release of multiple hormones by a direct action of interleukin-1 on pituitary cells. *Science* 238, 519–522.

Bernton, E.W., Bryant, H.U., and Holaday, J.W. (1991) Prolactin and immune function. In: R. Ader, D.L. Felten, and N. Cohen (Eds.), *Psychoneuroimmunology*. Academic Press, San Diego, pp. 403–428.

Besedovsky, H., Sorkin, E., Keller, M., and Miller, J. (1977) Hypothalamic changes during the immune response. *Eur J Immunol* 7, 323–325.

Besedovsky, H.O., delRey, A.E., and Sorkin, E. (1981) Lymphokine containing supernatants from Con A-stimulated cells increase corticosterone blood levels. *J Immunol* 126, 385–387.

Besedovsky, H.O., delRey, A.E., Sorkin, E., DaPrada, M., and Keller, H.A. (1979) Immunoregulation mediated by the sympathetic nervous system. *Cell Immunol* 48, 346.

Blalock, J.E. (1984) The immune system as a sensory organ. *J Immunol* 132, 1067–1070.

Blalock, J.E. (1994) The immune system: Our sixth sense. *Immunologist* 2, 8–15.

Blalock J.E., and Costa, O. (1990) Immune neuroendocrine interactions: Implications for reproductive physiology. *Ann N Y Acad Sci* 564, 261–266.

Blalock, J.E., and Smith, E.M. (1980) Human leukocyte interferon: Structural and biological relatedness to adrenocorticotropic hormone and endorphins. *Proc Natl Acad Sci USA* 77, 5972–5974.

Brooks, K.H. (1990) Adrenocorticotropin (ACTH) functions as a late-acting B cell growth factor and synergizes with interleukin 5. *J Mol Cell Immunol* 4, 327–335.

Bryant, P.A., Trinder, J., and Curtis, N. (2004) Sick and tired: Does sleep have a vital role in the immune system. *Nat Rev Immunol* 4, 457–467.

Carr, D.J., DeCosta, B.R., Jacobson, A.E., Rice, K.C., and Blalock, J.E. (1990) Corticotropin-releasing hormone augments natural killer cell activity through a naloxone-sensitive pathway. *J Neuroimmunol* 28, 53–61.

Carr, D.J.J. (1991) The role of endogenous opioids and their receptors in the immune system. *Soc Exp Biol Med* 198, 170–720.

Carr, D.J.J. (1992) Neuroendocrine peptide receptors on cells of the immune system. In: J.E. Blalock (Ed.), *Neuroimmunoendocrinology*. Karger, Basel, pp. 49–83.

Clarke, B.L., and Bost, K.L. (1989) Differential expression of functional adrenocorticotropic hormone receptors by subpopulations of lymphocytes. *J Immunol* 143, 464–469.

Clevenger, C.V., Sillman, A.L., Hanley-Hyde, J., and Prystowsky, M.B. (1992) Requirement for prolactin during cell cycle regulated gene expression in cloned T-lymphocytes. *Endocrinology* 130, 3216–3222.

Clevenger, C.V., Sillman, A.L., and Prystowsky, M.B. (1990) Interleukin-2 driven nuclear translocation of prolactin in cloned T-lymphocytes. *Endocrinology* 127, 3151–3159.

Cross, R.J., Markesbery, W.R., Brooks, W.H., and Roszman, T.L. (1980) Hypothalamic-immune interactions. I. The acute effect of anterior hypothalamic lesions on the immune response. *Brain Res* 196, 79–87.

DeBellis, A., Bizzarro, A., Pivonello, R., Lombardi, G., and Bellastella, A. (2005) Prolactin and autoimmunity. *Pituitary* 8, 25–30.

Delgado, M., Gonzalez-Rey, E., and Ganea, D. (2004) VIP/PACAP preferentially attract Th2 effectors through differential regulation of chemokine production by dendritic cells. *FASEB J* 18, 1453–1455.

DeSouza, E.B. (1993) Corticotropin-releasing factor and interleukin-1 receptors in the brain-endocrine-immune axis. *Ann N Y Acad Sci* 697, 9–27.

Dhabhar, F.S., and McEwen, B.S. (1999) Enhancing versus suppressive effects of stress hormones on skin immune function. *Proc Natl Acad Sci USA* 96, 1059–1064.

Dinarello, C.A. (2004) Infection, fever and exogenous and endogenous pyrogens: Some concepts have changed. *J Endotoxin Res* 10, 201–222.

Dobbs, C.M., Vasquez, M., Glaser, R., and Sheridan, J.F. (1993) Mechanisms of stress-induced modulation of viral pathogenesis and immunity. *J Neuroimmunol* 48, 151–160.

Dubuis, J.M., Dayer, J.M., Siegrist-Kaiser, C.A., and Burger, A.G. (1988) Human recombinant interleukin-1 beta decreases plasma thyroid hormone and thyroid stimulating hormone levels in rats. *Endocrinology* 123, 2175–2181.

Edwards, C.K., Ghiasuddin, S.M., Schepper, J.M., Yunger, L.M., and Kelley, K.W. (1988) A newly defined property of somatotropin: Priming of macrophages for production of superoxide anion. *Science* 239, 769–771.

Feistritzer, C., Clausen, J., Sturn, D.H., Djanani, A., Gunsilius, E., Wiedermann, C.J., and Kahler, C.M. (2003) Natural killer cell functions mediated by the neuropeptide substance P. *Regul Pept* 116, 119–126.

Felten, D.L., Felten, S.Y., Bellinger, D.L., Carlson, S.L., Ackerman K.D., Madden, K.S., Olschowka,J.A., and Livnat, S. (1987) Noradrenergic sympathetic neural interactions with the immune system: Structure and function. *Immunol Rev* 100, 225–260.

Fomicheva, E.E., Nemirovich-Danchenko, E.A., and Korneva, E.A. (2004) Immunoprotective effects of prolactin during stress-induced immune dysfunction. *Bull Exp Biol Med* 6, 544–547.

Foster, C.A., Mandak, B., Kromer, E., and Rot, A. (1992) Calcitonin gene-related peptide is chemotactic for human T lymphocytes. *Ann N Y Acad Sci* 657, 397–404.

Fujii, T., and Kawashima, K. (2000) Calcium signaling and c-fos gene expression via M3 muscarinic acetylcholine receptors in human T- and B-cells. *Jpn J Pharmacol* 84, 124–132.

Fujimoto, K., Matsui, M., Fujii, T., and Kawashima, K. (2001) Decreased acetylcholine content and choline acetyltransferase mRNA expression in circulating mononuclear leukocytes and lymphoid organs of the spontaneously hypertensive rat. *Life Sci* 69, 1629–1638.

Fujino, H., Kitamura, Y., Yada, T., Uehara, T., and Nomura, Y. (1997) Stimulatory roles of muscarinic acetylcholine receptors in T cell antigen receptor/CD3 complex-mediated interleukin-2 production in human peripheral blood lymphocytes. *Mol Pharmacol* 51, 1007–1114.

Fukushima, T., Ichinose, M., Shingai, R., and Sawada, M. (2001) Adrenocorticotropic hormone activates an outward current in cultured mouse peritoneal macropahges. *J Neuroimmunol* 113, 231–235.

Gaillard, R.C., Turnill, D., Sappino, P., and Muller, A.F. (1990) Tumor necrosis factor alpha inhibits the hormonal response of the pituitary gland to hypothalamic releasing factors. *Endocrinology* 127, 101–106.

Gala, R.R. (1991) Prolactin and growth hormone in the regulation of the immune system. *Proc Soc Exp Biol Med* 198, 513–527.

Gala, R.R., and Shevach, E.M. (1993) Identification by analytical flow cytometry of prolactin receptors on immunocompetent cell populations in the mouse. *Endocrinology* 133, 1617–1623.

Gan, X., Zhang, L., Solomon, G.F., and Bonavida, B. (2002) Mechanism of norepinephrine-mediated inhibition of human NK cytotoxic functions: Inhibition of cytokine secretion, target binding, and programming for cytotoxicity. *Brain Behav Immun* 16, 227–246.

Ganea, D., Rodriguez, R., and Delgado, M. (2003) Vasoactive intestinal peptide and pituitary adenylate cyclase-activating polypeptide: Players in innate and adaptive immunity. *Cell Mol Biol* 49, 127–142.

Geffner, M.E., Bersch, N., Lippe, B.M., Rosenfeld, R.G., Hintz, R.L., and Golde, D.W. (1990) Growth hormone mediates the growth of T-lymphoblast cell lines via locally generated insulin-like growth factor-I. *J Clin Endocrin Metab* 71, 464–469.

Glaser, R., Rice, J., Sheridan, J., Fertel, R., Stout, J., Speicher, C., Pinsky, D., Kotur, M., Post, A., Beck, M., and Kiecolt-Glaser, J.A. (1987) Stress-related immune suppression: Health implications. *Brain Behav Immun* 1, 7–20.

Gonzalez, M.C., Aguila, M.C., and McCann, S.M. (1991) In vitro effects of recombinant human gamma-interferon on growth hormone release. *Prog Neuroendocrin Immunol* 4, 222–227.

Gonzalez, M.C., Riedel, M., Rettori, V., Yu, W.H., and McCann, S.M. (1990) Effect of recombinant human gamma-interferon on the release of anterior pituitary hormones. *Prog Neuroendocrin Immunol* 3, 49–54.

Guarcello, V., Weigent, D.A., and Blalock, J.E. (1991) Growth hormone releasing hormone receptors on thymocytes and splenocytes from rats. *Cell Immunol* 136, 291–302.

Hadden, J.W., Hadden, E.M., and Middleton, E. (1970) Lymphocyte blast transformation. I. Demonstration of adrenergic receptors in human peripheral lymphocytse. *J Cell Immunol* 1, 583–595.

Hadden, J.W., Hadden, E.M., Middleton, E., and Good, R.A. (1971) Lymphocyte blast transformation. II. The mechanisms of action of alpha adrenergic receptors effects. *Int Arch Allergy Appl Immunol* 40, 526–539.

Harbour, D.V., and Hughes, T.K. Thyrotropin releasing hormone (TRH) induces gamma interferon release. *FASEB J* A5884. 1991.

Harbour, D.V., Leon, S., Keating, C., and Hughes, T.K. (1990) Thyrotropin modulates B-cell function through specific bioactive receptors. *Prog Neuroendocrin Immunol* 3, 266–276.

Hiruma, K., Koike, T., Nakamura, H., Sumida, T., Maeda, T., Tomioka, H., Yoshida, S, and Fujita, T. (1990) Somatostatin receptors on human lymphocytes and leukaemia cells. *Immunology* 71, 480–485.

Ho, W., Lai, J., Li, Y., and Douglas, S. (2002) HIV enhances substance P expression in human immune cells. *FASEB J* 16, 616–618.

Irwin, M. (2002) Effects of sleep and sleep loss on immunity and cytokines. *Brain Behav Immun* 16, 503–512.

Jain, R., Zwickler, D., Hollander, C.S., Brand, H., Saperstein, A., Hutchinson, B., Brown, C., and Audhya, T. (1991) Corticotropin-releasing factor modulates the immune response to stress in the rat. *Endocrinology* 128, 1329–1336.

Johnson, H.M., Downs, M.O., and Pontzer, C.H. (1992) Neuroendocrine peptide hormone regulation of immunity. *Chem Immunol* 52, 49–83.

Johnson, H.M., Farrar, W.L., and Torres, B.A. (1982a) Vasopressin replacement of interleukin 2 requirement in gamma interferon production: Lymphokine activity of a neuroendocrine hormone. *J Immunol* 129, 983–986.

Johnson, H.M., Smith, E.M., Torres, B.A., and Blalock, J.E. (1982b) Neuroendocrine hormone regulation of *in vitro* antibody formation. *Proc Natl Acad Sci USA* 79, 4171–4174.

Johnson, H.M., Torres, B.A., Smith, E.M., Dion, L.D., and Blalock, J.E. (1984) Regulation of lymphokine (gamma-interferon) production by corticotropin. *J Immunol* 132, 246–250.

Kakucska, I., Qi, Y., Clark, B.D., and Lechan, R.M. (1993) Endotoxin-induced corticotropin-releasing hormone gene expression in the hypothalamic paraventricular nucleus is mediated centrally by interleukin-1. *Endocrinology* 129, 2796.

Kalra, P.S., Sahu, A., and Kalra, S.P. (1990) Interleukin-1 inhibits the ovarian steroid-induced luteinizing hormone surge and release of hypothalamic luteinizing hormone- releasing hormone in rats. *Endocrinology* 126, 2145–2152.

Karanth, S., Aguila, M.C., and McCann, S.M. (1993) The influence of interleukin-2 on the release of somatostatin and growth hormone-releasing hormone by mediobasal hypothalamus. *Neuroendocrinology* 58, 185–190.

Karanth, S., and McCann, S.M. (1991) Anterior pituitary hormone control by interleukin 2. *Proc Natl Acad Sci USA* 88, 2961–2965.

Kawashima, K., and Fujii, T. (2003) The lymphocytic cholinergic system and its biological function. *Life Sci* 72, 2101–2110.

Kelley, K.W. (1989) Growth hormone, lymphocytes and macrophages [Review]. *Biochem Pharmacol* 38, 705–713.

Kelley, K.W., Arkins, S., and Li, Y.M. (1992) Growth hormone, prolactin, and insulin-like growth factors: New jobs for old players. *Brain Behav Immun* 6, 317–326.

Koff, W.C., and Dunnegan, M.A. (1986) Neuroendocrine hormones suppress macrophage-mediated lysis of herpes simplex virus-infected cells. *J Immunol* 136, 705–709.

Kooijman, R., and Coppens, A. (2004) Insulin-like growth facator-1 stimulates IL-10 production in human T cells. *J Leukoc Biol* 76, 862–867.

Kooijman, R., and Gerlo, S. (2002) Prolactin expression in the immune system. In: L. Matera and R. Rapaport (Eds.), *Growth and Lactogenic Hormones*. Elsevier Science, Amsterdam, pp. 147–159.

Kooijman, R., Hooghe-Peters, E.L., and Hooghe, R. (1996) Prolactin, growth hormone and insulin-like growth factor-1 in the immune system. *Adv Immunol* 63, 377–453.

Kruger, T.E., Smith, L.R., Harbour, D.V., and Blalock, J.E. (1989) Thyrotropin: An endogenous regulator of the in vitro immune response. *J Immunol* 142, 744–747.

Lai, J.-P., Ho, W.Z., Kilpatrick, L.E., Wang, X., Tuluc, F., Korchak, H.M., and Douglas, S.D. (2006) Full-length and truncated neurokinin-1 receptor expression and function during monocyte/macrophage differentiation. *Proc Acad Sci USA* 103, 7771–7776.

Laye, S., Bluthe, R.M., Kent, S., Combe, C., Medina, C., Parnet, P., Kelley, K., and Dantzer,R. (1995). Subdiaphragmatic vagotomy blocks induction of IL-1β mRNA in mice brain in response to peripheral LPS. Am J Physiol *268*, R1327-R1331.

Levite, M. (1998) Neuropeptides, by direct interaction with T cells, induce cytokine secretion and break the commitment to a distinct T helper phenotype. *Proc Natl Acad Sci USA* 95, 12544–12549.

Li, Y.M., Arkins, S., McCusker, R.H., Jr., Donovan, S.M., Liu, Q., Jayaraman, S., and Dantzer, R. (1996) Macrophages synthesize and secrete a 25 kDa protein that binds insulin-like growth factor-1. *J Immunol* 156, 64–72.

Lin, L., Hungs, M., and Mignot, E. (2001) Narcolepsy and the HLA region. *J Neuroimmunol* 117, 9–20.

Lipton, J. and Catania, A. (2003). Anti-inflammatory actions of the neuroimmunomodulator α-MSH. Immunol Today *18*, 140-145.

Luger, T.A., Scholzen, T.E., Brzoska, T., and Bohm, M. (2003) New insights into the functions of α-MSH and related peptides in the immune system. *Ann N Y Acad Sci* 994, 133–140.

Matera, L., Cesano, A., Bellone, G., and Oberholtzer, E. (1992) Modulatory effect of prolactin on the resting and mitogen-activity of T, B, and NK lymphocytes. *Brain Behav Immun* 6, 409–417.

Matera, L., Mori, M., and Galetto, A. (2001) Effect of prolactin on the antigen presenting function of monocyte-derived dendritic cells. *Lupus* 10, 728–734.

Mathew, R.C., Cook, G.A., Blum, A.M., Metwali, A., Felman, R., and Weinstock, J.V. (1992) Vasoactive intestinal peptide stimulates T lymphocytes to release IL-5 in murine schistosomiasis mansoni infection. *J Immunol* 148, 3572–3577.

McCann, S.M., Karanth, S., Kamat, A., Dees, W.L., Lyson, K., Gimeno, M., and Rettori, V. (1994) Induction of cytokines of the pattern of pituitary hormone secretion in infection. *Neuroimmunomodulation* 1, 2–13.

McGillis, J.P., Humphreys, S., and Reid, S. (1991) Characterization of functional calcitonin gene-related peptide receptors on rat lymphocytes. *J Immunol* 147, 3482–3489.

Metal'nikov, S., and Chorine, V. (1926) Role des reflexes conditionnels dans l'immunite. *Ann l'Institute Pasteur* 40, 893–900.

Milenkovic, L., Rettori, V., Snyder, G.E., Reutler, B., and McCann, S.M. (1989) Cachectin alters anterior pituitary hormone release by a direct action in vitro. *Proc Natl Acad Sci USA* 86, 2418–2422.

Moldofsky, H. (1993) Fibromyalgia, sleep disorder and chronic fatigue syndrome. *Ciba Found Symp* 173, 272–279.

Montgomery, D.W. (2001) Prolactin production by immune cells. *Lupus* 10, 665–675.

Montgomery, D.W., Shen, G.K., Ulrich, E.D., Steiner, L.L., Parrish, P.R., and Zukoski,C.F. (1992) Human thymocytes express a prolactin-like messenger ribonucleic acid and synthesize bioactive prolactin-like protein. *Endocrinology* 131, 3019–3026.

Mullington, J., Korth, C., Hermann, D.M., Orth, A., Galanos, C., Holsboer, F., and Pollmacher, T. (2000) Dose-dependent effects of endotoxin on human sleep. *Am J Physiol Regul Integr Comp Physiol* 278, R947–R955.

Naitoh, Y., Fukata, J., Tominaga, T., Nakai, Y., Tamai, S., Mori, K., and Imura, H. (1988) Interleukin-6 stimulates the secretion of adrenocorticotropic hormone in conscious, freely-moving rats. *Biochem Biophys Res Commun* 155, 1459–1463.

O'Neal, K.D., Schwarz, L.A., and Yu-Lee, L.Y. (1991) Prolactin receptor gene expression in lymphoid cells. *Mol Cell Endocrinol* 82, 127–135.

Obal, F., Jr., Fang, J., Payne, L.C., and Krueger, J. (1995) Growth-hormone-releasing hormone mediates the sleep-promoting activity of interleukin-1 in rats. *Neuroendocrinology* 61, 559–565.

Opp, M.R., and Imeri, L. (2001) Rat strains that differ in corticotropin-releasing hormone production exhibit different sleep-wake responses to interleukin-1. *Neuroendocrinology* 73, 272–284.

Opp, M.R., and Krueger, J.M. (1994) Anti-interleukin-1-beta reduces sleep and sleep rebound after sleep deprivation in rats. *Am J Physiol* 266, R688–R695.

Ottaway, C.A., Bernaerts, C., Chan, B., and Greenberg, G.R. (1983) Specific binding of vasoactive intestinal peptide to human circulating mononuclear cells. *Can J Phys Pharm* 61, 664–671.

Pascual, D.W., Xu-Amano, J.C., Kiyono, H., McGhee, J.R., and Bost, K.L. (1991) Substance P acts directly upon cloned B lymphoma cells to enhance IgA and IgM production. *J Immunol* 146, 2130–2136.

Pavlov, V.A., Wang, H., Czura, C.J., Friedman, S.G., and Tracey, K.J. (2003) The cholinergic anti-inflammatory pathway: A missing link in neuroimmunomodulation. *Mol Med* 9, 125–134.

Payne, L.C., Weigent, D.A., and Blalock, J.E. (1994) Induction of pituitary sensitivity to interleukin-1: A new function for corticotropin-releasing hormone. *Biochem Biophy Res Commun* 198, 480–484.

Pierpaoli, W., Baroni, C., Fabris, N., and Sorkin, E. (1969) Hormones and immunological capacity. II. Reconstitution of antibody production in hormonally deficient mice by somatotropic hormone, thyrotropic hormone and thyroxin. *J Immunol* 16, 217–230.

Provinciali, M., Di Stefano, G., and Fabris, N. (1992) Improvement in the proliferative capacity and natural killer cell activity of murine spleen lymphocytes by thyrotropin. *Int J Immunopharm* 14, 865–870.

Qiu, Y.-H., Peng, Y.-P., Jiang, J.-M., and Wang, J.-J. (2004) Expression of tyrosine hydroxlyase in lymphocytes and effect of endogenous catecholamines on lymphocyte function. *Neuroimmunomodulation* 11, 75–83.

Rasmussen, A.F., Jr., Marsh, J.T., and Brill, N.Q. (1957) Increased susceptibility to herpes simplex in mice subjected to avoidance-learning stress or restraint. *Proc Soc Exp Biol Med* 96, 183–189.

Renegar, K.B., Crouse, D., Floyd, R.A., and Krueger, J. (2000) Progression of influenza viral infection through the murine respiratory tract: The protective role of sleep deprivation. *Sleep* 23, 859–863.

Renier, G., Clement, I., Desfaits, A.C., and Lambert, A. (1996) Direct stimulatory effect of insulin-like growth factor-1 on monocyte and macarophage tumor necrosis factor-alpha production. *Endocrinology* 137, 4611–4618.

Rivest, S. (1995) Molecular mechanisms and neural pathways mediating the influence of interleukin-1 on the activity of neuroendocrine CRF motoneurons in the rat. *Int J Dev Neurosci* 13, 135.

Rouabhia, M., Chakir, J., and Deschaux, P. (1991) Interaction between the immune and endocrine systems: Immunomodulatory effects of luteinizing hormone. *Prog Neuroendocrin Immunol* 4, 86–91.

Roupas, P., and Herington, A.C. (1989) Cellular mechanisms in the processing of growth hormone and its receptor [Review]. *Mol Cell Endocrinol* 61, 1–12.

Ruzicka, B., and Huda, A.K.L. (1995) Differential cellular regulation of POMC by IL-1 and corticotrophin-releasing hormone. *Neuroendocrinology* 61, 136.

Sakami, S., Ishikawa, T., Kawakami, N., Haratani, T., Fukui, A., Kobayashi, F., Fujita, O., Araki, S., and Kawamura, N. (2002) Coemergence of insomnia and a shift in the Th1/Th2 balance toward Th2 dominance. *Neuroimmunomodulation* 10, 337–343.

Savard, J., Laroche, L., Simard, S., Ivers, H., and Morin, C.M. (2003) Chronic insomnia and immune functioning. *Psychosom Med* 65, 211–221.

Schobitz, B., Pezeshki, G., Pohl, T., Hemmann, U., Heinrich, P.C., Holsboer, F., and Reul, J.M.H.M. (1995) Soluble interleukin-6 (IL-6) receptor augments central effects of IL-6 in vivo. *FASEB J* 9, 659.

Siejka, A., Stepien, T., Lawnicka, H., Krupinski, R., Komorowski, J., and Stepien, H. (2005) Effect of growth hormone-releasing hormone [GHRH(1-44)NH2] on IL-6 and IL-8 secretion from human peripheral blood mononuclear cells in vitro. *Endocr Rev* 39, 7–11.

Singh, U. (1985) Lymphoiesis in the nude fetal mouse thymus following sympathectomy. *Cell Immunol* 93, 222–228.

Smith, A. (1992) In: R.J. Broughton and R.D. Ogilvie (Eds.), *Arousal and Performance*. Birkhauser, Boston, pp. 233–242.

Smith, E.M. (2003) Opioid peptides in immune cells. In: H. Machelska and C. Stein (Eds.), *Immune Mechanisms of Pain and Analgesia*. Kluwer Academic/Plenum Publishers, New York, pp. 51–68.

Smith, E.M., and Blalock, J.E. (1981) Human lymphocyte production of corticotropin and endorphin-like substances: Association with leukocyte interferon. *Proc Natl Acad Sci USA* 78, 7530–7534.

Smith, E.M., Hughes, T.K., Hashemi, F., and Stefano, G.B. (1992) Immunosuppressive effects of corticotropin and melanotropin and their possible significance in human immunodeficiency virus infection. *Proc Natl Acad Sci USA* 89, 782–786.

Spangelo, B.L., Judd, A.M., Isakson, P.C., and MacLeod, R.M. (1989) Interleukin-6 stimulates anterior pituitary hormone release in vitro. *Endocrinology* 125, 575–577.

Spangelo, B.L., and MacLeod, R.M. (1990) Regulation of acute phase response and neuroendocrine function by interleukin 6. *Prog Neuroendocrin Immunol* 3, 167–174.

Spiegel, K., Sheridan, J.F., and Van Cauter, E. (2002) Effect of sleep deprivation on response to immunization. *JAMA* 288, 1471–1472.

Stanisz, A.M., Befus, D., and Bienenstock, J. (1986) Differential effects of vasoactive intestinal peptide, substance P and somatostatin on immunoglobulin synthesis and proliferation by lymphocytes from Peyer's patches, mesenteric lymph nodes and spleen. *J Immunol* 136, 152.

Stein, M., Schiavi, R.C., and Camerino, M. (1976) Influence of brain and behavior on the immune system. *Science* 191, 435–440.

Suzuki, K., Suzuki, S., Saito, Y., Ikebuchi, H., and Terao, T. (1990) Human growth hormone-stimulated growth of human cultured lymphocytes (IM-9) and its inhibition by phorbol diesters through down-regulation of the hormone receptors. Possible involvement of phosphorylation of a 55,000 molecular weight protein associated with the receptor in the down-regulation. *J Biol Chem* 265, 11320–11327.

Tayebati, S.K., El-Assouad, D., Ricci, A., and Amenta, F. (2002) Immunochemical and immunocytochemical characterization of cholinergic markers in human peripheral blood lymphocytes. *J Neuroimmunol* 132, 147–155.

Torres, K.C.L., Antonelli, L.R.V., Souza, A.L.S., Teixeira, M.M., Dutra, W.O., and Gollob, K.J. (2005) Norepinephrine, dopamine and dexamethasone modulate discrete leukocyte subpopulations and cytokine profiles from human PBMC. *J Neuroimmunol* 166, 144–157.
Toth, L.A., and Krueger, J.M. (1988) Alteration of sleep in rabbits by Staphylococcus aureus infection. *Infect Immun* 56, 1785–1791.
Tracey, K. (2002) The inflammatory reflex. *Nature* 420, 853–859.
Tsao, C.W., Lin, Y.S., and Cheng, J.T. (1998) Inhibition of immune cell proliferation with haloperidol and relationship of tyrosine hydroxylase expression to immune cell growth. *Life Sci* 62, 335–344.
Vankelecom, H., Carmeliet, P., Heremans, H., Van Damme, J., Dijkmans, R., Billiau, A., and Denef, C. (1990) Interferon-gamma inhibits stimulated adrenocorticotropin, prolactin, and growth hormone secretion in normal rat anterior pituitary cell cultures. *Endocrinology* 126, 2919–2926.
Wang, H., Xing, L., Li, W., Hou, L., Guo, J., and Wang, X. (2002) Production and secretion of calcitonin gene-related peptide from human lymphocytes. *J Neuroimmunol* 130, 155–162.
Wang, H.C., and Klein, J.R. (2001) Immune function of thyroid stimulating hormone and receptor. *Crit Rev Immunol* 21, 323–337.
Warthan, M.D., Freeman, J., Loesser, K., Lewis, C., Hong, M., Conway, C., and Stewart, J. (2002) Phenylethanolamine N-methyl transferase expression in mouse thymus and spleen. *Brain Behav Immun* 16, 493–499.
Webster, E.L., Tracey, D.E., and De Souza, E.B. (1991) Upregulation of interleukin-1 receptors in mouse AtT-20 pituitary tumor cells following treatment with corticotropin-releasing factor. *Endocrinology* 129, 2796–2798.
Weigent, D.A., Baxter, J.B., and Blalock, J.E. (1992) The production of growth hormone and insulin-like growth factor-I by the same subpopulation of rat mononuclear leukocytes. *Brain Behav Immun* 6, 365–376.
Weigent, D.A., and Blalock, J.E. (1995) Associations between the neuroendocrine and immune systems. *J Leukoc Biol* 58, 137–150.
Weigent, D.A., Blalock, J.E., and LeBoeuf, R.D. (1991) An antisense oligodeoxynucleotide to growth hormone messenger ribonucleic acid inhibits lymphocyte proliferation. *Endocrinology* 128, 2053–2057.
Williams, J.M., Peterson, R.G., Shea, P.A., Schmedtje, J.R., Bauer, D.C., and Felten, D.L. (1981) Sympathetic innervation of murine thymus and spleen: Evidence for a functional link between the nervous and immune system. *Brain Res Bull* 6, 83.
Wilson, T., Yu-Lee, L., and Kelley, M. (1995) Coordinate gene expression of luteinizing hormone-releasing hormone (LHRH) and the LHRH receptor after prolactin stimulation in the rat Nb2 T cell line: Implications for a role in immunomodulation and cell cycle gene expression. *Mol Endocrinol* 9, 44–53.
Woloski, B.M., Smith, E.M., Meyer, W.J.3., Fuller, G.M., and Blalock, J.E. (1985) Corticotropin-releasing activity of monokines. *Science* 230, 1035–1037.
Wrona, D. (2006) Neural-immune interactions: An integrative view of the bidirectional relationship between the brain and immune systems. *J Neuroimmunol* 172, 38–58.
Zelazowski, P., Dohler, K.D., Stepien, H., and Pawlikowski, M. (1989) Effect of growth hormone-releasing hormone on human peripheral blood leukocyte chemotaxis and migration in normal subjects. *Neuroendocrinology* 50, 236–239.

2 Neuroimmunological Correlates of Circadian Rhythmicity in Oral Biology and Medicine

Francesco Chiappelli, Olivia S. Cajulis, Audrey Navarro, and David R. Moradi

2.1 Introduction

Sleep is a necessary behavior for physiological allostasis that is common to all vertebrates. Allostasis refers to the set of intertwined neuroendocrine-immune processes of bodily adaptation to stressful challenges during the wake cycle. The summative effects of these challenges, the allostatic load, signifies the total cost of wear and tear to the body. A finely regulated neuroimmunology of the sleep component of the circadian patterns is critical to the physiological repair mechanism required in allostasis (Solomon and Moos 1964; Solomon 1987; Sterling and Eyer 1988; Chrousos and Gold 1992; Kiecolt-Glaser, McGuire, Robles, and Glaser 2002; Irwin 2002; McEwen and Wingfield 2003; Schulkin 2003; Chiappelli and Cajulis 2004; Chiappelli et al. in press-c).

It was long suspected (e.g., Engle 1960), and is now widely documented by research findings from our laboratory and others that oral biology and systemic medicine are intimately intertwined. Certain prokaryotes that contribute the flora of the oral cavity are known to migrate into the cardiovascular system, to lodge in the aortic and the pulmonary valves, and to lead to severe cardiac malfunction and bacterial endocarditis. Serious tegumentary diseases, such as lichen planus, have their corresponding pathology in the oral mucosa (i.e., oral lichen planus, OLP, *vide infra*) (Chiappelli and Cajulis 2004). Several systemic neuroendocrine and immune responses to stress, to the allostatic response, and to bacteriokines, bacterial cytokine inducers that are produced by infectious pathogens and that contribute to the control of the pathological inflammatory response, are monitored in peripheral fluids (e.g., plasma, serum, urine), including total or parotid saliva. Case in point, the cold pressor challenge, whose physiological mechanism underlying the response consists of temperature-dependent constriction to blood flow of both superficial and deep tissues of the hand, and associated strong perception of pain. This challenge test is an adequate model to study psychobiological responses to stressful discomfort and the associated allostatic response, because cold-induced vasospasms are frequently present in syndromes of chronic pain. The caveat of this test is that it may have adverse effects in subjects with cardiovascular dysfunctions. In normal subjects, the cold pressor test alters the glucocorticoids-cytokine axis response driven by the hypothalamus-pituitary-adrenocortical (HPA) cell mediated immune (CMI) feedback system (*vide infra*, Prolo and Chiappelli in press). This leads to a time-dependent pattern of change in the circadian patterns of salivary interleukin (IL)-6. Salivary

IL-6 rises two- to threefold during the 90 min following the challenge. This effect is more pronounced in female compared to male subjects ($p < 0.05$), suggesting modulatory effects by the HPGonadal hormones, themselves under biologically important rhythmicity (Chiappelli et al. in press-c; Prolo and Chiappelli in press; Chiappelli, Prolo, Cajulis, Harper, Sunga, and Concepcion 2004).

IL-6 is a central neuroimmunological factor, and is one of the proinflammatory cytokines produced following immune challenges that contribute to stimulate the hypothalamic secretion of corticotropin releasing factor (CRF), the factor responsible for inducing secretion of adrenocorticotropin hormone (ACTH) by the anterior pituitary. ACTH regulates production and secretion of glucocorticoids by the adrenal cortex (i.e., cortisol in fish and Homo sapiens, corticosterone in rodents), which down-regulate both the HPA axis, and immune inflammatory events mediated by CMI. Autonomic neural activity, which includes palmar skin conductance, brachial artery systolic blood pressure, electrocardiogram, interbeat duration between R spikes, finger photoplethysmograph pulse peak amplitude, and peripheral finger photoplethysmograph peak during the final 60 s of a 15-min resting period, expressed as the SD of each indicator about its mean under each assessment condition, also plays a significant role in mediating CMI (Solomon 1987; Chrousos and Gold 1992; Kiecolt-Glaser et al. 2002; Chiappelli et al. in press-c; Prolo and Chiappelli in Press; Chiappelli et al. 2004). Saliva and plasma IL-6 levels are characterized by distinctive patterns of circadian fluctuation, and show a significant positive correlation (Chiappelli et al. in press-c).

In addition to its somnogenic properties, IL-6 administration or elevation of its endogenous levels results in sleep disturbance when associated with HPA axis activation (Chiappelli et al. 2004). Changes in sleep physiology associated with aging, including elevations of sleep-disturbing hormones and increased sensitivity of the sleep-controlling target organ to the actions of these hormones, markedly increase insomnia prevalence with aging. IL-6 peripheral levels correlate negatively with sex steroid levels, are decreased after a restful night of sleep, and are elevated in chronic pain/inflammatory syndromes (Irwin 2002).

The allostatic response consists of a set of psychobiological process, whose relevance to the health disease continuum is long lasting. The fundamental mechanisms of allostasis are defined by underlying psychoneuroendocrine-immune pathways. The psychoneuroimmune interactive system consists of the overlay of the complexity of the psyche with the short and long feedback loops that directly and indirectly regulate neuroendocrine responses, and the modulation of cellular and humoral, innate and antigen-dependent immunity, with the intertwined process of physiological rhythmicity (Solomon 1987; Kiecolt-Glaser et al. 2002; Chiappelli et al. in press-c; Chiappelli et al. 2004).

We present the case of oral mucositis (OM), the pervasive condition that afflicts the large majority of patients undergoing aggressive cancer treatments, as a significant allostatic threat. OM is a condition with established immunological mechanisms, with severe sequelae that include, among others, disturbed sleep. We discuss the fundamental psychobiology of the etiology of OM, and current treatment interventions. We present the case for chronomodulated cancer treatment that is one that respects biorhythmicity, as an effective measure to minimize or to prevent OM. We suggest avenues for future research to better characterize the fundamental neuroimmunology of OM, and the potential palliative effects of sleep for this condition.

2.2 Allostatic Response and Allostatic Intervention: Relevance to Oral Biology and Medicine

2.2.1 Allostasis

The nineteenth-century French physiologist, Claude Bernard (1813–1878) proposed that preservation of the internal milieu (*le milieu intérieur, 1856*) is a fundamental process of physiological regulation. The phrase "homeostasis" was coined. However, the concept of the benefits reaped by a state of physiological balance goes back millennia and across cultures, medical traditions and civilizations.

In Western civilization, we need to go beyond the classic Roman view of a healthy mind in a healthy body (*mens sana in corpore sano*), to the ancient Greek concept of Kalokagathia (from kalós kai agathós, Greek: καλòσ και αγατηòσ). In Greek mythology, Kalokagathia was the female form of the spirit (=Daimon), and represented nobility and goodness. She was related to Virtue (Arete) and to Excellence (Eukleia). In Greek philosophy (cf., Plato), it came to represent the ideal and harmonious bodily, moral and spiritual whole. Harmony, it was defended, was reached when the mind and the body dimensions of a human being were appropriately balanced.

Cannon (1871–1945) described the dynamic process that organisms use to maintain physiological balance in response to changing environmental demands. Hans Selye (1907–1982) labeled the concerted responses to stressful challenges as the "Generalized Stress Response." Contemporaneously, Polany, described consciousness and related perceptions of motivation and control as fundamental phenomena processed by the central nervous system (CNS).

The role of the psyche in the fine and dynamic balance that signifies physical health emerged as psychosomatic medicine, a discipline that evolved from Speransky's "neurodystrophic processes" theory. Today, this paradigm argues that all forms of human disease, oral and systemic, are related to alterations in the interaction among genetic, endocrine, nervous, immune, and psychic (including consciousness), behavioral, cognitive, and emotional factors (Chiappelli et al. in press-c; Prolo and Chiappelli in Press; Chiappelli et al. 2004; Vgontzas et al. 1999). The subdiscipline that specifically describes the cross talk among the central and peripheral nervous, the psychoneuroendocrine, and the immune systems refers to psychoneuroimmunology (Solomon and Moos 1964; Solomon 1987; Chrousos and Gold 1992; Kiecolt-Glaser et al. 2002; Irwin 2002; Chiappelli et al. in press-c; Prolo and Chiappelli in press; Chiappelli et al. 2004; Vgontzas et al. 1999).

The central process of the psychoneuroimmunologic response entails allostasis, which consists of the totality of intertwined psychosocial and psychophysiological responses that regulate the psychobiological adaptation to change in homeostasis (Sterling and Eyer 1988; McEwen and Wingfield 2003; Schulkin 2003; Chiappelli and Cajulis 2004; Chiappelli et al. in press-c). Allostasis pertains to the complex interaction of variable set points according to circadian and circa-mensual rhythmicity, and which are characterized by individual differences, themselves associated with anticipatory behavioral and physiological responses, and vulnerable to physiological overload and to the breakdown of regulatory capacities (Sterling and Eyer 1988; McEwen and Wingfield 2003; Schulkin 2003; Chiappelli and Cajulis 2004). In brief, allostasis represents the set of intertwined processes in the face of all possible challenges, including challenges to the immune system (e.g., bacterial and viral infection of the stoma) in the context of orderly biological fluctuations. In this framework, these challenges to psychobiological rhythmicity, including sleep, represent the allostatic load (Chiappelli et al. in press-c; Prolo and Chiappelli in press; Chiappelli et al. 2004; Vgontzas et al. 1999). The cumulative load to the allostatic process, the allostatic load, produces the pathological side effects of failed adaptation, which are commonly observed in a variety of states. These outcomes can lead to unease and *dis*-ease, endanger mental and physical health, and may result in serious sickness well described in psychosomatic medicine (Sterlin and Eyer 1988; McEwen and Wingfield 2003; Schulkin 2003; Chiappelli et al. in press-c).

In most situations, subjects position themselves along a spectrum of allostatic regulation, somewhere between allostasis (=toward regaining physiological balance), and allostatic overload (=toward physiological collapse, and associated potential onset of varied pathologies). In other words, allostasis refers to the entirety of the complexity of psychobiological processes that bring about stability through change of state consequential to stress or trauma (Sterling and Eyer 1988; McEwen and Wingfield 2003; Chiappelli et al. in press-c). Stress can be defined by a variety of models. For instance, stress may signify perceived lack of fit of one's perceived abilities and the demands of the environment (i.e., person/environment fit) (French, Rodgers, and Cobb 1974; Chiappelli, Manfrini, Edgerton, Rosenblum, Cajulis, and Prolo 2006b), or which is essentially equivalent to a serious threat to the perception of self (Gruenewald, Kemeny, and Aziz, 2006).

Regardless of its construct, psychoemotional or physiological stress leads to anxiety, irritability and anger, tension, depressed moods, fatigue, and a compendium of symptoms referred to as the "sickness behavior" (e.g., lethargy, nausea, fever). Sickness behavior is evident, among others, as an elevation in IL-6 and in tumor necrosis factor (TNF)-α (i.e., cachectin) in body fluids, including saliva. Stress results in a set of bodily manifestations, including perspiration, blushing or blanching of the face, increased heart beat or decreased blood pressure, intestinal cramps and discomfort, and suppressed immune resistance to antigenic threats (Solomon and Moos 1964; Solomon 1987; Chrousos and Gold 1992; Kiecolt-Glaser et al. 2002; Irwin 2002; Chiappelli et al. in press-c; Chiappelli et al. 2004). In the oral cavity, reactions to stress manifest as clenching of the teeth, with consequential pain, discomfort and trauma to the masticatory musculature, the temporomandibular joint, and various odontalgias with no clear etiology (Chiappelli et al. in press-c).

2.2.2 Allostatic Load

We noted that the process of allostatic regulation signifies the recovery and the maintenance of internal balance and viability amidst circadian rhythmicity and changing circumstances consequential to stress. This process encompasses a range of behavioral and physiological functions that direct the adaptive function of the regulating homeostatic systems in response to stress, trauma and other challenges. The allostatic load, in brief, refers to the set of physiological, psychological, and pathological *sequelae* of the allostatic process. Allostatic overload signifies the psychobiological collapse that is inevitably associated with a variety of health outcomes systemically (Schulkin 2003; Chiappelli and Cajulis 2004; Chiappelli et al. in press-c).

Type 1 allostatic load often utilizes the psychobiological responses to the challenge as a mean of self-preservation by developing and establishing temporary or permanent adaptation skills. The organism aims at surviving the perturbation in the best condition possible, and seeks to normalize its normal circadian life cycle. An example of Type 1 allostatic response in oral biology and medicine is, as noted elsewhere (Chiappelli and Cajulis 2004), recurrent aphthous stomatitis (RAS, i.e., canker sores). This is a stress-associated T cell-dependent pathology of the gingival mucosa. RAS lesions are characterized with increased TH1 patterns of cytokine (e.g., IL-2) in situ (Chiappelli and Cajulis 2004). Other examples of Type I allostatic response manifested in the oral cavity include teeth grinding, and tooth pain that cannot be attributed to carries, cracked tooth syndrome, pulpitis or other circum-dental inflammatory process (e.g., gum disease), temporomandibular disorder and pain, and pain at palpation of the masticatory muscles (*vide supra*). Stress and trauma to the dentition alters dental anatomy at the coronal surfaces such that, when subjects are followed up prospectively, changes in vertical distance of occlusion and the angle of Spee, the curved line that follows the occlusal surface of the posterior teeth of the mandible and is viewed from a sagittal plane, are clearly noted (Chiappelli et al. in press-c). Our data, cited elsewhere (Chiappelli et al. in press-c), on the relationship among dental pain, dental anxiety, and the perception of wellness show that dental patients can be dichotomized on the basis of pain perception: patients stratified for perceived pain intensity (high pain median 5.5 versus low pain: 3.0 [significant by t comparison of means, $p = 0.004$]), do not differ in terms of reported dental anxiety (medians: 1.6 versus 1.7 [not significant by Wilcoxon comparison of means]). Perception of pain impacts negatively (significant inverse correlation by Pearson coefficient: $r = -0.5$) upon dental patients' overall perception of wellness (medians: 2.3 versus 3.3 [significant by t comparison of means, $p = 0.049$]) (Chiappelli et al. in press-c).

In Type 2 allostatic load, the challenge is excessive, sustained, or continued, and drives allostasis chronically: an escape response cannot be found, and allostatic overload irremediably obtains. Type 2 drives allostasis beyond any possibility of return to homeostatic rhythmicity. The psychobiological and pathological responses to stress become permanent, with consequential erosion in health and immune surveillance mechanisms (McEwen and Wingfield 2003; Chiappelli and Cajulis 2004; Chiappelli et al. in press-c; Prolo and Chiappelli 2007 in press). Oral lichen planus (OLP, Fig. 2.1) is an example of a Type II allostatic response with evident mucosal inflammation of the oral cavity.

The typical pathology, which involves basal cell lysis, and lymphocyte transmigration into the epithelial compartment, favors a statistically significant increase in incidence of oral squamous cell carcinoma. This premalignant lesion manifests in most patients as a chronic inflammatory disorder, and affect in 2 to 3% of the US population. The prevalence of OLP is higher in women than in men, and increases with age with onset in middle age. Human hepatitis C virus may be a

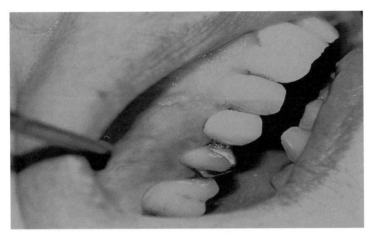

FIGURE 2.1. Oral lichen planus – Pathological manifestation (intraoral photograph) reticular stage.

predisposing factor, and there may be a genetic propensity for this condition since expression of HLA-Bw57 favors OLP, whereas HLA-Dql expression provides resistance against it. OLP is a life-long condition, which require serious commitment to long-term adaptability on the part of the afflicted patient, and of the significant others as well (Chiappelli and Cajulis 2004).

2.2.3 Allostatic Testing and Intervention

The process of testing for the allostatic response yields important information about the organism's psychobiological status. Therefore, several experimental protocols have been characterized for testing allostasis in animal and in human research. Human subjects under stress can be subjected to cognitive, sensory, or physiological challenges aimed at monitoring the allostatic process of immunephysiological adaptation to the experimental challenge in addition to basal states of the stress response.

Case in point the Stroop color–word interference test is an example of a cognitive challenge that was originally developed (1935) to examine patterns of cognitive decline. The Stroop challenge is measured as reaction time, and becomes evaluated in subjects who suffer from organic or stress-related cognitive dysfunction. One important *caveat* of this test is that patients with severe dementia show less Stroop interference effect, when the reaction time is adjusted for color naming performance, which suggests the need and the importance of partitioning out underlying deficits for the understanding of complex cognitive processes. Principal-components analyses, however, show that disruption of semantic knowledge and speeded verbal processing in certain patient (e.g., patients with Neuro-AIDS, patients with DAT) is a major contributor to the incongruent trial. The Stoop challenge is unquestionably an effective modality for testing the allostatic response of the glucocorticoids-cytokine (i.e., HPA-CMI) response in normal, healthy young men and women. Two groups of responders can be distinguished based on salivary and plasma glucocorticoids response to the challenge. The bottom 40% of the responders for cortisol responses to the challenge characteristically manifest significantly higher salivary and plasma levels of IL-6, but lower subjective experience of stress, compared to the high-glucocorticoids responders (top 40 percentile). This confirms that individual variations exist in the response of the HPA-CMI axis to a mild cognitive dissonance stress, and that these variations can have potentially adverse health outcomes (Chiappelli et al. in press-c; Prolo and Chiappelli in press; Kunz-Ebrecht, Mohamed-Ali, Feldman, Kirschbaum, and Steptoe 2003).

Another widely used stress paradigm for testing allostasis in human subjects is the dexamethasone-suppression test (DST), which tests the functional response of the HPA-CMI axis directly. Administration of 1 to 5 mg of dexamethasone (DEX) to normal subjects in the evening (23:00 h) results in a flattened cortisol circadian pattern, which lasts up to 24 h. The cortisol-suppression response is an indication of DEX-sensitivity and normal HPA response. A significant change in the migratory properties of circulating CD4+CD62L+T cells occurs in DEX-sensitive control subjects. This observation is significant since the CD4+CD62L+T lymphocyte subpopulation overlaps to a large extent with regulatory T cells (i.e., Tregs, CD4+CD25+FoxP3+), which play a central coordinating role in CMI (Chiappelli in press). The DST challenge is used in monitoring the HPA-CMI response in a variety of patient populations. In DEX-resistant patients who show failed suppression of cortisol following DST (e.g., most patients with anorexia nervosa, a large proportion of patients with Alzheimer's disease, or with major depression), the migratory properties of CD4+CD62L+T cells are also not altered. Therefore, this test provides a useful experimental tool to study the intimate relationship between the physiological regulation of the neuroendocrine and cellular immune systems in vivo in normal subjects and in certain patient populations. However, the DST is a static challenge to the HPA-CMI axis, and solely presents a reflection of the regulatory feedback response of the axis, rather than of the potential of the system per se. Moreover, a principal *caveat* of this test of the allostatic response is that it lacks of reliability, validity and specificity as a clinical diagnostic instrument. The test can be used in research protocols in oral biology and medicine, and we have shown that patients with OLP exhibit significant alterations in HPA-CMI regulation (Chiappelli and Cajulis 2004).

In brief, maladaptation of the person to the environment can cause psychological turmoil, with serious psychosomatic *sequelae*. A general "stress-coping-social support" model was proposed with the goal of identifying psychosocial risk factors for progression and to develop effective psychosocial modes of intervention (Mulder 1994). The model established social support as a beneficial mediating variable, and proposes certain underlying psychobiological "causal paths" among stress, coping, social support, and disease progression. Numerous traditional and alternative modes of interventions are beneficial in allostatic load and overload in psychosomatic medicine in general and in oral biology and medicine in particular. It is not the purpose of this chapter to discuss them, except to note that, within the context of allostatsis described above, their aim focuses at restoring the mind–body balance and harmony (e.g., mindfulness-based stress reduction program, Kabat-Zinn et al. 1992; Miller, Fletcher, and Kabat-Zinn 1995; Robert McComb, Tacon, Randolph, and Caldera 2004; Carlson, Speca, Patel, and Goodey 2004; Williams, Kolar, Reger, and Pearson 2001; Roth and Stanley 2002).

2.2.4 Allostasis Is Not Chaos-Or Is It?

The multitude of intertwined cellular and humoral events that characterize the complex process of allostasis behave according to ordered and organized randomness. At least two complex "strange attractors" may be hypothesized to determine and to characterize the chaotic behavior of psychobiological rhythmicity. Psychoneuroendocrine responses to basal homeostasis and to stress challenges may compound intrinsic immune regulatory events under basal condition and following antigenic threats in the form of a pathogen (=non-self) or autoimmune recognition (=self).

Georg Cantor (1845–1918) and Karl Theodor Wilhelm Weierstrass (1815–1897) first opened the field of fractal mathematics in the late nineteenth-century, but their work received little recognition until Benoit Mandelbrot (1924)[1] characterized a set of mathematical equations to represent graphs with self-similar properties in the mid-1970s. He introduced the terms "fractal" (Latin, *fractus* = broken into pieces; from the verb *frango, frangere, fregi, fractus*) to the scientific discourse, and defined fractals as graphical representations of relatively simple mathematical equations (Mandelbrot 1977, 1986). Computer visualization applied to fractal geometry offers a powerful visual argument by integrating into the scientific domain nonlinear dynamics and chaos theory (i.e., ordered nonlinear behavior), and object complexity.

"Strange attractors" have fractal structure. Fractal recurrence is one characteristic feature of inherent pattern of repetition of the fractal structure, which, together with certain attractors, which the MIT meteorologist Edward Lorenz named "*strange*," and which are now recognized to occur manifested by all biological systems (e.g., HPA-CMI axis) (Lorenz 1963). Chaotic motion can be represented as a phase diagram of that motion, with time as implicit, and each axis representing one dimension of the physiological state under study. This type of phase diagram depends on the initial state of the system as well as on a set of parameters, and certain forces toward which the system is attracted in its motion (cf., allostasis). Such forces are attractor*s*, and chaotic motion is driven and characterized by these "strange attractors," which of course can have great detail and complexity (Dalgleish 1999; Rew 1999; Glenny, Robertson, Yamashiro, and Bassingthwaighte 1991).

Fractal objects can be regular (or geometric) if they consist of large and small structures that are exact copies of each other, except for their size. Random fractals, by contrast, are characterized by the property of having the large-scale and the small-scale structures differ in the details (e.g., the shape of cell membranes). Fractal objects are either self-similar or self-affine, and are composed of subunits and of sub-subunits that are precisely or statistically similar (or affine) to the structure of the original.

---

[1] Sterling Professor Emeritus of Mathematics, Department of Mathematics, Yale University, benoit.mandelbrot@yale.edu

Deterministic fractals have the characteristic of precise similarity (or affinity), whereas statistical similarity (or affinity) characterizes statistical fractals. Self-similarity is proper to isotropic fractals, which are invariant under orthogonal scale transformation. Self-affinity pertains to fractals that are invariant under anisotropic (=affine) scale transformation (Nygard and Glattre 2003).

In brief, fractal analysis provides the means to quantify the irregularity of the contour of objects with a measurable value, the fractal dimension, which is rendered as:

$$\text{fractal dimension (FD)} = \frac{\log(\text{number of self-similar units})}{\log(\text{magnification factor})}$$

In mathematically stricter terms, for any fractal object of size P, and made up of smaller units of size p, the number of units (N) that fits into the larger object is equal to the size ratio (P/p) raised to the power of d, the Hausdorff dimension[2]:

$$d = \frac{\log N}{\log (P/p)}$$

Fractal analysis opens several areas of biological inquiry, including the study of the phenomenon of self-organized criticality, which refers to organisms continually driven out of balance striving to maintain or to regain a state of organized and poised critical balance (Rew 1999). Self-organized criticality corresponds to the characteristic state of criticality, or change that results from the process of self-organization during the time-dependent and transient period that lies between stability and chaos (Glenny et al. 1991). In the context of psychoneuroimmunology, of course, one example of self-organized criticality is the physiological events associated with allostasis noted above, and in this context, the fractal & recurrence analysis-based score (FRAS) may be an effective measuring tool. Case in point its use to quantify heart rate and pulse transit time in patients with probable Chronic Fatigue syndrome and in normal cohorts following the head-up tilt test (Naschitz et al. 2002). In the patient group, the challenge is likely to arise a Type II allostatic response due to the chronicity of the condition. By contrast, a Type I response is expected in normal control subjects. The data showed significant ($p < 0.0001$) elevation in FRAS in the patient compared to the control group (Naschitz et al. 2002). The FRAS index describes the reemergence of fractal structure within the context of the object under study—in this study, heart and pulse rate. It is timely to

---

[2] Named after Felix Hausdorff (1868–1942), but also referred to as Hausdorff-Besicovitch (Abram S. Besicovitch, 1891–1970) dimension, the capacity dimension, or the fractal dimension.

conceive of research paradigms that will incorporate the FRAS index in the context of investigations in oral biology and medicine. In that context, one possible patient population to be studied might be patients afflicted with oral mucositis.

2.3 Mucositis: A Case of Allostatic Load

2.3.1 Oral Mucositis

An important example of Type 2 allostasis in oral biology and medicine, which could benefit from fractal analysis, is the clinical condition of oropharyngeal mucositis (OM, Fig. 2.2). The term "mucositis" and "stomatitis" are often used interchangeably, but OM specifically pertains to pharyngeal-esophago-gastro-intestinal inflammation, that manifests as red, burn-like sore or ulcerations throughout the mouth.

Stomatitis is an inflammation of the oral tissues proper, which can present with or without sores, and is made worse by poor dental hygiene. Typically, stomatitis manifests 1 to 10 mm peripheral to a pathological lesion of the oral mucosa (Chiappelli 2005; Duncan and Grant 2003; Weissman, Janjan, and Byhardt 1989; Janjan, Weissman, and Pahule 1992).

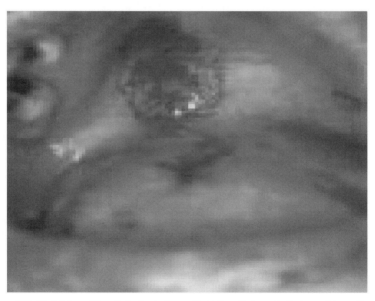

FIGURE 2.2. Oral Mucositis—Pathological manifestation (intraoral photograph)—soft palatal and lingual lesions.

OM is a common and treatment-limiting side effect of cancer therapy, which affect up to 100% of patients undergoing high-dose chemotherapy and hematopoietic stem cell transplantation, 80% of patients with malignancies of the head and neck receiving radiotherapy, and over 50% of patients receiving systemic chemotherapy. It is a critical condition since in 2004 alone, close to 1.5 million Americans were diagnosed with cancer and received some combination therapies for this disease, including surgery, radiation, chemotherapy, immunotherapy and/or cell transplantation. OM is the principal oral complication from cancer therapies. In most cancer patients, OM manifests as clinically observed atrophy (tissue damage) and telangiectasis (blood vessel, spider-like red spots) of the oral mucosa, with significant increase the risk for pain and necrosis (Chiappelli 2005; Duncan and Grant 2003).

OM is a limiting factor in the effectiveness of cancer therapy because it leads to significant impairment in quality of life (Weissman et al. 1989; Janjan et al. 1992). Patients with OM often report impaired eating, drinking, swallowing, and talking, as well as disturbed sleep. Increased predisposition to local and systemic infections is also serious *sequelae* commonly associated with OM (Cengiz, Ozyar, Ozturk, Akyol, Atahan, and Hayran 1999). Dry mouth, mouth sores and pain, taste changes, and sore throat are among the most frequently reported troublesome or debilitating side effects of OM. Together, they lead to significant weight loss, overall drop in energy level, fatigue, lethargy and muscle weakness (Rose-Ped, Bellm, Epstein, Trotti, Gwede, and Fuchs 2002). All in all, OM can be so severe as to lead to the need to interrupt or discontinue the cancer therapy protocol, and to hinder cure of the primary disease (Chiappelli 2005).

An important *caveat* to these reports, however, must be noted: many of the signs and symptoms listed above are clinically observed as well in many cancer patients who undergo radiation treatment alone or in combination with chemotherapy, and who do not manifest overt signs of OM. Therefore, it is often held that OM is one facet of the complex spectrum of side effects of cancer intervention, rather than the cause per se of the symptomatology described above.

What is certain is that OM is an acute oral mucosal inflammatory reaction secondary to cell death of the basal epithelial cell lining. The oral mucosa is a rapidly replacing body tissue: one that has received little attention in terms of defining its cell kinetics and cellular organization, relative to other domains of cell biology. The tissue is sensitive to the effects of cytotoxic agents, the consequence of which can be progenitor cell death, with the subsequent development of the symptoms of cell death and, consequentially, of OM. Radiation and chemotherapeutic agents act by blocking mitosis of rapidly dividing cell, including cellular toxicity and apoptotic cell death, and by decreasing their regeneration. These effectors kill cancer cells and fast-proliferating normal cells, such as oro-gastro-intestinal epithelial cells, thus generating lesions. Immune cells are henceforth actively recruited to migrate at the site of these emerging lesions, and an inflammatory state and alteration of oral flora ensue. An inflammatory lesion is what is commonly recognized as OM. The involvement of immunity in OM can involve systemic immune suppression, significantly increased risk of local and systemic infection, opportunistic infections and mortality due to sepsis. In brief, the loss of rapidly dividing epithelial progenitor cells triggers the onset of OM, but its severity and duration are determined by changes in local oral immunity (Chiappelli 2005; Duncan and Grant 2003; Potten et al. 2002; Alvarado, Bellm, and Giles 2002; Blijlevens, Donnelly, and DePauw 2005).

Oral microorganisms play an important role in aggravating the pathology of the impaired epithelium, and individual oral hygiene has considerable influence on symptom presence and severity. Common dental care products that offer several different active ingredients to reduce bacterial counts and plaque, such as fluorides, have antibacterial (sodium fluoride, amine fluoride, tin fluoride) and plaque decreasing effects (amine fluoride and tin fluoride), and substantially improve the prognosis of OM (Chiappelli 2005). Chlorhexidine is antimicrobial and very effective in inhibiting plaque growth, but, because of side effects, which include mucosal irritation, is not recommended as a viable for long-term use for OM. Smoking, and alcohol use and abuse, and psychoemotional stress are reported to contribute to exacerbate OM (Chiappelli 2005; Janjan et al. 1992; Cengiz et al. 1999; Alvarado et al. 2002; Kwong 2004).

Certain cytokines play an important role in OM. Evidence-based research has established that applications of granulocyte-macrophage colony-stimulating factor (GM-CSF) plays a critical role in preventing the onset or the exacerbation of oropharyngeal mucositis (risk ratio, RR = 0.51, CI[95]: 0.29–0.91; NNT = 3, CI[95]: 2–20) (Clarkson, Worthington, and Eden 2003). Clinical trials suggest that GM-CSF has clinical benefits beyond enhancing neutrophil recovery, including shortening the duration of mucositis and diarrhea (Nemunaitis et al. 1995), stimulating dendritic cells, preventing infection, acting as an adjuvant vaccine agent, and facilitating antitumor activity (Buchsel, Forgey, Grape, and Hamann 2002). GM-CSF belongs to the family of colony-stimulating factors (CSFs), which are central to the development and maturation of cells of the immune system, the modulation of their functional responses, as well as the maintenance of immune homeostasis and overall immunity. This group of glycoproteins consists of the macrophage-CSF (M-CSF), granulocyte-CSF (G-CSF), granulocyte/macrophage-CSF (GM-CSF), and multi-CSF (interleukin[IL]-3). GM-CSF functions at early stages of lineage commitment regulating the expansion and maturation of hematopoietic progenitor cells (Barreda, Hanington, and Belosevic 2004; Chiappelli in press).

GM-CSF is a proinflammatory cytokine, which stimulates proliferation and differentiation of neutrophilic, eosinophilic and monocytic lineages of cellular immunity. In immunosuppressed patients, and in murine models of therapeutic immune suppression, GM-CSF administration is effective in boosting the innate immune response, while continuing to suppress the adaptive immune response to prevent graft rejection (Chiappelli 2005; Xu, Lucas, and Wendel 2004). GM-CSF belongs to a pattern of cytokines distinct to the traditionally recognized TH1 and TH2 patterns. By its action on dendritic cells, GM-CSF modulates the production of IL-23 by these cells. IL-23 belongs to the IL-12 cytokine family, which shares p40, and binds to the IL12β1/IL-23R receptor complex to play a critical role in end-stage inflammation via signaling of the JAK/STAT pathway. Administration of IL-11 down-regulates IL-12. IL-23 augments production of IL-17 by T cells, which becomes the significant cytokine in the involvement of T cells in inflammation. IL-17 leads to decreased neutrophilia in animal models. In vitro administration of IL-17 synergizes the effects of TNF-α for GM-CSF induction, for increased intercellular adhesion molecule (ICAM)-1 (CD54) expression by CD34+ progenitors, for increased neutrophil maturation, and for increased proinflammatory IL-6 and prostaglandin(PG)E2 and G-CSF growth factor production by keratinocytes and endothelial cells. IL-17 also favors increased secretion of the migration-inducing cytokine, IL-8. In human U937 cells, and probably in all lymphoid and myeloid cells, the JAK/STAT signaling pathway is responsible for transducing signals from the IL-17 receptor. Human vascular embryonic cells (HUVEC) express a homologue of IL-17R, as do ductal epithelial cells of human salivary glands, suggesting a significant role for IL-17 in immunity in the oral mucosa. In fact, in inflamed gingival tissue, IL-17 is the predominant cytokine in tissue proximal to 4 to 5 mm diseased periodontal pockets. IL-17 plays a critical role in immunity and mucosal inflammation, by modulating the effects of GM-CSF and other growth factors (Watford, Hissong, Bream, Kann, Muul, and O'Shea 2004; Chiappelli in press).

In an evidence-based research study on the comparison of utilization of complementary and alternative (CAM) treatment for OM to traditional pharmacological intervention (e.g., GM-CSF), we have examined 424 reports. Critical analysis suggests that traditional medical intervention is not inferior to CAM in the treatment of mucositis. Issues in the quality of the research methodology were identified that hamper the overall quality of the data, including the inadequate use of double blinding, randomization, and placebo-controlled, and the lack of satisfactory statistical analysis of the data. Research protocols in this area often suffer from significant weaknesses, which hamper their efficacy. No meta-analysis was generated because of the nonhomogeneity of measures, and the lack of statistical stringency.

2.3.2 Neuroendocrine-Immune Orderly Chaos in OM

The immune system is a sensory organ for external stimuli, such as local cellular trauma caused by histo- and cyto-pathology and infectious agents that cannot be detected directly by the nervous system (*vide supra*). The bidirectional nature of the communication between the immune and nervous systems is responsible for directing the organism's response to infection and to inflammation (Solomon 1987; Chrousos and Gold 1992; Kiecolt-Glaser et al. 2002; Irwin 2002; Chiappelli et al. in press-c; Engle 1960; Prolo and Chiappelli in press; Chiappelli et al. 2004). The physiological challenge that ensues following sensory perceptions, including sympathetic, parasympathetic and hormonal responses, plays a crucial role in regulating neuroendocrine-immunity, thus ensuring the functionality and the relevance of the psychoneuroimmune loop. Because both the nervous and the immune systems have the ability to learn, and the capacity for memory, and because cells of both systems are endowed with receptors for neural and immune factors, the lymphocyte was termed the "mobile brain" (Pierpaoli and Maetroni 1988).

The intercellular communication between the nervous and immune systems via common receptors and signal molecules yields a conceptual framework for this cross talk. Soluble immune regulatory factors, including cytokines (e.g., proinflammatory, TH1, TH2), growth factors (e.g., GM-CSF) modulate neurobiological responses at rest and during the allostatic response. Neurotransmitters (e.g., epinephrine, acetylcholine), neuropeptides (e.g., endogenous opioids), and hormones (e.g., glucocorticoids) have specific receptors on and in cells of the immune system, that contribute to mediating immune events involved in inflammation, surveillance and healing (i.e., during allostatic load and overload). This cross talk has important implications in maintaining the health of tissues, including the oral mucosa (Chiappelli et al. in press-c; Chiappelli in press; Blalock 1994). "Hardware" connections exist between the nervous system and immune organs (Bulloch, Cullen, Schwartz, and Longo 1987; Felten et al. 1987) that are relevant to the physiological cross talk between the psychoneuroendocrine and the immune system (Sterling and Eyer 1988; McEwen and Wingfield 2003; Chiappelli and Cajulis 2004; Chiappelli et al. in press-c; Chiappelli et al. 2004; Besedovsky and del Rey 1996).

It is possible and even probable that the severity of tissue damage in OM, as well as the pattern and rate of recovery of the oral mucosa depend in large part on its neurobiological structure. The oral mucosa is richly innervated by sympathetic and parasympathetic fibers, which establish an intimate dialogue with resident and invading myeloid and lymphoid cells in the healthy tissue as well as in pathological oral lesions (e.g., OLP, 41). Our studies of the glossopharyngeal nerve demonstrated the implications of the psychoneuroendocrine-immune network within the oral mucosa (Romeo, Tio, Rahman, Chiappelli, and Taylor 2001).

As noted above, fractal analysis has emerged in the biological sciences in the last decade to provide a new powerful look into physiological processes that follow nonlinear dynamics, as well as a reliable approach to quantify the complexity of cells and cellular components and organelles. Since function is interdependent with structure, fractal analysis, by elucidating the fundamental morphological changes in immune cells at the cellular and the nuclear levels following psychoneuroendocrine modulation, will contribute to our mounting understanding and knowledge of immunophysiology. The fractal characterization of human T lymphocytes, with emphasis on the Tregs subpopulation will be critical in a variety of clinical situations, including oral pathology (Chiappelli et al. 2005), and OM in particular.

The analysis of fractal dimension yields a noninteger measure of how "complicated" a self-similar object is, and is strictly greater than the integer derived from a topological Euclidean dimension.[3] Fractals have a noninteger dimension, which represents the measure of space-filling ability of the object under study. The fractal dimension allows the quantitative comparison and categorization of fractals. By contrast, the traditional Euclidean dimension is the number of coordinates required to specify the object. A straight line, for example, has a Euclidean dimension of 1, but a plane has an Euclidean dimension of 2. A jagged line takes up more space than a straight line, and has a fractal dimension between 1 and 2 (e.g., 1.56). Fractal dimensions are, therefore, usually, albeit not always, about 20% larger than Euclidean dimensions, and are all the more informative, in that they represent the complexity of the object under measurement (e.g., T lymphocytes invading the oral mucosa in OLP, Chiappelli et al. 2005).

---

[3] In the traditional sense of the word, "dimension" refers to a measure of how an object fills space, a topological space. This measure, the Lebesgue (Henri Lebesgue, 1875–1941) covering dimension, is the topological dimension, and is rendered by the archetypal example of the Euclidean n-space with a topological dimension n. (Lebesgue H. 1904. *Leçons sur l'Intégration et la Recherche des Fonctions Primitives, Professées au Collège de France.* Paris, Gauthier-Villars).

Lacunarity analysis is another a multiscaled method of this approach to object analysis, typically a sliding box algorithm, for describing patterns of spatial dispersion, which is used with both binary and quantitative data in one, two, and three dimensions. It is a parsimonious analysis of the overall fraction of a map, or transect, covered by the attribute of interest, the presence of self-similarity, the presence and scale of randomness, and the existence of hierarchical structure. For self-similar objects such as cells, lacunarity analysis can be used to determine the fractal dimension by readily interpretable graphic results. Lacunarity, λ, is a counterpart to fractal dimension in that it describes the texture of a fractal. In more practical terms, lacunarity is concerned with the size distribution of the holes that compose a fractal. High lacunarity pertains to a fractal that has large gaps or holes. A fractal that is translationally invariant has characteristically low lacunarity. Different objects could have the same fractal dimension, but appear largely different because they differ in lacunarity (Mandelbrot 1994).

The use of fractal dimension and lacunarity has increasingly permeated the biological sciences. Case in point, fractal dimension was shown to be a reliable measure of atypical nuclei in dysplastic lesions of the *cervix uteri* (Sedivy, Windischberger, Svozil, Moser, and Breitenecker 1999). In that study, fractal dimension was obtained by the box-counting method. This approach involves starting with an image of which the fractal dimension is sought. The image is translated into a binary version where the pixels above a given cut off brightness are set to one, and the others are set to zero. An image of contours around the bright sections is generated, and divided up into boxes of a given size. The count of the number of boxes required to contain the contour is obtained, and the plot of the log of the box size vs. the numbers obtained for that box size provides the best-fit slope estimate, which corresponds to the fractal dimension of the image. That study impressed the notion that whereas the same fractal dimension could characterize certain nuclei, they differed substantially in their porosity, their lacunarity (Sedivy et al. 1999).

Box-counting fractal analysis was used to compare the hepatocyte nuclei obtained from normal liver and from hepatocellular carcinoma. In that study, the number of grid squares (boxes, tiles) that the cell border-image contacts, regardless of the number of pixels contained within the border, was counted. The log number of squares was plotted against the log of the box edge length, rendered as

$$D_{\mathrm{B}} = \lim_{\varepsilon \to 0} \frac{\log N(\varepsilon)}{\log N(1/\varepsilon)}$$

where D_{B} represented the box-counting fractal dimension of the nuclear image in binary form, and $N(\varepsilon)$ the smallest number of boxes with side length ε that completely covered the outline of the object under study. The log–log plot of the $(1/\varepsilon)$ (X-axis) versus $N(\varepsilon)$ (Y-axis) was derived to render the linear approximation of the first-degree polynomial equation ($Y = a + mX$). The slope of the least square fitted line, m, thus obtained is taken to represent D_{B}, the box-counting fractal dimension. Self-similarity can be established by means of the evaluation of the fit of regression lines. Higher values for the coefficient of determination are representative of better fit. The closer the coefficient of determination is to 1.0, the larger the proportion of random error accounted for, and the greater the extent of self-similarity of the objects under study (Kerenji, Bozovic, Tasic, Budimlija, Klem, and Polozovic 2000).

Common approaches of fractal analysis of self-similarity include, in addition to the box analysis, the Hurst[4] exponent. This value measures the smoothness of fractal time series and is based on the asymptotic behavior of the rescaled range of the process, the perimeter-area dimension analysis, the information dimension analysis, the mass dimension analysis, and the ruler dimension analysis. Common fractal dimension calculation methods for self-affine traces can also include rescaled range analysis, power-spectral analysis, roughness length analysis, variogram, and wavelets. Fractal dimension of size–frequency data is most commonly estimated by the method of fragmentation dimension. It is unquestionable that the assessment of the fractal dimension has become a useful tool in quantitative histology and cytology over the past decade. This measurement is reliably obtained from computerized image analysis systems. It is generally accurate, with errors of <1.5%, and repro-ducible (reliability coefficient, $r = 0.972$, CI^{95}: 0.868–0.987) (Cross, Cotton, and Underwood 1994).

Fractal dimension analysis adequately distinguishes cell populations with complex plasma membranes at diverse stages of maturation. Oligodendrocytes in culture, for example, were analyzed by the box-counting method. In order to generate estimates on the ascending part of the curve, which encompasses the range in which cells express the property of self-similarity, box size was restricted between 1 and 20 µM. Larger boxes provide no additional information on cell morphology, and smaller boxes are well within the lower-end sensitivity range (0.1 µM) of the typical analysis software,[5] but provide no further information about the cells filamentous projections, typically in the 1 µM range. Below this range, cells typically loose the property of self-similarity[6] (Takeda, Ishikawa, Ohtomo, Kobayashi, and Matsuoka 1992; Bernard, Bossu, and Gaillard 2001). In brief, future studies aimed at providing lacunarity and box-counting fractal analysis assessments of invading lymphocyte subpopulations at different stages of OM pathology and treatment will be most revealing of OM immunopathology.

A principal *caveat* of this analytical protocol, however, lies in the fact that errors in fractal quantification arise because estimates are derived from pixel counts, a function of the number of pixels cut by the boundary of the object. Cellular contours with a fractal dimension greater than the planar Euclidean dimension of 1.0 will

---

[4] British hydrologist, H.E. Hurst, who systematically characterized the hydrologic properties of the Nile (Hurst, H.E., Black, R.P. and Simaika, Y.M. 1965. Long-Term Storage: An Experimental Study. Constable, London).

[5] e.g., Benoit, Trusoft

[6] Prof. S. Gaillard, pers. commun.

appear more complex as the pixel size decreases and the resolution increases. By contrast, the more contorted boundaries will have higher fractal dimensions, with less error of estimate. Flawed interpretations of pixel-based quantification of fractal dimension also arise as a function of the distribution of the pixels: a compact distribution yields typically more reliable results and less error, compared to more widely dispersed, diffuse, or patchy pixel phenomenon. Typically, a scattered phenomenon yields the following relationship:

Scatteredness: (SD/dimension mean) × 100 = (dimension mean − 1.00)/2

Lymphocytes are typically smaller and less complex than oligodendrocytes. Nevertheless, the perimeters of the surface membranes of different lymphocyte populations can be digitized from electron microphotographs, and the data analyzed to derive the fractal perimeter dimension. Values range from 1.02 to 1.34, independently of the magnification, and are consistently obtained across the lymphocyte population of the same type (Chiappelli et al. 2005; Keough, Hyam, Pink, and Quinn 1991).

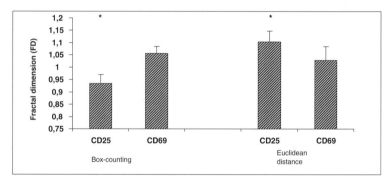

FIGURE 2.3. Fractal dimension in CD25+ and CD69+ human T lymphocytes (measured by box counting or by Euclidean distance map): oral lichen planus: fractal dimension in CD25+ and CD69+ human T lymphocytes measured by the box-counting method, compared to Euclidean distance dimension map. Biopsies from the oral mucosa from patients with oral lichen planus were obtained, fixed and process for standard immunohistochemistry. T cells at different stages of activation were identified with either monoclonal antihuman CD25 or antihuman CD69, and visualized by horseraddish peroxidase staining. Micrographs (60× + 10× eyepiece magnification), were obtained, and the stained cells isolated and adjusted the brightness by means of computer-aided software (Micrografx picture publisher). Euclidean estimates of size (i.e., relative estimates of cell radius) as well as fractal dimension were obtained by box-counting (i.e., relative complexity of the staining pattern on the cell membrane) with the FractalFox software (Qichang Li, Univ. Memphis).

Immunological studies have confirmed that when blood cells are projected into an image plane their contours appear as borderlines of irregular shape with the property of the statistical self-similarity. The fractal perimeter dimension of normal peripheral blood mononuclear cells and mature T lymphocytes fall in the median range of 1.23 to 1.17, and the same dimension obtained from T cells of hairy-cell leukemia with highly convoluted morphology range between 1.32 and 1.36. By contrast, blast cells obtained of acute lymphoblastic leukemia and generated from activation of normal T cells in vitro appeared larger with a less complex membrane—stretched, as it were. They yielded lower fractal perimeter dimensions in the range of 1.11 to 1.16 (Losa, Baumann, and Nonnenmacher 1992). Human T lymphocytes from normal donors, and hairy leukemic cells exhibit distinct fractal dimension (mean ± SD: 1.15 ± 0.03 for normal T cells, compared to 1.34 ± 0.04 for hairy leukemic T cells, $p < 0.05$) (Nonnenmacher, Baumann, Barth, and Losa 1994). In our studies, active T lymphocytes that invade the oral mucosa in OLP have similar dimensions (Chiappelli et al. 2005), which suggests that immune cells that will be found in OM pathological tissue ought to be characterized by contours in that same range.

Specifically, our estimates of scatterdness, based on the relationship provided above, suggested an even distribution of the staining pattern around the circumference of the cell membrane, as opposed to compacted and clumped patterns of staining. Our data showed a reliability of measurement of 0.92 ± 0.11 (Fig. 2.3).

The data summarized in Fig. 2.3 show that the box-counting (fractal dimension) and Euclidean distance was not significantly different among CD69-stained cells, and was close to 1.0 (mean ± SD: 1.056 ± 0.029 and 1.029 ± 0.056, respectively, $p >$ 0.05), suggesting that the configuration of these cells upon the slide was rather planar and linear. The fractal dimension of CD25-stained cells was significantly lower than their CD69 counterpart (0.934 ± 0.036 and 1.056 ± 0.29, respectively, $p < 0.0001$), suggesting that the membrane contours of CD25+ T cells was overall smoother than that of CD69+ T cells. By contrast, the Euclidean dimension of CD25-stained cells was significantly larger than that of CD69-stained cells (1.104 ± 0.043 and 1.029 ± 0.056, respectively, $p < 0.028$), indicating that CD25+ T cells were overall larger (larger relative estimate of cell radius), compared to CD69+ T cells.

One interpretation of these observations suggests the possibility that CD25+ T cells show signs of stretching of the plasma membrane as they enlarge toward the blast stage of cell activation. Euclidean dimension for both CD25+ and CD69+ T cells was freer of random error, with 98% of the variance accounted for. Box-counting assessments, by contrast, suffered from more random error across both cell populations with 75% of the variance accounted for. Such stretching is precisely what has been observed in cellular immune studies of T cell activation: early following antigenic trigger, T cells express CD69. Expression is rather transient, and leads to continued signaling pathways, which result in the expression of IL-2 and the α chain of its receptor, CD25 within a few hours. During this time, the cell advances through G1 and S of the cell cycle, and enlarges toward its blast state, which is fully attained several hours later (Chiappelli in press). In brief, fractal dimension analysis appears to be sensitive enough to distinguish CD69+ from CD25+ activated T cells.

It is possible and even probable, therefore, that this approach can be reliably utilized to identify subpopulations of Tregs.

Fractal dimension or lacunarity, as well as the fractal dimension of the epithelial-connective tissue, are associated with functional changes of transformed lymphocytic cells (Sedivy et al. 1999). The Zipf law[7] can serve to characterize the T lymphocyte repertoire. In a study of murine CD8+ T cells, the fractal dimension derived from the Zipf plots correlated significantly ($p > 0.05$) with the nature of the repertoire (Burgos and Moreno-Tovar 1996). In a similar vein, distinct CD8+ cytotoxic T cell clonotypes were established on the basis of the unique DNA sequence of the third complementarity-determining region of the TcR β-chain. The frequency distribution of the clonotypes indicated a complex population pattern probably dependent upon mechanisms for maintaining a large number of antigen-specific clonotypes at a low frequency in the memory pool. The clonotypes were ranked in terms of population characteristics similar to power law distribution. When the repertoire was divided into specific subsets, based on their immunogenic properties, the resulting clonotype frequencies described a power law-like distribution, which indicates self-similarity of the repertoire, in which smaller pieces are slightly altered copies of the larger piece. The power law-like description was found to be stable in time, and was replicable across donors.

Self-similarity and power laws are associated with fractal systems, demonstrating the applicability of fractal analysis to the characterization of memory CD8 T cell repertoires (Naumov, Naumova, Hogan, Selin, and Gorski 2003), and could serve to characterize diverse repertoires of invading immune cells during the OM immunopathological process. This is particularly true and applicable to the Tregs subpopulation because Tregs easily migrate from the peripheral blood, to the lymphatic system, to the oro-pharyngo-gastro-intestinal mucosa (Chiappelli in press).

In brief, it is possible and even probable that fractal analysis will generate new insights about neuroendocrine-immune interaction in OM. Fractal analysis will also undoubtedly emerge as a novel and reliable experimental approach for the

---

[7] Relations of the form $f(x) = k\ x^b$ are called power laws, but do not imply fractal structure. Power-laws imply that small occurrences are extremely common, whereas large instances are extremely rare, and are also referred to as Zipf or as Pareto relations. The Zipf law (George Kingsley Zipf, Harvard linguistics professor, 1920–1950) refers to the "size" y of an occurrence of an event relative to it's rank r, and seeks to determine the "size" of the 3rd or 8th or 100th most common occurrence. Size denotes the frequency of occurrence. Zipf's law states that the size of the rth largest occurrence of the event is inversely proportional to it's rank: $y \sim r^{-b}$, with b close to unity. By contrast, Pareto's law (Vilfredo Pareto, Italian economist, 1948–1923) is given in terms of the cumulative distribution function, that is the number of events larger than x is an inverse power of x: $P[X > x] \sim x^{-k}$. Pareto's distribution reports the number of events at any given x, and represents the probability distribution function derived from Pareto's law.

elucidation of the genomic, proteonomic and interactomic molecular cartography (*vide infra*) that defines and characterizes the phenotypic and the functional properties of Tregs and other immune cell subpopulations involved in OM and other stomatological pathologies.

2.3.3 Circadian Neurobiology and Neuroimmunopathology in OM

According to the guidelines for prophylaxis and therapy of mucositis (http://www3.interscience.wiley.com/cgi-bin/fulltext/108069518 [2005-05-04]), the risk to develop a grade 3 (severe) or grade 4 (life threatening) OM lies between 2 and 66%, depending on the type of chemotherapy (without additional radiotherapy), and between 1 and 53% depending on the type of tumor. In over 30% of the patients who develop a grade 3 or 4 OM, the start of the next chemotherapy cycle needs to be delayed, resulting in impaired cancer treatment (http://www3.interscience. wiley.com/cgi-bin/fulltext/108069518 [2005-05-04]).

Chemotherapeutic interventions that use 5-fluorouracil (FU) commonly lead to OM. The relationship between FU, one of the oldest and most widely used anticancer drugs, pharmacokinetics and patient response support a link between systemic drug exposure, physiology, and drug response, and consequentially survival. Hence, the maximal tolerated exposure is derived from pharmacokinetic follow-up and individual dose adjustment. Maximal tolerated exposure is dependent upon FU individual catabolic rate (Milano and Chamorey 2002).

The reported circadian pattern in epithelial cell response to insult following radiation or chemotherapy is likely consequential to neurobiological regulation, as well as biological circadian pattern inherent to the biochemical and molecular properties of certain populations of cells (e.g., expression of cell-autonomous Dec or Per genes). This hypothesis has barely been tested in isolated cell systems, and more in-depth and rigorously focused research is needed.

Data to date indicate that the activity of dihydropyrimidine dehydrogenase (DPD[8]), the rate-limiting enzyme in FU catabolism, varies according to the time of day in pattern reminiscent of the circadian variation of cortisol in normal subjects (Raida et al. 2002). This observation suggests that glucocorticoids and glucocorticoids-responsive elements in the nuclear compartment may mediate the regulation of the DPD gene. Whereas physiologic data from both human and animal investigations confirm circadian rhythmicity in DPD activity (Milano and Chamorey 2002), genomic and epigenetic (i.e., by noncoding DNA, *vide infra*) studies on the regulation of the DPD gene in the chromatin must now be designed.

These investigations will be particularly important because recent evidence indicates fine circadian regulation of DPD. For example, in an in vivo murine model of OM, the toxicity of FU on the oral mucosa was decreased by two to eightfold if

---

[8] EC 1.3.1.2.

this drug was injected near the middle of the day (rest span) rather than in the middle of the night (activity span). Since the rhythm in FU toxicity appears to be linked to the sleep-wakefulness endogenous circadian cycle across species, the least toxic time in patients would correspond to 4:00 AM (Levi et al. 1995). Furthermore, in an experimental model of neonatal male rats, DPD activity in liver increased 48 h following glucocorticoids injection (Fujimoto, Kikugawa, Kaneko, and Tamaki 1992).

Experiments in vitro have demonstrated that DPD activity is expressed in most tissues, including epithelial cells and leukocytes, and is highest in hepatocytes. Peripheral blood mononuclear cells (PBMC) are routinely used to monitor clinically DPD activity, and a significant, albeit weak, correlation between PBMC and liver DPD activity has been reported (Milano and Chamorey 2002). Among circulating leukocytes, myeloid derivatives (e.g., monocytes, macrophages) express greater activity than their lymphoid counterpart (i.e., T & B cells). A significant positive correlation was reported between PBMC–DPD activity and the differential percentage of monocytes/macrophages, which contributes to the observed inter- and intrapatient variability in the activity of DPD among the patients (van Kuilenburg, Meinsma, and van Gennip 2004).

Oral mucosa samples taken from healthy subjects show circadian variations in DPD activity. These findings were obtained by measuring DPD activity by HPLC. The data revealed a change in enzymatic activity from 0.004 to 0.13 nmol/min/mg protein at 10:00 h and from 0.07 to 0.16 nmol/min/mg protein at 24:00 h, with a 30% average relative increase from morning to midnight (range: -20% to $+100\%$; $p < 0.073$). This pattern mirrors that of DPD activity observed in leukocytes. In brief, DPD rhythmicity in the oral mucosa cells peaks at 01:00 h, which is an important determinant of FU tolerability in the context of OM. These findings have considerable potential clinical relevance for the pattern of chronomodulated delivery of FU chemotherapy (Barrat, Renee, Mormont, Milano, and Levi 2003).

However, the pattern of circadian DPD activity levels can vary considerably, particularly among subjects. In some cancer patients, and for biological reasons that are not fully elucidated as of yet, this circadian pattern of DPD activity vanishes. The diurnal cortisol pattern showed the expected consistent circadian rhythm in the control subjects, and a blunted circadian cortisol pattern in a group of patients with advanced gastrointestinal carcinomas treated with FU. A trend towards a circadian rhythm of DPD mRNA expression in PBMC was also observed that corresponded roughly to the cortisol curve (i.e., peak of DPD mRNA expression at 05:00 h, trough at 14:00 h, $p < 0.005$ Mann–Whitney–Wilcoxon test). The DPD mRNA circadian pattern could be fitted to a cosine wave ($p = 0.001$, 0.014, Cosinor analysis) in 40% of the control subjects, but in not in the cancer patients. In brief, in these cancer patients, PBMC-DPD mRNA expression showed no trend toward consistent circadian rhythmicity, whereas circadian endogenous cortisol secretion pattern was maintained (Raida et al. 2002).

The principal *caveats* to the study of the circadian physiology of DPD in the context of OM lie in the fact that the relationship between PBMC-DPD activity and FU systemic clearance is weak ($r^2 = 0.10$). Thus, simply determining PBMC-DPD is not sufficient to predict accurately FU clearance in general, and OM in particular. Evidently several as-of-yet undetermined intervening variables contribute to the maximal tolerated exposure. Case in point, population pharmacokinetic studies identify patient covariables that influence FU clearance, including increased age, high serum alkaline phosphatase, duration of drug infusion, in addition to low PBMC-DPD (Milano and Chamorey 2002).

The observations about the circadian pattern of DPD have brought forth the realization that optimal maximal tolerated exposure to FU could be obtained by carefully controlling the timing of FU administration in a chronomodulated mode. Research data overall support this hypothesis. Plasma pharmacokinetics of FU (600 mg/m^2/day) differs among patients with advanced colorectal cancer, when the drug is administered with a programmable-in-time pump by continuous infusion for 5 days, either at a constant rate of delivery or with a chronomodulated rate. The peak of FU appears at 04:00 h in the chronomodulated schedule of administration. When FU is administered at a constant rate, mean plasma concentration varies in a circadian manner each treatment day, with a peak at 04:00 h (approximately 800 ng/ml) and a trough at 13:00 h (approximately 100 ng/ml). Severe OM was invariably observed by all patients on a flat schedule of FU chemotherapy, but only in a small proportion of patients on the chronomodulated schedule ($p < 0.008$). OM was generally less pronounced among patients with circadian rhythms in FU, which suggests that one mode of control of OM may be, in certain groups of cancer patients at least, via amplification or induction of FU circadian rhythms (Metzger et al. 1994).

In a related study, in patients receiving protracted low dose FU infusion, the circadian rhythm in FU plasma concentration was observed to peak at 11:00 h and to be lowest at 23:00 h, on average. The inverse relationship observed between the circadian profile of FU plasma concentration and PBMC-DP activity in these same patients indicated a link between DPD activity and FU pharmacokinetics. When the impact of the biological time of drug administration was also studied with short venous infusions; clearance was 70% greater at 13:00 h than at 01:00h. Similarly, peak drug concentration occurred in the first half of the night in patients receiving constant rate FU infusion for 2 to 5 days. Wide interindividual variations in the timing of the peak and trough of the 24 h rhythm in DPD activity were observed (Milano and Chamorey 2002).

Consequently, a rationale for FU chronomodulated therapy emerged based on the apparent circadian rhythm in host drug tolerance, which is greatest during the nighttime when the proliferation of normal target tissue is least. A randomized study of chronomodulated FU therapy with maximal delivery rate at 04:00 h further showed to be significantly more effective and less toxic than control flat FU therapy (Milano and Chamorey 2002). Other studies have shown mixed results of chronomodulated chemotherapy regimens with variable-rate infusions of FU have shown mixed results (Raida et al. 2002).

The novel tools of evidence-based research in medicine and of systematic review were employed to critically review the pertinent available research. We evaluated the evidence in support of the hypothesis that chronomodulated delivery of FU could be recommended, compared to control flat FU administration for therapeutic cancer intervention, in terms of avoiding the side effects of OM. The process for the search of the literature, critical evaluation and analysis of the evaluations were as described elsewhere (Chiappelli, Navarro, Moradi, Manfrini, and Prolo 2006a). The number needed top treat could not be computed because the design of the studies failed to include appropriate control groups, when considering the efficacy of a particular chemotherapy regimen. A X^2 analysis was used to test if the difference between the percentage of patients who developed mucositis was statistically significant between the groups. From a total of 291 papers, 63 were included in this study, of which 33 implemented a continuous treatment of FU, and 30 studies reported chrono-modulated administration of FU. Of a total of 1395 patients, 766 (54.9%) experienced mucositis with a continuous infusion of FU, compared to 532 (34.4%) out of 1545 patients receiving chronotherapy. Among patients in the treatment group that received constant administration of FU, the ratio of those who developed mucositis to those who did not (766/629 = 1.2178) was significantly higher than the same ratio obtained in the group following chronomodulated administration of FU (532/1013 = 0.5252) (X^2 = 123.83, p < 0.0001). These findings indicate that the published research evidence supports chronomodulated FU therapy as leading to significantly fewer patients to develop mucositis compared to the continuous infusion regimen of FU.

This systematic evaluation of the literature confirms that research toward understanding and controlling the biochemical and molecular variables that direct and regulate the circadian pattern of FU catabolism by DPD requires immediate attention. Specifically, it has been proposed that future research focus on easy-to-obtain markers of specific rhythms to individualize the chronomodulated FU delivery (Milano and Chamorey 2002).

2.4 Directions for the Future

Pyrimidine antagonists, such as FU, capecitabine, a tumor-selective pre-prodrug of FU, cytarabine (ara-C) and gemcitabine (dFdC), are widely used in chemotherapy regimes for a variety of solid and circulating cancers. Extensive metabolism is a prerequisite for conversion of these pyrimidine prodrugs into active compounds. Interindividual variation in the activity of metabolizing enzymes can affect the extent of prodrug activation and, as a result, act on the efficacy of chemotherapy treatment. Genetic factors at least partly explain interindividual variation in antitumor efficacy and toxicity of pyrimidine antagonists. The expression of three enzymes is involved in the metabolism and antitumor activity of FU:

> thymidine synthase (TS) binds FU and utilizes it in spurious DNA synthesis;
> thymidine phosphorylase (TP) catalyzes the reversible phosphorylation of thymidine to thymine and 2-deoxyribose-1-phosphate and is an essential rate-limiting enzyme for activation of FU by conversion to its active metabolites, thus playing an important role in the inhibition of TS;
> DPD catabolizes FU into its inactive metabolites.

DPD and pyrimidine nucleoside phosphorylase (PyNPase) are the first and rate-limiting enzymes that regulate FU catabolism, and tumoral DPD activity appears to be a promising predictor of FU sensitivity (Ueda et al. 2000). The measurement of the intratumoral level of these enzymatic activities may be useful in predicting tumor sensitivity to FU, and may shed light on the loss of DPD circadian pattern in some patients (*vide supra*).

Certain newly available chemotherapeutic choices may take the TP:DPD ratio into consideration (Fujiwaki, Hata, Nakayama, Fukumoto, and Miyazak 2000). For instance, the mRNA levels of DPD and TS was examined in 28 oral squamous cell carcinomas (SCC), and 22 salivary gland tumors by semiquantitative reverse transcription polymerase chain reaction. The correlation of the responsiveness of the patients with oral SCC to FU with the intratumoral levels of DPD and TS mRNA was also studied. Biopsy specimens were obtained before treatment. Patients received oral FU (UFT), irradiation of cobalt-60 (up to 60 Gy), and injection of an immuno-potentiator (OK-432). Intratumoral levels of DPD mRNA in the patients who showed complete response and partial response were significantly lower than the levels in the patients who showed no effect. Intratumoral levels of DPD mRNA did not correlate with the local recurrence of the tumor. Taken together, these data indicate that intratumoral levels of DPD mRNA may predict the tumor response to FU-based chemo-immuno-radiation therapy in the patients with oral SCC (Hoque et al. 2001). However, and based on the previous discussion, fractal analyses could have significantly increased the relevance of these observations.

Together, TP and DPD influence the activities of fluoropyrimidine anticancer drugs. The sensitivity of cancer cells to capecitabine, for example, correlate better to the TP/DPD ratio than to levels of either enzyme alone (Megyeri, Bacso, Shields, and Eliason 2005), and certain newly available chemotherapeutic choices may take the TP:DPD ratio into consideration (Tuchman, Ramnaraine, and O'Dea 1985). Deficiency or down-modulation of DPD can lead to elevated levels of unmetabolized FU, and consequential toxicity. Indeed, the field of molecular diagnostic has recently established the pivotal role of the catabolic pathway of FU in the determination of toxicity towards FU. Patients with a partial DPD deficiency proved to be at risk of developing severe toxicity after the administration of FU, and partial DPD deficiency has now proven to be a novel pharmacogenetic disorder associated with severe FU toxicity (van Kuilenburg et al. 2004).

The Michaelis constant (Km) for the catabolism of FU in vitro by rat liver DPD is 3.49 ± 0.41 (SE) uM, about one and a half that of uracil (2.26 ± 0.28 uM) or thymine (2.23 ± 0.34 uM). The specific activity of DPD (nmol/min/mg of protein) is 0.82 for FU, but 0.68 and 0.56 for uracil and thymine respectively. The reduction of FU is most sensitive to the inhibitory effects of increased substrate concentration; but uridine is the most potent noncompetitive inhibitor of pyrimidine base degradation in vitro, with an inhibition constant (Ki) for FU of 0.71 uM. Total inhibition of FU degradation is obtained at 10 uM uridine, and 24 uM thymidine (Tuchman et al. 1985, 1986). By contrast, thymine appears to be the more potent modulator of FU catabolism in hepatocytes, compared to uracil (Sommadossi, Gewirtz, Cross, Goldman, Cano, and Diasio 1985). These distinctions are important since the pharmacological inactivation of DPD, for example with oral eniluracil, represents one strategy to improve FU therapy (Baker 2000). These biochemical data, and others that indicate negative regulation of DPD by its substrate, FU, and inhibition of DPD activity results shortly following FU administration in both colorectal cancer patients and an animal model, are critical to our understanding of the genomic underpinnings of the biochemistry of OM. Since more than 80% of the administered FU dose is catabolized by DPD, elucidation of the biochemistry and the molecular biology responsible for the regulation of this dehydrogenase is most important to the characterization of the determinants of the maximal tolerated exposure (Milano and Chamorey 2002).

Molecular cartography is the science of recognizing and identifying the multifaceted and intricate array of interacting genes and gene products that characterize the function and specialization of each individual cell in the context of cell–cell interaction, tissue and organ function. The relevance of molecular epigenetics pertains to every domain of physiology, including immunity.

Epigenetics is the study of DNA at the chromatin level (control of gene expression) and also the study of what was termed junk DNA by molecular biologists. The classical genes, exons, occupy a minor part of our total genome, perhaps less than 2 to 3%. The rest is noncoding DNA of unknown function. It contains regions with many repeat sequences, and signatures that are only currently being elucidated with advanced computer technologies. Noncoding DNA is believed to modulate chromatin structure and function in a complex series of molecular processes that are the domain of study of epigenetics.

Molecular cartography will provide to the study of the circadian pattern in OM the fundamental knowledge and understanding of the genomic, the proteonomic and interactomic processes that regulate the emergence, stability and function of immune cell populations, and the mechanisms by which the psychoneuroendocrine system regulates immunity. It holds the promise of new fundamental knowledge with respect to the processes that signify the etiology of the condition, the neuroimmunological mechanisms that tissue pathology and associated immune suppression, and the basic circadian regulatory.

Case in point, human hepatoma (HepG2) cells were treated in vitro with 0 to 150 ng/ml of consensus interferon (c-IFN). The treated cells showed a significant time-dependent reduction in the level of DPD mRNA following c-IFN pretreatment ($p <$ 0.05). HepG2 proliferation was blunted by treatment with FU and c-IFN, and this effect was significantly increased upon pre-treatment with c-IFN for 72 h (Dou et al. 2005), confirming that treatment with IFN augments the antitumor effects of FU (Milano et al. 1994), and suggesting underlying genomic and epigenetic regulation.

In a similar vein, epidermal growth factor (EGF) and TNF-α led to a concentration-dependent increase tumor cell growth in vitro of the EGF receptor-expressing uterine cervical carcinoma SKG-IIIb cells and of PyNPase activity. Tumoral DPD activity was inhibited under these experimental conditions, and FU sensitivity of the tumor cells was increased (Ueda et al. 2000; Kashima, Ueda, and Kanazawa 2003). In human T98G and A172 glioblastoma cells, both type I and type II IFNs up-regulated TP mRNA and protein expression, and inhibited cell proliferation. IFN-induced TP mRNA accumulation was not inhibited by the protein synthesis inhibitor cycloheximide, but was strongly blocked by the transcription inhibitor actinomycin-D, and by transcription factor decoy oligodeoxynucleotides containing the putative IFN response element or the activated sequence in the TP promoter. The Janus kinase (JAK) inhibitor AG-490 blocked both IFN-induced STAT1 (signal transducers and activators of transcription 1) phosphorylation and TP expression. All IFNs increased the stability of TP mRNA as well. In addition, IFN-evoked TP enzyme activity enhanced the cytotoxicity of FU. TP expression was up-regulated by IFNs via the JAK-STAT signaling pathway and both transcriptional and posttranscriptional mechanisms (Yao, Kubota, Sato, Takeuchi, Kitai, and Matsukawa 2005). In a normal male mouse model, TS activity was monitored in response to IFN administration. Synthetase activity in plasma at 24 h following IFN-α (10 mU/kg, I.V.) was significantly higher for injections given at 9:00 AM (rest cycle), compared to injections given at 9:00 PM (activity cycle) ($p < 0.05$). The uptake of ^{3}H-thymidine by lymphocytes after 24-h incubation with IFN was significantly higher in cells obtained at 9:00 AM than in the cells obtained at 9:00 PM ($p < 0.01$) (Ohdo et al. 2000). Taken together, these data confirm that environmental factors may exert critical regulation of DPD and PyNPase activities, and of FU sensitivity and toxicity (Ueda et al. 2000; Kashima et al. 2003).

Topoisomerase II is a target of common chemotherapeutic agents such as doxorubicin and etoposide, because these drugs that induce DNA damage by altering the activity of this enzyme. Topoisomerase II expression in gastrointestinal epithelium does have a significant circadian variation, confirming that administration of chemotherapeutic agents at the proper time of day might reduce their mucositis side effects (Clayton, Tessnow, Fang, Holden, and Moore 2002). In contrast to DPD, Topoisomerase II-α, and TS, TP was found not to have a circadian period of rhythmicity (Naguib, Soong, and el Kouni 1993).

At the molecular level, it has been shown that the E2F family of transcription factors (e.g., E2F1) regulates the transcription of genes that encode proteins required for DNA synthesis, including TS, TP, and DPD (Kubota 2005). However, little is more known about the regulation of Topoisomerase II-α, TS, TP and DPD mRNA expression to date, if not that preliminary data suggest that they mat be regulated at the levels of both transcription and translation. These observations establish an urgent need to characterize the apparent link between the expression of these three enzymes and the regulation of cell proliferation, specifically at the G_1 phase of the cell cycle. One among the principal roles of the phosphorylation of the retinoblastoma tumor suppressor protein (pRb) is the temporal relationships between transition events in G_1 phase, which includes release of E2F1 from pRb, and accumulation of cyclin E (Martinsson, Starborg, Erlandsson, and Zetterberg 2005).

It is possible and even probable that these growth factors exercise their effect by modulating coding and, putatively, noncoding DNA. These changes are "translated" at the molecular level as altered genomic and epigenetic outcomes, which molecular cartography, including genomics, proteonomic and interactomic are now actively pursuing. Since biological structure and function are intimately related, it is important to develop new insights as to how circadian signals alter the structure and the functional circadian regulation of cells. Here, we have examined these relationships in the context of oral biology and medicine, and specifically of the neuroimmunological underpinnings of OM. We have discussed the potential contributions of two cutting-edge domains in contemporary experimental biology, which pertain to chaos theory and fractal analysis on the one hand, and genomes and epigenetic on the other.

The future of clinical and translational oral biology and medicine for the treatment of OM unquestionably depends upon advances in our grasp of the fundamental neuroimmunological processes that concern the pathological process in OM. This progress will rest on traditional cellular, biochemical and molecular biology studies, as well as on fractal analyses of the immune cells invading the oral mucosa and the oral epithelium that undergoes apoptosis and on the genomic, proteonomic, interactomic, and epigenetic characterization of the fundamental events. The quality of the evidence will be monitored by means of systematic reviews, meta-analysis and other evidence-based research protocols. Taken together, these findings will signify a more complete knowledge and understanding of the neuroimmunological correlates of circadian rhythmicity in oral biology and medicine in general and in OM in particular.

References

Alvarado, Y., Bellm, L.A., and Giles, F.J. (2002) Oral mucositis: Time for more studies. *Hematology* 7, 281–289.

Baker, S.D. (2000) Pharmacology of fluorinated pyrimidines: Eniluracil. *Invest New Drugs* 18, 373–381.

Barrat, M.A., Renee, N., Mormont, M.C., Milano, G., and Levi, F. (2003) Circadian variations of dihydropyrimidine dehydrogenase (DPD) activity in oral mucosa of healthy volunteers. *Pathol Biol (Paris)* 51, 191–193.

Barreda, D.R., Hanington, P.C., and Belosevic, M. (2004) Regulation of myeloid development and function by colony stimulating factors. *Dev Comp Immunol* 28, 509–554.

Bernard, F., Bossu, J.L., and Gaillard, S. (2001) Identification of living oligodendrocyte developmental stages by fractal analysis of cell morphology. *J Neurosci Res* 65, 439–445.

Besedovsky, H.O., and del Rey, A. (1996) Immune-neuro-endocrine interactions: Facts and hypotheses. *Endocrine Rev* 17, 64–102.

Blalock, J.E. (1994) The syntax of immune-neuroendocrine communication. *Immunol Today* 15, 504–511.

Blijlevens, N.M., Donnelly, J.P., and DePauw, B.E. (2005) Inflammatory response to mucosal barrier injury after myeloablative therapy in allogeneic stem cell transplant recipients. *Bone Marrow Transplant* 36, 703–707.

Buchsel, P.C., Forgey, A., Grape, F.B., and Hamann, S.S. (2002) Granulocyte macrophage colony-stimulating factor: Current practice and novel approaches. *Clin J Oncol Nurs* 6, 198–205.

Bulloch, K., Cullen, M.R., Schwartz, R.H., and Longo, D.L. (1987) Development of innervation within syngeneic thymus tissue transplanted under the kidney capsule of the nude mouse: A light and ultrastructural microscope study. *J Neurosci Res* 18, 16–27.

Burgos, J.D., and Moreno-Tovar, P. (1996) Zipf-scaling behavior in the immune system. *Biosystems* 39, 227–232.

Carlson, L.E., Speca, M., Patel, K.D., and Goodey E. (2004) Mindfulness-based stress reduction in relation to quality of life, mood, symptoms of stress and levels of cortisol, dehydroepiandrosterone sulfate (DHEAS) and melatonin in breast and prostate cancer outpatients. *Psychoneuroendocrinology* 29, 448–474.

Cengiz, M., Ozyar, E., Ozturk, D., Akyol, F., Atahan, I.L., and Hayran, M. (1999) Sucralfate in the prevention of radiation-induced oral mucositis. *J Clin Gastroenterol* 28, 40–43.

Chiappelli, F. (2005) The molecular immunology of mucositis: Implications for evidence-based research in alternative and complementary palliative treatments. *Evidence-Based Complement Altern Med* 2, 489–494.

Chiappelli, F. (in press) Immunity. In: S. Vanstone, G. Chrousos, I. Craig, R. de Kloet, G. Feuerstein, B. McEwen, N. Rose, R. Rubin, and A. Steptoe (Eds.), *Encyclopeadia of Stress II*. Elsevier, Amsterdam.

Chiappelli, F., Prolo, P, Cajulis, E, Harper, S., Sunga, E., and Concepcion E. (2004) Consciousness, emotional self-regulation, and the psychosomatic network: Relevance to oral biology and medicine. In: M. Beauregard (Ed.), *Consciousness, Emotional Self-regulation and the Brain. Advances in Consciousness Research*. John Benjamins, Philadelphia, Ch. 9, pp. 253–274.

Chiappelli, F., and Cajulis, O.S. (2004) Psychobiological views on "stress-related oral ulcers". *Quint Int* 35, 223–227.

Chiappelli, F., Alwan, J., Prolo, P., Christensen, R., Fiala, M., Cajulis, O.S., and Bernard, G. (2005) Neuro-immunity in stress-related oral ulcerations: A fractal analysis. *Front Biosci* 10, 3034–3041.

Chiappelli, F., Manfrini, E., Edgerton, M., Rosenblum, M., Cajulis, K.D., and Prolo, P. (2006a) Clinical evidence and evidence-based dental treatment of special populations: Patients with Alzheimer's disease. *California Dental Assoc J* 34, 439–447.

Chiappelli, F., Navarro, A.M., Moradi, D.R., Manfrini, E., and Prolo P. (2006b) Evidence-based research in complementary and alternative medicine II: Treatment of patients with Alzheimer's disease. *Evidence-Based Complement Altern Med* 34, 439–447.

Chiappelli, F., Prolo, P, Fiala, M., Cajulis, O.S., Cajulis, J., Iribarren, J., Panerai, A., Neagos, N. Younai, F., and Bernard, G. (in press) Allostasis in HIV infection and AIDS. In: P.A. Minagar and P. Shapshak (Eds.), *Neuro-AIDS*. Nova Science Publ, Hauppauge, NY.

Chrousos, G.P., and Gold, P.W. (1992) The concepts of stress and stress system disorders. Overview of physical and behavioral homeostasis. *JAMA*, 267, 1244–1252.

Clarkson, J.E., Worthington, H.V., and Eden, O.B. (2003) Interventions for preventing oral mucositis for patients receiving treatment. *Cochrane Database Syst Rev* 3, CD000978.

Clayton, F., Tessnow, K.A., Fang, J.C., Holden, J.A., and Moore, J.G. (2002) Circadian variation of topoisomerase II-alpha in human rectal crypt epithelium: Implications for reduction of toxicity of chemotherapy. *Mod Pathol* 15, 1191–1196.

Cross, S.S., Cotton, D.W., and Underwood, J.C. (1994) Measuring fractal dimensions. Sensitivity to edge-processing functions. *Anal Quant Cytol Histol* 16, 375–379.

Dalgleish, A. (1999) The relevance of non-linear mathematics (chaos theory) to the treatment of cancer, the role of the immune response and the potential for vaccines. *QJM* 92, 347–359.

Dou, J., Iwashita, Y., Sasaki, A., Kai, S., Hirano, S., Ohta, M., and Kitano, S. (2005) Consensus interferon enhances the anti-proliferative effect of 5-fluorouracil on human hepatoma cells via downregulation of dihydropyrimidine dehydrogenase expression. *Liver Int* 25, 148–152.

Duncan, M., and Grant, G. (2003) Oral and intestinal mucositis––Causes and possible treatments. *Aliment Pharmacol Ther* 18, 853–874.

Engle, G.L. (1960) A unified concept of health and disease. *Perspect Biol Med* 3, 459–485.

Felten, D.L., Felten, S.Y., Bellinger, D.L., Carlson, S.L., Ackerman, K.D., Madden K.S., Olschowki, J.A., and Livnat S. (1987) Noradrenergic sympathetic neural interactions with the immune system: Structure and function. *Immunol Rev* 100, 225–260.

French, J.R.P., Rodgers, W., and Cobb, S. (1974) Adjustment as person-environment fit. In: G.V. Coelho, D.A. Hamburg, and J.E. Adams (Eds.), *Coping and Adaptation*. Basic Books, New York, Ch. 11.

Fujimoto, S., Kikugawa, M., Kaneko, M., and Tamaki, N. (1992) Role of dihydropyrimidine dehydrogenase in the uridine nucleotide metabolism in the rat liver. *J Nutr Sci Vitaminol* 38, 39–48.

Fujiwaki, R., Hata, K., Nakayama, K., Fukumoto, M., and Miyazak, K. (2000). Gene expression for dihydropyrimidine dehydrogenase and thymidine phosphorylase influences outcome in epithelial ovarian cancer. *J Clin Oncol* 18, 3946–3951.

Glenny, R.W., Robertson, H.T., Yamashiro, S., and Bassingthwaighte, J.B. (1991) Applications of fractal analysis to physiology. *J Appl Physiol* 70, 2351–2367.

Gruenewald, T.L., Kemeny, M.E., and Aziz, N. (2006) Subjective social status moderates cortisol responses to social threat. *Brain Behav Immun* 20, 410–419.

Hoque, M.O., Kawamata, H., Nakashiro, K.I., Omotehara, F., Shinagawa, Y., Hino, S., Begum, N.M., Uchida, D., Yoshida, H., Sato, M., and Fujimori, T. (2001) Dihydropyrimidine dehydrogenase mRNA level correlates with the response to 5-

fluorouracil-based chemo-immuno-radiation therapy in human oral squamous cell cancer. *Int J Oncol* 19, 953–958.

Irwin, M. (2002) Effects of sleep and sleep loss on immunity and cytokines. *Brain Behav Immun* 16, 503–512.

Janjan, N.A., Weissman, D.E., and Pahule, A. (1992) Improved pain management with daily nursing intervention during radiation therapy for head and neck carcinoma. *Int J Radiat Oncol Biol Phys* 23, 647–652.

Kabat-Zinn, J., Massion, A.O., Kristeller, J., Peterson, L.G., Fletcher, K.E., Pbert, L., Lenderking, W.R., and Santorelli, S.F. (1992) Effectiveness of a meditation-based stress reduction program in the treatment of anxiety disorders. *Am J Psychiatry* 149, 936–943.

Kashima, N., Ueda, M., and Kanazawa, J. (2003) Effect of 5-fluorouracil and epidermal growth factor on cell growth and dihydropyrimidine dehydrogenase regulation in human uterine cervical carcinoma SKG-IIIb cells. *Cancer Sci* 94, 821–825.

Keough, K.M., Hyam, P., Pink, D.A., and Quinn, B. (1991) Cell surfaces and fractal dimensions. *J Microsc* 163, 95–99.

Kerenji, A.S., Bozovic, Z.L., Tasic, M.M., Budimlija, Z.M., Klem, I.A., and Polozovic, A.F. (2000) Fractal dimension of hepatocytes' nuclei in normal liver vs. hepatocellular carcinoma (HCC) in human subjects--Preliminary results. *Arch Oncol* 8, 47–50.

Kiecolt-Glaser, J.K., McGuire, L., Robles, T.F., and Glaser, R. (2002) Psychon-euroimmunology and psychosomatic medicine: Back to the future. *Psychosom Med* 64, 15–28.

Kubota, T. (2005) Real-time RT-PCR (TaqMan) of tumor mRNA to predict sensitivity of specimens to 5-fluorouracil. *Methods Mol Med* 111, 257–265.

Kunz-Ebrecht, S.R., Mohamed-Ali, V., Feldman, P.J., Kirschbaum, C., and Steptoe, A. (2003) Cortisol responses to mild psychological stress are inversely associated with proinflammatory cytokines. *Brain Behav Immun* 17, 373–383.

Kwong, K.K. (2004) Prevention and treatment of oropharyngeal mucositis following cancer therapy: Are there new approaches? *Cancer Nurs* 27, 183–205.

La Recherche, Mars 1986 N° 175, 420–424.

Levi, F., Soussan, A., Adam, R., Caussanel, J.P, Metzger, G., Jasmin, C., Bismuth, H., Smolensky, M., and Misset, J.L. (1995) A phase I-II trial of five-day continuous intravenous infusion of 5-fluorouracil delivered at circadian rhythm modulated rate in patients with metastatic colorectal cancer. *J Infus Chemother* 5, 153–158.

Lorenz, E.N. (1963) Deterministic nonperiodic flow. *J Atmos Sci* 20, 130–141.

Losa, G.A., Baumann, G., and Nonnenmacher, T.F. (1992) Fractal dimension of pericellular membranes in human lymphocytes and lymphoblastic leukemia cells. *Pathol Res Pract* 188, 680–686.

McEwen, B., and Wingfield, J.C. (2003) The concept of allostasis in biology and biomedicine. *Hormones Behav* 43, 2–15.

Mandelbrot, B. (1977) The Fractal Geometry of Nature (1st Edn.). Freeman, New York.

Mandelbrot, B. (1986) Comment j'ai decouvert les fractales. *La Recherche*, 420–424.

Mandelbrot B. (1994) A Fractal's lacunarity, and how it can be tuned and measured. In: T.F. Nonnenmacher, G.A. Losa, E.R. Weibel (Eds.), *Fractals in Biology and Medicine.* Birkhäuser Verlag, Basel & Boston.

Martinsson, H.S., Starborg, M., Erlandsson, F., and Zetterberg A. (2005) Single cell analysis of G1 check points--The relationship between the restriction point and phosphorylation of pRb. *Exp Cell Res* 305, 383–391.

Megyeri, A., Bacso, Z., Shields, A., and Eliason, J.F. (2005) Development of a stereological method to measure levels of fluoropyrimidine metabolizing enzymes in tumor sections using laser scanning cytometry. *Cytometry A* 64, 62–71.

Metzger, G., Massari, C., Etienne, M.C., Comisso, M., Brienza, S., Touitou, Y., Milano, G., Bastian, G., Misset, J.L., and Levi, F. (1994) Spontaneous or imposed circadian changes in plasma concentrations of 5-fluorouracil coadministered with folinic acid and oxaliplatin:

Relationship with mucosal toxicity in patients with cancer. *Clin Pharmacol Ther* 56, 190–201.

Milano, G., Fischel, J.L., Etienne, M.C., Renee, N., Formento, P., Thyss, A., Gaspard, M.H., Thill, L., and Cupissol, D. (1994) Inhibition of dihydropyrimidine dehydrogenase by alpha-interferon: Experimental data on human tumor cell lines. *Cancer Chemother Pharmacol* 34, 147–152.

Milano, G., and Chamorey, A.L. (2002) Clinical pharmacokinetics of 5-fluorouracil with consideration of chronopharmacokinetics. *Chronobiol Int* 19, 177–189.

Miller, J.J., Fletcher, K., and Kabat-Zinn, J. (1995) Three-year follow-up and clinical implications of a mindfulness meditation-based stress reduction intervention in the treatment of anxiety disorders. *Gen Hosp Psychiatry* 17, 192–200.

Mucositis Study Section of the Multinational Association of Supportive Care in Cancer and the International Society for Oral Oncology: Mucositis–Perspectives and Clinical Practice Guidelines–Perspectives on Cancer Therapy-Induced Mucosal Injury [online]. (2005) URL: http://www3.interscience.wiley.com/cgi-bin/fulltext/108069518 [2005-05-04].

Mulder, C.L. (1994) Psychosocial correlates and the effects of behavioral interventions on the course of human immunodeficiency virus infection in homosexual men. *Patient Educ Couns* 24, 237–247.

Naguib, F.N., Soong, S.J., and el Kouni, M.H. (1993) Circadian rhythm of orotate phosphoribosyltransferase, pyrimidine nucleoside phosphorylases and dihydrouracil dehydrogenase in mouse liver. Possible relevance to chemotherapy with 5-fluoropyrimidines. *Biochem Pharmacol* 45, 667–673.

Naschitz, J.E., Sabo, E., Naschitz, S., Rosner, I., Rozenbaum, M., Priselac, R.M., Gaitini, L., Zukerman, E., and Yeshurun, D. (2002) Fractal analysis and recurrence quantification analysis of heart rate and pulse transit time for diagnosing chronic fatigue syndrome. *Clin Auton Res* 12, 264–272.

Naumov, Y.N., Naumova, E.N., Hogan, K.T., Selin, L.K., and Gorski, J. (2003) A fractal clonotype distribution in the CD8+ memory T cell repertoire could optimize potential for immune responses. *J Immunol* 170, 3994–4001.

Nemunaitis, J., Rosenfeld, C.S., Ash, R., Freedman, M.H., Deeg, H.J., Appelbaum, F., Singer, J.W., Flomenberg, N., Dalton, W., Elfenbein, G.J., Refkin, R., Rubin, A., Agosti, J., Hayes, F.A., Holcenberg, J., Shadduck, R.K. (1995) Phase III randomized, double-blind placebo-controlled trial of rhGM-CSF following allogeneic bone marrow transplantation. *Bone Marrow Transplant* 15, 949–954.

Nonnenmacher, T.F., Baumann, G., Barth, A., and Losa, G.A. (1994) Digital image analysis of self-similar cell profiles. *Int J Biomed Comput* 37, 131–138.

Nygard, J.F., and Glattre, E. (2003) Fractal analysis of time series in epidemiology: Is there information hidden in the noise? *Norsk Epidemiol* 13, 303–308.

Ohdo, S., Wang, D.S., Koyanagi, S., Takane, H., Inoue, K., Aramaki, H., Yukawa, E., and Higuchi, S. (2000) Basis for dosing time-dependent changes in the antiviral activity of interferon-alpha in mice. *J Pharmacol Exp Ther* 294, 488–493.

Pierpaoli, W., and Maetroni, G.J. (1988) Neuroimmunomodulation: Some recent views and findings. *Int J Neurosci* 39, 165–175.

Potten, C.S., Booth, D., Cragg, N.J., Tudor, G.L., O'Shea, J.A., Appleton, D., Barthel, D., Gerike, T.G., Meineke, F.A., Loeffler, M., and Booth, C. (2002) Cell kinetic studies in the murine ventral tongue epithelium: Thymidine metabolism studies and circadian rhythm determination. *Cell Prolif* 35S 1, 1–15.

Prolo, P., and Chiappelli, F. (in press). Immune suppression. In: S. Vanstone, G. Chrousos, I. Craig, R. de Kloet, G. Feuerstein, B. McEwen, N. Rose, R. Rubin, and A. Steptoe (Eds.), *Encyclopeadia of Stress II*. Elsevier, Amsterdam.

Raida, M., Kliche, K.O., Schwabe, W., Hausler, P., Clement, J.H., Behnke, D., and Hoffken, K. (2002) Circadian variation of dihydropyrimidine dehydrogenase mRNA expression in leukocytes and serum cortisol levels in patients with advanced gastrointestinal carcinomas compared to healthy controls. *J Cancer Res Clin Oncol* 128, 96–102.

Rew, D.A. (1999) Tumour biology, chaos and non-linear dynamics. *Eur J Surg Oncol* 25, 86–89.

Robert McComb, J.J., Tacon, A., Randolph, P., and Caldera, Y. (2004) A pilot study to examine the effects of a mindfulness-based stress-reduction and relaxation program on levels of stress hormones, physical functioning, and submaximal exercise responses. *J Altern Complement Med* 10, 819–827.

Romeo, H.E., Tio, D.L., Rahman, S.U., Chiappelli, F., and Taylor, A.N. (2001) The glossopharyngeal nerve as a novel pathway in immune-to-brain communication: Relevance to neuroimmune surveillance of the oral cavity. *J Neuroimmunol* 115, 91–100.

Rose-Ped, A.M., Bellm, L.A., Epstein, J.B., Trotti, A., Gwede, C., and Fuchs, H.J. (2002) Complications of radiation therapy for head and neck cancers. The patient's perspective. *Cancer Nurs* 25, 461–467.

Roth, B., and Stanley, T.W. (2002) Mindfulness-based stress reduction and healthcare utilization in the inner city: Preliminary findings. *Altern Ther Health Med* 8, 64–66.

Schulkin, J. (2003) Allostasis: A neural behavioral perspective. *Hormones Behav* 43, 21–27.

Sedivy, R., Windischberger, C., Svozil, K., Moser, E., and Breitenecker, G. (1999) Fractal analysis: An objective method for identifying atypical nuclei in dysplastic lesions of the cervix uteri. *Gynecol Oncol* 75, 78–83.

Solomon, G.F. (1987) Psychoneuroimmunology: Interactions between central nervous system and immune system. *J Neuroscien Res* 18, 1–9.

Solomon, G.F., and Moos, R.H. (1964) Emotions, immunity, and disease. *Arch Gen Psychiat* 11, 657–674.

Sommadossi, J.P., Gewirtz, D.A., Cross, D.S., Goldman, I.D., Cano, J.P., and Diasio, R.B. (1985) Modulation of 5-fluorouracil catabolism in isolated rat hepatocytes with enhancement of 5-fluorouracil glucuronide formation. *Cancer Res* 45, 116–121.

Sterling, P., and Eyer, J. (1988) Allostasis: A new paradigm to explain arousal pathology. In: S. Fisher and J. Reason (Eds.), *Handbook of Life Stress, Cognition, and Health.* Wiley, New York.

Takeda, T., Ishikawa, A., Ohtomo, K., Kobayashi, Y., and Matsuoka, T. (1992) Fractal dimension of dendritic tree of cerebellar Purkinje cell during onto- and phylogenetic development. *Neurosci Res* 13, 19–31.

Tuchman, M., Ramnaraine, M.L., and O'Dea, R.F. (1985) Effects of uridine and thymidine on the degradation of 5-fluorouracil, uracil, and thymine by rat liver dihydropyrimidine dehydrogenase. *Cancer Res* 45, 5553–5556.

Tuchman, M., Ramnaraine, M.L., and O'Dea, R.F. (1986) In vitro degradation of pyrimidine bases: studies of rat liver dihydropyrimidine dehydrogenase. *Adv Exp Med Biol* 195B, 245–248.

Ueda, M., Kitaura, K., Kusada, O., Mochizuki, Y., Yamada, N., Terai, Y., Kumagai, K., Ueki, K., and Ueki, M. (2000) Regulation of dihydropyrimidine dehydrogenase and pyrimidine nucleoside phosphorylase activities by growth factors and subsequent effects on 5-fluorouracil sensitivity in tumor cells. *Jpn J Cancer Res* 91, 1185–1191.

T.C., Kasperkovitz, P.V., Verbeet, N., and Verweij, C.L. (2004) Genomics in the immune system. *Clin Immunol* 111, 175–185.

van Kuilenburg, A.B., Meinsma, R., and van Gennip, A.H. (2004) Pyrimidine degradation defects and severe 5-fluorouracil toxicity. *Nucleosides Nucleotides Nucleic Acids* 23, 1371–1375.

Vgontzas, A.N., Papanicolaou, D.A., Bixler, E.O., Lotsikas, A., Zachman, K., Kales, A., Wong, M.-L., Licinio, J., Prolo, P., Gold, P.W., Hermida, R.C., Mastorakis, G., and Chrousos, G.P. (1999) Circadian interleukin-6 secretion and quantity and depth of sleep. *J Clin Endocrinol Metabol* 84, 2603–2607.

Watford, W.T., Hissong, B.D., Bream, J.H., Kanno, Y., Muul, L., and O'Shea, J.J. (2004) Signaling by IL-12 and IL-23 and the immunoregulatory roles of STAT4. *Immunol Rev.* 202, 139-156.

Weissman, D.E., Janjan, N.A., and Byhardt, R.W. (1989) Assessment of pain during head and neck irradiation. *J Pain Symptom Manage* 4, 90–95.

Williams, K.A., Kolar, M.M., Reger, B.E., and Pearson, J.C. (2001) Evaluation of a wellness-based mindfulness stress reduction intervention: A controlled trial. *Am J Health Promot* 15, 422–432.

Xu, J., Lucas, R., and Wendel, A. (2004) The potential of GM-CSF to improve resistance against infections in organ transplantation. *Trends Pharmacol Sci* 25, 254–258.

Yao, Y., Kubota, T., Sato, K., Takeuchi, H., Kitai, R., and Matsukawa, S. (2005) Interferons upregulate thymidine phosphorylase expression via JAK-STAT-dependent transcriptional activation and mRNA stabilization in human glioblastoma cells. *J Neurooncol* 72, 217–223.

Zou, L., Barnett, B., Safah, H., Larussa, V.F., Evdemon-Hogan, M., Mottram, P., Wei, S., David, O., Curiel, T.J., and Zou, W. (2004) Bone marrow is a reservoir for CD4+CD25+ regulatory T cells that traffic through CXCL12/CXCR4 signals. *Cancer Res* 64, 8451–8455.

3 Circadian Organization of the Immune Response

Lessons from the Adjuvant Arthritis Model

Daniel P. Cardinali, Ana I. Esquifino, Georges J.M. Maestroni, and Seithikurippu R. Pandi-Perumal

3.1 The Circadian Clock Is One of the Most Indispensable Biological Functions

Organisms populating the Earth are under the steady influence of daily and seasonal changes resulting from the planet's rotation and orbit around the sun. This periodic pattern is most prominently manifested by the light–dark cycle and has led to the establishment of endogenous circadian timing systems that synchronize biological functions to the environment. This is the basis of predictive homeostasis (Moore-Ede 1986), evolving as an adaptation to anticipate predictable changes in the environment, such as light and darkness, temperature, food availability or predator activity. Therefore, the circadian clock is one of the most indispensable biological functions for living organisms that acts like a multifunctional timer to adjust the homeostatic system, including sleep and wakefulness, hormonal secretions and various other bodily functions, to the 24-h cycle (Buijs, van Eden, Goncharuk, and Kalsbeek 2003; Collins and Blau 2006; Hastings, Reddy, and Maywood 2003). In mammals, the circadian system is composed of many individual, tissue-specific cellular clocks. To generate coherent physiological and behavioral responses, the phases of this multitude of cellular clocks are orchestrated by a master circadian pacemaker residing in the suprachiasmatic nuclei (SCN) of the hypothalamus.

At a molecular level, circadian clocks are based on clock genes, some of which encode proteins able to feedback and inhibit their own transcription. These cellular oscillators consist of interlocked transcriptional and posttranslational feedback loops that involve a small number of core clock genes (about 12 genes identified currently). The positive drive to the daily clock is constituted by two, basic helix—loop–helix, PAS-domain containing transcription factor genes, called Clock and Bmal1. The protein products of these genes form heterodimeric complexes that control the transcription of other clock genes, notably three Period (Per1/Per2/Per3) genes and two Cryptochrome (Cry1/Cry2) genes, which in turn provide the negative feedback signal that shuts down the Clock/Bmal drive to complete the circadian cycle (Okamura 2003). Per and Cry messenger RNAs peak in the SCN in mid-to-late circadian day, regardless of whether an animal is nocturnal or diurnal. Other clock genes provide additional negative and positive transcriptional/translational feedback loops to form the rest of the core clockwork, which has been characterized in rodents by a transgenic gene deletion methodology. Clock gene expression oscillates because of the delay in the feedback loops, regulated in part by phosphorylation of the clock proteins that control their stability, nuclear reentry and transcription complex formation (Collins and Blau 2006; Lakin-Thomas 2006). Clock genes are expressed in a tissue-specific fashion, often with unknown function. Although a substantial number of genes are rhythmic (about 10% in the SCN or peripheral tissues), the rhythmic genes tend to be different in the different tissues. For example, in comparisons between heart and liver, or between the SCN and liver, only a 10% coincidence is seen (Okamura 2003). The phase of the peripheral clock oscillations is delayed by 3 to 9 h as compared to that of SCN cells, suggesting that the peripheral tissues might be receiving timing cues from the master SCN oscillator. Furthermore, oscillations in isolated peripheral tissues dampen rapidly, unlike the persistent rhythms in isolated SCN neurons (Albrecht and Eichele 2003; Buijs et al. 2003; Fukuhara and Tosini 2003; Hastings et al. 2003; Okamura 2003).

Sorting of the cycling transcripts into functional groups has revealed that the major classes of clock-regulated genes are implicated in processes specific to the tissue in which they are found. For example, many cycling transcripts in the liver are involved in nutrient metabolism. It is also of interest that many of the regulated transcripts correspond to rate limiting steps in their respective pathways, indicating that control is selective and very efficient. Indeed, about 10% of the genome is under control of the circadian clock (Ueda et al. 2005). As noted, the trillions of cellular clocks in primates are synchronized by a few thousand neurons located in the SCN. It is remarkable that such a small group of neurons display the properties of a central clock. Indeed, these "neuronal oligarchies," like the human ones, control trillions of cells in the body by (a) taking control of the major communication channels (the endocrine and autonomic nervous systems); (b) concentrating the relevant information in a private way (i.e., light information arriving via the retinohypothalamic tract). Thus, it is not surprising that anatomical studies have showed that the SCN projects to at least three different neuronal targets: endocrine neurons, autonomic neurons of the paraventricular nucleus (PVN) of the hypothalamus, and other hypothalamic structures that transmit the circadian signal to other brain regions (Buijs et al. 2003). The SCN projections are generally indirect, via the sub-PVN zone (Saper, Lu, Chou, and Gooley 2005). Through autonomic nervous system projections involving the superior cervical ganglia the SCN controls the release of a major internal synchronizer, the pineal substance melatonin (Cardinali 1981; Hardeland, Pandi-Perumal, and Cardinali 2006).

Recordings from single dispersed SCN neurons have demonstrated that the circadian mechanism is not an emergent property of the SCN neuronal network but it is expressed in each individual cell. Multisynaptic links of SCN through the hypothalamic sub-PVN zone outflow to the adrenocorticotropic and other neuroendocrine axes and to autonomic ganglia that innervate the viscera including all the immune system, whereas innervation of the dorsomedial hypothalamus contributes to circadian control of the orexin/hypocretin system, that participates in wakefulness (Buijs et al. 2003; Saper et al. 2005). By projecting to areas outside the hypothalamus, such as the lateral geniculate bodies and the paraventricular nucleus of the thalamus, the SCN neurons can synchronize hypothalamic-induced behavior (e.g., feeding) and locomotor activity. The circadian control of rest/activity cycles also involves SCN paracrine signaling, e.g., the secretion of transforming growth factor-α and prokineticin-2 (Hastings et al. 2003; Saper et al. 2005).

In the case of the immune system, our own work concentrated on the role of the autonomic nervous system (parasympathetic and sympathetic) to provide the anatomical basis for the circadian control of lymph node function (Cardinali and Esquifino 1998; Esquifino and Cardinali 1994). These studies were the continuation of our former studies on the role of sympathetic and parasympathetic nerves in thyroid follicular and C cell and parathyroid cell regulation (Cardinali and Romeo 1991; Cardinali and Stern 1994). The concept that autonomic nerves are a very efficient avenue to convey time of day information to the periphery has been since then generalized to tissues like the adrenal, pancreas, liver, ovaries and many other organs (Buijs et al. 2003). Although circadian rhythms are anchored genetically, they are synchronized by and maintain certain phase relationships to external factors (Murphy and Campbell 1996). These rhythms will persist with a period different from 24 h when external time cues are suppressed or removed, such as during complete social isolation or in constant light or darkness. Research in animals and humans has shown that only a few such environmental cues, like light–dark cycles, are effective entraining agents for the circadian oscillator ("Zeitgebers").

An entraining agent can actually reset, or phase shift, the internal clock. Depending on when an organism is exposed to such an entraining agent, circadian rhythms may be advanced, delayed, or not shifted at all. Therefore, involved in adjusting the daily activity pattern to the appropriate time of day is a rhythmic variation in the influence of the Zeitgeber as a resetting factor (Murphy and Campbell 1996). In humans, light exposure during the first part of the night delays the phase of the cycle; a comparable light change near the end of the night, advances it. At other times during the day light exposure has no phase-shifting influence (Lewy, Ahmed, and Sack 1996; Pandi-Perumal et al. in press). Melatonin, a chemical code of the night in most species, showed an opposite phase response curve to light, producing phase advances during the first half of the night and phase delays during the second (Lewy et al. 1996; Pandi-Perumal et al. in press).

3.2 The Immune System Shows Circadian Organization

Light and daily rhythms have a profound influence on immune function. Many studies have described circadian variations of different immune parameters such as lymphocyte proliferation, antigen presentation, and cytokine gene expression. The number of lymphocytes and monocytes in the human blood reach maximal values during the night and are lowest after waking. Natural killer (NK) cells, by contrast, reach their highest level in the afternoon, with a normal decrease in number and activity around midnight.

Immune cells have been checked for the presence of clock genes. In a study aimed to investigate whether circadian clock genes function in human peripheral blood mononuclear cells, circadian clock genes human Per1, Per2, Per3 and Dec1 were found to be expressed in a circadian manner in human peripheral blood mononuclear cells, with the a peak level occurring during the second part of the active phase (Boivin et al. 2004). Studies on clock gene oscillations in splenic-enriched NK cells have also been reported. In addition, the circadian expression of granzyme B, perforin, interferon (IFN)-γ, and tumor necrosis factor (TNF)-α found in NK cells underlines the circadian nature of NK cell function (Arjona and Sarkar 2005). Thus, the existence of molecular clock machinery may be conserved across different lymphocyte subsets and peripheral blood cells. Moreover, they may share common entrainment signals. Emerging data in the human and animal literature suggest that circadian regulation may be crucial for the host defenses against cancer. Virtually all immunological variables investigated to date in animals and humans have been shown to display biological periodicity. In most of its components, the immune system shows regularly recurring, rhythmic variations in numerous frequencies, the circadian being the best known (Haus and Smolensky 1999; Pelegri, Vilaplana, Castellote, Rabanal, Franch, and Castell 2003; Petrovsky and Harrison 1998).

Both the humoral arm and the delayed (cellular) arm of the immune system function in a rhythmic manner. Indeed, circadian variations in immunocompetent cells in peripheral blood are of a magnitude to require attention in medical diagnostics (Mazzoccoli et al. 1999; Niehaus, Ervin, Patel, Khanna, Vanek, and Fagan 2002; Undar, Ertugrul, Altunbas, and Akca 1999). Circadian changes in the circulation of T, B, or natural killer NK lymphocyte subsets in peripheral blood and in the density of epitope molecules at their surface, which may be related to cell reactivity to antigen exposure, have been reported (Cardinali, Brusco, Garcia, and Esquifino 1998b). Changes in lymphocyte subset populations can depend on time of day-associated changes in cell proliferation in immunocompetent organs and/or on diurnal modifications in lymphocyte release and traffic among lymphoid organs. Circadian rhythmicity is revealed in circulating cells, lymphocyte metabolism and transformability, circulating hormones and other substances that may exert various actions on different targets of the immune system, cytokines, receptors, and adhesion molecules, cell cycle events in health and cancer, reactions to antigen challenge, and disease etiology and symptoms. It must be noted that the role of the SCN, the central circadian pacemaker, in entrainment of lymphocyte function and in coordinating signals by which circadian information is conveyed to the immune cells remains unsettled. Rhythms in the number of circulating T cells persisted in rats with disrupted circadian outputs (Kobayashi, Oishi, Hanai, and Ishida 2004). Similarly, SCN ablation did not affect the 24-h rhythms in cell cycle phase distribution in bone marrow cells (Filipski et al. 2004), suggesting that some rhythms in the immune system may be SCN-independent. It is known that circadian gene expression can be maintained in vitro (Yoo et al. 2004). Thus, some peripheral clocks may be able to independently generate circadian oscillations and this could be also the case for lymphocytes. Rather than a mere rhythm generator for the periphery, the SCN should be envisioned as a transducer for light entrainment. However, there are entrainment signals other than light that may be coordinating the rhythm in NK cell function and other immunological parameters. For example, feeding is an important zeitgeber for peripheral clock gene expression (Kobayashi et al. 2004), and interestingly enough, internal desynchronization produced by restricted feeding during the light period slowed down tumor progression in mice (Wu, Li, Xian, and Levi 2004). Daily activity rhythms are also considered to act as entrainment cues for peripheral tissues (Schibler, Ripperger, and Brown 2003) and may as well influence the molecular clock in lymphocyte cells. In addition, intrinsic immunological outputs such a cytokine secretion could function as entrainment factors for immune cells. Indeed, interleukin (IL)-6 has been shown to induce Per1 expression in vitro (Motzkus, Albrecht, and Maronde 2002).

Several studies have investigated the changes in cytokine levels that occur during the sleep–wake cycle; however, it is difficult to measure these changes because endogenous cytokine levels are low. Plasma TNF levels peak during sleep, and the circadian rhythm of TNF release is disrupted by obstructive sleep apnea. Plasma IL-1β levels also have a diurnal variation, being highest at the onset of non-REM sleep (Obal and Krueger 2005). The levels of other cytokines (including IL-2, IL-6, IL-10, and IL-12) and the proliferation of T cells in response to mitogens also change during the sleep–wake cycle. Although the production of macrophage-related cytokines (such as TNF) increases during sleep (in response to in vitro stimulation), this occurs in parallel with the rise in monocyte numbers in the blood. The production of T-cell-related cytokines (such as IL-2) increases during sleep, independent of migratory changes in T-cell distribution (Obal and Krueger 2005).

All of these observed diurnal changes could be specific to the effects of sleep or associated with the circadian oscillator. To dissociate the effects that result from the sleep–wake cycle from those due to the endogenous circadian oscillator, experimental procedures such as constant routine or forced desynchrony need to be used. At present, there are no reports of studies using these methods to elucidate the effects of sleep on immunity.

Sleep and the immune system share regulatory molecules. These are involved in both physiological sleep and sleep in the acute-phase response to infection. This supports the view that sleep and the immune system are closely interconnected (Obal and Krueger 2005). It is probable that sleep influences the immune system through the action of centrally produced cytokines that are regulated during sleep. These endogenous cytokines are known to function through the autonomic nervous system and the neuroendocrine axis, although other pathways might be involved.

During the last years, we have examined the regulation of circadian rhythmicity of lymph cell proliferation in a number of experimental models in rat submaxillary lymph nodes. The bilateral anatomical location of submaxillary lymph nodes and their easily manipulable autonomic innervation allowed us to dissect some of humoral and neural mechanisms regulating the lymphoid organs and their interaction. A significant diurnal variation of rat submaxillary lymph node ornithine decarboxylase activity (ODC), an index of cell proliferation in immunocompetent organs (Neidhart and Larson 1990) and endocrine glands (Scalabrino, Ferioli, Modena, and Fraschini 1982), was uncovered, displaying maximal activity at early afternoon (Cardinali, Della Maggiore, Selgas, and Esquifino 1996a). Such a maximum coincided with peak mitotic responses to lipopolysaccharide (LPS) and concanavalin A (Con A) in incubated lymph node cells (Esquifino, Selgas, Arce, Della Maggiore, and Cardinali 1996). A purely neural pathway including as a motor leg the autonomic nervous system innervating the lymph nodes was identified (Cardinali and Esquifino 1998). The combined sympathetic–parasympathetic denervation of the lymph node suppressed circadian variation in lymph cell proliferation. In addition, a hormonal pathway involving the circadian secretion of melatonin also plays a role to induce rhythmicity (Cardinali, Brusco, Cutrera, Castrillon, and Esquifino 1999; Cardinali et al. 1998b).

3.3 "Sickness Behavior" Includes Changes in Circadian Rhythms

Circadian neuroimmune connections imply a very important feedback component provided by the immune cells to the brain. Indeed, there are several mechanisms by which the immune system can modify central clock structures (Dantzer 2001; Johnson 2002; Larson 2002). Peripheral inflammation results in production of cytokines, which can signal the SCN. In the case of rheumatoid arthritis, inflammation is characterized by increased local synovial and systemic levels of the proinflammatory cytokines IL-1, IL-6, IFN, and TNF-α, which are directly involved in disease's pathophysiology (Feldmann, Brennan, Foxwell, and Maini 2001). Such increased cytokine production plays a key role in neuroendocrine activation pathways in arthritis (Cardinali and Esquifino 2003). As large, hydrophilic proteins, cytokines can only cross the blood–brain barrier at leaky points (the circumventricular organs) or via specific active transport mechanisms (Kastin, Pan, Maness, and Banks 1999). Cytokines act at the level of the organum vasculosum laminae terminalis, a circumventricular organ located at the anterior wall of the third ventricle. IL-1 binds to cells located on the vascular side of this circumventricular structure, thereby inducing synthesis and release of second messenger systems, such as the nitric oxide (NO) synthase/NO and the cyclooxygenase/prostaglandin systems (McCann, Kimura, Karanth, Yu, and Rettori 2002). It must be noted that a central compartment for cytokines exists and that there are data indicating that an increase in peripheral cytokines can evoke a mirror increase in brain levels of cytokines (for ref. see Dantzer 2001). Inflammatory stimuli can also induce CNS stress response through afferent peripheral neural signaling. This was shown mainly for cytokines from the peritoneum that can cause early rapid activation of the nucleus of tractus solitarius in the brainstem via the vagus nerve (Bluthe et al. 1994). Experimental evidence suggests that symptomatology after antigen administration, like anorexia and depressed activity, is a part of a defense response to antigenic challenge and is mediated by the neural effects of cytokines. These changes are known generally as "sickness behavior," that is, the "nonspecific" symptoms (anorexia, depressed activity, loss of interest in usual activities, disappearance of body care activities) that accompany the response to infection. These "nonspecific" symptoms of infection include fever and profound psychological and behavioral changes, among them in their circadian structure (Johnson 2002). Sick individuals experience weakness, malaise, listlessness, and inability to concentrate (Larson 2002). They consistently show indication of decreased amplitude of circadian rhythmicity, like superficial sleep at night and hypersomnia, loss of interest and depressed activity during the day.

One of the most studied physiological roles of immune variables in the central nervous system is the regulation of sleep by pro and anti-inflammatory cytokines. It is now clear that proinflammatory cytokines induce sleep while anti-inflammatory cytokines prevent sleep induction (Majde and Krueger 2005; Obal and Krueger 2005). LPS injections produce similar results to those of the proinflammatory cytokines on sleep regulation and exert differential effects on EEG activity in rats depending on the time of administration (Lancel, Mathias, Faulhaber, and Schiffelholz 1996). Although there is a substantial amount of information regarding the circadian modulation of many immunological variables, there are relatively few data about the possible effect of immune factors on the circadian system itself. Several reports suggest a possible immune feedback regulation of the circadian clock. For example, immunosuppressant drugs such as cyclosporin A affect the phase of locomotor activity (Marpegan et al. 2004) and of hormone secretion (Esquifino, Selgas, Vara, Arce, and Cardinali 1999b; Selgas, Pazo, Arce, Esquifino, and Cardinali 1998) in laboratory animals. Moreover, immune-related transcription factors are present and active in the SCN and its activity is partially necessary for light-induced phase shifts (Marpegan et al. 2004).

Introduction of gram-negative bacteria into the body causes the liberation of toxic, soluble products of the bacterial cell wall, such as LPS, also known as endotoxin. Peripheral administration of LPS exerts profound effects on the sleep–wake cycle and sleep architecture and may produce, at higher doses, fever and a characteristic "sickness behavior" observed during inflammatory diseases, including sleep pattern changes and fever oscillations along the day (Krueger, Obal, Fang, Kubota, and Taishi 2001). In mice, susceptibility to lethal doses of endotoxin increase dramatically during the resting period (Halberg, Johnson, Brown, and Bittner 1960) and a similar temporal pattern of induced mortality has also been established for tumor necrosis factor α (TNF-α) (Hrushesky, Langevin, Kim, and Wood 1994).

Results in hamsters indicate that LPS treatment induces changes in the phase of locomotor activity rhythms in a manner similar to light-induced phase delays (Marpegan, Bekinschtein, Costas, and Golombek 2005). The phase-shifting response to LPS was reduced when the activation of NF-κB, a transcription factor reported to play a role in the photic input of the circadian system (Marpegan et al. 2004), was prevented by sulfasalazine. LPS treatment stimulates the dorsal area of the SCN as assessed by c-Fos activation (Marpegan et al. 2005). Data from our laboratory indicate that melatonin, administered in the drinking water, has the capacity to counteract the effect of LPS on body temperature in hamsters, when injected at "Zeitgeber" time (ZT) 0 (ZT12 defined as the time of light off) (Bruno, Scacchi, Pérez Lloret, Esquifino, Cardinali, and Cutrera 2005). Evidence that melatonin improves survival from endotoxin shock has also been published (Crespo et al. 1999; Maestroni 1996; Wu, Chiao, Hsiao, Chen, and Yen 2001).

Therefore, one possible mechanism through which infection-related changes in circadian rhythms can occur is by modifying directly the activity of cells in the SCN. Cytokine receptors, e.g., IFN-γ receptors, have been detected in neuronal elements of ventrolateral SCN (Lundkvist, Robertson, Mhlanga, Rottenberg, and Kristensson 1998). Expression of SCN IFN-γ receptors followed a 24-h rhythm, coinciding with the expression of janus kinase 1 and 2 as well as the signal transducer and activator of transcription factor 1, the main intracellular signaling pathway for IFN-γ. In an ontogenic study, SCN IFN-γ receptors were found to reach their adult pattern between postnatal day 11 and 20, at a time when capacity for photic entrainment of the pacemaker became established (Lundkvist, Andersson, Robertson, Rottenberg, and Kristensson 1999). Indeed, high doses of an IFN-γ–TNF-α cocktail disrupt electrical activity of SCN neurons (Lundkvist, Hill, and Kristensson 2002).

The capacity of intracerebroventricular administration of IFN-γ to modify 24-h wheel running activity was assessed in golden hamsters (Boggio et al. 2003). Animals received IFN-γ or saline at ZT 6 or ZT 18. Intracerebroventricular administration of IFN-γ at ZT 6 produced a significant phase advance in acrophase of rhythm, an effect not seen with injection at ZT 18. IFN-γ depressed mesor value of rhythm significantly; the effect was seen both with ZT 6 and ZT 18 injections

(Boggio et al. 2003). IFN-γ was very effective to disrupt circadian rhythmicity of pituitary hormone release (Cano, Cardinali, Jimenez, Alvarez, Cutrera, and Esquifino 2005). The results supported the view that the circadian sequels arising during the immune reaction can rely partly on central effects of IFN-γ (Boggio et al. 2003). A disruptive effect of systemic administration of IFN-α on the circadian rhythm of locomotor activity, body temperature and clock-gene mRNA expression in SCN has also been documented in mice (Ohdo, Koyanagi, Suyama, Higuchi, and Aramaki 2001). Moreover, LPS incubation modified the circadian arginine–vasopressin release from SCN cultures (Nava, Carta, and Haynes 2000). Motzkus et al. (2002) demonstrated that IL-6-induced murine Per1 expression in SCN cell cultures.

In recent years we examined the circadian disruption of hormone release and immune-related mechanisms in several animal models including alcoholism (Jimenez, Cardinali, Alvarez, Fernandez, Boggio, and Esquifino 2005; Jimenez, Cardinali, Cano, Alvarez, Reyes Toso, and Esquifino 2004), calorie restriction (Cano, Cardinali, Fernandez, Reyes Toso, and Esquifino 2006; Chacon, Cano, Jimenez, Cardinali, Marcos, and Esquifino 2004; Chacon, Esquifino, Perelló, Cardinali, Spinedi, and Alvarez 2005; Esquifino, Chacon, Cano, Marcos, Cutrera, and Cardinali 2004b), and social isolation (Cano et al. 2006; Esquifino, Alvarez, Cano, Chacon, Reyes Toso, and Cardinali 2004a; Esquifino, Chacon, Jimenez, Reyes Toso, and Cardinali 2004c; Perelló, Chacon, Cardinali, Esquifino, and Spinedi 2006). Indeed, severe immune challenges such as animal models of sepsis (Bauhofer, Witte, Celik, Pummer, Lemmer, and Lorenz 2001) or infection with blood-borne parasites such as *Trypanosoma cruzi* or *T. brucei* (Bentivoglio, Grassi-Zucconi, Peng, and Kristensson 1994) or HIV-infected animals or patients (Bourin, Mansour, Doinel, Roue, Rouger, and Levi 1993; Vagnucci and Winkelstein 1993) display different levels of circadian disruption, including complete arrhythmicity, suggesting that circadian rhythms can be considered a good quality-of-health indicator.

3.4 The Rat Adjuvant Arthritis Is an Experimental Model of Rheumatoid Arthritis

Rheumatoid arthritis is a systemic inflammatory disorder that mainly affects the diarthrodial joint. It is the most common form of inflammatory arthritis and affects about 1% of the population, in a female/male ratio of 2.5/1. The disease can occur at any age, but it is most common among those aged 40 to 70 years. The geographic distribution of rheumatoid arthritis is worldwide, with a notably low prevalence in rural areas (Goodson and Symmons 2002; Reginster 2002). Although it initially presents as a symmetrical polyarticular synovitis with prominent hand involvement, rheumatoid arthritis has multiple potential systemic manifestations. The clinical course of the disorder is extremely variable, ranging from mild, self-limiting arthritis to rapidly progressive multisystem inflammation with profound morbidity and mortality. Fever and weight loss can be part of the acute symptoms, while splenomegaly, vasculitis, neutropenia, and amyloidosis are some of the disease's complications, which may occur in patients with long-standing disease (Bowman 2002; Reginster 2002; Youssef and Tavill 2002).

Rheumatoid arthritis is a T-cell-driven autoimmune process associated with the production of autoantibodies. Rheumatoid arthritis is initiated by CD4$^+$ T cells, which amplify the immune response by stimulating other mononuclear cells, synovial fibroblasts, chondrocytes, and osteoclasts. The release of cytokines, especially TNF-α, IL-1, and IL-6, causes synovial inflammation. In rheumatoid arthritis the inflammatory process, usually tightly regulated by mediators that initiate and maintain inflammation and mediators that shut the process down, becomes imbalanced leaving inflammation unchecked and resulting in the destruction of cartilage and bone. The etiology of rheumatoid arthritis (RA) is not known. Genetic susceptibility involving the HLA system along with environmental factors play a role. It is possible that the environmental factors are infectious agents such as Epstein Barr virus, cytomegalovirus, mycoplasmas, and parvovirus, or a response to microbial products. RA is a chronic multisystem disease with inflammatory synovitis of the joints serving as a major feature. Besides the pain and swelling in joints, morning stiffness, fatigue, and generalized weakness may occur and the clinical course may be variable. Therapy may consist of nonsteroidal anti-inflammatory drugs, Cox-2-specific inhibitors, glucocorticoids, disease-modiftying antirheumatic agents, the TNF-neutralizing agents, and immunosuppressive and cytotoxic drugs. In most studies of RA, sleep disturbances are often associated with pain. However, disease activity may lead to the release of cytokines affecting multiple neurobiological mechanisms.

Efforts to develop safer and more effective treatments for rheumatoid arthritis rely heavily on the availability of suitable animal models (Bendele 2001). Among these models, the rat's adjuvant arthritis is widely employed (Whitehouse 1988). Hallmarks of this rat model are reliable onset and progression of easily measurable, polyarticular inflammation, marked bone resorption and periosteal bone proliferation. Induction of adjuvant disease can be done with either Freund's complete adjuvant (FCA) supplemented with mycobacterium or by injecting synthetic adjuvants (Bendele 2001; Whitehouse 1988). The pathogenesis for development of adjuvant disease following injection of mycobacterial preparations is not fully understood, although a cross-reactivity of mycobacterial wall antigens with cartilage proteoglycans occurs.

After FCA injection to rats, the inflammatory disease of the joints shows four stages in its time-course: preclinical (first week), acute (weeks 2 to 4), post-acute (weeks 5 to 8), and recovery (weeks 9 to 11) (Calvino, Crepon-Bernard, and Le Bars 1987). The preclinical stage of FCA arthritis (first week) is characterized by discrete radiological lesions of the forepaws and slight increase in the threshold for struggle triggered by foot pressure, presumably due to an impending, initially painless, stiffness.

The acute stage or arthritis (weeks 2 to 4) is defined by signs of hyperalgesia, lack of mobility and a pause in body weight gain; during the acute period, hindpaw and forepaw joint diameters increase (Calvino et al. 1987). In the later, acute, stages of disease (day 12+), adjuvant arthritis rats are often relatively immobile due to severity of paw swelling.

At day 18th, an increase in scratching behavior and signs of hyperalgesia are clearly established as compared with the adjuvant's vehicle-injected group. FCA arthritis is induced most easily in inbred Lewis rats; it is also produced, to a milder extent, in Wistar and Sprague–Dawley rats (Holoshitz, Matiau, and Cohen 1984; Knight et al. 1992; Pearson 1956; Stenzel-Poore, Vale, and Rivier 1993; Tanaka, Ueta, Yamashita, Kannan, and Yamashita 1996).

The use of the adjuvant's arthritis model offers an opportunity to study pathological changes in a variety of tissues other than the joints. Among these, central nervous system (CNS) changes are most relevant (Dantzer 2001; Johnson 2002; Larson 2002).

3.5 Adjuvant Arthritis Disrupts Normal Chronobiological Organization

We have examined a number of immune and neuroendocrine circadian rhythms in FCA-injected rats by looking for changes in the preclinical phase of arthritis (2 to 3 days after FCA injection) as well as in the acute phase of the disease (18 days after FCA injection) (Tables 3.1 to 3.3). Generally, changes in circadian rhythms in lymph node immune function tended to be more profound at the preclinical phase of the disease. For example, B-cell- and T-cell-mediated mitogenic activity of lipopolysaccharide (LPS) and concanavalin (Con A), respectively, were modified in amplitude or acrophase during the preclinical phase (Esquifino, Castrillón, Chacon, Cutrera, and Cardinali 2001) while exhibiting few or none changes during the acute phase of experimental arthritis (Garcia Bonacho, Cardinali, Castrillon, Cutrera, and Esquifino 2001) (Table 3.1). Similarly, 24-h variations of B and T cells, as well as of $CD4^+$ (T helper) and $CD8^+$ (T cytolytic) cells became significantly changed during the preclinical phase (Castrillon, Esquifino, Varas, Zapata, Cutrera, and Cardinali 2000; Esquifino et al. 2001), with absence of changes during the acute phase (Garcia Bonacho et al. 2001). In the case of lymph node cell proliferation and local autonomic nerve activity, the increase in amplitude and mesor of rhythms found in the preclinical phase of arthritis was higher than that observed as the disease progressed (Cardinali, Brusco, Selgas, and Esquifino 1998a). Therefore, the results suggest that some sort of homeostatic compensation of initial changes in circadian rhythmicity of immune changes occurs with the development of arthritis (Table 3.1). As far as changes in neuroendocrine rhythmicity during rat's arthritis, early data had indicated in FCA-injected rats that the 24-h organization of the biologic responses was altered. For example, morning–evening differences in circulating ACTH and corticosterone disappeared by days 7 to 21, and between days 6 and 8 after FCA injection a loss of the adrenocortical ODC circadian rhythm of activity was found (Neidhart 1996).

TABLE 3.1. Summary of changes in circadian rhythms of submaxillary lymph node immune responses during the preclinical (3rd day) and acute (18th day) phases of Freund's adjuvant arthritis in rats. (Results from Cardinali et al. 1996a, 1997a, 1998a, 1998b; Castrillon et al. 2000; Esquifino et al. 1996, 2001; Garcia Bonacho et al. 2001.)

24-h rhythms	Amplitude	Acrophase	Mean
Preclinical phase			
Cell proliferation[a]	Increased	Unchanged	Increased
Mitotic response to Con A	Decreased	Unchanged	Unchanged
Mitotic response to LPS	Unchanged	Changed	Unchanged
NK cells	Absence of rhythm	Absence of rhythm	Unchanged
B cells	Increased	Unchanged	Increased
T cells	Unchanged	Unchanged	Unchanged
B-T cells	Induction of rhythm	Induction of rhythm	Increased
CD4+ cells	Induction of rhythm	Induction of rhythm	Unchanged
CD8+ cells	Suppression of rhythm	Suppression of rhythm	Unchanged
CD4+–CD8+ cells	Induction of rhythm	Induction of rhythm	Unchanged
Noradrenergic activity[b]	Increased	Unchanged	Increased
Cholinergic activity[c]	Increased	Unchanged	Increased
Acute phase			
Cell proliferation[a]	Increased	Unchanged	Increased
Mitotic response to Con A	Unchanged	Unchanged	Unchanged
Mitotic response to LPS	Unchanged	Unchanged	Unchanged
B cells	Unchanged	Unchanged	Unchanged
T cells	Unchanged	Unchanged	Unchanged
CD4+ cells	Unchanged	Unchanged	Unchanged
CD8+ cells	Unchanged	Unchanged	Unchanged
Noradrenergic activity[b]	Increased	Unchanged	Increased
Cholinergic activity[c]	Increased	Unchanged	Increased
5-HT	Absence of rhythm	Absence of rhythm	Unchanged
Inhibitory amino acids[d]	Unchanged	Unchanged	Unchanged
Excitatory amino acids[e]	Unchanged	Unchanged	Unchanged

[a] Estimated from ornithine decarboxylase activity changes.
[b] Estimated from tyrosine hydroxylase activity and neuronal NE uptake.
[c] Estimated from ^{3}H-acetylcholine synthesis and neuronal choline uptake.
[d] Aspartate, glutamate.
[e] GABA, taurine.

In our own studies conducted during the preclinical phase of arthritis, a significant effect of immune-mediated inflammatory response on diurnal rhythmicity of circulating ACTH, growth hormone (GH), prolactin and thyrotropin (TSH) release was found, and was partially sensitive to immunosuppression by cyclosporine (Selgas et al. 1998) (Table 3.2). Further experiments indicated that hypothalamic levels of corticotrophin-releasing hormone (CRH), thyrotropin-releasing hormone (TRH), GH-releasing hormone (GHRH), and somatostatin were altered in the preclinical phase of arthritis (Esquifino et al. 1999). In the median eminence, adjuvant's vehicle-injected rats exhibited significant 24-h variations for the four hypophysiotropic hormones examined, with maxima at noon.

TABLE 3.2. Summary of changes in circadian rhythms of hypothalamic and hypophysial hormones and neurotransmitters, pineal melatonin and plasma proteins during the preclinical (3rd day) phases of Freund's adjuvant arthritis in rats.

24-h rhythms	Amplitude	Acrophase	Mean
Preclinical phase			
Serum			
ACTH	Unchanged	Unchanged	Augmented
Prolactin	Unchanged	Unchanged	Augmented
GH	Absence of rhythm	Absence of rhythm	Decreased
TSH	Suppression of rhythm	Suppression of rhythm	Decreased
LH	Suppression of rhythm	Suppression of rhythm	Decreased
Albumin	Absence of rhythm	Absence of rhythm	Unchanged
α-1 globulin	Decreased	Unchanged	Unchanged
α-2 globulin	Decreased	Unchanged	Decreased
β-globulin	Absence of rhythm	Absence of rhythm	Unchanged
γ-globulin	Absence of rhythm	Absence of rhythm	Unchanged
Median eminence			
CRH	Decreased	Unchanged	Decreased
TRH	Suppression of rhythm	Suppression of rhythm	Decreased
GHRH	Decreased	Unchanged	Decreased
Somatostatin	Decreased	Unchanged	Decreased
NE	Decreased	Unchanged	Unchanged
DA turnover	Unchanged	Changed	Unchanged
5-HT turnover	Suppression of rhythm	Suppression of rhythm	Decreased
Anterior hypothalamus			
CRH	Suppression of rhythm	Suppression of rhythm	Decreased
TRH	Suppression of rhythm	Suppression of rhythm	Decreased
GHRH	Unchanged	Changed	Unchanged
Somatostatin	Unchanged	Changed	Unchanged
NE	Decreased	Unchanged	Decreased
DA turnover	Decreased	Unchanged	Unchanged
5-HT turnover	Decreased	Unchanged	Decreased
Medial hypothalamus			
CRH	Unchanged	Unchanged	Unchanged
TRH	Suppression of rhythm	Suppression of rhythm	Decreased
GHRH	Unchanged	Unchanged	Unchanged
Somatostatin	Unchanged	Unchanged	Unchanged
NE	Decreased	Unchanged	Decreased
DA turnover	Decreased	Unchanged	Decreased
5-HT turnover	Decreased	Unchanged	Unchanged
Posterior hypothalamus			
CRH	Unchanged	Unchanged	Unchanged
TRH	Unchanged	Unchanged	Unchanged
GHRH	Unchanged	Unchanged	Unchanged
Somatostatin	Unchanged	Unchanged	Unchanged
NE	Unchanged	Changed	Unchanged
DA turnover	Decreased	Unchanged	Decreased
5-HT turnover	Unchanged	Changed	Unchanged
Pineal gland			
Melatonin	Unchanged	Unchanged	Unchanged

These 24-h rhythms were inhibited or suppressed 3 days after FCA injection. The administration of the immunosuppressant drug cyclosporine impaired the depressing effect of FCA injection on CRH, TRH, and somatostatin content in median eminence, but not that on GHRH. The activity of cyclosporine was less evident in other hypothalamic regions examined (Esquifino et al. 1999). Generally, a decrease amplitude or mesor of transmitter rhythms were detectable, mainly in anterior and medial hypothalamic regions (Castrillón, Cardinali, Pazo, Cutrera, and Esquifino 2001) (Table 3.2). We also examined the changes in circadian rhythms of CNS and hypophysial hormones and neurotransmitters during the acute phase of Freund's adjuvant arthritis (i.e., 18 days after FCA administration) (Table 3.3).

TABLE 3.3. Summary of changes in circadian rhythms of hypothalamic and hypophysial hormones and neurotransmitters, pineal melatonin and plasma proteins during the acute (18th day) phase of Freund's adjuvant arthritis in rats.

24-h rhythms	Amplitude	Acrophase	Mean
Acute phase			
Serum			
ACTH	Suppression of rhythm	Suppression of rhythm	Augmented
FSH	Unchanged	Unchanged	Unchanged
LH	Suppression of rhythm	Suppression of rhythm	Decreased
Prolactin	Unchanged	Unchanged	Unchanged
Testosterone	Suppression of rhythm	Suppression of rhythm	Decreased
GH	Absence of rhythm	Absence of rhythm	Decreased
TSH	Suppression of rhythm	Suppression of rhythm	Unchanged
Anterior hypothalamus			
NE	Suppression of rhythm	Suppression of rhythm	Unchanged
DA turnover	Suppression of rhythm	Suppression of rhythm	Unchanged
Medial hypothalamus			
NE	Unchanged	Unchanged	Unchanged
DA turnover	Induction of rhythm	Induction of rhythm	Unchanged
Posterior hypothalamus			
NE	Unchanged	Unchanged	Unchanged
DA turnover	Suppression of rhythm	Suppression of rhythm	Unchanged
Adenohypophysis			
DA	Suppression of rhythm	Suppression of rhythm	Unchanged
Neurointermediate lobe			
NE	Unchanged	Unchanged	Unchanged
DA	Unchanged	Unchanged	Unchanged
Pineal gland			
Melatonin	Decreased	Unchanged	Decreased
NE	Unchanged	Unchanged	Unchanged
5-HT turnover	Decreased	Unchanged	Decreased

Differing from the relative compensation of circadian immune changes seen at this time of arthritis, changes in 24-h rhythms of neuroendocrine parameters persisted during the clinical phase of the disease (Garcia Bonacho, Esquifino, Castrillon, Reyes Toso, and Cardinali 2000).

Daily rhythms in plasma luteinizing hormone (LH), testosterone and TSH became suppressed or disrupted in arthritic rats. Concerning GH, the depressed mean values found in the preclinical phase of arthritis also persisted during the acute phase, as it was the case for the changes in catecholamine transmitter activity (Cano, Cardinali, Castrillón, Reyes Toso, and Esquifino 2001).

Twenty-four hours of variation in dopamine (DA) content were blunted in the anterior hypophysial lobe but remained unaltered in the neurointermediate lobe (Cano et al. 2001).

Disruption of endocrine circadian rhythms of plasma prolactin, insulin-like growth factor-1, LH and testosterone and of pituitary prolactin mRNA was recently reported in male Long Evans rats injected with FCA 23 days earlier (Roman, Seres, Herichova, Zeman, and Jurcovicova 2003).

3.6 Melatonin as a Circadian Immunoregulatory Signal in Adjuvant Arthritis

Pineal melatonin is normally synthesized and secreted during the dark phase of the day in all species studied to date. The primary physiological function of melatonin is to convey information about daily cycles of light and darkness to body physiology. By its pattern of secretion during darkness, melatonin indicates the length of the scotophase, thus representing the chemical code of the night. This information is used for the organization of functions, which respond to changes in photoperiod such as circadian and seasonal rhythms (Arendt, Middleton, Stone, and Skene 1999; Cardinali and Pevet 1998).

Melatonin is secreted by the pineal gland by simple diffusion and its lipophilicity contributes to an easy passive diffusion across cell membranes as well as through cell layers. This assures a rapid distribution of melatonin throughout the body and effects on virtually every cell type of the organism. Radioactive melatonin administered intravenously rapidly disappears from the blood with a half-life of about 30 min; about 60 to 70% of melatonin in plasma is bound to albumin (Cardinali, Lynch, and Wurtman 1972).

The first experiments on brain melatonin receptor sites were carried out in the late seventies by employing ^{3}H-melatonin as a ligand indicating the existence of melatonin acceptor sites in bovine brain (Cardinali 1981). By using a 2-^{125}I-iodomelatonin analog, melatonin binding was then detected in several brain areas, the choroid plexus and in some brain arteries as well as in peripheral organs, like primary and secondary lymphoid organs, the Harderian glands, the adrenals, heart and lungs, the gastrointestinal tract, the mammary glands, the kidney and male reproductive organs. Indeed, to have an organ devoid of melatonin binding site may constitute an exception rather than the rule (Brydon et al. 1999).

A first classification of putative melatonin receptor sites into the ML1 and ML2 was based on kinetic and pharmacological differences of 2-^{125}I-iodomelatonin binding and a major achievement in the field was the cloning of the ML1 melatonin receptors (Brydon et al. 1999; Reppert, Weaver, and Ebisawa 1994). A nomenclature of melatonin receptors has been proposed by the International Union of Pharmacology (IUPHAR) (Dubocovich et al. 2000). Besides membrane acceptor sites, evidence has accumulated on nuclear binding of melatonin as well (Acuña-Castroviejo, Reiter, Menendez-Pelaez, Pablos, and Burgos 1994). Melatonin appears to interact with the orphan nuclear hormone receptor superfamily RZR/ROR (Missbach, Jagher, Sigg, Nayeri, Carlberg, and Wiesenberg 1996). Among presumptive second messengers for melatonin action, the cAMP generating system has received paramount attention. The main signal transduction pathway of high affinity MT_1 receptors in both neuronal and nonneuronal tissues is the inhibition of cAMP formation through a pertussis toxin sensitive inhibitory Gi protein.

Coupling of the high affinity melatonin receptors to other signaling pathways has also been reported. Melatonin modifies cGMP levels, decreases Ca^{2+} influx and inhibits arachidonate acid conversion and prostaglandin synthesis and the NO synthase/NO system. Direct effects of melatonin on calmodulin and other intracellular proteins, nuclear receptor activity for melatonin, and the free radical scavenging properties of melatonin should be also considered (for ref. see Cardinali, Golombek, Rosenstein, Cutrera, and Esquifino 1997b; Reiter, Tan, and Burkhardt 2002).

Pineal ablation, or any other experimental procedure that inhibits melatonin synthesis and secretion, induces a state of immunodepression which is partly counteracted by melatonin in several species (Beskonakli, Palaoglu, Aksaray, Alanoglu, Turhan, and Taskin 2001; Fraschini, Demartini, Esposti, and Scaglione 1998; Liebmann, Wolfler, Felsner, Hofer, and Schauenstein 1997; Maestroni 2001; Martins et al. 2001; Skwarlo-Sonta 2002). Melatonin treatment increases T-cell proliferation (Konakchieva, Kyurkchiev, Kehayov, Taushanova, and Kanchev 1995; Pioli, Caroleo, Nistico, and Doria 1993), enhances antigen presentation by macrophages to T cells by increasing the expression of MHC class II molecules (Pioli et al. 1993), activates splenic, lymph node, and bone marrow cells (Drazen and Nelson 2001; Wajs, Kutoh, and Gupta 1995), stimulates antibody-dependent cellular cytotoxicity (Giordano and Palermo 1991) and augments natural and acquired immunity (Bonilla et al. 2001; Negrette et al. 2001; Poon, Liu, Pang, Brown, and Pang 1994). Melatonin also stimulates NK cell activity (Currier, Sun, and Miller 2000; del Gobbo, Libri, Villani, Calio, and Nistico 1989), activates monocytes (Morrey, McLachlan, Serkin, and Bakouche 1994), increases the number of Th-2 lymphocytes (Lissoni et al. 1995), augments $CD4^+$ lymphocytes and decreased $CD8^+$ lymphocytes in submaxillary lymph nodes (Castrillón et al. 2001), restores impaired Th cell activity in immuno-depressed mice (Maestroni, Conti, and Pierpaoli 1988) and augments antibody responses in vivo (Akbulut, Gonul, and Akbulut 2001; Fraschini et al. 1998; Maestroni 1995; Pioli et al. 1993; Poon et al. 1994). Concerning B cells, there is information on melatonin inhibition of apoptosis during early B-cell development in mouse bone marrow (Yu, Miller, and Osmond 2000).

Besides the release of proinflammatory Th-1 cytokines, such as IFN-γ and IL-2, administration of melatonin to antigen-primed mice increased the production of IL-10, indicating that melatonin also activates anti-inflammatory Th-2-like immune responses (Raghavendra, Singh, Kulkarni, and Agrewala 2001a; Raghavendra, Singh, Shaji, Vohra, Kulkarni, and Agrewala 2001b). It is not yet clear whether melatonin acts only on Th-1 cells or also on Th-2 cells. This is an important subject as the Th-1/Th-2 balance is significant for the immune response (Maestroni 2001). Relevant to this, melatonin treatment suppressed the subsequent in vitro stimulation by the mitogenic agents LPS (that stimulates B cells) and Con A (that stimulates T cells) in submaxillary lymph nodes (Castrillón et al. 2001). In addition, an inhibitory influence of melatonin on parameters of the immune function has also been demonstrated, i.e., in human lymphocytes NK cell activity, DNA, IFN-γ, and TNF-α synthesis, as well as the proliferation of T lymphocytes and lymphoblastoid cell lines were depressed by melatonin (Arzt et al. 1988; Lewinski, Zelazowski, Sewerynek, Zerek-Melen, Szkudlinski, and Zelazowska 1989; Persengiev and Kyurkchiev 1993). In our own studies, melatonin treatment exerted an inhibitory influence on submaxillary lymph node cytolytic, CD8$^+$ cells (Esquifino et al. 2001).

Melatonin possesses anti-inflammatory activity (Cuzzocrea and Reiter 2002). It reduces tissue destruction during inflammatory reactions by a number of means, among them scavenging of free radicals (Reiter et al. 2002). Additionally, melatonin prevents the translocation of nuclear factor-kappa B to the nucleus and its binding to DNA, thereby reducing the up-regulation of a variety of proinflammatory cytokines (for ref. see Guerrero and Reiter 2002). Finally, there is evidence that melatonin inhibits the production of adhesion molecules that promote the sticking of leukocytes to endothelial cells, attenuating transendothelial cell migration and edema (Sasaki et al. 2002).

The subcellular mechanisms involved in the immunomodulatory activity of melatonin remain to be elucidated. There is evidence of MT$_1$ receptors for melatonin, mainly in human circulating CD4$^+$ T helper lymphocytes with few in CD8$^+$ T cytolytic cells and none in B lymphocytes (Garcia-Maurino et al. 1997). Such a predominant effect on CD4$^+$ cells is supported by the observations on melatonin efficacy to augment CD4$^+$ cells in submaxillary lymph nodes (Castrillón et al. 2001). However, expression of Mel1a-melatonin receptor was found in rat thymus and spleen, melatonin receptor mRNA being expressed in all the thymic lymphocyte subpopulations (CD4$^+$,CD8$^+$, doubled positive, doubled negative, and B cells), indicating possible effects of melatonin on all these cells (Pozo et al. 1997; Yu et al. 2000). Nuclear melatonin receptors may also mediate immunomodulation, since drugs that bind to RZR/ROR receptors are active in experimental models of autoimmune diseases (Missbach et al. 1996). Melatonin is also a potent antioxidant, acting by itself rather than through specific binding sites (Reiter 1998; Reiter et al. 1999). In addition, melatonin could affect centrally the release of hormone in the hypothalamic-hypophyseal unit (Cardinali et al. 1997b) as well as the activity of autonomic pathways to the lymphoid organs (Brusco, Garcia Bonacho, Esquifino, and Cardinali 1998).

Our studies on melatonin role in arthritis have been mainly addressed to examine the participation of melatonin in regulation of circadian rhythmicity of immune parameters in rats (Castrillón et al. 2001). Pretreatment for 11 days with a pharmacological dose of melatonin (100 μg) affected some aspects of the early phase of the immune response elicited by FCA injection, at the preclinical phase of disease. Cell proliferation in rat submaxillary lymph nodes and spleen during the immune reaction (as assessed by ODC activity) exhibited a pineal-dependent diurnal rhythmicity, as it was reduced by pinealectomy or pineal sympathetic denervation (Cardinali, Cutrera, Castrillon, and Esquifino 1996b; Cardinali, Cutrera, Garcia Bonacho, and Esquifino 1997a). This effect was counteracted by a pharmacological melatonin dose (100 μg/day). Exogenous melatonin also restored the reduced amplitude in diurnal rhythms of lymph node or splenic tyrosine hydroxylase (TH) activity and lymph node acetylcholine synthesis (Cardinali et al. 1996b, 1997a).

Further examination of melatonin activity on circadian rhythmicity of cell proliferation in submaxillary lymph nodes and spleen at the clinical phase of arthritis was conducted in young and old Sprague–Dawley rats (Cardinali et al. 1998b). Pineal melatonin content was measured, as well as the efficacy of melatonin treatment to recover modified circadian rhythmicity of submaxillary lymph node and splenic ODC and TH activities and of lymph node ^{3}H-acetylcholine synthesis. After 17 daily injections of 10 or 100 μg of melatonin at the evening, the treatment restored the inflammatory response in old rats (assessed plethysmographically in hind paws) to the level found in young animals. In young rats, an inflammation-promoting effect of 100 μg melatonin could be demonstrated. As a consequence of the immune reaction, submaxillary lymph node and splenic lymph cell proliferation augmented significantly, with acrophases of 24-h rhythms at the afternoon for lymph nodes or in the morning for spleen. Mesor and amplitude of ODC rhythm were lowest in old rats, while melatonin injection generally augmented its amplitude. Lymph node and splenic TH activity attained maximal values at early night while maxima in lymph node. ^{3}H-acetylcholine synthesis occurred at the afternoon. Amplitude and mesor of these rhythms were lowest in old rats, an effect generally counteracted by melatonin treatment. The results were compatible with an age-dependent, significant depression of pineal melatonin synthesis during adjuvant-induced arthritis and with decreased amplitude of circadian rhythms in immune cell proliferation and autonomic activity in lymph nodes and spleen at the clinical phase of the disease. This picture was generally counteracted by melatonin injection, mainly in old rats (Cardinali et al. 1998b).

A number of studies were carried out to examine the participation of melatonin in altered 24-h rhythms of serum ACTH, GH, prolactin, LH and insulin in rats at the preclinical phase of Freund's adjuvant arthritis (Esquifino, Castrillon, Garcia Bonacho, Vara, and Cardinali 1999a). The data indicated that several early changes in levels and 24-h rhythms of circulating ACTH, PRL and LH in FCA-injected rats were sensitive to treatment with melatonin (100 μg). An effect of melatonin treatment on 24-h variations in hypothalamic 5-HT and DA turnover during the preclinical phase of Freund's adjuvant arthritis was also apparent (Pazo et al. 2000). FCA injection suppressed circadian rhythmicity of 5-HT turnover in the anterior hypothalamus, an effect prevented by the previous injection of melatonin. Melatonin decreased 5-HT turnover rate in the anterior hypothalamus. Melatonin also prevented the changes in 5-HT turnover of medial hypothalamus evoked by Freund's adjuvant. As far as hypothalamic DA turnover, the preventing effect of melatonin was less clear, sometimes synergizing with the mycobacterial adjuvant to modify the normal 24-h pattern detected in hypothalamic regions (Pazo et al. 2000).

Physiological circulating levels of melatonin at midnight in rats are about 90 pg/ml in rats (Chan, Pang, Tang, and Brown 1984) while the melatonin levels achieved within 15 min after the systemic administration of a 100 μg-dose are about 30 or 200 ng/ml plasma (Raynaud, Mauviard, Geoffriau, Claustrat, and Pevet 1993), with a half-life of about 20 min (Chan et al. 1984). We addressed this subject by examining whether the administration of melatonin to pinealectomized rats in a way that reproduced the plasma values and daily rhythm of endogenous melatonin could affect immune responses during arthritis development (Cardinali, García, Cano, and Esquifino 2004). Pinealectomized rats exhibited a significantly less pronounced inflammatory response, which was restored to normal by physiological melatonin administration. The physiological doses of melatonin employed were effective to counteract the impaired response of lymph node ODC seen in pinealectomized rats.

It must be thus noted that the pharmacological effect of melatonin on the immune response may not always be beneficial, particularly in young subjects. In autoimmune arthritis developed in mice with type II rat collagen melatonin administration (1 mg/kg) induced a more severe arthritis. Accordingly, pinealectomy in two strains of mice immunized with rat type II collagen reduced severity of the arthritis as shown by a slower onset of the disease, a less severe course of the disease (reduced clinical scores) and reduced serum levels of anti- collagen II antibodies (Hansson, Holmdahl, and Mattsson 1992, 1993). Using a 100-µg dose of melatonin an inflammation-promoting effect could be demonstrated in young rats injected with FCA. In contrast, melatonin administration (10 or 100 µg) to old rats restored the inflammatory response in hind paws of FCA-injected rats to levels found in young rats (Cardinali et al. 1998b). Therefore, high levels of melatonin in young animals may stimulate the immune system and cause exacerbation of both autoimmune collagen II and mycobacterial arthritis. Indeed, recent data indicate that rheumatoid arthritis patients have increased nocturnal plasma levels of melatonin and their synovial macrophages respond to melatonin with an increased production of IL-12 and NO (Maestroni, Sulli, Pizzorni, Villaggio, and Cutolo 2002; Sulli et al. 2002). In these patients, inhibition the antagonism of melatonin synthesis or effect could be therapeutically desirable.

Acknowledgments. Work in authors' laboratories was supported in part by DGES, Spain, Agencia Nacional de Promoción Científica y Tecnológica, Argentina, and the University of Buenos Aires and Consejo Nacional de Investigaciones Científicas y Técnicas (CONICET), Argentina.

References

Acuña-Castroviejo, D., Reiter, R.J., Menendez-Pelaez, A., Pablos, M.I., and Burgos, A. (1994) Characterization of high-affinity melatonin binding sites in purified cell nuclei of rat liver. *J Pineal Res* 16, 100–112.

Akbulut, K.G., Gonul, B., and Akbulut, H. (2001) The effects of melatonin on humoral immune responses of young and aged rats. *Immunol Invest* 30, 17–20.

Albrecht, U., and Eichele, G. (2003) The mammalian circadian clock. *Curr Opin Genet Dev* 13, 271–277.

Arendt, J., Middleton, B., Stone, B., and Skene D. (1999) Complex effects of melatonin: Evidence for photoperiodic responses in humans? *Sleep* 22, 625–635.

Arjona, A., and Sarkar, D.K. (2005) Circadian oscillations of clock genes, cytolytic factors, and cytokines in rat NK cells. *J Immunol* 174, 7618–7624.

Arzt, E.S., Fernandez-Castelo, S., Finocchiaro, L.M., Criscuolo, M.E., Diaz, A., Finkielman, S., and Nahmod, V.E. (1988) Immunomodulation by indoleamines: Serotonin and melatonin action on DNA and interferon-gamma synthesis by human peripheral blood mononuclear cells. *J Clin Immunol* 8, 513–520.

Bauhofer, A., Witte, K., Celik, I., Pummer, S., Lemmer, B., and Lorenz, W. (2001) Sickness behaviour, an animal equivalent to human quality of life, is improved in septic rats by G-CSF and antibiotic prophylaxis. *Langenbeck's Arch Surg* 386, 132–140.

Bendele, A.M. (2001) Animal models of rheumatoid arthritis. *J Musculoskel Neuron Interact* 1, 377–385.

Bentivoglio, M., Grassi-Zucconi, G., Peng, Z.C., and Kristensson, K. (1994) Sleep and timekeeping changes, and dysregulation of the biological clock in experimental trypanosomiasis. *Bull Soc Pathol Exotique* 87, 372–375.

Beskonakli, E., Palaoglu, S., Aksaray, S., Alanoglu, G., Turhan, T., and Taskin, Y. (2001) Effect of pinealectomy on immune parameters in rats with Staphylococcus aureus infection. *Neurosurg Rev* 24, 26–30.

Bluthe, R.M., Walter, V., Parnet, P., Laye, S., Lestage, J., Verrier, D., Poole, S., Stenning, B.E., Kelley, K.W., and Dantzer, R. (1994) Lipopolysaccharide induces sickness behaviour in rats by a vagal mediated mechanism. *C R Acad Sci III* 317, 499–503.

Boggio, V., Castrillon, P., Pérez Lloret, S., Riccio, P., Esquifino, A.I., Cardinali, D.P., and Cutrera, R.A. (2003) Cerebroventricular administration of interferon-gamma modifies locomotor activity in the golden hamster. *NeuroSignals* 12, 89–94.

Boivin, D.B., James, F.O., Wu, A., Cho-Park, P.F., Xiong, H., and Sun, Z.S. (2003) Circadian clock genes oscillate in human peripheral blood mononuclear cells. *Blood* 102, 4143–4145.

Bonilla, E., Rodon, C., Valero, N., Pons, H., Chacin-Bonilla, L., Garcia, T.J., Rodriguez, Z., Medina-Leendertz, S., and Anez, F. (2001) Melatonin prolongs survival of immunodepressed mice infected with the Venezuelan equine encephalomyelitis virus. *Trans R Soc Trop Med Hyg* 95, 207–210.

Bourin, P., Mansour, I., Doinel, C., Roue, R., Rouger, P. and Levi, F. (1993) Circadian rhythms of circulating NK cells in healthy and human immunodeficiency virus-infected men. *Chronobiol Int* 10, 298–305.

Bowman, S.J. (2002) Hematological manifestations of rheumatoid arthritis. *Scand J Rheumatol* 31, 251–259.

Bruno, V.A., Scacchi, P., Pérez Lloret, S., Esquifino, A.I., Cardinali, D.P. and Cutrera, R.A. (2005) Melatonin treatment counteracts the hyperthermic effect of lipopolysaccharide injection in the syrian hamster. *Neurosci Lett* 389, 169–172.

Brusco, L.I., Garcia Bonacho, M., Esquifino, A.I., and Cardinali, D.P. (1998) Diurnal rhythms in norepinephrine and acetylcholine synthesis of sympathetic ganglia, heart and adrenals of aging rats. Effect of melatonin. *J Auton Nerv Syst* 74, 49–61.

Brydon, L., Petit, L., de Coppet, P., Barrett, P., Morgan, P.J., Strosberg, A.D., and Jockers, R. (1999) Polymorphism and signalling of melatonin receptors. *Reprod Nutr Dev* 39, 315–324.

Buijs, R.M., van Eden, C.G., Goncharuk, V.D., and Kalsbeek, A. (2003) The biological clock tunes the organs of the body: Timing by hormones and the autonomic nervous system. *J Endocrinol* 177, 17–26.

Calvino, B., Crepon-Bernard, M.O., and Le Bars, D. (1987) Parallel clinical and behavioural studies of adjuvant-induced arthritis in the rat: Possible relationship with 'chronic pain'. *Behav Brain Res* 24, 11–29.

Cano, P., Cardinali, D.P., Castrillón, P., Reyes Toso, C., and Esquifino, A.I. (2001) Age-dependent changes in 24-hour rhythms of catecholamine content and turnover in hypothalamus, corpus striatum and pituitary gland of rats injected with Freund's adjuvant. *BMC Physiol* 1, 14.

Cano, P., Cardinali, D.P., Fernandez, P., Reyes Toso, C., and Esquifino, A.I. (2006) 24-Hour rhythms of splenic mitogenic responses, lymphocyte subset populations and interferon γ release after calorie restriction or social isolation of rats. *Biol Rhythm Res* 37(3), 255–263.

Cano, P., Cardinali, D.P., Jimenez, V., Alvarez, M.P., Cutrera, R.A., and Esquifino, A.I. (2005) Effect of interferon-gamma treatment on 24-hour variations in plasma ACTH, growth hormone, prolactin, luteinizing hormone and follicle-stimulating hormone of male rats. *Neuroimmunomodulation* 12, 146–151.

Cardinali, D., Della Maggiore, V., Selgas, L., and Esquifino, A. (1996a) Diurnal rhythm in ornithine decarboxylase activity and noradrenergic and cholinergic markers in rat submaxillary lymph nodes. *Brain Res* 711, 153–162.

Cardinali, D.P. (1981) Melatonin. A mammalian pineal hormone. *Endocr Rev* 2, 327–346.

Cardinali, D.P., Brusco, L.I., Selgas, L., and Esquifino, A.I. (1998a) Diurnal rhythms in ornithine decarboxylase activity and norepinephrine and acetylcholine synthesis in submaxillary lymph nodes and spleen of young and aged rats during Freund's adjuvant-induced arthritis. *Brain Res* 789, 283–292.Cardinali, D.P., Brusco, L.I., Cutrera, R.A., Castrillon, P., and Esquifino, A.I. (1999) Melatonin as a time-meaningful signal in circadian organization of immune response. *Biol Signals Recept* 8, 41–48.

Cardinali, D.P., Brusco, L.I., Garcia, B.M., and Esquifino, A.I. (1998b) Effect of melatonin on 24-hour rhythms of ornithine decarboxylase activity and norepinephrine and acetylcholine synthesis in submaxillary lymph nodes and spleen of young and aged rats. *Neuroendocrinology* 67, 349–362.

Cardinali, D.P., Cutrera, R.A., Castrillon, P., and Esquifino, A.I. (1996b) Diurnal rhythms in ornithine decarboxylase activity and norepinephrine and acetylcholine synthesis of rat submaxillary lymph nodes: Effect of pinealectomy, superior cervical ganglionectomy and melatonin replacement. *Neuroimmunomodulation* 3, 102–111.

Cardinali, D.P., Cutrera, R.A., Garcia Bonacho, M., and Esquifino, A.I. (1997a) Effect of pinealectomy, superior cervical ganglionectomy and melatonin treatment on 24-hour rhythms in ornithine decarboxylase and tyrosine hydroxylase activities of rat spleen. *J Pineal Res* 22, 210–220.

Cardinali, D.P., and Esquifino, A.I. (1998) Neuroimmunoendocrinology of the cervical autonomic nervous system. *Biomed Rev* 9, 47–59.

Cardinali, D.P., and Esquifino, A.I. (2003) Circadian disorganization in experimental arthritis. *NeuroSignals* 12, 267–282.

Cardinali, D.P., García, A.P., Cano, P., and Esquifino, A.I. (2004) Melatonin role in experimental arthritis. *Curr Drug Targets—-Immune-Endocr Metab Dis* 4, 1–22.

Cardinali, D.P., Golombek, D.A., Rosenstein, R.E., Cutrera, R.A., and Esquifino, A.I. (1997b) Melatonin site and mechanism of action: Single or multiple? *J Pineal Res* 23, 32–39.

Cardinali, D.P., Lynch, H.J., and Wurtman, R.J. (1972) Binding of melatonin to human and rat plasma proteins. *Endocrinology* 91, 1213–1218.

Cardinali, D.P., and Pevet, P. (1998) Basic aspects of melatonin action. *Sleep Med Rev* 2, 175–190.

Cardinali, D.P., and Romeo, H.E. (1991) The autonomic nervous system of the cervical region as a channel of neuroendocrine communication. *Front Neuroendocrinol* 12, 278–297.

Cardinali, D.P., and Stern, J.E. (1994) Peripheral neuroendocrinology of the cervical autonomic nervous system. *Braz J Med Biol Res* 27, 573–599.

Castrillón, P., Cardinali, D.P., Pazo, D., Cutrera, R.A., and Esquifino, A.I. (2001) Effect of superior cervical ganglionectomy on 24-h variations in hormone secretion from anterior hypophysis and in hypothalamic monoamine turnover, during the preclinical phase of Freund's adjuvant arthritis in rats. *J Neuroendocrinol* 13, 288–295.

Castrillon, P.O., Esquifino, A.I., Varas, A., Zapata, A., Cutrera, R.A., and Cardinali, D.P. (2000) Effect of melatonin treatment on 24-h variations in responses to mitogens and lymphocyte subset populations in rat submaxillary lymph nodes. *J Neuroendocrinol* 12, 758–765.

Chacon, F., Cano, P., Jimenez, V., Cardinali, D.P., Marcos, A., and Esquifino, A.I. (2004) 24-Hour changes in circulating prolactin, follicle-stimulating hormone, luteinizing hormone and testosterone in young male rats subjected to calorie restriction. *Chronobiol Int* 21, 393–404.

Chacon, F., Esquifino, A.I., Perelló, M., Cardinali, D.P., Spinedi, E., and Alvarez, M.P. (2005) 24-Hour changes in ACTH, corticosterone, growth hormone and leptin levels in young male rats subjected to calorie restriction. *Chronobiol Int* 22, 253–265.

Chan, M.Y., Pang, S.F., Tang, P.L., and Brown, G.M. (1984) Studies on the kinetics of melatonin and N-acetylserotonin in the rat at mid-light and mid-dark. *J Pineal Res* 1, 227–236.

Collins, B., and Blau, J. (2006) Keeping time without a clock. *Neuron* 50, 348–350.

Crespo, E., Macias, M., Pozo, D., Escames, G., Martin, M., Vives, F., Guerrero, J.M. and Acuña-Castroviejo, D. (1999) Melatonin inhibits expression of the inducible NO synthase

II in liver and lung and prevents endotoxemia in lipopolysaccharide-induced multiple organ dysfunction syndrome in rats. *FASEB J* 13, 1537–1546.

Currier, N.L., Sun, L.Z., and Miller, S.C. (2000) Exogenous melatonin: Quantitative enhancement in vivo of cells mediating non-specific immunity. *J Neuroimmunol* 104, 101–108.

Cuzzocrea, S., and Reiter, R.J. (2002) Pharmacological actions of melatonin in acute and chronic inflammation. *Curr Top Med Chem* 2, 153–165.

Dantzer R. (2001) Cytokine-induced sickness behavior: Mechanisms and implications. *Ann N Y Acad Sci* 933, 222–234.

del Gobbo, V., Libri, V., Villani, N., Calio, R., and Nistico, G. (1989) Pinealectomy inhibits interleukin-2 production and natural killer activity in mice. *Int J Immunopharmacol* 11, 567–573.

Drazen, D.L., and Nelson, R.J. (2001) Melatonin receptor subtype MT_2 (Mel 1b) and not mt_1 (Mel 1a) is associated with melatonin-induced enhancement of cell-mediated and humoral immunity. *Neuroendocrinology* 74, 178–184.

Dubocovich, M.L., Cardinali, D.P., Delagrange, P., Krause, D.N., Strosberg, D., Sugden, D., and Yocca, F.D. (2000) Melatonin receptors. In: IUPHAR Ed. *The IUPHAR Compendium of Receptor Characterization and Classification, 2nd. Edn.* IUPHAR Media, London, pp. 271–277.

Esquifino, A., Selgas, L., Arce, A., Della Maggiore, V., and Cardinali, D. (1996) Twenty four hour rhythms in immune responses in rat submaxillary lymph nodes and spleen. Effect of cyclosporine. *Brain Behav Immun* 10, 92–102.Esquifino, A.I., Alvarez, M.P., Cano, P., Chacon, F., Reyes Toso, C.F., and Cardinali, D.P. (2004a) 24-Hour pattern of circulating prolactin and growth hormone levels and submaxillary lymph node immune responses in growing male rats subjected to social isolation. *Endocrine* 25, 41–48.

Esquifino, A.I., and Cardinali, D.P. (1994) Local regulation of the immune response by the autonomic nervous system. *Neuroimmunomodulation* 1, 265–273.

Esquifino, A.I., Castrillón, P., Chacon, F., Cutrera, R.A., and Cardinali, D.P. (2001) Effect of local sympathectomy on 24-hour changes in mitogenic responses and lymphocyte subset populations in rat submaxillary lymph nodes during the preclinical phase of Freud's adjuvant arthritis. *Brain Res* 888, 227–234.

Esquifino, A.I., Castrillon, P., Garcia Bonacho, M., Vara, E., and Cardinali, D.P. (1999a) Effect of melatonin treatment on 24-hour rhythms of serum ACTH, growth hormone, prolactin, luteinizing hormone and insulin in rats injected with Freund's adjuvant. *J Pineal Res* 27, 15–23.

Esquifino, A.I., Chacon, F., Cano, P., Marcos, A., Cutrera, R.A., and Cardinali, D.P. (2004b) 24-Hour rhythms of mitogenic responses, lymphocyte subset populations and amino acid content in submaxillary lymph nodes of growin male rats subjected to calorie restriction. *J Neuroimmunol* 156, 66–73.

Esquifino, A.I., Chacon, F., Jimenez, V., Reyes Toso, C., and Cardinali, D.P. (2004c) 24-Hour changes in circulating prolactin, follicle-stimulating hormone, luteinizing hormone and testosterone in male rats subjected to social isolation. *J Circad Rhythms* 2, 1.

Esquifino, A.I., Selgas, L., Vara, E., Arce, A., and Cardinali, D.P. (1999b) Twenty-four hour rhythms of hypothalamic corticotropin-releasing hormone, thyrotropin-releasing hormone, growth hormone-releasing hormone and somatostatin in rats injected with Freund's adjuvant. *Biol Signals Recept* 8, 178–190.

Feldmann, M., Brennan, F.M., Foxwell, B.M., and Maini, R.N. (2001) The role of TNF alpha and IL-1 in rheumatoid arthritis. *Curr Dir Autoimmun* 3, 188–199.

Filipski, E., King, V.M., Etienne, M.C., Li, X., Claustrat, B., Granda, T.G., Milano, G., Hastings, M.H., and Levi, F. (2004) Persistent twenty-four hour changes in liver and bone marrow despite suprachiasmatic nuclei ablation in mice. *Am J Physiol Regul Integr Comp Physiol* 287, R844–R851.

Fraschini, F., Demartini, G., Esposti, D., and Scaglione, F. (1998) Melatonin involvement in immunity and cancer. *Biol Signals* 7, 61–72.

Fukuhara, C., and Tosini, G. (2003) Peripheral circadian oscillators and their rhythmic regulation. *Front Biosci* 8, d642–d651.

Garcia Bonacho, M., Cardinali, D.P., Castrillon, P., Cutrera, R.A., and Esquifino, A.I. (2001) Aging-induced changes in 24-h rhythms of mitogenic responses, lymphocyte subset populations and neurotransmitter and amino acid content in rat submaxillary lymph nodes during Freund's adjuvant arthritis. *Exp Gerontol* 36, 267–282.

Garcia Bonacho, M., Esquifino, A.I., Castrillon, P., Reyes Toso, C., and Cardinali, D.P. (2000) Age-dependent effect of Freund's adjuvant on 24-hour rhythms in plasma prolactin, growth hormone, thyrotropin, insulin, follicle-stimulating hormone, luteinizing hormone and testosterone in rats. *Life Sci* 66, 1969–1977.

Garcia-Maurino, S., Gonzalez-Haba, M.G., Calvo, J.R., Rafii-el-Idrissi, M., Sanchez-Margalet, V., Goberna, R., and Guerrero, J.M. (1997) Melatonin enhances IL-2, IL-6, and IFN-gamma production by human circulating CD4$^+$ cells: A possible nuclear receptor-mediated mechanism involving T helper type 1 lymphocytes and monocytes. *J Immunol* 159, 574–581.

Giordano, M., and Palermo, M.S. (1991) Melatonin-induced enhancement of antibody-dependent cellular cytotoxicity. *J Pineal Res* 10, 117–121.

Goodson, N., and Symmons, D. (2002) Rheumatoid arthritis in women: Still associated with an increased mortality. *Ann Rheum Dis* 61, 955–956.

Guerrero, J.M., and Reiter, R.J. (2002) Melatonin-immune system relationships. *Curr Top Med Chem* 2, 167–179.

Halberg, F., Johnson, E.A., Brown, B.W., and Bittner, J.J. (1960) Susceptibility rhythm to E. coli endotoxin and bioassay. *Proc Soc Exp Biol Med* 103, 142–144.

Hansson, I., Holmdahl, R., and Mattsson, R. (1992) The pineal hormone melatonin exaggerates development of collagen-induced arthritis in mice. *J Neuroimmunol* 39, 23–30.

Hansson, I., Holmdahl, R., and Mattsson, R. (1993) Pinealectomy ameliorates collagen II-induced arthritis in mice. *Clin Exp Immunol* 92, 432–436.

Hardeland, R., Pandi-Perumal, S.R., and Cardinali, D.P. (2006) Melatonin. *Int J Biochem Cell Biol* 38, 313–316.

Hastings, M.H., Reddy, A.B., and Maywood, E.S. (2003) A clockwork web: Circadian timing in brain and periphery, in health and disease. *Nat Rev Neurosci* 4, 649–661.

Haus, E., and Smolensky, M.H. (1999) Biologic rhythms in the immune system. *Chronobiol Int* 16, 581–622.

Holoshitz, J., Matiau, J., and Cohen, I. (1984) Arthritis induced in rats by cloned T lymphocytes responsive to mycobacteria but not to collagen type II. *J Clin Invest* 73, 211–215.

Hrushesky, W.J.M., Langevin, T., Kim, Y.J., and Wood, P.A. (1994) Circadian dynamics of tumor necrosis factor α (cachectin) lethality. *J Exp Med* 180, 1059–1065.

Jimenez, V., Cardinali, D.P., Alvarez, M.P., Fernandez, M.P., Boggio, V., and Esquifino, A.I. (2005) Effect of chronic ethanol feeding on 24-hour rhythms of mitogenic responses and lymphocyte subset populations in thymus and spleen of peripubertal male rats. *Neuroimmunomodulation* 12 (6), 357-365.

Jimenez, V., Cardinali, D.P., Cano, P., Alvarez, M.P., Reyes Toso, C., and Esquifino, A.I. (2004) Effect of ethanol on 24-hour hormonal changes in peripubertal male rats. *Alcohol* 34, 127–132.

Johnson, R.W. (2002) The concept of sickness behavior: A brief chronological account of four key discoveries. *Vet Immunol Immunopathol* 87, 443–450.

Kastin, A.J., Pan, W., Maness, L.M., and Banks, W.A. (1999) Peptides crossing the blood-brain barrier: Some unusual observations. *Brain Res* 848, 96–100.

Knight, B., Katz, D.R., Isenberg, D.A., Ibrahim, M.A., Le Page, S., Hutchings, P., Schwartz, R.S., and Cooke, A. (1992) Induction of adjuvant arthritis in mice. *Clin Exp Immunol* 90, 459–465.

Kobayashi, H., Oishi, K., Hanai, S., and Ishida, N. (2004) Effect of feeding on peripheral circadian rhythms and behaviour in mammals. *Genes Cells* 9, 857–864.

Konakchieva, R., Kyurkchiev, S., Kehayov, I., Taushanova, P., and Kanchev, L. (1995) Selective effect of methoxyindoles on the lymphocyte proliferation and melatonin binding to activated human lymphoid cells. *J Neuroimmunol* 63, 125–132.

Krueger, J.M., Obal, F., Fang, J., Kubota, T., and Taishi, P. (2001) The role of cytokines in physiological sleep regulation. *Ann N Y Acad Sci* 933, 211–221.

Kusanagi, H., Mishima, K., Satoh, K., Echizenya, M., Katoh, T., and Shimizu, T. (2004) Similar profiles in human period 1 gene expression in peripheral mononuclear and polymorphonuclear cells. *Neurosci Lett* 365, 124–127.

Lakin-Thomas, P.L. (2006) Transcriptional feedback oscillators: Maybe, maybe not. *J Biol Rhythms* 21, 83–92.

Lancel, M., Mathias, S., Faulhaber, J., and Schiffelholz, T. (1996) Effect of interleukin-1 β on EEG power density during sleep depends on circadian phase. *Am J Physiol Regul Integr Comp Physiol* 270, R830–R837.

Larson, S.J. (2002) Behavioral and motivational effects of immune-system activation. *J Gen Psychol* 129, 401–414.

Lewinski, A., Zelazowski, P., Sewerynek, E., Zerek-Melen, G., Szkudlinski, M., and Zelazowska, E. (1989) Melatonin-induced suppression of human lymphocyte natural killer activity in vitro. *J Pineal Res* 7, 153–164.

Lewy, A.J., Ahmed, S., and Sack, R.L. (1996) Phase shifting the human circadian clock using melatonin. *Behav Brain Res* 73, 131–134.

Liebmann, P.M., Wolfler, A., Felsner, P., Hofer, D., and Schauenstein, K. (1997) Melatonin and the immune system. *Int Arch Allergy Immunol* 112, 203–211.

Lissoni, P., Barni, S., Brivio, F., Rossini, F., Fumagalli, L., Ardizzoia, A., and Tancini, G. (1995) A biological study on the efficacy of low-dose subcutaneous interleukin- 2 plus melatonin in the treatment of cancer-related thrombocytopenia. *Oncology* 52, 360–362.

Lundkvist, G.B., Andersson, A., Robertson, B., Rottenberg, M.E., and Kristensson, K. (1999) Light-dependent regulation and postnatal development of the interferon-gamma receptor in the rat suprachiasmatic nuclei. *Brain Res* 849, 231–234.

Lundkvist, G.B., Hill, R.H., and Kristensson, K. (2002) Disruption of circadian rhythms in synaptic activity of the suprachiasmatic nuclei by African trypanosomes and cytokines. *Neurobiol Dis* 11, 20–27.

Lundkvist, G.B., Robertson, B., Mhlanga, J.D., Rottenberg, M.E., and Kristensson, K. (1998) Expression of an oscillating interferon-gamma receptor in the suprachiasmatic nuclei. *Neuroreport* 9, 1059–1063.

Maestroni, G.J. (1995) T-helper-2 lymphocytes as a peripheral target of melatonin. *J Pineal Res* 18, 84–89.

Maestroni, G.J. (1996) Melatonin as a therapeutic agent in experimental endotoxic shock. *J Pineal Res* 20, 84–89.

Maestroni, G.J. (2001) The immunotherapeutic potential of melatonin. *Exp Opin Invest Drugs* 10, 467–476.

Maestroni, G.J., Conti A., and Pierpaoli, W. (1988) Pineal melatonin, its fundamental immunoregulatory role in aging and cancer. *Ann N Y Acad Sci* 521, 140–148.

Maestroni, G.J., Sulli, A., Pizzorni, C., Villaggio, B., and Cutolo, M. (2002) Melatonin in rheumatoid arthritis: Synovial macrophages show melatonin receptors. *Ann N Y Acad Sci* 966, 271–275.

Majde, J.A., and Krueger, J.M. (2005) Links between the innate immune system and sleep. *J Allergy Clin Immunol* 116, 1188–1198.

Marpegan, L., Bekinschtein, T.A., Costas, M.A., and Golombek, D.A. (2005) Circadian responses to endotoxin treatment in mice. *J Neuroimmunol* 160, 102–109.

Marpegan, L., Bekinschtein, T.A., Freudenthal, R., Rubio, M.F., Ferreyra, G.A., Romano, A., and Golombek, D.A. (2004) Participation of transcription factors from the Rel/NF-[kappa]B family in the circadian system in hamsters. *Neurosci Lett* 358, 9–12.

Martins E., Ligeiro De Oliveira A.P., Fialho De Araujo A.M., Tavares D.L., Cipolla-Neto J. and Costa Rosa L.F. (2001) Melatonin modulates allergic lung inflammation. J. Pineal Res. 31, 363–369.

Mazzoccoli, G., Balzanelli, M., Giuliani, A., De Cata, A., La Viola, M., Carella, A.M., Bianco, G., and Tarquini, R. (1999) Lymphocyte subpopulations anomalies in lung cancer patients and relationship to the stage of disease. *In Vivo* 13, 205–209.

McCann, S.M., Kimura, M., Karanth, S., Yu, W.H., and Rettori, V. (2002) Role of nitric oxide in the neuroendocrine response to cytokines. *Front Horm Res* 29, 117–129.

Missbach, M., Jagher, B., Sigg, I., Nayeri, S., Carlberg, C., and Wiesenberg, I. (1996) Thiazolidine diones, specific ligands of the nuclear receptor retinoid Z receptor/retinoid acid receptor-related orphan receptor alpha with potent antiarthritic activity. *J Biol Chem* 271, 13515–13522.

Moore-Ede, M.C. (1986) Physiology of the circadian system: Predictive versus reactive homeostasis. *Am J Physiol* 250, R737–R752.

Morrey, K.M., McLachlan, J.A., Serkin, C.D., and Bakouche, O. (1994) Activation of human monocytes by the pineal hormone melatonin. *J Immunol* 153, 2671–2680.

Motzkus, D., Albrecht, U., and Maronde, E. (2002) The human PER1 gene is inducible by interleukin-6. *J Mol Neurosci* 18, 105–109.

Murphy, P.J., and Campbell, S.S. (1996) Physiology of the circadian system in animals and humans. *J Clin Neurophysiol* 13, 2-16.

Nava, F., Carta, G., and Haynes, L.W. (2000) Lipopolysaccharide increases arginine-vasopressin release from rat suprachiasmatic nucleus slice cultures. *Neurosci Lett* 288, 228–230.

Negrette, B., Bonilla, E., Valero, N., Pons, H., Garcia, T.J., Chacin-Bonilla, L., Medina-Leendertz, S., and Anez, F. (2001) Melatonin treatment enhances the efficiency of mice immunization with Venezuelan equine encephalomyelitis virus TC-83. *Neurochem Res* 26, 767–770.

Neidhart, M., and Larson, D. (1990) Freund's complete adjuvant induces ornithine decarboxylase activity in the central nervous system of male rats and triggers the release of pituitary hormones. J Neuroimmunol 26, 97–105.

Neidhart, M. (1996) Effect of Freund's complete adjuvant on the diurnal rhythms of neuroendocrine processes and ornithine decarboxylase activity in various tissues of male rats. *Experientia* 52, 900–908.

Niehaus, G.D., Ervin, E., Patel, A., Khanna, K., Vanek, V.W., and Fagan, D.L. (2002) Circadian variation in cell-adhesion molecule expression by normal human leukocytes. *Can J Physiol Pharmacol* 80, 935–940.

Obal, F., and Krueger, J.M. (2005) Humoral mechanisms of sleep. In: P.L. Parmeggiani and R. Velluti (Eds.), *The Physiological Nature of Sleep*. Imperial College Press, London, pp. 23–44.

Ohdo, S., Koyanagi, S., Suyama, H., Higuchi, S., and Aramaki, H. (2001) Changing the dosing schedule minimizes the disruptive effects of interferon on clock function. Nat Med 7, 356–360.

Okamura, H. (2003) Integration of mammalian circadian clock signals: From molecule to behavior. *J Endocrinol* 177, 3–6.

Pandi-Perumal, S.R., Smits, M., Spence, W., Srinivasan, V., Cardinali, D.P., Lowe, A.D., and Kayumov L. (in press) Dim light melatonin onset (DLMO): A tool for the analysis of circadian phase in human sleep and chronobiological disorders. *Prog Neuro-Psychopharmacol Biol Psychiatry*.

Pazo, D., Cardinali, D.P., Garcia Bonacho, M., Reyes Toso, C., and Esquifino A.I. (2000) Effect of melatonin treatment on 24-hour variations in hypothalamic serotonin and dopamine turnover during the preclinical phase of Freund's adjuvant arthritis in rats. *Biol Rhythm Res* 32, 202–211.

Pearson, C.M. (1956) Development of arthritis, periarthritis and periostitis in rats given adjuvant. *Proc Soc Exp Biol Med* 91, 95–103.

Pelegri, C., Vilaplana, J., Castellote, C., Rabanal, M., Franch, A., and Castell, M. (2003) Circadian rhythms in surface molecules of rat blood lymphocytes. *Am J Physiol Cell Physiol* 284, C67–C76.

Perelló, M., Chacon, F., Cardinali, D.P., Esquifino, A.I., and Spinedi, E. (2006) Effect of social isolation on 24-hour pattern of stress hormones and leptin in rats. *Life Sci* 78, 1857–1862.

Persengiev, S.P., and Kyurkchiev, S. (1993) Selective effect of melatonin on the proliferation of lymphoid cells. *Int J Biochem* 25, 441–444.

Petrovsky, N., and Harrison, L.C. (1998) The chronobiology of human cytokine production. *Int Rev Immunol* 16, 635–649.

Pioli, C., Caroleo, M.C., Nistico, G., and Doria, G. (1993) Melatonin increases antigen presentation and amplifies specific and non-specific signals for T-cell proliferation. *Int J Immunopharmacol* 15, 463–468.

Poon, A.M., Liu, Z.M., Pang, C.S., Brown, G.M., and Pang, S.F. (1994) Evidence for a direct action of melatonin on the immune system. *Biol Signals* 3, 107–117.

Pozo, D., Delgado, M., Fernandez-Santos, J.M., Calvo, J.R., Gomariz, R.P., Martin-Lacave, I., Ortiz, G.G. and Guerrero, J.M. (1997) Expression of the Mel 1a-melatonin receptor mRNA in T and B subsets of lymphocytes from rat thymus and spleen. *FASEB J* 11, 466–473.

Raghavendra, V., Singh, V., Kulkarni, S.K., and Agrewala, J.N. (2001a) Melatonin enhances Th2 cell mediated immune responses: Lack of sensitivity to reversal by naltrexone or benzodiazepine receptor antagonists. *Mol Cell Biochem* 221, 57–62.

Raghavendra, V., Singh, V., Shaji, A.V., Vohra, H., Kulkarni, S.K., and Agrewala, J.N. (2001b) Melatonin provides signal 3 to unprimed CD4$^+$ T cells but failed to stimulate LPS primed B cells. *Clin Exp Immunol* 124, 414–422.

Raynaud, F., Mauviard, F., Geoffriau, M., Claustrat, B., and Pevet, P. (1993) Plasma 6-hydroxymelatonin, 6-sulfatoxymelatonin and melatonin kinetics after melatonin administration to rats. *Biol Signals* 2, 358–366.

Reginster, J.Y. (2002) The prevalence and burden of arthritis. *Rheumatol (Oxford)* 41(Suppl 1), 3–6.

Reiter, R.J. (1998) Cytoprotective properties of melatonin: Presumed association with oxidative damage and aging. *Nutrition* 14, 691–696.

Reiter, R.J., Tan, D., Cabrera, J., D'Arpa, D., Sainz, R.M., Mayo, J.C., and Ramos, S. (1999) The oxidant/antioxidant network: Role of melatonin. *Biol Signals Recept* 8, 56–63.

Reiter, R.J., Tan, D.X., and Burkhardt, S. (2002) Reactive oxygen and nitrogen species and cellular and organismal decline: Amelioration with melatonin. *Mech Ageing Dev* 123, 1007–1019.

Reppert, S.M., Weaver, D.R., and Ebisawa, T. (1994) Cloning and characterization of a mammalian melatonin receptor that mediates reproductive and circadian responses. *Neuron* 13, 1177–1185.

Roman, O., Seres, J., Herichova, I., Zeman, M., and Jurcovicova, J. (2003) Daily profiles of plasma prolactin (PRL), growth hormone (GH), insulin-like growth factor-1 (IGF-1), luteinizing hormone (LH), testosterone, and melatonin, and of pituitary PRL mRNA and GH mRNA in male long evans rats in acute phase of adjuvant arthritis. *Chronobiol Int* 20, 823–836.

Saper, C.B., Lu, J., Chou, T.C., and Gooley, J. (2005) The hypothalamic integrator for circadian rhythms. *Trends Neurosci* 28, 152–157.

Sasaki, M., Jordan, P., Joh, T., Itoh, M., Jenkins, M., Pavlick, K., Minagar, A., and Alexander, S.J. (2002) Melatonin reduces TNF-α induced expression of MAdCAM-1 via inhibition of NF-kB. *BMC Gastroenterol* 2, 9.

Scalabrino, G., Ferioli, M.E., Modena, D., and Fraschini, F. (1982) Enhancement of ornithine decarboxylase activity in rat adenohypophysis after pinealectomy. *Endocrinology* 111, 2132–2134.

Schibler, U., Ripperger, J., and Brown, S.A. (2003) Peripheral circadian oscillators in mammals: Time and food. *J Biol Rhythms* 18, 250–260.

Selgas, L., Pazo, D., Arce, A., Esquifino, A.I., and Cardinali, D.P. (1998) Circadian rhythms in adenohypophysial hormone levels and hypothalamic monoamine turnover in mycobacterial-adjuvant-injected rats. *Biol Signals Recept* 7, 15–24.

Skwarlo-Sonta, K. (2002) Melatonin in immunity: Comparative aspects. *Neuroendocrinol Lett* 23(Suppl 1), 61–66.

Stenzel-Poore, M., Vale, W.W., and Rivier, C. (1993) Relationship between antigen-induced immune stimulation and activation of the hypothalamic-pituitary-adrenal axis in the rat. *Endocrinology* 132, 1313–1318.

Sulli, A., Maestroni, G.J., Villaggio, B., Hertens, E., Craviotto, C., Pizzorni, C., Briata, M., Seriolo, B., and Cutolo, M. (2002) Melatonin serum levels in rheumatoid arthritis. *Ann N Y Acad Sci* 966, 276–283.

Tanaka, H., Ueta, Y., Yamashita, U., Kannan, H., and Yamashita, H. (1996) Biphasic changes in behavioral, endocrine, and sympathetic systems in adjuvant arthritis in Lewis rats. *Brain Res Bull* 39, 33-7X.

Ueda, H.R., Hayashi, S., Chen, W., Sano, M., Machida, M., Shigeyoshi, Y., Iino, M., and Hashimoto, S. (2005) System-level identification of transcriptional circuits underlying mammalian circadian clocks. *Nat Genet* 37, 187–192.

Undar, L., Ertugrul, C., Altunbas, H., and Akca, S. (1999) Circadian variations in natural coagulation inhibitors protein C, protein S and antithrombin in healthy men: A possible association with interleukin-6. *Thromb Haemost* 81, 571–575.

Vagnucci, A.H., and Winkelstein, A. (1993) Circadian rhythm of lymphocytes and their glucocorticoid receptors in HIV-infected homosexual men. *J Acquir Immun Defic Syndr* 6, 1238–1247.

Wajs, E., Kutoh, E., and Gupta, D. (1995) Melatonin affects proopiomelanocortin gene expression in the immune organs of the rat. *Eur J Endocrinol* 133, 754–760.

Whitehouse, M.W. (1988) Adjuvant-induced polyarthritis in rats. In: R.A. Greenwald and H.S. Diamad (Eds.), *Handbook of Animal Models for the Rheumatic Diseases*. CRC Press, New York, pp. 3–16.

Wu, C.C., Chiao, C.W., Hsiao, G., Chen, A., and Yen, M.H. (2001) Melatonin prevents endotoxin-induced circulatory failure in rats. *J Pineal Res* 30, 147–156.

Wu, M.W., Li, X.M., Xian, L.J., and Levi, F. (2004) Effects of meal timing on tumor progression in mice. *Life Sci* 75, 1181–1193.

Yoo, S.H., Yamazaki, S., Lowrey, P.L., Shimomura, K., Ko, C.H., Buhr, E.D., Siepka, S.M., Hong, H.K., Oh, W.J., Yoo, O.J., Menaker, M., and Takahashi, J.S. (2004) Inaugural Article: PERIOD2::LUCIFERASE real-time reporting of circadian dynamics reveals persistent circadian oscillations in mouse peripheral tissues. *PNAS* 101, 5339–5346.

Youssef, W.I., and Tavill, A.S. (2002) Connective tissue diseases and the liver. *J Clin Gastroenterol* 35, 345–349.

Yu, Q., Miller, S.C., and Osmond, D.G. (2000) Melatonin inhibits apoptosis during early B-cell development in mouse bone marrow. *J Pineal Res* 29, 86–93.

**Part II
Basic Research**

4 The Biological Clock in Inflammation and Sleep Switch Alterations

Marina Bentivoglio, Mikael Nygård, Maria Palomba, and Krister Kristensson

4.1 Introduction

The word inflammation derives from Latin *inflammare*, meaning to set in flame, and signifies the heat and redness related to an increased blood flow and vasodilation in the affected tissue. Although vasodilation, edema, and extravasation of white blood cells are the hallmarks of inflammation, its definition has become broader and has been more loosely applied to tissue reactions to injuries. With the discovery of molecules that mediate inflammation and, in particular, the extensive number of cytokines that regulate the cellular response to inflammation, an inflammatory component is now frequently ascribed to neurodegenerative diseases that were traditionally considered as noninflammatory. This is due to the production of proinflammatory cytokines in activated microglial cells and astrocytes in such diseases. The classical hallmarks of inflammation are, however, lacking.

On the other hand, events considered as components of systemic inflammation occur also in normal brain function, e.g., increase in local blood flow related to increased brain activities, and release of certain proinflammatory cytokines, including those that may be involved in mediating sleep–wakefulness.

In this overview we will focus on recent observations on how mediators of vascular responses in inflammation outside the nervous system (e.g., histamine) and proinflammatory cytokines may affect the biological clock in the brain and be involved in the regulation of the switch between sleep and wakefulness, and hence cause disturbances in the function of this switch during diseases.

4.2 Interactions Between the Biological Clock and the Sleep Switch

The suprachiasmatic nucleus (SCN) in the anterior hypothalamus is also referred to as the biological clock, since it acts as a major pacemaker in the mammalian brain to drive various circadian rhythms. One important feature of circadian rhythmicity is the sleep–wake cycle, which is greatly influenced by signals originating in the SCN. In spite of the close relation between the circadian timing and sleep–wake regulating systems, their physiological, anatomical, and possibly humoral, interconnections have started only recently to be elucidated (for reviews, see Pace-Schott and Hobson 2002; Saper, Scammell, and Lu 2005).

Transitions between sleep and wakefulness have to be rapid and flawless. Such transitions have been suggested to be controlled by wake-promoting and sleep-promoting nuclei in the hypothalamus and brainstem that are mutually inhibitory, forming a bistable switch that ensures rapid transitions between sleep and wakefulness with no in-between states (Saper et al. 2005).

On one leg of the seesaw switch (Fig. 4.1), are the wake-promoting neural centers, such as histaminergic tuberomammillary nucleus (TMN) in the posterior hypothalamus, and the serotoninergic dorsal raphe nucleus (DRN), and noradrenergic locus coeruleus (LC) in the brainstem. At all these sites, neurons fire during wakefulness, but slow down or stop firing during slow-wave sleep and rapid eye movement (REM) sleep, respectively. On the other leg of the seesaw, these wake-promoting groups of neurons are counteracted by a sleep-promoting cell group, which is located in the ventrolateral preoptic nucleus (VLPO). The GABA- and galanin-containing inhibitory VLPO neurons project to wake-promoting groups of neurons (TMN, DRN, and LC), and are therefore thought to suppress arousal when activated. Thus, they can switch a wake state into a non-REM sleep state. The VLPO in turn is thought to be inhibited during wakefulness by the wake-promoting neurons through their anatomical input. The TMN neurons contain, in addition to the excitatory histamine, also GABA and galanin (Haas and Panula 2003) which could both inhibit the VLPO; an inhibitory function could also be played by serotonin in the DRN and noradrenaline in LC neurons (Pace-Schott et al. 2002).

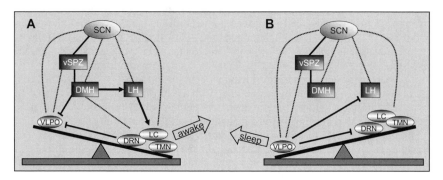

FIGURE 4.1. Transmission of circadian signals to components of the sleep switch, depicted as the flip-flop model proposed by Saper et al. (2005). Circadian signals from the SCN to sleep-regulatory nuclei are relayed via the ventral SPZ and the DMH. During wakefulness (A), the DMH is active and a subset of DMH neurons excites the orexinergic neurons in the LH, while a GABAergic subset of DMH neurons instead inhibits the VLPO. Orexinergic cells activate the wake-promoting monoaminergic nuclei (TMN, DRN, LC), which also inhibit the VLPO thereby stabilizing the awake state. During sleep (B), the VLPO inhibits the LH and the monoaminergic nuclei, and this releases the VLPO from the inhibition exerted by monoaminergic nuclei. Excitatory connections are shown as arrows and inhibitory connections as blunt ends. Dashed lines indicate some of the anatomical interconnections, that are yet of unclear functional significance, between the biological clock and sleep switch structures.

The sleep switch is influenced by the SCN through indirect polysynaptic pathways, and also to some extent through direct, though relatively sparse, connections
(Fig. 4.1). The main bulk of SCN efferents terminate within the hypothalamus, targeting especially the subparaventricular zone (SPZ) and the dorsomedial nucleus of the hypothalamus (DMH). These areas integrate SCN-derived signals with noncircadian signals, and convey information to downstream target nuclei, which are involved in the circadian regulation of feeding, body temperature, sleep, and corticosteroid release. The DMH plays a key role in sleep regulation, and lesions of the DMH cause altered circadian rhythms in sleep–wakefulness behavior (Chou, Scammell, Gooley, Gaus, Saper, and Lu 2003). The DMH projects to all components of the sleep switch, as well as to the lateral hypothalamus (LH), which is the exclusive site of neurons containing the peptide orexin. In addition, the SCN has some sparse direct projections to the TMN and the VLPO (Abrahamson and Moore 2001; Chou, Bjorkum, Gaus, Lu, Scammell, and Saper 2002).

Orexin-containing neurons could play a strategic role in the cross talk between the circadian system and the sleep switch. Orexin exists in two forms (orexin-A and orexin-B, also known as hypocretin-1 and hypocretin-2), which are derived from the same precursor (prepro-orexin) and are colocalized in the same neurons. Orexin release in the cerebrospinal fluid has a daily oscillation which is under SCN control (Zhang et al. 2004). Besides other physiological functions, orexin has been especially implicated in arousal (review in Sutcliffe and de Lecea (2002)). In particular, orexin is suggested to stabilize the sleep-switch by activating the TMN and the other wake-promoting nuclei, thereby preventing untimely transitions from wakefulness to sleep. The orexin-containing neurons receive sparse direct innervation from the SCN, and are targets of polysynaptic pathways relayed via intrahypothalamic targets of SCN efferents, which, as mentioned above, are represented by the SPZ and DMH, and also include the median preoptic area (Abrahamson, Leak, and Moore 2001; Aston-Jones, Chen, Zhu, and Oshinsky 2001; Chou et al. 2003). Through indirect projections via hypothalamic relays, the SCN, therefore, reaches both sleep-promoting preoptic nuclei and wake-regulatory cell groups.

Recent physiological evidence indicates that sleep states can alter the activity of the SCN, indicative of a feedback mechanism (Deboer, Vansteensel, Detari, and Meijer 2003). The anatomical pathways involved in this circuit remain, however, to be determined. The homeostatic sleep process can also feed back on the orexin-containing cells in the LH, indicating that sleep homeostatic and circadian signals converge on these cells (Deboer et al. 2004).

In addition to the switch which regulates sleep–wakefulness described above, a putative switch for the regulation of REM/non-REM sleep has been recently identified in defined brainstem structures (the sublaterodorsal nucleus, the precoeruleus nucleus, lateral pontine tegmentum, and the periaqueductal grey matter) (Lu, Sherman, Devor, and Saper 2006).

Furthermore, the potential role of diffusible factors on sleep switch structures is of special interest in the context of inflammatory signaling. Besides classical synaptic transmission, there is now substantial evidence that the SCN can regulate state-dependent behavior also via diffusible molecules reviewed in Mistlberger (2005). The two candidate molecules discovered so far are transforming growth factor (TGF)-α and prokineticin-2, which are expressed in the SCN. A remarkable finding is that astrocytes are the source of TGF-α in the SCN (Li, Sankrithi, and Davis 2002), thus implicating glial cells in diffusible outputs of the biological clock. In view of the main role played by activation of glia (both astrocytes and microglia) in inflammatory signaling, diffusible factors released by SCN glia could potentially play an important role in inflammatory conditions. This is an unexplored issue that remains open for future investigations.

4.3 Effects of Inflammatory Molecules on the SCN and Sleep Switch Structures

Certain molecules originally defined on the basis of their role in inflammation display activities in the sleep–wakefulness-promoting structures, namely histamine as well as interleukin (IL)-1β and tumor necrosis factor (TNF)-α.

During peripheral inflammation, the biogenic amine histamine is released from mast cells and is a major mediator of vasodilation and permeability, but in the brain it acts mainly as an excitatory neurotransmitter. Histamine is produced by neurons located in the TMN, which send widespread projections in the brain, including a relatively dense projection to the SCN (Moga and Moore 1997). Histamine plays an essential role in arousal, an effect that may be mediated by direct excitation of cortical neurons or by stimulatory effects on sleep-regulatory centers such as the cholinergic basal forebrain neurons (Haas et al. 2003). It is interesting to note that the rodent SCN contains neurons immunoreactive to histamine, which they receive via TMN fibers through a not yet characterized uptake mechanism (Michelsen et al. 2005). These data suggest that local uptake, storage and release of histamine within the SCN could potentially play a novel role in the function of the biological clock.

The other source of histamine in the brain is provided by a limited number of resident mast cells, which in the rat brain parenchyma are concentrated in the thalamus (Florenzano and Bentivoglio 2000). It is not clear, however, if and how brain histamine is involved in sleep–wakefulness dysregulation occurring during inflammation in the central nervous system (CNS).

IL-1β is released from macrophages/microglial cells during inflammation. It activates T cells, and is the principal cytokine involved in induction of the so-called "sickness behavior," i.e., sleepiness, loss of appetite, and fever. IL-1β is a very potent somnogen and when administered at the anterior preoptic area, which includes the VLPO, IL-1β activates sleep-related neurons in the VLPO; this may, therefore, mediate, at least in part, the non-REM sleep-inducing properties of this cytokine (Alam et al. 2004). In addition, blocking of IL-1β with neutralizing antibodies inhibits spontaneous non-REM sleep in normal rats, indicating that this cytokine is endogenously produced in the brain and affects its normal function. In line with this, both IL-1β and TNF-α, which are also usually associated with release from activated macrophages/microglial cells, are expressed in the brain with a day/night variation (for review, see Vitkovic, Bockaert, and Jacque (2000)). Systemic administration of TNF-α also increases the time spent in non-REM sleep (Fang, Wang, and Krueger 1997). However, mice deficient in the genes encoding either TNF-α or its receptor show no changes in the amount of non-REM sleep, and show instead a reduction in REM sleep episode frequency (Deboer, Fontana, and Tobler 2002).

In contrast to well-documented roles of histamine and of the cytokines IL-1β and TNF-α in sleep–wakefulness regulation, effects of these molecules and other cytokines on circadian rhythms are less well known. Through the above-mentioned direct projections of TMN neurons to the SCN, which may be excitatory or inhibitory, histamine may have effects on neuronal activities in the biological clock (Liou, Shibata, Yamakawa, and Ueki 1983). Direct application of histamine to the SCN can phase shift the circadian rhythm of the neuronal firing rate (Cote and Harrington 1993), but whether this biogenic amine may modulate the biological clock in relation to arousal activities is not clear.

The response of the SCN to inflammatory mediators such as lipopolysaccharide (Marpegan, Bekinschtein, Costas and Golombek 2005), interferon (IFN)-α (Ohdo, Koyanagi, Suyama, Higuchi, and Aramaki 2001), or an IFN-γ/TNF-α cocktail displays day/night variation (Sadki, Bentivoglio, Kristensson, and Nygard 2006). Interestingly, the sensitivity of SCN neurons to IFN-γ coincides in time with the expression of an immunopositive IFN-γ-receptor-like molecule in the SCN (Lundkvist, Robertson, Mhlanga, Rottenberg, and Kristensson 1998). That IFN-γ may affect the SCN is indicated by induction of Fos protein (a marker of neuronal activity) in SCN neurons following intracerebroventricular injection of the cytokine (Robertson, Kong, Peng, Bentivoglio, and Kristensson 2000), and by the observation that IFN-γ can cause reduced synaptic activity in the ventral part of the SCN after application on slice preparations (Lundkvist, Hill, and Kristensson 2002). The IFN-γ-receptor-like molecule is localized to the ventrolateral part of the rat SCN, which is also the site of innervation from the TMN. In turn, the TMN contains a molecule that cross-reacts with monoclonal antibodies directed to different epitopes of IFN-γ (Bentivoglio, Florenzano, Peng, and Kristensson 1994) but the nature of this IFN-γ-immunopositive molecule in the TMN and its potential effects on the SCN remain to be determined. Neither has any role for an IL-6-immunopositive molecule in SCN neurons been determined (Gonzalez-Hernandez et al. 2006).

Thus, although no role has been demonstrated so far for inflammatory cytokines in the normal functioning of the SCN, such molecules may potentially affect SCN synaptic activity during inflammatory diseases, and hence influence the circadian rhythms and sleep–wakefulness states.

In addition, the nuclear factor (NF)-κB, a transcription factor which plays a major role in inflammation, was found to be expressed in the SCN of hamsters, and NF-κB inhibition blocked the light-induced phase advance, suggesting that NF-κB family of proteins could serve a role in the entrainment of circadian rhythms (Marpegan et al. 2004).

In terms of susceptibility to inflammation, a number of recent data suggest that the orexinergic system could be a site of vulnerability in the sleep switch network. Substantial evidence has indicated in the last years that alterations of the orexinergic system play a key role in idiopathic narcolepsy. This disease, which is usually sporadic, is characterized by chronic sleepiness and intrusions into wakefulness of manifestations of REM sleep, cataplexy, sleep paralysis, and hypnagogic hallucinations (for reviews, see Sutcliffe et al. (2002), Scammell (2003), and Baumann and Bassetti (2005)). Impaired orexin signaling (with loss of orexin-producing neurons in the LH, low or undetectable levels of orexin in the cerebrospinal fluid, and orexin ligand deficiency) is considered to hallmark narcolepsy. The cause of such degeneration of orexin neurons is still unknown. Interestingly, the disease is associated with strong linkage to the human leukocyte antigen (HLA) complex, since most narcoleptic patients have a HLA-DR2 haplotype (Lin, Hungs and Mignot 2001). TNF-α signaling has also been implicated in the disease since polymorphisms in the TNF-α and TNF receptor 2 genes are associated with some cases of narcolepsy (Scammell 2003). This has led to the proposal of an autoimmune pathogenesis of loss of orexin-producing neurons in narcolepsy, but no immune changes have been found in the patients. Experimentally, chronic infusion of lipopolysaccharide in the LH of rats resulted in neuronal loss which included orexinergic neurons but it was not selective for these cells, suggesting that also other factors may be involved in the pathology of narcolepsy (Gerashchenko and Shiromani 2004).

We are at present examining experimentally the vulnerability of orexin-containing neurons in systemic and CNS inflammatory conditions.

4.4 Intracellular Inflammatory Regulators

In the preceding sections, the potential role of certain proinflammatory molecules in sleep–wakefulness regulation has been pointed out. However, the brain is an immune-privileged site and has a powerful intrinsic capacity to down-regulate inflammatory processes (for review, see Niederkorn (2006)). In addition, a number of molecules that regulate the intracellular effects of cytokines have been recently defined in the immune system. Some of these molecules have been detected in the nervous system and a short survey of these is given below.

4.4.1 SHP, PIAS, and SOCS

Most cytokines activate membrane receptors that signal via the janus kinase (JAK)/signal transducer and activator of transcription (STAT) pathway. There are several strategies for a cell to attenuate cytokine signaling to avoid excessive or insufficient signaling, and at least three main classes of intracellular negative regulators that affect components of the JAK/STAT pathway have been identified (Fig. 4.2). These include tyrosine phosphatases (e.g., SH2-containing phosphatases (SHP) that dephosphorylate and thereby deactivate cytokine receptors), protein inhibitors of activated STATs (PIAS), and suppressors of cytokine signaling (SOCS) (for review, see Rakesh and Agrawal (2005)). The SHP and PIAS molecules are expressed constitutively, whereas SOCS molecules are induced in response to cytokines and act as a negative feedback by inhibiting the JAK-STAT signaling cascade.

4.4.2 Intracellular Regulation of Inflammation and the Biological Clock

Given the many effects exerted by cytokines on various cellular elements and regions of the brain, the control of these effects deserves special attention. PIAS1, an inhibitor of activated STAT1 that is a component of the intracellular signaling pathway for IFN-γ, is constitutively expressed in the brain but its cellular localization remains to be determined (Maier, Kincaid, Pagenstecher, and Campbell 2002). The SH2-containing phosphatase SHP-1 is expressed in oligodendroglial cells (Massa, Saha, Wu, and Jarosinski 2000) and deficiency of SHP-1 is associated with increased levels of IL-1β in the brain (Zhao and Lurie 2004). A role for SHP-1 in limiting CNS inflammation following injury and disease has therefore been suggested (Zhao et al. 2004). Since cytokine levels can increase dramatically in the brain during inflammatory diseases, the negative feedback regulators SOCS could also play an important role in limiting the response to cytokines. Induction of SOCS has been described in mononuclear cells infiltrating the brain in experimental allergic encephalomyelitis (Campbell 2005), which provides a model for multiple sclerosis, while induction of transcript for SOCS in glial cells and neurons has only rarely been reported (Rosell, Akama, Nacher, and McEwen 2003).

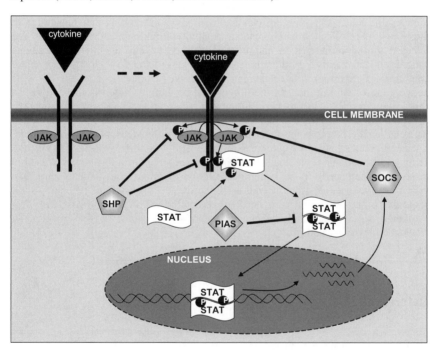

FIGURE 4.2. Regulation of JAK/STAT signaling. Ligand (cytokine) binding leads to receptor phosporylation and subsequent activation of STAT molecules which dimerize; the complex thereafter translocates to the nucleus and activates transcription. SOCS is one class of molecules induced in response to specific cytokines and it down-regulates signaling mainly via inhibition of JAKs. SHP and PIAS are constitutive inhibitors that act on JAK and STAT molecules, respectively.

In the context of the present survey, it is interesting to note that SOCS1 and SOCS3 are expressed in the SCN, where the level of their transcripts displays a circadian fluctuation (Sadki et al. 2006).

This could represent a potent regulatory mechanism accounting for day/night differences in sensitivity of the biological clock to inflammatory molecules. We are presently studying whether SOCS can be expressed in neurons and whether this occurs also in brain areas involved in sleep regulation after an inflammatory challenge.

4.5 Concluding Remarks

Inflammatory cytokines are of interest in the context of sleep and circadian rhythms for several reasons. These molecules usually operate on much longer timescales, i.e., minutes and hours, than those of classical neurotransmitters, and they concord with those of sleep switches and circadian rhythms. Proinflammatory cytokines are released during inflammation and infectious diseases, and are therefore suitable candidates for mediating alterations in sleep and circadian rhythms in these conditions. Furthermore, since therapies based on manipulations of the cytokine network are of current interest, the timing of drug administration in these therapies should be considered in order to minimize adverse effects on sleep and circadian rhythms. Considering the complexity of circadian rhythms and sleep regulation and the relationship between the two processes, a more detailed understanding of the roles and actions played by cytokines in these neural circuits is required. Thus, the cellular localization of cytokines and cytokine receptors in neural centers involved in these processes needs to be better mapped. Furthermore, the potential circadian variation in their expression, as well as in the intracellular systems regulating their responses in the nervous system, needs to be determined. Novel knowledge on how cytokines can regulate and dysregulate sleep and circadian rhythms is a main challenge for the near future.

Acknowledgment.   This work was supported by EC grant LSHM-CT-518189.

References

Abrahamson, E.E., Leak, R.K., and Moore, R.Y. (2001) The suprachiasmatic nucleus projects to posterior hypothalamic arousal systems. *Neuroreport* 12, 435–440.

Abrahamson, E.E., and Moore, R.Y. (2001) Suprachiasmatic nucleus in the mouse: Retinal innervation, intrinsic organization and efferent projections. *Brain Res* 916, 172–191.

Alam, M.N., McGinty, D., Bashir, T., Kumar, S., Imeri, L., Opp, M. R., and Szymusiak, R. (2004) Interleukin-1beta modulates state-dependent discharge activity of preoptic area and basal forebrain neurons: Role in sleep regulation. *Eur J Neurosci* 20, 207–216.

Aston-Jones, G., Chen, S., Zhu, Y., and Oshinsky, M.L. (2001) A neural circuit for circadian regulation of arousal. *Nat Neurosci* 4, 732–738.

Baumann, C.R., and Bassetti, C.L. (2005) Hypocretins (orexins) and sleep-wake disorders. *Lancet Neurol* 4, 673–682.

Bentivoglio, M., Florenzano, F., Peng, Z.C., and Kristensson, K. (1994) Neuronal IFN-gamma in tuberomammillary neurones. *Neuroreport* 5, 2413–2416.

Campbell, I.L. (2005) Cytokine-mediated inflammation, tumorigenesis, and disease-associated JAK/STAT/SOCS signaling circuits in the CNS. *Brain Res Brain Res Rev* 48, 166–177.

Chou, T.C., Bjorkum, A.A., Gaus, S.E., Lu, J., Scammell, T.E., and Saper, C.B. (2002) Afferents to the ventrolateral preoptic nucleus. *J Neurosci* 22, 977–990.

Chou, T.C., Scammell, T.E., Gooley, J.J., Gaus, S.E., Saper, C.B., and Lu, J. (2003) Critical role of dorsomedial hypothalamic nucleus in a wide range of behavioral circadian rhythms. *J Neurosci* 23, 10691–106702.

Cote, N.K., and Harrington, M.E. (1993) Histamine phase shifts the circadian clock in a manner similar to light. *Brain Res* 613, 149–151.

Deboer, T., Fontana, A., and Tobler, I. (2002) Tumor necrosis factor (TNF) ligand and TNF receptor deficiency affects sleep and the sleep EEG. *J Neurophysiol* 88, 839–846.

Deboer, T., Overeem, S., Visser, N.A., Duindam, H., Frolich, M., Lammers, G.J., and Meijer, J.H. (2004) Convergence of circadian and sleep regulatory mechanisms on hypocretin-1. *Neuroscience* 129, 727–732.

Deboer, T., Vansteensel, M.J., Detari, L., and Meijer, J.H. (2003) Sleep states alter activity of suprachiasmatic nucleus neurons. *Nat Neurosci* 6, 1086–1090.

Fang, J., Wang, Y., and Krueger, J.M. (1997) Mice lacking the TNF 55 kDa receptor fail to sleep more after TNFalpha treatment. *J Neurosci* 17, 5949–5955.

Florenzano, F., and Bentivoglio, M. (2000) Degranulation, density, and distribution of mast cells in the rat thalamus: A light and electron microscopic study in basal conditions and after intracerebroventricular administration of nerve growth factor. *J Comp Neurol* 424, 651–669.

Gerashchenko, D., and Shiromani, P.J. (2004) Effects of inflammation produced by chronic lipopolysaccharide administration on the survival of hypocretin neurons and sleep. *Brain Res* 1019, 162–169.

Gonzalez-Hernandez, T., Afonso-Oramas, D., Cruz-Muros, I., Barroso-Chinea, P., Abreu, P., del Mar Perez-Delgado, M., Rancel-Torres, N., and del Carmen Gonzalez, M. (2006) Interleukin-6 and nitric oxide synthase expression in the vasopressin and corticotrophin-releasing factor systems of the rat hypothalamus. *J Histochem Cytochem* 54, 427–441.

Haas, H., and Panula, P. (2003) The role of histamine and the tuberomamillary nucleus in the nervous system. *Nat Rev Neurosci.* 4(2):121–130.

Li, X., Sankrithi, N., and Davis, F.C. (2002) Transforming growth factor-alpha is expressed in astrocytes of the suprachiasmatic nucleus in hamster: Role of glial cells in circadian clocks. *Neuroreport* 13, 2143–2147.

Lin, L., Hungs, M., and Mignot, E. (2001) Narcolepsy and the HLA region. *J Neuroimmunol* 117, 9–20.

Liou, S.Y., Shibata, S., Yamakawa, K., and Ueki, S. (1983) Inhibitory and excitatory effects of histamine on suprachiasmatic neurons in rat hypothalamic slice preparation. *Neurosci Lett* 41, 109–113.

Lu, J., Sherman, D., Devor, M., and Saper, C.B. (2006) A putative flip-flop switch for control of REM sleep. *Nature* 441, 589–594.

Lundkvist, G.B., Hill, R.H., and Kristensson, K. (2002) Disruption of circadian rhythms in synaptic activity of the suprachiasmatic nuclei by African trypanosomes and cytokines. *Neurobiol Dis* 11, 20–27.

Lundkvist, G.B., Robertson, B., Mhlanga, J.D., Rottenberg, M.E., and Kristensson, K. (1998) Expression of an oscillating interferon-gamma receptor in the suprachiasmatic nuclei. *Neuroreport* 9, 1059–1063.

Maier, J., Kincaid, C., Pagenstecher, A., and Campbell, I.L. (2002) Regulation of signal transducer and activator of transcription and suppressor of cytokine-signaling gene expression in the brain of mice with astrocyte-targeted production of interleukin-12 or experimental autoimmune encephalomyelitis. *Am J Pathol* 160, 271–288.

Marpegan, L., Bekinschtein, T.A., Costas, M.A., and Golombek, D.A. (2005) Circadian responses to endotoxin treatment in mice. *J Neuroimmunol* 160, 102–109.

Marpegan, L., Bekinschtein, T.A., Freudenthal, R., Rubio, M.F., Ferreyra, G.A., Romano, A., and Golombek, D.A. (2004) Participation of transcription factors from the Rel/NF-kappa B family in the circadian system in hamsters. *Neurosci Lett* 358, 9–12.

Massa, P.T., Saha, S., Wu, C., and Jarosinski, K.W. (2000) Expression and function of the protein tyrosine phosphatase SHP-1 in oligodendrocytes. *Glia* 29, 376–385.

Michelsen, K.A., Lozada, A., Kaslin, J., Karlstedt, K., Kukko-Lukjanov, T.K., Holopainen, I., Ohtsu, H., and Panula, P. (2005) Histamine-immunoreactive neurons in the mouse and rat suprachiasmatic nucleus. *Eur J Neurosci* 22, 1997–2004.

Mistlberger, R.E. (2005) Circadian regulation of sleep in mammals: Role of the suprachiasmatic nucleus. *Brain Res Brain Res Rev* 49, 429–454.

Moga, M.M., and Moore, R.Y. (1997) Organization of neural inputs to the suprachiasmatic nucleus in the rat. *J Comp Neurol* 389, 508–534.

Niederkorn, J.Y. (2006) See no evil, hear no evil, do no evil: The lessons of immune privilege. *Nat Immunol* 7, 354–359.

Ohdo, S., Koyanagi, S., Suyama, H., Higuchi, S., and Aramaki, H. (2001) Changing the dosing schedule minimizes the disruptive effects of interferon on clock function. *Nat Med* 7, 356–360.

Pace-Schott, E.F., and Hobson, J.A. (2002) The neurobiology of sleep: Genetics, cellular physiology and subcortical networks. *Nat Rev Neurosci.* 2002 Aug, 3(8):591–605.

Rakesh, K., and Agrawal, D.K. (2005) Controlling cytokine signaling by constitutive inhibitors. *Biochem Pharmacol* 70, 649–657.

Robertson, B., Kong, G., Peng, Z., Bentivoglio, M., and Kristensson, K. (2000) Interferon-gamma-responsive neuronal sites in the normal rat brain: Receptor protein distribution and cell activation revealed by Fos induction. *Brain Res Bull* 52, 61–74.

Rosell, D.R., Akama, K.T., Nacher, J., and McEwen, B.S. (2003) Differential expression of suppressors of cytokine signaling-1, -2, and -3 in the rat hippocampus after seizure: Implications for neuromodulation by gp130 cytokines. *Neuroscience* 122, 349–358.

Sadki, A., Bentivoglio, M., Kristensson, K., and Nygard, M. (2006) Suppressors, receptors and effects of cytokines on the aging mouse biological clock. *Neurobiol Aging.*

Saper, C.B., Scammell, T.E., and Lu, J. (2005) Hypothalamic regulation of sleep and circadian rhythms. *Nature* 437, 1257–1263.

Scammell, T.E. (2003) The neurobiology, diagnosis, and treatment of narcolepsy. *Ann Neurol* 53, 154–166.

Sutcliffe, J.G., and de Lecea, L. (2002) The hypocretins: Setting the arousal threshold. *Nat Rev Neurosci.* 3(5):339–349.

Vitkovic, L., Bockaert, J., and Jacque, C. (2000) "Inflammatory" cytokines: neuromodulators in normal brain? *J Neurochem* 74, 457–471.

Zhang, S., Zeitzer, J.M., Yoshida, Y., Wisor, J.P., Nishino, S., Edgar, D.M., and Mignot, E. (2004) Lesions of the suprachiasmatic nucleus eliminate the daily rhythm of hypocretin-1 release. *Sleep* 27, 619–627.

Zhao, J., and Lurie D.I. (2004) Loss of SHP-1 phosphatase alters cytokine expression in the mouse hindbrain following cochlear ablation. Cytokine. 2004 Oct 7;28(1):1-9.

Zhao, J., Lurie D.I. (2004) Cochlear ablation in mice lacking SHP-1 results in an extended period of cell death of anteroventral cochlear nucleus neurons. Hear Res. 2004 Mar; 189(1–2):63–75.

5 Vasoactive Intestinal Polypeptide and Prolactin Cytokines: Role in Sleep and Some Immune Aspects

Fabio García-García, Tatiana Olivares-Bañuelos, and René Drucker-Colín

5.1 Introduction

Vasoactive intestinal polypeptide (VIP) was originally isolated from the small intestine and the lung as a peptide with potent vasodilator effects (Said and Mutt 1970) and has recently been suggested to play a major regulatory role in a large number of acute and chronic diseases (Delgado, Abad, Martinez, Leceta, and Gomariz 2001; Delgado and Ganea 2001). The name VIP is derived from the profound and long-lasting gastrointestinal smooth muscle relaxation that this peptide produces following systemic administration. The primary structure of VIP is closely related to pituitary adenylate cyclase activating polypeptide (PACAP). VIP contains 28 amino acid residues with a molecular weight of 3326 Da. The amino acid sequence of VIP in the rat, human and other mammalians species is identical (Henning and Sawmiller 2001).

VIP has recently been attributed to function not only as a vasodilator and bronchodilator but also as a potent immunomodulator. VIP is a critical mediator of neuronal development and survival, hormone secretion, smooth muscle function, glandular secretion, cellular migration, adhesion and production of cytokines, and other proteins. VIP also regulates negatively and positively the generation of cytokines by macrophages and lymphocytes (Said and Mutt 1988). Furthermore, several studies indicated that VIP has a significant role in sleep regulation.

On the other hand, prolactin (PRL) is a hormone expressed in mammals, birds, reptiles, amphibians, and fishes, and has a wide spectrum of effects. In fact, more than 300 distinct biological activities of PRL have been recorded (Nicoll 1980). PRL is expressed in many extrapiuitary sites, particularly within the female and male reproductive organs (uterus, mammary gland, prostate), cells of the immune system (T cells, B cells, NK cells), and certain brain regions, acting locally as an autocrine or paracrine cytokine in diverse tissues and cells. It is produced by lactotrophs and somatolactotrophs of the anterior pituitary. PRL has a wide range of cellular and physiological effects. The effects are cell and context dependent.

The regulatory neuropeptide VIP, and cytokines such as PRL, are multifunctional pleiotropic proteins that play crucial roles in cell-to-cell communication and cellular activation. Both molecules are involved not only in the immune response but also in a variety of physiological and pathological process, including the sleep regulation.

In the present chapter we describe the biological activities of VIP and PRL on the central nervous system (CNS), with special emphasis on role they have on sleep and some immune system regulations.

5.2 Vasoactive Intestinal Polypeptide (VIP)

5.2.1 Receptors

VIP is specifically recognized by members of a subfamily of G protein-coupled receptors (GPCRs). VIP binds to $VPAC_1$ and $VPAC_2$ with similar affinities ($K_d = 3$ to 10 nM) and PAC1 with much lower affinity ($K_d = 1$ μM). $VPAC_1$, $VPAC_2$, and PAC_1 show a high level of homology to each other and share structural features, as a long N-terminus and short loops. As for most subfamilies of GPCRs, the greatest degree of homology is within the transmembrane domains and least in the N- and C-terminal segments. $VPAC_1$ and $VPAC_2$ are often expressed with reciprocal densities in one type of cell, and both show high levels of inducibility and repressibility. VIP receptors are presumed to transduce all of the effects of VIP but often act as mediators of other cytokines (Goetzl, Voice, and Dorsam 2001).

5.2.2 Gene Regulation

VIP is a 28 amino acid linear peptide with a C-terminal amide and without any complex substituents. Prepro VIP (VIP precursor) is also the source of biologically active peptide histidine isoleucine (PHI) or peptide histidine metionine (PHM). Human prepro VIP gene encoding is 8.8 Kb and has been localized in chromosome 6p24. Expression of prepro VIP is regulated with cell type specificity by intracellular cyclic AMP (cAMP) and Ca^{2+}, phorbol esters, and several members of the neuropoietic family of cytokines (Yamagami et al. 1988; Fink, Verhave, Walton, Mandel, and Goodman 1991). Neural cell-specific full expression of the prepro VIP gene in mediated by a tissue-specific element consisting of a 452-bp domain with two AT-rich octamer-like sequences between –4.7 and 4.2 Kb upstream of the transcription start site (Hahm and Eiden 1998). Ca^{2+}-mediated gene regulation increases in prepro VIP mRNA levels in human neuroblastoma cell lines contrast with those evoked by cAMP in being cAMP-independent, slower in time course and requiring de novo protein synthesis. Although increases in Ca^{2+} induce differential rises in several transcription factors relative to cAMP, no promoter element was unequivocally coupled to the effects of Ca^{2+} (Symes, Gearan, Eby, and Fink 1997).

VIP is expressed at levels predominantly determined by the balance between transcriptionally regulated synthesis and susceptibility to peptidolysis, and it exerts many different types of effects in numerous organ systems (Goetzl et al. 2001). It has been demonstrated that light affects VIP expression (Marshall and Born 2002) and that the prepro VIP mRNA display diurnal variations in the suprachiasmatic nucleus (SCN) after exogenous administration of VIP (Bredow, Kacsoh, Obál, Fang, and Krueger 1994). As a consequence, VIP affects the circadian clock, resulting in prolonged sleep cycles and earlier occurrence of the cortisol nadir (Steiger 2003).

It is known that VIP mRNA is fairly broadly distributed within the SCN, with the exception of its most rostral or most caudal sections. Short photoperiod exposure inhibited VIP mRNA expression in the SCN of hamsters for at least several hours, and this effect does not require the pineal gland. Thus, the pineal gland may chronically and differentially modulate the expression of VIP mRNA in the SCN where the activity of VIP neurons is decreased after the short photoperiod. Evidence suggest that VIP neurons of the SCN project to the paraventricular nuclei, which is a necessary component of the neural pathway generating the circadian pacemaker, and also modulates the endocrine functions with which these neurons have been associated (Duncan 1998).

5.3 VIP: Brain Distribution

VIP has been localized in neuronal cell bodies, axons and dendrites, and presynaptic nerve terminals from which VIP is released as a neurotransmitter. In the CNS, VIP is present in the cerebral cortex, the hypothalamus, amygdala, hippocampus, corpus striatum, and the vagal centers of the medulla oblongata (Usdin, Bonner, and Mezey 1994; Wei and Mojsov 1996). VIP receptors are widely distributed in cerebral cortex, amygdaloid nuclei, hippocampus, olfactory lobes, thalamus, and SCN. In contrast, VIP receptors are present in lower concentrations in hippocampus, brainstem, spinal cord and dorsal root ganglia. VIP is also involved in the release of other hormones such as corticotrophin-releasing factor (CRH), PRL, oxytocin, and vasopressin (Table 5.1).

5.4 VIP and Sleep Regulation

Several experimental studies have suggested a role for VIP as a sleep-inducing factor. During the 80s Riou, Cespuglio, and Jouvet (1981) demonstrated that i.c.v. administration of 100 ng of VIP promotes rapid eye movements sleep (REMS) in normal rats and in rats pretreated with *para*-chlorophenylalanine (PCPA). This study was the first evidence that demonstrate that VIP is capable of enhancing sleep and specifically REMS. Subsequent studies showed that i.c.v. injection of VIP also promotes REMS in other species, for example, cats (Drucker-Colín, Bernal-Pedraza, Fernandez-Cancino, and Oksenberg 1984; Drucker-Colín, Aguilar-Roblero, and Arankowsky-Sandoval 1986) and rabbits (Obál 1986; Obál, Opp, Cady, Johannsen, and Krueger 1989). The REMS enhancement observed after VIP administration is independent of brain temperature increases (Obál, Sary, Alfoldi, Rubicsek, and Obal 1986).

In 1986, Prospero-García, Morales, Arankowsky-Sandoval, and Drucker-Colín (1986) reported that both the cerebral fluid (CSF) of sleep-deprived cats (24 h), and VIP, were capable of restoring REMS in PCPA insomniac cats, suggesting that the CSF of sleep-deprived cats contains a VIP-like sleep factor possibly involved in triggering REMS. Subsequent studies demonstrated that i.c.v. administration of CSF in cats incubated with VIP antibodies blocked its REMS-inducing effects (Drucker-Colín, Prospero-García, Perez-Montfort, and Pacheco-Cano 1988), and that VIP also induces REMS recovery in insomniac forebrain lesioned cats (Pacheco-Cano, García-Hernandez, Prospero-García, and Drucker-Colín 1990).

TABLE 5.1. Tissue distribution of both VIP- and PRL-binding sites in vertebrates.

	VIP	PRL
Central nervous system		
Cortex	+	+
Hippocampus	+	+
Thalamus	+	−
Suprachiasmatic nucleus	+	−
Spinal cord	+	−
Dorsal root ganglia	+	−
Hypothalamus	−	+
Choroid plexus	−	+
Corpus callosum	−	+
Astrocytes	−	+
Glial cells	−	+
Striatrum	−	+
Olfactory bulb	−	+
Lung	+	+
Gastrointestinal tract	+	+
Liver	+	+
Prostate	+	+
Macrophages and lymphocytes	+	+
Skeletal muscle	+	+
Cardiac muscle	+	+
Kidney	+	+
Pancreas	−	+
Adipose tissue	+	+
Bone tissue	−	+
Skin	−	+
Ganglia	−	+
Ovary	−	+
Testes	+	+
Mammary gland	−	+
Submandibular and submaxillary gland	−	+

Furthermore, it has been observed that VIP-immunoreactivity in different areas of the rat brain after 24 and 48 h of REMS deprivation are not changing in any analyzed brain structure, suggesting that VIP is not involved in REMS homeostatic process (Morin, Denoroy, and Jouvet 1992). However, other studies have shown that in both the SCN and periventricular nucleus the concentration of VIP-immunoreactivity increases during the dark period, when rodents are mostly awake, and decreases during the subsequent light period, when most of the time is spent in sleep (Morin, Denoroy, and Jouvet 1993; Schilling and Nurnberger 1998). VIP-immunoreactivity exhibits daily variations in the locus coeruleus nucleus, the periaqueductal gray matter and the paraventricular nucleus. These daily variations suggest that VIP acts as an endogenous hypnogenic factor, which progressively accumulates during the waking period, until it reaches a critical level where it could trigger REMS (Jimenez-Anguiano, Baez-Saldana, Drucker-Colín 1993).

Additionally, it has been demonstrated that the brain distribution of VIP receptors changes following REMS deprivation, since there is an increase in the density of receptors in several brainstem and forebrain structures after 24 and 72 h of REMS deprivation (Fig. 5.1). The increase of VIP receptors occurred mainly in several of the brainstem areas that have persistently been suggested to be involved in the generation and maintenance of REMS.

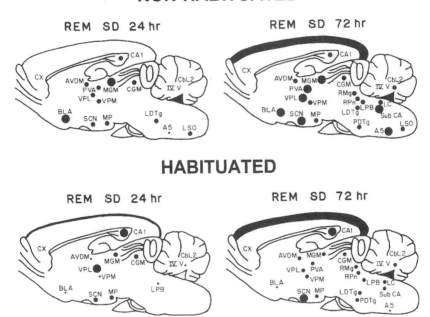

FIGURE 5.1. This representation illustrates the distribution of areas within the diencephalon and brainstem where increases in VIP receptor density were observed in comparison to control. The black dots indicate the structures where the changes occurred and size of the dots portrays areas where large changes were observed, while small dots portray smaller changes (Jimenez-Anguiano, García-García, Mendoza-Ramirez, Duran-Vazquez, and Drucker-Colín 1996) (*Abbreviations*: A5, noradrenaline cells; AVDM, anteroventral thalamic nucleus; BLA, basolateral amygdaloid nucleus; CA1, field CA1 of Ammon's horn; CbL2, cerebellar lobule 2; CGM, central gray medial; Cx, cortex; IV V, choroid plexus of IV ventricle; LDTg, laterodorsal tegmental nucleus; LPB, lateral parabrachial nucleus; LR, lateral recess of 4th ventricle; LSO, lateral superior olive; MGM, medial geniculate nucleus; MP, mammillary peduncle; PDTg, posterodorsal tegmental nucleus; PVA, paraventricular thalamic nucleus; RMg, raphe magnus nucleus; RPn, raphe pontis nucleus; SubCA, subcoeruleus nucleus alpha; VPL, ventral posterolateral thalamic nucleus; VPM, ventral posteromedial thalamic nucleus).

5.5 How VIP Induces REMS: A Hypothesis

Several studies have demonstrated that injection of VIP in areas of the brain related with REMS generation induced REMS for several days (Bourgin et al. 1997). The most remarkable result is that VIP, possibly in association with cholinergic mechanisms, plays a critical role in the long-term regulation of REMS. Furthermore, the coexistence of VIP with acetylcholine (ACh) has been reported (Whittaker 1989; Magistretti 1990).

It has been suggested that the oral pontine reticular nucleus (PnO) could be a common target for the induction of REMS by VIP, and that the interaction between VIP-ACh at the PnO level is responsible for REMS induction. This hypothesis is supported by the observation that the injection the VIP into PnO in combination with atropine, a cholinergic muscarinic antagonist, prevented REMS increase (Bourgin, Ahnaou, Laporte, Hamon, and Adrien 1990). Also, it has been reported that injection of VIP into medial pontine reticular formation (mPRF) of rats induced REMS and neuronal depolarization (Kohlmeier and Reiner 1999). In addition, the microinjection of a cholinergic agonists into the mPRF also produces a state that is behaviorally indistinguishable from naturally occurring REMS (Baghdoyan, Spotts, and Snyder 1993). Together these results support that VIP plays an important role in the induction and maintenance of the REMS via its interaction with ACh. The mechanisms could be prolonged depolarization of mPRF cells during REMS period; however, more studies are necessary to elucidate this hypothesis.

On the other hand, there are other studies that suggest that the effect of VIP on REMS is independent of cholinergic mechanisms. For instance, VIP administered in combination with atropine in PCPA-treated cats insomnia induced an increase in total time REMS (Prospero-García, Jimenez-Anguiano, and Drucker-Colín 1993). Additionally, it has been observed that the administration of PRL-antiserum blocked the increase in circulating levels of PRL and prevented VIP-enhanced REMS (Obál, Payne, Kacsoh, Opp, Kapas, Grosvenor, and Krueger 1994). VIP injections stimulate PRL secretion in plasma and i.c.v. injection of VIP also increase mRNA PRL levels in the hypothalamus of the rat (Bredow et al. 1994; Abe, Engler, Molitch, Bollinguer-Gruber, and Reinchlin 1985). According to these results it is possible to suggest that the REMS-promoting activity of VIP may be mediated by endogenous PRL.

5.6 VIP Clinical Implications

Currently, there are no clinical studies where VIP alterations (overexpression, mutations, lack of receptors, etc.) are related with human sleep disorders. For this reason in the present section, we will be rely mention clinical studies where VIP is associated with other kind of human pathologies.

Transgenic up-regulation of expression of VIP in pancreatic islet β-cells resulted in greater secretion of insulin and improved glucose tolerance in mice (Kato et al. 1994). On the other hand, deficiencies of VIPergic nerves, assessed by morphometric criteria, have been observed on the affected tissue sites of patients with *esophageal achalasia* and *Hirschsprung's disease* of the colon, where absence of VIP may contribute to each characteristic disorder of gastrointestinal motility. Similarly, patients with *cystic fibrosis* or *asthma* have diminished respiratory airway content of VIP that may account in part for the respective abnormalities of exocrine secretion and bronchial reactivity. The absence of inactivating genetic anomalies of any of $VPAC_1$, $VPAC_2$, or PAC_1 receptors precludes specific assignment of each neuropeptide effect to one type of receptor (Said et al. 1980; Goetzl, Adelman, and Sreedharan 1990).

5.7 Prolactin (PRL)

The secretion of PRL, unlike that of other pituitary hormones, is under tonic inhibition by hypothalamic dopamine. THR, VIP, and PRL-releasing peptide are potent stimulatory peptides (Hinuma et al. 1998; Thorner, Vance, Laws, Horvath, and Kovacs 1998). In addition, estrogens are strong stimulators of pituitary PRL secretion (Day, Koike, Sakai, and Maurer 1990). Besides, steroid hormones such as estrogen and progesterone, from the placenta, frequently modulate the effects of PRL. However, interactions between PRL and steroids are highly dependent of cell type and hormonal milieu. PRL and glucocorticoids appear to have a synergistic antiapoptotic effect in differentiated mammary gland. In addition, interactions have been described with polypeptide growth factors such as EGF, IGF-1, and insulin.

At the cellular level, PRL regulates the growth, survival, differentiation, and activation state of target cells. At the physiological level, the effects of PRL may be divided into seven partially overlapping areas: (1) reproduction, (2) immune function, (3) water/electrolyte balance (osmoregulation), (4) stress adaptation, (5) behavior/brain/psychology, (6) metabolism, and (7) skin function.

5.8 PRL Molecular Aspects

After the processing of a 28 residue signal peptide, the mature human PRL molecule is secreted as a polypeptide with 199 amino acid residues, whereas mouse PRL is two residues shorter (Cooke 1989). In both species, six cysteines form three intramolecular disulfide bridges. The molecular weight of human PRL is approximately 23 kDa, but a 26-kDa glycosylated form is also produced. PRL circulates in blood as monomers of 23 to 26 kDa (Lewis, Sigh, Sinha, and Vanderlaan 1985). Larger "macroprolactins" (big-prolactin and big–big prolactin) represent both homo-oligomeric aggregates and immunoglobulin-complexed PRL. Furthermore, the physiological proteolysis of PRL to a C-terminally truncated 16 kDa variant results in a molecule with distinct biological activities that may activate unique receptors (Mittra 1980; Clapp and Weiner 1992). PRL is a tetrahelical cytokine most closely related to growth hormone and placental lactogenes. It binds to specific prolactin receptors that belong to the WS-motif cytokine receptor family.

The pituitary PRL secretory rate is approximately 18.6 nmol/day (400 µg/day). The hormone is cleared by the liver (75%) and the kidney (25%), and its half-life in plasma is approximately 50 min (Thorner et al. 1998).

5.8.1 Gene Regulation

Human PRL gene is present as a single copy on chromosome 6 (Owerbach, Rutter, Cooke, Martial, and Shows 1981). It is located close proximity to the HLA complex. This colocalization is of interest because of a possible association between prolactin-secreting adenomas and specific HLA alleles (Farid, Noel, Sampson, and Russell 1980). Meanwhile, the mouse PRL gene maps to chromosome 13, clustered with genes encoding mouse placental lactogenes and other prolactin-like genes (Jackson-Grusby, Pravtcheva, Ruddle, and Linzer 1988).

The human PRL gene is more than 15 Kb long and contains six exons (Truong, Duez, Belayew, Renard, Pictet, Bell, and Martial 1984). It is known that transcriptional control of the distal nonpituitary start site in endometrial stromal cells is linked to decidual differentiation during the secretory phase of the ovulatory cycle (DiMattia, Gekkersen, Duckworth, and Freisen 1990). Two consensus-binding sites for CCAAT/enhancer-binding proteins (C/EBP) mediate the cAMP/PKA-induced activation of this nonpituitary PRL gene promoter in human decidual cells (Pohnke, Kempf, and Gellersen 1999). Cyclic AMP, alone or in synergy with PHA, also stimulates the activation of this upstream PRL gene promoter in Jurkat T cells, possibly through activation of C/EBP proteins (Reem, Ray, and Davis 1999).

PRL activates the transcription factor STAT5 in most target cells and tends to interact positively and in a redundant manner with other cytokines that also activate STAT5 as IL-2, IL-3, IL-5, IL-7, IL-9, IL-15, GM-CSF, erythropoietin, thrombopoietin, and growth hormone (Kirken, Rui, Malabarba, and Farrar 1994).

5.8.2 Posttranslational Modifications

PRL protein undergoes posttranslational modifications: oxidation, proteolysis, glycosylation, and phosphorylation. Mature prolactin, formed by proteolytic removal of a 28-kDa signal peptide, can be further modified by proteases. Cathepsin-D proteolysis at position 133 generates two fragments of 16 and 8 kDa respectively, which may exist as both disulfide-linked heterodimers and monomers (Mittra et al. 1980; Cole, Nichols, Lauziere, Edmunds, and McPeherson 1991). The six cystein residues in PRL undergo oxidation and form stable, successive intramolecular disulfide bonds. Besides, a proportion of pituitary and circulating PRL is glycosylated in most species. Approximately 20% of circulating human PRL is glycosylated through N-linkage at position 31 (Lewis et al. 1985; Champier, Claustrat, Sassolas, and Berger 1987). The physiological function of glyscosylation of PRL may be to reduce biological potency, while extending the half-life of the molecule (Hoffmann, Penel, and Ronin 1993). Finally, a significant proportion of PRL molecules are phosphorylated on their serine and threonine residues. Phosphorylation of PRL is associated with reduced bioactivity, but does not affect the biological half-life (Wang and Walker 1993). Because PRL is phosphorylated in secretory granules during its release from pituitary lactotrophs, it is possible that phosphorylation serves to reduce local bioactivity during secretion.

PRL could be considered to be a monogamous cytokine in that it binds exclusively to PRL receptors. The PRL receptor (PRLR) exists in all vertebrates in many isoforms, soluble or membrane-bound. It is expressed in a wide variety of tissues where PRL activates PRLR by inducing its homodimerization, the first step required for triggering signaling cascades. No other membrane chain is required for signaling. PRLR is a single-pass transmembrane chain, with the N-terminus outside the cell. In mammals, the overall length of the PRLR varies from 200 amino acids for the soluble-binding protein up to 600 residues for the long membrane isoforms (Goffin and Kelly 2001). The gene encoding the human PRLR is located on chromosome 5(p13–14) and contains at least 10 exons for an overall length >100 kb (Arden, Boutin, Djiane, Kelly, and Cavenee 1990).

It is usually observed that the soluble binding protein (PRLbp) has a higher affinity (10 times) than the membrane-bound PRLR for a given ligand (Postel-Vinay et al. 1991). The affinity of the PRLR will vary depending on the type and species of origin of the ligand considered. It is usually in the range of $K_d = 10^{-9}$ to 10^{-10} M. The PRLR is activated by dimerization (Goffin, Shiverick, Kelly, and Martial 1996), which involves two regions (called binding sites 1 and 2) of the ligands, each interacting with one molecule of PRLR. It is known that both binding sites interact with virtually overlapping epitopes within the receptor (Goffin et al. 2001). The level of expression of the PRLR varies from 10 to 2000 fmol/mg of membrane protein. The expressions of short and long forms of receptor have been shown to vary as a function of the stage of the estrous cycle, pregnancy, and lactation. PRL is able to both up- and down-regulate its receptor, the latter process probably being due to an acceleration of internalization of hormone/receptor complexes. The effect of PRL on its receptor is also a function of the hormone concentration and time of exposure of the tissue. Growth hormone is also able to up-regulate the PRLR (Kelly, Djiane Postel-Vinay, and Edery 1991). The JAK/STAT signaling pathway is the most widely described cascade for the PRLR. Activation of the JAK2 occurs very rapidly after hormonal stimulation (within 1 min), suggesting that this Janus kinase occupies a central and very upstream role in the activation of several signaling pathways of the PRLR (Chang, Ye, and Clevenger 1998). However, many other signaling proteins were found to be activated by the PRLR. The well-known MAP kinase pathway involves the She/SOS/Grb2/Ras/Raf/MAP kinase cascade and this pathway has been demonstrated to be activated by the PRLR in various cell systems (Piccoletti, Marioni, Bendinelli, Bernelli-Zazzera 1994; Erwin, Kirken, Malabarba, Farrar, and Rui 1996). Although the JAK/STAT and the MAP kinase cascades were initially regarded as independent pathways, recent data rather suggest that these pathways are interconnected (Chida, Wakao, Yoshimura, and Miyajima 1998) (Fig. 5.2).

5.9 PRL and Sleep Regulation

In 1986 Michel Jouvet showed that the systemic administration of PRL enhances the total time of REMS in cats (Jouvet et al. 1986). Studies in hypoprolactinemic rats showed that REMS duration was decreased, also circadian rhythm of REMS disappeared, while that of nonrapid eye movements sleep (NREMS) remains unchanged (Valatx and Jouvet 1988).

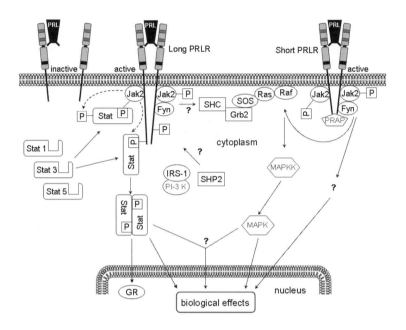

FIGURE 5.2. Representation of PRL and its receptor (PRLR) signaling pathways. Long and short isoforms of the PRLR are represented. PRLR activates STAT1, and STAT3. The MAP kinase pathway involves the Shc, Grb2, Sos, Ras, Raf cascade and is presumably activated by both PRLR isoforms. Connections between the JAK/STAT and MAP kinase pathways have been suggested.

These studies were the first to evidence that involvement of PRL in sleep regulation. Further studies have shown that the administration of exogenous PRL in different animal species including rabbits and rats induced REMS (Obál et al. 1989; Roky, Valatx, and Jouvet 1993; Zhang, Kimura, and Inoue 1999).

Additionally, administration of anti-PRL antibodies and antiserum in rats reduced REMS (Obál et al. 1992; Roky, Valatx, Paut-Pagano, and Jouvet 1994). Likewise, it has been observed that the effect of PRL on REMS is photoperiod-dependent, because the amount of REMS decreased or increased when PRL is administered during the dark or light period, respectively (Obál et al. 1989; Roky et al. 1993). Furthermore, it has been reported that PRL-releasing peptide (PrRP) also promotes REMS when it is administered to rats (Zhang, Inoue, and Kimura 2001; Zhang, Kimura, and Inoue 2000). These results suggest that PRL stimulated by i.c.v. injection of PrRP could be responsible to induce REMS.

PRL administered locally into the dorsolateral hypothalamus, an area that contain PRL-immunoreactive neurons (Paut-Pagano, Roky, Valatx, Kitahama, and Jouvet 1993), either increase REMS when is given diurnally or decreases REMS when is given nocturnally (Roky et al. 1994). Also, it has been demonstrated that in hypoprolactinemic (IPL) rats under light–dark conditions, the circadian rhythms of

NREMS and REMS display an alteration of their phase relation (Valatx et al. 1988). This result suggests that the promoting effect of REMS by PRL could be regulated by circadian factors and probably through the relationship between other hormones, for example melatonin, since melatonin is involved in the regulation of circadian PRL secretion (Bispink, Zimmermann, Weise, and Leidenberger 1990), and sleep (Tobler, Jaggi, and Borbely 1994). It is possible that PRL could have its effect on REMS through a circadian mechanism (Roky et al. 1995).

Furthermore, it has been observed that the firing of frequency of mesopontine tegmental neurons, an area involved in REMS generation (Baghdoyan et al. 1993) was increased after PRL injection (Takahashi, Koyama, Kayama, and Yamamoto 2000). These results suggest that PRL is capable of inducing REMS and modulate cholinergic activity, but the mechanisms remain unclear. Additionally, there are other studies that show that PRL injection into the central nucleus of the amygdale, an area where microinjections of cholinergic agonist produced REMS (Calvo, Simon-Arceo, and Fernandez-Mas 1996) and also containing high concentration of PRL-immunoreactive fibers and receptors (Siaud, Manzoni, Balmefrezol, Barbanel, Assenmacher, and Alonso 1989; Roky et al. 1996), decrease NREMS (Sanford, Nassar, Ross, Schulkin, and Morrison 1998).

In addition, it has been shown that high secretion of PRL occurs during NREMS and that PRL secretion is coupled to delta waves in humans, in contrast alpha and beta bands frequencies were inversely proportional to PRL secretion (Spiegel et al. 1995). Thus, it has been suggested that endogenous PRL accumulation during NREMS is responsible to induce the subsequent REMS periods. For example, it is well established that plasma PRL concentrations exhibit a sleep-dependent pattern, the high levels occurring during sleep and the low levels during waking (Spiegel et al. 1995; Sassin, Frantz, Weitzman, and Kapen 1972; Sassin, Frantz, Kapen, and Weitzman 1973; Van Cauter, L'Hermite, Copinschi, Refetoff, Desir, and Robyn 1981; Linkowski et al. 1998; Appelberg, Katila, and Rimon 2002). However, some reports show that occasionally poor sleep does not influence PRL secretion in normal humans (Spiegel, Follenius, Simon, Saini, Ehrhart, and Brandenberger 1994).

5.10 Relationship Between PRL and Stress

On the other hand, PRL has been associated with stress regulation and sleep. Several studies have demonstrated that different types of stressors may have different effects on sleep (García-García, Beltrán-Parrazal, Jimenez-Anguiano, Vega-González, and Drucker-Colín 1998; Palma, Suchecki, and Tufik 2000). One and two hours stress by immobilization applied at the beginning of the dark period is followed by a significant REMS rebound during 10 consecutive hours after stress (Rampin, Cespuglio, Chastrette, and Jouvet 1991). Although, the response to immobilization is often limited to increase in REMS, but sometimes clear increases in NREMS could be observed (Gonzalez, Debilly, Valatx, and Jouvet 1995; Bouyer, Deminiere, Mayo, and Le Moal 1997). In addition, the activation of the hypothalamo-pituitary-adrenocortical axis by stress elicited PRL secretion (Gala 1990; Akema, Chiba, Oshida, Kimura, and Toyoda 1995). Bodosi, Obál, Gardi, Komlodi, Fang, and Krueger (2000) showed that ether exposition as a stressor increases REMS and PRL secretion in rats, suggesting that PRL secretion could be involved in the mediation of the stress-induced increase in REMS. Spontaneous REMS rebound disappear after exposure to ether stress in HYPOX rats and with the administration of PRL-antiserum. However, PRL-antiserum decreases spontaneous REMS in absence of exposure to ether vapor. These authors suggest that the i.c.v. injections of PRL-antiserum may modify REM sleep propensity independently of PRL secretion (Bodosi et al. 2000). These findings suggest that PRL plays an important role in stress regulation and that PRL could be the hormone responsible of the increase REMS after stress. Other data support this hypothesis because in different strains of mice the PRL plasma levels was significantly increased after restraint stress (Meerlo, Easton, Bergmann, and Turek 2001). Nevertheless, it has been reported that after 24-h of sleep disturbance, the mRNA expression levels of prolactin did not change (Ogawa et al. 2005).

5.11. Stress and REMS: Role of PRL

There are different possible explanations for a parallel increase in PRL and REMS after stress. One possibility is the relationship between PRL and CRH. CRH is responsible to regulate the stress response through of adrenocorticotropic hormone (ACTH) release, which in turn regulates cortisol production (Menzaghi, Heinrichs, Pich, Weiss, and Koob 1993). Some studies have been demonstrated that CRH induces PRL release during stress conditions (Akema et al. 1995; Morel et al. 1989), and that the i.c.v injection of alpha-helical 9–41 a CRF-receptor antagonist before stress by immobilization in rats abolished REMS rebound after stress (Gonzalez and Valatx 1997; García-García 2000). These results suggest that PRL could be modulating REMS increase after stress via CRH.

Another possibility is the relationship that exists between PrRP and CRH. Recently, it has been demonstrated that i.c.v. injections of PrRP increases CRH release (Maruyama et al. 2001; Matsumoto et al. 2000). Probably, PrRP can cause release of PRL via CRH receptor activation, possibly by the direct activation of CRH neurons in the PVN. Matsumoto and coworkers (2000) provided the functional support for this model by demonstrating that when PrRP is injected i.c.v. into rat brains, a dramatic induction of c-Fos expression is found in the paraventricular nucleus (PVN), and some of these neurons appear to be CRH positive. Double labeling of CRH with PrRP found PrRP fibers making contact with a few CRH cells, supporting the possibility that PrRP may affect CRH release. Indeed, i.c.v. injection of 10 nmol PrRP causes a significant increase in serum ACTH, which is blocked when coadministered with alpha-helical CRH. Furthermore, it has been demonstrated that PrRP mRNA levels are increased in PVN and other nuclei involved in the processing of visceral sensory information, for example the nucleus of the solitary tract, and have been implicated in affecting adaptive responses to stress (Morales, Hinuma, and Sawchenko 2000).

In summary, these results clearly support the hypothesis that PRL possesses REMS-promoting activity and also that PRL could be involved in REMS rebound after stress, possibly via PrRP or CRH. Nevertheless, there is evidence showing that PRL is related to other sleep-inducing substances that play an important role in sleep regulation, for example VIP.

5.12. PRL Expression and REMS

In a mutant mouse model, mice lacking an interferon type I receptor have very low levels of hypothalamic PRL mRNA and approximately a 30% reduction in REMS during both the light and the dark period (Bohnet, Traynor, Majde, Kacsoh, and Krueger 2004). Recently, it has been reported that total time in REMS is reduced in mice that produce a biologically inactive 11 kDa PRL-immunoreactive polypeptide because of a targeted disruption of the PRL gene. In addition, the same study suggested that these mice are not responding neither to stress nor VIP administration but REMS is restored after 14 days of PRL therapy (Obál et al. 2005). According of this findings it is plausible to suggest that the actions of PRL on REMS mechanism are indirect, involving perhaps the brainstem structures and neurotransmitters such as acetylcholine.

5.13. PRL Clinical Implications

It has been shown that rats chronically hyperprolactinemic with anterior pituitary grafts present an augmentation in duration and frequency of REMS during the 12-h light period (Obál, Kacsoh, Bredow, Guha-Thakurta, and Krueger 1997). However, patients with prolactinoma did not show significant differences in REMS quantity, in contrast an NREMS increase was observed (Frieboes, Murck, Stalla, Antonijevic, and Steiger 1998). Conversely, rats with hypophysectomy (HYPOX) exhibit a decrease in NREMS and REMS (Obál, Floyd, Kapas, Bodosi, and Krueger 1996). However, there is also evidence showing rats with HYPOX have decreased or normal REMS and decrease, normal, or increased NREMS (Valatx, Chouvet, and Jouvet 1975). Perhaps, the time lapse between HYPOX rats and recording might contribute to the divergent findings, because it has been demonstrated that the amount of REMS is practically compensated 30 days after HYPOX surgery (Valatx et al. 1975).

Among pituitary tumors, 60% secrete PRL and cause a state of chronic hyperprolactinemia. In certain asymptomatic subjects with hyperprolactinemia, stable circulating complexes of immunoglobulin and PRL have been identified (Bonhoff, Vuille, Gomez, and Gallersen 1995; Hattori and Inagaki 1998), suggesting that antiprolactin autoantibodies may occasionally neutralize the activity of the hormone. Pituitary adenomas reveal pathophysiologic effects of PRL in men by inducing decreased libido, impotence, gynecomastia, galactorrhea, hypospermia, and occasionally reduced beard growth (Thorner et al. 1998). In premenopausal women, the cardinal effects of hyperprolactinemia are amenorrhea, a cessation of the normal cyclic ovarian function, galactorrhea, decreased libido, and an increased long-term risk of osteoporosis (Thorner et al. 1998; Palermo, Albano, Mangione, and Napoli 1994).

Data suggest that an elevated PRL level represents a risk factor for certain autoimmune diseases in humans and rodents. These include adjuvant arthritis in rats, collagen type II-induced arthritis in rats and mice, type I diabetes in mice, and systemic lupus erythematosis (SLE) in mice and humans. Furthermore, a connection between an elevated PRL level and rheumatoid arthritis (RA) has also been suggested (Brennan, Ollier, Worthington, Hajeer, and Silman 1996), as well an involvement of PRL in the development and progression of leukemia and lymphoma. PRL may promote the growth of hematopoietic cancers. In fact, elevated serum PRL was detected in more than 50% of patients with acute myeloid leukemia (Hatfill, Kirby, Hanley, Rybicki, and Bohm 1990), although this observation might be the result of an associated stress response. There is, to date, no disease known to cause by mutation of the PRLR, or even of its ligand. Interestingly, in vitro studies have shown that antihuman PRL antibodies can prevent proliferation of breast cancer cell lines induced by locally produced PRL (Ginsburg and Vonderhaar 1995).

In knockout mice, the data suggested that PRL is critical for fertility and mammary development in female, whereas PRL deficiency can be compensated with regard to male fertility and general hematopoiesis and immune system development. Besides, PRL receptor knockout mice presented multiple reproductive defects in female mice as sterility caused by complete failure of embryonic implantation, and almost complete loss of lactation after the first but not subsequent pregnancies. Furthermore, half of the male PRL receptors knockout mice were infertile or showed reduced fertility. Knockout mice for PRLR indicated that this receptor is not absolute requirement for fetal development. The main phenotypes of PRLR knockout mice are linked to sterility of double negative females, due to a failure of embryo implantation. No immunological phenotype was observed in either PRLR or PRL knockout mice. This suggests that PRL- or PRLR knockout mice can probably compensate for the lack of receptor functions via redundancy of other cytokine(s) (Goffin et al. 2001).

Meanwhile, transgenic PRL overexpression led to an increased rate of mammary tumor formation in females (Wennbo, Gebre-Medhin, Gritli-Linde, Ohlsson, Isaksson, and Tornell 1997). In addition, advanced prostate hyperplasia was observed in male mice, an effect that was associated with a moderate elevation of circulating testosterone level (Wennbo, Kindblom, Isaksson, and Tornell 1997).

In conclusion, both VIP and PRL cytokines are two of the most relevant sleep regulatory substances. Those cytokines are related to REMS regulation in mammals. However, the relationship between VIP, PRL, and the immune system is not clear. Thereby more studies are necessary to understand their role in both immune diseases and sleep disorders.

Acknowledgments.   This work was supported in part by Programa del Mejoramiento del Profesorado (PROMEP) UVER-PTC-118, CONACYT 25122-M and Fideicomiso to R.D.C.

References

Abe, H., Engler, D., Molitch, M.E., Bollinguer-Gruber, J., and Reinchlin, S. (1985) Vasoactive intestinal peptide is a physiological mediator of prolactin release in the rat. *Endocrinology* 1116, 1383–1390.
Akema, T., Chiba, A., Oshida, M., Kimura, F., and Toyoda, J. (1995) Permissive role of corticotropin releasing factor in the acute stress-induced prolactin release in female rats. *Neurosci Lett* 198, 146–148.

Appelberg, B., Katila, H., and Rimon, R. (2002) REM sleep and prolactin in patients with non-affective psychoses. *Psychoneuroendocrinology* 27, 661–669.

Arden, K.C., Boutin, J.M., Djiane, M., Kelly, P.A., and Cavenee, W.K. (1990) The receptors for prolactin and growth hormone are localized in the same region of human chromosome 5. *Cytogenet Cell Genet* 53, 161–165.

Baghdoyan, H.A., Spotts, J.L., and Snyder, S.G. (1993) Simultaneous pontine and basal forebrain microinjections of carbachol suppress REM sleep. *J Neurosci* 13, 229–242.

Bispink, G., Zimmermann, R., Weise, H.C., and Leidenberger, F. (1990) Influence of melatonin on the sleep-independent component of prolactin secretion. *J Pineal Res* 8, 97–106.

Bodosi, B., Obál, F., Jr., Gardi, J., Komlodi, J., Fang, J., and Krueger, J.M. (2000) An ether stressor increases REM sleep in rats: Possible role of prolactin. *Am J Physiol Regul Integr Comp Physiol* 279, R1590–R1598.

Bohnet, S.G., Traynor, T.R., Majde, J.A., Kacsoh, B., and Krueger, J.M. (2004) Mice deficient in the interferon type I receptor have reduced REM sleep and altered hypothalamic hypocretin, prolactin and 2',5'-oligoadenylate synthetase expression. *Brain Res* 1027, 117–125.

Bonhoff, A.J., Vuille, C., Gomez, F., and Gallersen, B. (1995) Identification of macroprolactin in a patient with asymptomatic hyperprolactinemia as a stable PRL-IgG complex. *Exp Clin Endocrinol Diabetes* 103, 252–255.

Bourgin, P., Ahnaou, A., Laporte, A.M., Hamon, M., and Adrien, J. (1999) Rapid eye movement sleep induction by vasoactive intestinal peptide infused into the oral pontine tegmentum of the rat may involve muscarinic receptors. *Neuroscience* 89, 291–302.

Bourgin, P., Lebrand, C., Escourrou, P., Gaultier, C., Franc, B., Hamon, M., and Adrien, J. (1997) Vasoactive intestinal polypeptide microinjections into the oral pontine tegmentum enhance rapid eye movement sleep in the rat. *Neuroscience* 77, 351–360.

Bouyer, J.J., Deminiere, J.M., Mayo, W., and Le Moal, M. (1997) Inter-individual differences in the effects of acute stress on the sleep-wakefulness cycle in the rat. *Neurosci Lett* 225, 193–196.

Bredow, S., Kacsoh, B., Obál, F., Jr., Fang, J., and Krueger, J.M. (1994) Increase of prolactin mRNA in the rat hypothalamus after intracerebroventricular injection of VIP or PACAP. *Brain Res* 660, 301–308.

Brennan, P., Ollier, B., Worthington, J., Hajeer, A., and Silman, A. (1996) Are both genetic and reproductive associations with rheumatoid arthritis linked to prolactin? *Lancet* 348, 106–109.

Calvo, J.M., Simon-Arceo, K., and Fernandez-Mas, R. (1996) Prolonged enhancement of REM sleep produced by carbachol microinjection into the amygdale. *Neuroreport* 31, 577–580.

Champier, J., Claustrat, B., Sassolas, G., and Berger, M. (1987) Detection and enzymatic deglycosylation of a glycosylated variation of prolactin in human plasma. *FEBS Lett* 212, 220–224.

Chang, W.P., Ye, Y., and Clevenger, C.V. (1998) Stoichiometric structure–function analysis of the prolactin receptor signaling domain by receptor chimeras. *Mol Cell Biol* 18, 896–905.

Chida, D., Wakao, H., Yoshimura, A., and Miyajima, A. (1998) Transcriptional regulation of the beta-casein gene by cytokines: Cross talk between STAT5 and other signaling molecules. *Mol Endocrinol* 12, 1972–1806.

Clapp, C., and Weiner, R.I. (1992) A specific, high affinity, saturable binding site for the 16-kilodalton fragment of prolactin on capillary endothelial cells. *Endocrinology* 139, 609–616.

Cole, E.S., Nichols, E.H., Lauziere, K., Edmunds, T., and McPeherson, J.M. (1991) Characterization of the microheterogeneity of recombinant primate prolactin: Implications for post-translational modifications of the hormone *in vivo*. *Endocrinology* 129, 2639–2646.

Cooke, N.E. (1989) Prolactin: Normal synthesis, regulation, and actions. In: *Endocrinology*. W.B. Saunders, Philadelphia, USA.

Day, R.N., Koike, S., Sakai, M., and Maurer, R.A. (1990) Both Pit-1 and the estrogen receptor are required for estrogen responsiveness of rat prolactin gene. *Mol Endocrinol* 4, 1964–1971.

Delgado, M., and Ganea, D. (2003) Vasoactive intestinal peptide prevents activated microglia-induced neurodegeneration under inflammatory conditions: Potential therapeutic role in brain trauma. *FASEB J* 17, 1922–1924.

Delgado, M., Abad, C., Martinez, C., Receta, J., and Gomariz, R.P. (2001) Vasoactive intestinal peptide prevents experimental arthritis by downregulating both autoimmune and inflammatory components of the disease. *Nat. Med.* 7(5), 563–568.

DiMattia, G.E., Gekkersen, B., Duckworth, M.L., and Freisen, H.G. (1990) Human prolactin gene expression: The use of an alternative noncoding exon in decidual and the IM-9-P3 lymphoblast cell line. *J Biol Chem* 265, 16412–16421.

Drucker-Colín, R., Aguilar-Roblero, R., and Arankowsky-Sandoval, G. (1986) Sleep factors released from brain of unrestrained cats: A critical appraisal. *Ann N Y Acad Sci* 473, 449–460.

Drucker-Colín, R., Bernal-Pedraza, J., Fernandez-Cancino, F., and Oksenberg, A. (1984) Is vasoactive intestinal polypeptide (VIP) a sleep factor? *Peptides* 5, 837–840.

Drucker-Colín, R., Prospero-García, O., Perez-Montfort, R., and Pacheco-Cano, M.T. (1988) Vasoactive intestinal polypeptide (VIP) a sleep factor? *Ann N Y Acad Sci* 527, 627–630.

Duncan, M.J. (1998) Photoperiodic regulation of hypothalamic neuropeptide messenger RNA expression: Effect of pinealectomy and neuroanatomical location. *Mol Brain Res* 57, 142–148.

Erwin, R.A., Kirken, R.A., Malabarba, M.G., Farrar, W.L., and Rui, H. (1996) Prolactin activates ras via signaling proteins Shc, growth factor receptor bound 2 and Son of Sevenless. *Endocrinology* 136, 3512–3518.

Farid, N.R., Noel, E.P., Sampson, L., and Russell, N.A. (1980) Prolactin-secreting adenomata are possibly associated with HLA-B8. *Tissue Antigen* 15, 333–335.

Fink, J.S., Verhave, M., Walton, K., Mandel, G., and Goodman, R.H. (1991) Cyclic AMP- and phorbol ester- induced transcriptional activation are mediated by the same enhancer element in the vasoactive intestinal peptide gene. *J Biol Chem* 266, 3882–3887.

Frieboes, R.M., Murck, H., Stalla, G.K., Antonijevic, I.A., and Steiger, A. (1998) Enhanced slow wave sleep in patients with prolactinoma. *J Clin Endocrinol Metab* 83, 2706–2710.

Gala, R.R. (1990) The physiology and mechanisms of the stress-induced changes in prolactin secretion in the rat. *Life Sci* 46, 1407–1420.

García-García, F. (2000) La vigilia como regulador del sueño: estudio sobre los péptidos involucrados. Doctoral thesis, UNAM, Mexico.

García-García, F., Beltrán-Parrazal, L., Jimenez-Anguiano, A., Vega-González, A., and Drucker-Colín, R. (1998) Manipulations during forced wakefulness have differential impact on sleep architecture, EEG power spectrum, and Fos induction. *Brain Res Bull* 47, 317–324.

Ginsburg, E., and Vonderhaar, B.K. (1995) Prolactin synthesis and secretion by human breast cancer cells. *Cancer Res* 55, 2591–2595.

Goetzl, E.J., Adelman, D.C., and Sreedharan, S.P. (1990) Neuroimmunology. *Adv Innunol* 48, 147–155.

Goetzl, E.J., Voice, J.K., and Dorsam, G. (2001) VIP and PACAP. PACAP and VIP receptors. In: *Cytokine Reference*. Academic Press, USA, pp. 1396–1405, 2249–2253.

Goffin, V., Binart, N., Touraine, P., Kelly, P.A. Prolactin: the new biology of an old hormone. Annu Rev Physiol. 2002; 64: 47–67.

Goffin, V., and Kelly, P.A. (2001) Prolactin receptor. *Cytokine Reference.* Academic Press, USA, pp. 1547–1562.

Goffin, V., Shiverick, K.T., Kelly, P.A., and Martial, J.A. (1996) Sequence–function relationships within the expanding family of prolactin, growth hormone, placental lactogen and related proteins in mammals. *Endocr Rev* 17, 385–410.

Gonzalez, M.M., and Valatx, J.L. (1997) Effect of intracerebroventricular administration of alpha-helical CRH (9–41) on the sleep/waking cycle in rats under normal conditions or after subjection to an acute stressful stimulus. *J Sleep Res* 6, 164–170.

Gonzalez, M.M., Debilly, G., Valatx, J.L., and Jouvet, M. (1995) Sleep increase after immobilization stress: Role of the noradrenergic locus coeruleus system in the rat. *Neurosci Lett* 202, 5–8.

Hahm, S.H., and Eiden, L.E. (1998) Five discrete *cis*-active domains direct cell type-specific transcription of the vasoactive intestinal peptide (VIP) gene. *J Biol Chem* 273, 17086–17094.

Hatfill, S.J., Kirby, R., Hanley, M., Rybicki, E., and Bohm, L. (1990) Hyperprolactinemia in acute myeloid leukemia and indication of ectopic expression of human prolactin in blast cells of a patient of subtype M4. *Leuk Res* 14, 57–62.

Hattori, N., and Inagaki, C. (1998) Immunological aspects of human growth hormone and prolactin. *Domest Anim Endocrinol* 15, 371–375.

Henning, R.J., and Sawmiller, D.R. (2001) Vasoactive intestinal peptide: Cardiovascular effects. *Cardiovasc Res* 49, 27–37.

Hinuma, S., Habata, Y., Fujii, R., Kawamata, Y., Hosoya, M., Fukusumi, S., Kitada, C., Masuo, Y., Asano, T., Matsumoto, H., Sekiguchi, M., Kurokawa, T., Nishimura, O., Onda, H., and Fujino, M. (1998) A prolactin-releasing peptide in the brain. *Nature* 393, 272–276.

Hoffmann, T., Penel, C., and Ronin, C. (1993) Lycosilation of human prolactin regulates hormone bioactivity and metabolic clearance. *J Endocrinol Invest* 16, 807–816.

Jackson-Grusby, L.L., Pravtcheva, D., Ruddle, F.H., and Linzer, D.I. (1988) Chromosomal mapping of prolactin/growth hormone gene family in the mouse. *Endocrinology US* 122, 2462–2466.

Jimenez-Anguiano, A., García-García, F., Mendoza-Ramirez, J.L., Duran-Vazquez, A., and Drucker-Colín, R. (1996) Brain distribution of vasoactive intestinal peptide receptors following REM sleep deprivation. *Brain Res* 728, 37–46.

Jimenez-Anguiano, A., Baez-Saldana, A., and Drucker-Colín, R. (1993) Cerebrospinal fluid (CSF) extracted immediately after REM sleep deprivation prevents REM rebound and contains vasoactive intestinal peptide (VIP). *Brain Res* 63, 345–348.

Jouvet, M., Buda, C., Cespuglio, R., Chastrette, N., Denoyer, M., Sallanon, M., and Sastre, J.P. (1986) Hypnogenic affects of some hypothalamo-pituitary peptides. *Clin Neuropharmacol* 9, 465–467.

Kato, I., Suzuki, Y., Akabane, A., Yonekura, H., Tanaka, O., Kondo, H., Takasawa, S., Yoshimoto, T., and Okamoto, H. (1994) Transgenic mice overexpressing human vasoactive intestinal peptide (VIP) gene in pancreatic beta cells. Evidencefor improved glucose tolerance and enhanced insulin secretion by VIP and PHM-27 in vivo. *J Biol Chem* 269, 21223–21228.

Kelly, P.A., Djiane, J., Postel-Vinay, M.C., and Edery, M. (1991) The prolactin/growth hormone receptor family. *Endocr Rev* 12, 235–251.

Kirken, R.A., Rui, H., Malabarba, M.G., and Farrar, W.L. (1994) Identification of interleukin-2 receptor-associated tyrosine kinase p116 as novel leukocyte-specific Janus kinase. *J Biol Chem* 269, 19136–19141.

Kohlmeier, K.A., and Reiner, P.B. (1999) Vasoactive intestinal polypeptide excites medial pontine reticular formation neurons in the brainstem rapid eye movement sleep-induction zone. *J Neurosci* 19, 4073–4081.

Lewis, U.J., Sigh, R.N., Sinha, Y.N., and Vanderlaan, W.P. (1985) Glycosylated human prolactin. *Endocrinology* 116, 359–363.

Linkowski, P., Spiegel, K., Kerkhofs, M., L'Hermite-Baleriaux, M., Van Onderbergen, A., Leproult, R., Mendlewicz, J., and Van Cauter, E. (1998) Genetic and environmental influences on prolactin secretion during wake and during sleep. *Am J Physiol* 274, E909–E919.

Magistretti, P.J. (1990) VIP neurons in the cerebral cortex. *Trends Pharmacol Sci* 11, 250–254.

Marshall, L., and Born, J. (2002) Brain-immune interactions in sleep. *Int Rev Neurobiol* 52, 93–131.

Maruyama, M., Hirokazu, M., Fujiwara, K., Noguchi, J., Kitada, C., Fujino, M., and Inoue, K. (2001) Prolactin-releasing peptide as a novel stress mediator in the central nervous system. *Endocrinology* 142, 2032–2038.

Matsumoto, H., Maruyama, M., Noguchi, J., Horikoshi, Y., Fujiwara, K., Kitada, C., Hinuma, S., Onda, H., Nishimura, O., Inoue, K., and Fujino, M. (2000) Stimulation of corticotropin-releasing hormone-mediated adrenocorticotropin secretion by central administration of prolactin-releasing peptide in rats. *Neurosci Lett* 285, 234–238.

Meerlo, P., Easton, A., Bergmann, B.M., and Turek, F.W. (2001) Restraint increases prolactin and REM sleep in C57BL/6J mice but not in BALB/cJ mice. *Am J Physiol Regul Integr Comp Physiol* 281, R846–R854.

Menzaghi, F., Heinrichs, S.C., Pich, E.M., Weiss, F., and Koob, G.F. (1993) The role of limbic and hypothalamic corticotropin-releasing factor in behavioral responses to stress. *Ann N Y Acad Sci* 697, 142–154.

Mittra, I. (1980) A novel cleaved prolactin in the rat pituitary. I: Biosynthesis, characterization and regulation control. *Biochem Biophys Res Commun* 95, 1750–1759.

Mittra, I. (1984) Somatomedins and proteolytic bioactivation of prolactin and growth hormone. *Cell.* 38(2):347–348.

Morales, T., Hinuma, S., and Sawchenko, P.E. (2000) Prolactin-releasing peptide is expressed in afferents to the endocrine hypothalamus, but not in neurosecretory neurons. *J Neuroendocrinol* 12, 131–140.

Morel, G., Enjalbert, A., Proulx, L., Pelletier, G., Barden, N., Grossard, F., and Dubois, P.M. (1989) Effect of corticotropin-releasing factor on the release and synthesis of prolactin. *Neuroendocrinology* 49, 669–675.

Morin, A.J., Denoroy, L., and Jouvet, M. (1992) Effect of paradoxical sleep deprivation on vasoactive intestinal peptide-like immunoreactivity in discrete brain areas and anterior pituitary of the rat. *Brain Res Bull* 28, 655–661.

Morin, A.J., Denoroy, L., and Jouvet, M. (1993) Daily variations in concentration of vasoactive intestinal peptide immunoreactivity in hypothalamic nuclei of rats rendered diurnal by restricted-schedule feeding. *Neurosci Lett* 152, 121–124.

Nicoll, C.S. (1980) Ontogeny and evolution of prolactin's functions. *Fed Proc* 39, 2563–2566.

Obál, F., Jr. (1986) Effects of peptides (DSIP, DSIP analogues, VIP, GRF and CCK) on sleep in the rat. *Clin Neuropharmacol* 4, 459–461.

Obál, F., Jr., Opp, M., Cady, A.B., Johannsen, L., and Krueger, J.M. (1989) Prolactin, vasoactive intestinal peptide, and peptide histidine methionine elicit selective increases in REM sleep in rabbits. *Brain Res* 490, 292–300.

Obál, F., Jr., Payne, L., Kacsoh, B., Opp, M., Kapas, L., Grosvenor, C.E., and Krueger, J.M. (1994) Involvement of prolactin in the REM sleep-promoting activity of systemic vasoactive intestinal peptide (VIP). *Brain Res* 645, 143–149.

Obál, F. Jr., Sary, G., Alfoldi, P., Rubicsek, G., and Obal, F. (1986) Vasoactive intestinal polypeptide promotes sleep without effects on brain temperature in rats at night. *Neurosci Lett* 64, 236–240.

Obál, F., Jr., Floyd, R., Kapas, L., Bodosi, B., and Krueger, J.M. (1996) Effects of systemic GHRH on sleep in intact and hypophysectomized rats. *Am J Physiol* 270, E230–E237.

Obál, F., Jr., Garcia-Garcia, F., Kacsoh, B., Taishi, P., Bohnet, S., Horseman, N.D., and Krueger, J.M. (2005) Rapid eye movement sleep is reduced in prolactin-deficient mice. *J Neurosci* 25, 10282–10289.

Obál, F., Jr., Kacsoh, B., Alfoldi, P., Payne, L., Markovic, O., Grosvenor, C., and Krueger, J.M. (1992) Antiserum to prolactin decreases rapid eye movement sleep (REM sleep) in the male rat. *Physiol Behav* 52, 1063–1068.

Obál, F., Jr., Kacsoh, B., Bredow, S., Guha-Thakurta, N., and Krueger, J.M. (1997) Sleep in rats rendered chronically hyperprolactinemic with anterior pituitary grafts. *Brain Res* 755, 130–136.

Ogawa, T., Kiryu-Seo, S., Tanaka, M., Konishi, H., Iwata, N., Saido, T., Watanabe, Y., and Kiyama, H. (2005) Altered expression of neprilysin family members in the pituitary gland of sleep-disturbed rats, an animal model of severe fatigue. *J Neurochem* 95, 1156–1166.

Owerbach, D., Rutter, W.J., Cooke, N.E., Martial, J.A., and Shows, T.B. (1981) The prolactin gene is located on chromosome 6 in humans. *Science* 212, 815–816.

Pacheco-Cano, M.T., García-Hernandez, F., Prospero-García, O., and Drucker-Colín, R. (1990) Vasoactive intestinal polypeptide induces REM recovery in insomniac forebrain lesioned cats. *Sleep* 13, 297–303.

Palermo, R., Albano, C., Mangione, D., and Napoli, P. (1994) Chronic anovulation due to prolactin hypersecretion *Acta Eur Fertil* 25, 161–172.

Palma, B.D., Suchecki, D., and Tufik, S. (2000) Differential effects of acute cold and footshock on the sleep of rats. *Brain Res* 861, 97–104.

Paut-Pagano, L., Roky, R., Valatx, J.L., Kitahama, K., and Jouvet, M. (1993) Anatomical distribution of prolactin-like immunoreactivity in the rat brain. *Neuroendocrinology* 58, 682–695.

Piccoletti, R., Marioni, P., Bendinelli, P., and Bernelli-Zazzera, A. (1994) Rapid stimulation of mitogen-activated protein kinase of rat liver by prolactin. *Biochem J* 303, 429–433.

Pohnke, Y., Kempf, R., and Gellersen, B. (1999) CCAAT/enhancer-binding proteins are mediators in the protein kinase A-dependent activation of the decidual prolactin promoter. *J Biol Chem* 274, 24808–24818.

Postel-Vinay, M.C., Belair, L., Kayser, C., Kelly, P.A., Djiane, J. (1991) Identification of prolactin and growth hormone binding proteins in rabbit milk. *Proc Natl Acad Sci.* 88(15): 6687–6690.

Prospero-García, O., Jimenez-Anguiano, A., and Drucker-Colín, R. (1993) The combination of VIP and atropine induces REM sleep in cats rendered insomniac by PCPA. 18. *Neuropsychopharmacology* 8, 387–390.

Prospero-García, O., Morales, M., Arankowsky-Sandoval, G., and Drucker-Colín, R. (1986) Vasoactive intestinal polypeptide (VIP) and cerebrospinal fluid (CSF) of sleep-deprived cats restores REM sleep in insomniac recipients. *Brain Res* 385, 169–173.

Rampin, C., Cespuglio, R., Chastrette, N., and Jouvet, M. (1991) Immobilisation stress induces a paradoxical sleep rebound in rat. *Neurosci Lett* 126, 113–118.

Reem, G.H., Ray, D.W., and Davis, J.R. (1999) The human prolactin gene upstream promoter is regulated in lymphoid cells by activators of T-cells and by camp. *J Mol Endocrinol* 22, 285–292.

Riou, F., Cespuglio, R., and Jouvet, M. (1981) Hypnogenic properties of the vasoactive intestinal polypeptide in rats. *C R Seances Acad Sci III* 293, 679–682.

Roky, R., Obál, F., Jr., Valatx, J.L., Bredow, S., Fang, J., Pagano, L.P., and Krueger, J.M. (1995) Prolactin and rapid eye movement sleep regulation. *Sleep* 18, 536–542.

Roky, R., Paut-Pagano, L., Goffin, V., Kitahama, K., Valatx, J.L., Kelly, P.A., and Jouvet, M. (1996) Distribution of prolactin receptors in the rat forebrain. Immunohistochemical study. *Neuroendocrinology* 63, 422–429.

Roky, R., Valatx, J.L., and Jouvet, M. (1993) Effect of prolactin on the sleep-wake cycle in the rat. *Neurosci Lett* 156, 117–120.

Roky, R., Valatx, J.L., Paut-Pagano, L., and Jouvet, M. (1994) Hypothalamic injection of prolactin or its antibody alters the rat sleep-wake cycle. *Physiol Behav* 55, 1015–1019.

Said, S.I., and Mutt, V. (1970) Polypeptide with broad biological activity: Isolation from small intestine. *Science* 169, 1217–1218.

Said, S.I., and Mutt, V. (1988) Vasoactive intestinal peptide and related peptides. *Ann N Y Acad Sci* 527, 1–691.

Sanford, L.D., Nassar, P., Ross, R.J., Schulkin, J., and Morrison, A.R. (1998) Prolactin microinjections into the amygdalar central nucleus lead to decreased NREM sleep. *Sleep Res Online* 1, 109–113.

Sassin, J.A., Frantz, S., Kapen, S., and Weitzman, E. (1973) The nocturnal rise of human prolactin is dependent on sleep. *J Clin Endrocrinol Metab* 37, 436–440.

Sassin, J.A., Frantz, S., Weitzman, E., and Kapen, S. (1972) Human prolactin: 24-hour patterns with increased release during sleep. *Science* 177, 1205–1207.

Schilling, J., and Nurnberger, F. (1998) Dynamic changes in the immunoreactivity of neuropeptide systems of the suprachiasmatic nuclei in golden hamsters during the sleep-wake cycle. *Cell Tissue Res* 294, 233–241.

Siaud, P., Manzoni, O., Balmefrezol, M., Barbanel, G., Assenmacher, I., and Alonso, G. (1989) The organization of prolactin-like-immunoreactive neurons in the rat central nervous system. Light- and electron-microscopic immunocytochemical studies. *Cell Tissue Res* 255, 107–115.

Spiegel, K., Follenius, M., Simon, C., Saini, J., Ehrhart, J., and Brandenberger, G. (1994) Prolactin secretion and sleep. *Sleep* 17, 20–27.

Spiegel, K., Luthringer, R., Follenius, M., Schaltenbrand, N., Macher, J.P., Muzet, A., and Brandenberger, G. (1995) Temporal relationship between prolactin secretion and slow-wave electroencephalic activity during sleep. *Sleep* 18, 543–548.

Steiger, A. (2003) Sleep and endocrinology. *J Int Med* 254, 13–22.

Symes, A., Gearan, T., Eby, J., and Fink, J.S. (1997) Integration of Jak-Stat and AP-1 signaling pathways at the vasoactive intestinal peptide cytokine response element regulates ciliary neurotrophic factor-dependent transcription. *J Biol Chem* 272, 9648–9654.

Takahashi, K., Koyama, Y., Kayama, Y., and Yamamoto, M. (2000) The effects of prolactin on the mesopontine tegmental neurons. *Psychiatry Clin Neurosci* 54, 257–278.

Thorner, M.O., Vance, M.L., Laws, E.R., Jr., Horvath, E., and Kovacs, K. (1998) The anterior pituitary. *Williams Textbook of Endocrinology, 9th Edn.* W.B. Saunders, Philadelphia, USA.

Tobler, I., Jaggi, K., and Borbely, A.A. (1994) Effects of melatonin and the melatonin receptor agonist S-20098 on the vigilance states, EEG spectra, and cortical temperature in the rat. *J Pineal Res* 16, 26–32.

Truong, A.T., Duez, C., Belayew, A., Renard, A., Pictet, R., Bell, G.I., and Martial, J.A. (1984) Isolation and characterization of the human prolactin gene. *EMBO J* 3, 429–437.

Usdin, T.B., Bonner, T.I., and Mezey, E. (1994) Two receptors for vasoactive intestinal polypeptide with similar specificity and complementary distributions. *Endocrinology* 135, 2662–2680.

Valatx, J.L., and Jouvet, M. (1988) Circadian rhythms of slow-wave sleep and paradoxical sleep are in opposite phase in genetically hypoprolactinemic rats. *C R Acad Sci III* 307, 789–794.

Valatx, J.L., Chouvet, G., and Jouvet, M. (1975) Sleep-waking cycle of the hypophysectomized rat. *Prog Brain Res* 42, 115–120.

Van Cauter, E., L'Hermite, M., Copinschi, G., Refetoff, S., Desir, D., and Robyn, C. (1981) Quantitative analysis of spontaneous variations of plasma prolactin in normal man. *Am J Physiol* 241, E355–E363.

Wang, Y.F., and Walker, A.M. (1993) Dephosphorylation of standard prolactin produces a more biologically active molecule: Evidence for antagonism between nonphosphorylated and phosphorylated prolactin in the stimulation of Nb2 cell proliferation. *Endocrinology* 133, 2156–2160.

Wei, Y., and Mojsov, S. (1996) Tissue specific expression of different human receptor types for pituitary adenylate cyclase activating polypeptide and vasoactive intestinal polypeptide: Implications for their role in human physiology. *J Neuroendocrinol* 8, 811–817.

Wennbo, H., Gebre-Medhin, M., Gritli-Linde, A., Ohlsson, C., Isaksson, O.G., and Tornell, J. (1997) Activation of the prolactin receptor but not the growth hormone receptor is important for induction of mammary tumors in transgenic mice. *J Clin Invest* 100, 2744–2751.

Wennbo, H., Kindblom, J., Isaksson, O.G., and Tornell, J. (1997) Transgenic mice overexpressing the prolactin gene develop dramatic enlargement of the prostate gland. *Endocrinology* 138, 4410–4415.

Whittaker, V.P. (1989) Vasoactive intestinal polypeptide (VIP) as a cholinergic co-transmitter: Some recent results. *Cell Biol Int Rep* 13, 1039–1051.

Yamagami, T., Ohsawa, K., Nishizawa, M., Inoue, C., Gotoh, E., Yanaihara, N., Yamamoto, H., and Okamoto, H. (1988) Complete nucleotide sequence of human vasoactive intestinal peptide/PHM-27 gene and its inducible promoter. *Ann N Y Acad Sci* 527, 87–102.

Zhang, S.Q., Inoue, S., and Kimura, M. (2001) Sleep-promoting activity of prolactin-releasing peptide (PrRP) in the rat. *Neuroreport* 12, 3173–3176.

Zhang, S.Q., Kimura, M., and Inoue, S. (1999) Effects of prolactin on sleep in cyclic rats. Psychiatry. *Clin Neurosci* 53, 101–103.

Zhang, S.Q., Kimura, M., and Inoue, S. (2000) Effects of prolactin-releasing peptide (PrRP) on sleep regulation in rats. *Psychiatry Clin Neurosci* 54, 262–264.

6 Immune Signaling to Brain: Mechanisms and Potential Pathways Influencing Sleep

Lisa Goehler and Ronald Gaykema

6.1 Introduction

Immune activation modulates sleep by influencing brain neurons involved in the regulation and maintenance of sleep–wake cycles and arousal states. The precise mechanisms by which this occurs are not completely established, but seem to follow from direct influences of immune-derived mediators, such as cytokines, on sleep/wake neurons (Opp 2005), and/or via activation of peripheral viscerosensory neurons driving brain neural circuits that modulate arousal states in response to internal and external challenges. In this chapter we will review the organization of visceral sensory pathways that carry immune-related information to the brain, and recent findings regarding the brain neurocircuitry connecting immune–brain interfaces with arousal network nuclei that influence sleep–wake states.

6.2 Mechanisms of Immune Signaling to the Brain

Signals generated in response to immune challenge, like other viscerosensory stimuli, take multiple pathways to the brain. These pathways can be broadly subdivided into two categories: neural pathways in which immune-derived signals, such as cytokines interact with peripheral nerves (Goehler, Gaykema, Hansen, Maier, and Watkins 2000), and endocrine-like signals, in which the immune- or pathogen-derived signals circulate in the blood to reach specialized immune-sensitive regions of the brain directly (Banks 2005).

6.2.1 Neural Pathways

Peripheral sensory neurons that respond to inflammation or immune activation are predominantly unmyelinated, small diameter Aδ and C fibers. Functionally these immune-responsive fibers can be divided into a group of viscerosensory neurons, associated primarily with the glossopharyngeal and vagus cranial nerves, and nociceptive C fibers associated with somatic and visceral sensory spinal nerves and the trigeminal cranial nerve (Kobierski, Srivastava, and Borsook 2000; Scafers, Svensson, Sommer, and Sorkin 2003; Zhang, Li, Liu, and Brull 2002). The glossopharyngeal and vagus nerves innervate most of the alimentary canal, as well as many other important visceral tissues including lungs and lymph nodes. These tissues are notable as major points of entry for diverse pathogens, and the immune responses nerves that innervate them function to initiate physiological and behavioral responses to homeostatic (visceral) challenges, such as infection. In contrast, activation of somatic nociceptors are involved in exaggerated pain states following inflammation in trigeminal terminal fields such as the eyes, nose, and mouth, and inflammatory mediators produced in damaged or inflamed spinal nerves modulate spinal mechanisms of pain transmission. Immune-responsive somatic nociceptors can then influence arousal states via modulation of central (e.g., spinothalamic) pain pathways.

The glossopharyngeal nerve (the ninth cranial nerve) innervates the posterior two-thirds of the tongue as well as other posterior oral structures. Specialized immune structures including the tonsils are located in this region, thus the glossopharyngeal nerve is well positioned for a role in immunosensory surveillance. In support of this idea, application of either lipopolysaccharide (LPS), an immune stimulant derived from bacterial cell walls, or interleukin-1 (IL-1), a proinflammatory cytokine, to the soft palate (receptive field of the glossopharyngeal nerve) induces a fever that can be blocked by prior section of the glossopharyngeal nerve (Romeo, Tio, Rahman, Chiappelli, and Taylor 2001). In addition to innervating the oral cavity, sensory fibers of the glossopharyngeal nerve innervate the carotid bodies, which consist of a very large collection of chemosensory glomus cells that are sensitive to blood gasses and likely other chemical stimuli in the general circulation. The carotid bodies express IL-1 receptor type 1 immunoreactivity (Wang et al. 2002), indicating that in addition to monitoring stimuli relevant to respiratory reflexes, these structures may also participate in signaling systemic immune-related signals.

The vagus nerve (the tenth cranial nerve), like the glossopharyngeal nerve, is well positioned to interact with pathogen products and immune-derived mediators. *Vagus* means wanderer and this nerve innervates every internal structure, from the larynx to the colon except the spleen. Internal tissues commonly in contact with pathogens, notably the lungs, gastrointestinal tract, and liver are richly supplied with vagal afferents capable of signaling immune activation in these tissues (Berthoud and Neuhuber 2000). The cell bodies of vagal sensory neurons occupy two ganglia, the nodose (or inferior vagal), and jugular (or superior vagal) ganglia, which lie just outside the caudal cranium. These two ganglia form a complex with the petrosal ganglion, which contains the cell bodies of sensory neurons contributing to the glossopharyngeal nerve. The central projections of these neurons terminate in the dorsal vagal complex of the caudal brainstem (see below). In this way, the vagus nerve is positioned to detect immune signals generated in response to local infection or inflammation in tissues commonly in contact with pathogens.

If vagal sensory nerve fibers function as sentinels for the early activation of brain-mediated host defense responses, then one would expect these fibers to innervate immune tissues, such as spleen and lymph nodes that process early signals from pathogens. Surprisingly, the spleen, which is located in the abdomen, and might be expected to receive vagal innervation, clearly does not (Cano, Sved, Rinaman, Rabin, and Card 2001; Nance and Burns 1989). This may be related to the fact that the spleen is not quite on the first line of defense, but rather operates as a filter for circulating immune and pathogen components. Lymph nodes, however, provide a site of early immune activation, as these are the major locations in which antigen presenting cells interface with the T cells that serve to coordinate immune responses. The lymphatic system comprises an interconnecting network of conducting vessels that carry immune cells and antigens, including microorganisms, from lymph node to node, progressively to the heart, and thereafter into the general circulation. Lymph nodes are innervated by both sympathetic and sensory neuropeptide containing nerve fibers (Felton, Livnat, Felton, Carlson, Bellinger, and Yeh 1984; Popper, Mantyh, Vigna, Maggio, and Mantyh 1988), as well as vagal neurons in the nodose and jugular ganglia (Goehler et al. 2000). These observations underline the idea that a major role for vagal afferents involves monitoring early stage immune activation.

Although the lymphatic system is an important early player in host defense, inhaled or ingested pathogens first interact with epithelial barrier tissue in the lung and gut. The gastrointestinal tract contains abundant immune type tissue and cells including specialized immune tissue, such as lymphoid nodules (which are organized somewhat like lymph nodes) and Peyers patches of the small intestine, which reside directly beneath the epithelium. In addition, macrophages and dendritic cells line the epithelium, and overlie the Peyers patches (Nagura, Ohtani, Masuda, Kimura, and Nakamura 1991). Vagal sensory fibers were found in close association with mast cells (Berthoud and Neuhuber 2000) as well, indicating that vagal sensory neurons occupy a position in which they can respond to proinflammatory mediators produced in response to local infections. This arrangement allows vagal sensory fibers to be rapidly activated in response to pathogens in the gastrointestinal system.

In addition to activating vagal afferents directly, immune-related signals may activate vagal immunosensitive pathways via the chemoreceptive cells located in the vagal paraganglia (see below), and/or similar, vagally innervated structures, the neuroepithelial bodies, which are found in lung airways (Adriaensen, Timmermans, Brouns, Berthoud, Neuhuber, and Scheuermann 1998). The vagal paraganglia are collections of glomus cells interspersed throughout the vagus nerve, which are innervated by vagal sensory neurons (Berthoud, Kressel, and Neuhuber 1995). Vagal paraganglia are penetrated by blood and lymph vessels, suggesting that these structures are likely monitoring substances circulating in body fluids. Glomus cells of vagal paraganglia, like those of the carotid bodies, express IL-1 receptors (Goehler et al. 1997; Wang, Wang, Duan, Liu, and Ju 2000). Immune cells expressing LPS-induced IL-1 immunoreactivity codistribute with these glomus cells expressing IL-1 receptors (Goehler et al. 1999), providing an alternative arrangement whereby vagal sensory nerves may monitor immune-related stimuli circulating in either blood or lymph.

Experimental evidence supporting a functional role for the vagus in immunosensory signaling initially relied on studies that involved cutting the vagus nerve in the abdomen, below the diaphragm (subdiaphragmatic vagotomy). Unfortunately, this is a partial lesion, leaving thoracic structures, notably lung and lymph nodes with intact vagal innervation. However, sectioning the vagus above these structures is only feasible in anesthetized terminal preparations. Consequently, results from vagotomy studies need to be interpreted carefully. Findings from these studies have shown that vagotomy can block or attenuate a wide range of illness responses (Goehler et al. 2000 for review), including hypersomnolence (Hansen and Krueger 1997; Opp and Toth 1998). In general, the effects of vagotomy are most pronounced when the immune stimulus is presented to peritoneal cavity (Bluthe, Michaud, Kelly, and Dantzer, 1996), and when the dose of stimulant is low (Hansen, O'Conner, Goehler, Watkins, and Maier 2001). These findings suggest that the vagus nerve may contribute to the signaling of immune activity locally in visceral tissues, and that higher doses of immune stimulants such as cytokines recruit additional immunosensory pathways associated with the brain, e.g., brain barrier tissues (see below). Additionally, or alternatively, vagal sensory nerves left intact (innervating thoracic structures including lung and lymph nodes) may contribute to immune-to-brain signaling following subdiaphragmatic vagotomy.

Further evidence for a vagal immune to brain signaling pathways is provided by findings that vagal sensory neurons respond to injections of cytokines such IL-1 (Ek, Kurosawa, Lundeberg, and Ericsson 1998; Goehler, Gaykema, Hammack, Maier, and Watkins 1998; Niijima 1996), as well as LPS (Gaykema et al. 1998), staphylococcus enterotoxin B (SEB), a product of gram positive bacteria (Goehler et al. 2000), and live bacteria in the lower gut (Goehler, Gaykema, Opitz, Reddaway, Badr, and Lyte 2005). Interestingly, cells in the nodose ganglia express mRNA and protein corresponding to the TOLL-like receptor 4 (TLR4), a pathogen receptor sensitive to bacterial LPS (Hosoi, Okuma, Matsuda, and Nomura 2005). From the study it was not possible to know the cell type expressing TLR4, which could be satellite cells, resident immune cells, or vagal sensory neurons. Nonetheless, taken together these studies indicate that vagal sensory neurons are likely to be sensitive to a wide variety of pathogenic or immune-derived stimuli.

6.2.2 Endocrine-Like Pathways: Immune Sensing Within the Brain

Although the brain is protected by the blood–brain barrier from diffusion of large lipophobic molecules, such as cytokines, several mechanisms exist that allow circulating cytokines to interact with specialized brain tissue and thereby signal the brain regarding the status of peripheral immune activation. Because cytokines are relatively large, lipophobic proteins, they cannot diffuse passively into the brain. Consequently, most of the pathways identified to date that support direct action of peripherally generated cytokines on the brain involve interactions of cytokines with receptors on cells outside of the blood–brain barrier. Such brain-associated immunosensory tissues include meninges, choroid plexus and endothelium, associated with the blood–brain barrier, as well as specialized brain regions called circumventricular organs (CVOs) and the ventricular ependyma.

The meninges form the outer coverings of the brain, as well as provide the outer structure of sinuses and the fourth ventricle. Meninges contain immune cells, primarily dendritic-like cells and macrophages (McMenamin 1999). These cells can respond directly to pathogens, via their expression of receptors for pathogen products or structural components, or indirectly via their expression of receptors for cytokines. Meningeal cells respond to both central and peripheral immune activation with the expression of cytokines, notably IL-1 (Vernet-der Garabedian, Lemaigre-Dubreuil, and Mariani 2000). The choroid plexus is best known for its role in the production of CSF, but it also contains abundant immune cells. Like those within the meninges, these cells are primarily dendritic cells and macrophages (McMenamin 1999), and express both cytokines and cytokine receptors. Choroid plexus is distributed throughout the ventricular system, and with the meninges, likely provides a significant contribution to the production of cytokines that may access brain regions involved in the regulation of sleep via circulation in the cerebrospinal fluid, via volume transmission in the brain (Konsman, Tridon, and Dantzer 2000) or via accessing the CVOs (see below).

The ventricular ependyma consists of cells lining the walls of the ventricular system. As such they are in contact with substances, including cytokines, present within the cerebrospinal fluid throughout the brain and spinal cord. The cells possess cilia, a typical characteristic of sensory cells (e.g., hair cells, photoreceptors, taste cells and olfactory receptors). They express IL-1 receptors (Ericsson, Liu, Hart, and Sawchenko 1995), and respond to intracerebroventricular injection of LPS or IL-1 with the expression of c-Fos protein (Goehler 2003), an activation marker for many cells. These observations are consistent some kind of immunosensory role for these cells. The mechanisms by which they interface with neural, or glial, elements are unknown, but may involve direct interactions with hypothalamic neurons (Carpenter and Sutin 1983).

Endothelium consists of the luminal lining of cells within blood vessels, rendering them in contact with virtually every circulating substance, including those derived from immune activation such as cytokines. Endothelial cells express a diverse array of substances and receptors, including receptors for IL-1, IL-6, and interferons, which have been implicated in modulation of sleep (Opp 2005) and they also express enzymes necessary for producing prostaglandins. Prostaglandin E_2 (PGE_2) is strongly implicated in the mediation of several illness responses, including fever and activation of the HPA axis, based on the observations that anti-inflammatory drugs that prevent the synthesis of prostaglandins prevent these responses to systemic cytokine treatment. Compelling evidence indicates that PGE_2 produced in endothelial cells contributes to actions of circulating IL-1 (Ericsson, Arias, and Sawchenko 1997). For instance, IL-1 receptors are codistributed in endothelial cells with synthetic enzymes for PGE_2 and this codistribution occurs in areas of the brain in which nearby neurons respond to IL-1 (as assessed by induction of activation markers), express PGE_2 receptors (EP3), and have been previously implicated in the mediation of illness responses. These regions include the area postrema, nucleus of the solitary tract, and ventrolateral medulla of the caudal brainstem, as well as midline thalamic nuclei and the preoptic area in the forebrain. Thus, immunosensory endothelial cells that express IL-1 receptors can release PGE_2 to into the adjacent brain parenchyma to activate neurons that mediate brain responses to immune activation. In this way, prostaglandins represent important diffusible substances that are released via the actions of cytokines circulating within the brain vasculature.

Circumventicular organs, the area postrema, organum vasculosum lamina terminalis (OVLT), subfornical organ (SFO), and median eminence, are situated along the brain ventricles (hence the name) at different locations throughout the neuraxis. The blood–brain barrier is weak within these structures, allowing access to circulating substances excluded from the brain parenchyma. Like the meninges and choroid plexus, CVOs contain numerous immune cells. However, in CVOs, immune cells intersperse among neurons. This arrangement allows for the possibility that immune cells in CVOs respond to circulating substances not accessible to neurons in the brain parenchyma, and signal adjacent neurons via a local, or paracrine kind of mechanism. In addition, IL-1-expressing dendritic like cells in the area postrema make direct contact with axons and dendrites of neurons in the area postrema, providing a direct pathway by which those neurons could activate brain neurocircuitry subserving illness responses (Goehler, Erisir, and Gaykema 2006). This type of interaction may allow for a more specific kind of activation, in that activated immune cells could interface with specific neurons within the CVO that project to and activate specific regions of the brain, allowing for high selectivity in illness responses.

Although cytokines cannot passively diffuse into the brain, an active transport mechanism for circulating cytokines carries them across the blood–brain barrier (Banks 2005). Thus, active transport constitutes an additional mechanism for signal transduction of circulating cytokines, and because its relevance may be selective for certain cytokines, such as IL-6, may provide one mechanism by which different cytokines exert effects on the brain.

6.3 Neurocircuitry Mediating Sleep and Waking States

Illness exerts a strong influence on sleep–waking cycles, with the most common result being an increase in non-REM (NREM), particularly slow wave sleep (SWS) (Opp 2005). In order to determine the mechanisms by which illness influences sleep, it is necessary to understand the brain substrates that control transitions between wakefulness and sleep. Recent evidence indicates that sleep wake cycles are controlled by a "switch" mechanism that is exerted by interactions of hypothalamic nuclei (reviewed in Saper, Scammell, and Lu 2005; see Fig. 6.1).

Histaminergic neurons of the tuberomammillary posterior hypothalamus serve an activating role in the brain, becoming functionally quiescent during sleep when behavioral activity is minimal, and are believed to be particularly important in regulating transitions between sleep and arousal (Saper et al. 2005). In contrast, a compact cluster of neurons in the ventrolateral preoptic nucleus (VLPO) promotes sleep, via inhibitory projections to the histaminergic, as well as all other monoaminergic nuclei contributing to arousal, such as the noradrenergic locus coeruleus and the serotonergic raphe nuclei.

The sleep-promoting neurons of the VLPO are under recurrent inhibitory influence by the monoaminergic arousal systems (Chou, Bjorkum, Gaus, Lu, Scammell, and Saper 2002; see Fig. 6.1). The arrangement of this circuitry indicates that any increased drive to sleep due to inflammation or illness would be executed by inhibition of critical monoaminergic arousal systems, including the histaminergic neurons, which as a consequence would release their inhibitory control over the sleep-active neurons within the VLPO.

6.4 What Are the Links Between Immune–Brain Interfaces and Sleep Neurocircuitry?

In order for peripherally generated immune signals to influence sleep, immuno-sensory interfaces must connect with the arousal-related neurocircuitry outlined above. The following describes immune-responsive brain regions that likely relay immune-related information to the brain regions that control arousal and the pathways by which they target these regions.

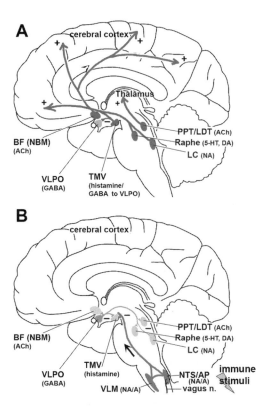

FIGURE 6.1. Schematic representations (modified after Saper et al. 2005) of key components of the "ascending reticular arousal system" mediating wakefulness and arousal states (panel A) and the inhibitory control on these components by the ventrolateral preoptic nucleus (VLPO) situated at the bottom of the anterior hypothalamus, as well as by ascending projections from the lower brainstem (panel B). The histaminergic neurons in the ventral tuberomammillary nucleus (TMV) at the bottom of the posterior hypothalamus provide a strong inhibitory influence on the VLPO (panel A) with the cotransmitter GABA, as the VLPO lacks histamine receptors. Activation of ascending projections from the lower brainstem, as indicated with the arrow in B, may represent a key mechanism for enhancement of sleep as a result of peripheral immune challenge. Inhibition of the histaminergic neurons in the TMV would subsequently disinhibit the VLPO, thus facilitating the transition from waking to sleep and suppressing the reverse switch. The components of the ascending reticular activating system further include the raphe nuclei (serotonergic and dopaminergic neurons), the locus coeruleus (LC, noradrenergic neurons), the pedunculopontine and laterodorsal tegmenti (PPT/LDT), and the basal forebrain (BF, cholinergic neurons), including the nucleus basalis of Meynert (NBM).

6.4.1 Primary Sensory Relay Nuclei

Sensory fibers associated with the vagal and glossopharyngeal ganglia collect signals from the tissues that they innervate, and via sensory neurotransmitter glutamate (Sykes, Spyer, and Izzo 1997), convey this information to brainstem dorsal vagal complex (DVC) (Beckstead and Norgren 1979; Chernicky, Barnes, Ferrario, and Conomy 1984) which consists of the nucleus of the solitary tract (NTS) and the area postrema (a circumventricular organ). Both of these brainstem structures express activation markers (e.g., c-Fos) following peripheral and central administration of a variety of immune stimulants (Elmquist, Scammell, Jacobson, and Saper 1996; Goehler et al. 2005; Wan, Janz, Vriend, Sorensen, Greenberg, and Nance 1993), and both IL-1 and LPS induce the release of glutamate into the DVC (Mascarucci, Perego, Terrazzino, and DeSimoni 1998). These nuclei coordinate local, protective reflexes, such as emesis and gastric retention, and relay immune-related viscerosensory signals to forebrain regions concerned with integration of visceral information with ongoing behavior and other sensory inputs.

In addition to neural input, immune-derived activation of dorsal vagal complex also occurs via humoral routes. The area postrema is a circumventricular organ, in which the blood–brain barrier is weak (hence it is sensitive to circulating signals unavailable to the brain parenchyma), and it contains immune cells that respond to peripherally administered LPS by expressing IL-1 immunoreactivity (Goehler et al. 2006). In addition, it has been suggested that the NTS may respond to cytokine signals directly (Banks 2005). Thus, the dorsal vagal complex is situated to function as a crossroads for converging immune-related signaling.

A functional role of the DVC in brain responses to illness is supported by findings that lesions of the area postrema can attenuate HPA axis responses or hypothalamic norepinephrine increase to systemic immune activation (Ishizuka, Ishida, Kunitake, Kato, Hanamori, Mitsuyama, and Kannan 1997; Lee, Whiteside, and Herkenham 1998), and that inactivation of the DVC, using a local anesthetic, blocks sickness-induced behavioral depression, psychomotor retardation and c-Fos expression in the brain (Marvel, Chen, Badr, Gaykema, and Goehler 2004). A role for the NTS and area postrema in the regulation of sleep is supported by findings that stimulation of the caudal NTS synchronizes EEG (Golanov and Reis 2001), and that activity in the area postrema corresponds to that in the cortex during SWS (Bronzino, Stern, Leahy, and Morgane 1976).

Whereas the DVC is uniquely situated to transduce both neural and humoral signals derived from immune activation, to convey these signals to forebrain regions that regulate arousal, other brain regions likely also play a role. However, the neural pathways by which other immunosensory interfaces modulate arousal or sleep are less well documented. For example, neurons residing in forebrain CVOs, such as the OVLT or SFO may project to hypothalamic regions that control sleep–waking states, and thus may contribute to immune-related influences on them.

In addition, or alternatively, sleep-modulating cytokines could be produced in the brain (by local microglia) or transported across the blood–brain barrier (Banks 2005). It should be noted however, that local production of cytokines is unlikely to play a role in the induction of somnolence and NREM/SWS states, based on the time course of cytokine induction in the brain parenchyma (Quan, Whiteside and

Herkenham 1998), which lags the induction of somnolescence and sleep, but may play a role in the maintenance of sleep states.

6.4.2 Ascending Pathways Influencing Arousal

Ascending pathways arising from the dorsal vagal complex and VLM, which are primarily catecholaminergic (Riche, De Pommery, and Menetrey 1990), respond to peripheral immune stimulation (Buller, Xu, Dayas, and Day 2001). The association of increased catecholamine release in the hypothalamic and preoptic regions with immune activation is well established.

Lower brainstem projection neurons provide a significant source of these catecholamines, based on findings that noradrenergic and adrenergic neurons within the nTS and ventrolateral medulla (VLM) target the hypothalamus and respond to IL-1 (Buller et al. 2001), and that ascending catecholaminergic projections originating in the nTS and VLM appear crucial in the initiation of HPA axis responses to immune challenges (Ericsson, Kovacs and Sawchenko 1994). These findings support an important role for brainstem-derived catecholamines as molecules mediating behavioral symptoms of illness, although a role for cotransmitter molecules (such as neuropeptide Y and ATP) cannot be ruled out.

Given the important role of tuberomammillary histamine neurons in arousal associated with behavior, and in the transitions between arousal and sleep, it was important to determine whether these histamine neurons are influenced by immune challenge, and whether caudal medullary projection neurons target these neurons. To address this issue, we recently demonstrated that activation of histaminergic neurons concomitant with behaviors, such as exploration of a novel environment or social interaction, is inhibited by the immune stimulant LPS (Fig. 6.2A and B). Moreover, these tuberomammillary neurons are targeted by LPS-responsive catecholamine neurons in the VLM and NTS (Gaykema et al. submitted; see Fig. 6.2C to E).

Thus, during illness, caudal brainstem projection neurons may act to inhibit tuberomamillary neurons that normally serve as a brake on sleep-inducing VLPO neurons. This disinhibition would allow VLPO neurons to inhibit neural systems that maintain wakefulness, resulting in somnolence and increased SWS.

6.5 Conclusions and Perspectives

Mediators expressed in immune cells in the periphery or in the area postrema activate immunosensory neurons in brainstem autonomic relay nuclei. Catecholaminergic projection neurons activate forebrain autonomic network nuclei that mediate host defense responses and which may also act to suppress the activity of key forebrain regions that mediate arousal, notably the histaminergic neurons of the tuberomammillary nuclei. Alternatively, brainstem projection neurons may directly inhibit some of these regions, or may act via yet unidentified relay nuclei. In this way, immune-driven brainstem neurons, through catecholaminergic ascending projections, impinge upon and modulate the neurocircuitry that mediates arousal. This system is organized such that early local signals can be conveyed from immune

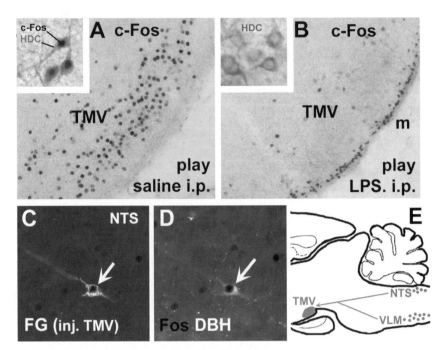

FIGURE 6.2. (A and B) The strong activation of rat histaminergic system is during arousal and behavioral activities are blocked by proinflammatory challenge with lipopolysaccharide (LPS). Active behavior of young adult rats in a playful interaction coincides with strong activation in histaminergic neurons in the ventral tuberomammillary nucleus (TMV) as shown with staining for c-Fos immunoreactivity (A), which is blocked in an LPS-treated rat introduced in the same play arena (B). Inserts show colocalization of nuclear c-Fos staining in neurons immunoreactive for the histamine-synthesizing enzyme histidine decarboxylase (HDC) in saline- (insert A), but not LPS-treated rats (insert B), which treatment lead to meningeal activation (m). (C and D) Application of the retrograde tracer Fluorogold (FG) into the TMV labels neurons in the nucleus of the solitary tract (NTS) and the ventrolateral medulla that project to the TMV (in C), are (nor-)adrenergic, as they stain positively for dopamine beta-hydroxylase (DBH), and are activated (nuclear c-Fos staining) by LPS challenge (in D). Panel E provides a schematic representation of the TMV and lower brainstem afferents in the rat brain.

tissue in contact with invading pathogens via peripheral nerves to brain regions that modulate arousal. In addition systemic, humoral signals can influence the brain directly. This arrangement can allow for appropriate responses to a variety of immune challenges, including serious illness, as well as for subtle, local perturbations to affect ongoing behavior.

Acknowledgment. This work was supported by National Institutes of Health grant MH068834.

References

Adriaensen, D., Timmermans, J.-P., Brouns, I., Berthoud H.-R., Neuhuber, W.L., and Scheuermann D.W. (1998) Pulmonary intraepithelial vagal nodose afferent nerve terminals are confined to neuroepithelial bodies: An anterograde tracing and confocal microscopy study in adult rats. *Cell Tiss Res* 293, 395–405.

Banks, W.A. (2005) Blood–brain barrier transport of cytokines: A mechanism for neuropathology. *Curr Pharm Des* 11, 973–84.

Beckstead, R.M., and Norgren, R. (1979) An autoradiographic examination of the central distribution of the trigeminal, facial, glossopharyngeal, and vagal nerves in the monkey. *J Comp Neurol* 184, 455–472.

Berthoud, H.-R., Kressel, M., and Neuhuber, W.L. (1995) Vagal afferent innervation of the rat abdominal paraganglia as revealed by anterograde DiI-tracing and confocal microscopy. *Acta Anat* 152, 127–132.

Berthoud, H.-R., and Neuhuber, W.L. (2000) Functional and chemical anatomy of the afferent vagal system. *Auton Neurosci* 85, 1–17.

Bluthe, R.-M., Michaud, B., Kelly, K.W., and Dantzer, R. (1996) Vagotomy blocks behavioral effects of interleukin-1 injected via the intraperitoneal route but not by other systemic routes. *Neuroreport* 7, 2823–2827.

Bronzino, J.D., Stern, W.C., Leahy, J.P., and Morgane, P.J. (1976) Power spectral analysis of EEG activity obtained from cortical and subcortical sites during vigilance states in the cat. *Brain Res Bull* 1, 285–294.

Buller, K., Xu, Y., Dayas, C., and Day, T. (2001) Dorsal and ventral medullary catecholamine cell groups contribute differentially to systemic interleukin-1beta-induced hypothalamic pituitary adrenal axis responses. *Neuroendocrinology* 73, 129–138.

Cano, G., Sved, A.F., Rinaman, L., Rabin, B.S., and Card, J.P. (2001) Characterization of the central nervous system innervation of the rat spleen using viral transneuronal tracing. *J Comp Neurol* 439, 1–18.

Carpenter, M.B., and Sutin J (1983) *Human Neuroanatomy*. 8th Edn. Williams &Wilkins, Baltimore.

Chernicky, C.L., Barnes, K.L., Ferrario, C.M., and Conomy, J.P. (1984) Afferent projections of the cervical vagus and nodose ganglion in the dog. *Brain Res Bull* 13, 401–411.

Chou, T.C., Bjorkum, A.A., Gaus, S.E., Lu, J., Scammell, T.E., and Saper, C.B. (2002) Afferents to the ventrolateral preoptic nucleus. *J Neurosci* 22, 977–990.

Ek, M., Kurosawa, M., Lundeberg, T., and Ericsson, A. (1998) Activation of vagal afferents after intravenous injection of interleukin-1β: Role of endogenous prostaglandins. *J Neurosci* 18, 9471–9479.

Elmquist, J.K., Scammell, T.E., Jacobson, C.D., and Saper, C.B. (1996) Distribution of Fos-like immunoreactivity in the rat brain following intravenous lipopolysaccharide administration. *J Comp Neurol* 371, 85–103.

Ericsson A., Arias, C., and Sawchenko, P.E. (1997) Evidence for an intramedullary prostaglandin-dependent mechanism in the activation of stress-related neuroendocrine circuitry by intravenous interleukin-1. *J Neurosci* 17, 7166–7179.

Ericsson, A., Kovacs, K.J., and Sawchenko, P.E. (1994) A functional anatomical analysis of central pathways subserving the effects of interleukin-1 on stress-related neuroendocrine neurons. *J Neurosci* 14, 897–913.

Ericsson, A., Liu, C., Hart, R.P., and Sawchenko, P.E. (1995) Type 1 interleukin-1 receptor in the rat brain: Distribution, regulation, and relationship to sites of IL-1-induced cellular activation. *J Comp Neurol* 361, 681–698.

Felton, D.L., Livnat, S., Felton, S.Y., Carlson, S.L., Bellinger, D.L., and Yeh, P. (1984) Sympathetic innervation of lymph nodes in mice. *Brain Res Bull* 13, 693–696.

Gaykema, R.P.A., Goehler, L.E., Tilders, F.J.H., Bol, J.G.M., McGorry, M.M., Maier, S.F., and Watkins, L.R. (1998) Bacterial endotoxin induces Fos immunoreactivity in primary afferent neurons of the vagus nerve. *Neuroimmunomodulation* 5, 234–240.

Goehler, L.E. (2003) Immunsensory signaling: Role of cytokines. In: Z. Kronfol (Ed.), *Cytokines and Mental Health.* Kluwer Academic, Boston, pp. 15–38.

Goehler, L.E., Erisir, A., and Gaykema, R.P.A. (2006) Neural-immune interface in the rat area postrema. *Neuroscience* 140, 1415–1434.

Goehler, L.E., Gaykema, R.P.A., Hammack, S.E., Maier, S.F., and Watkins L.R. (1998) Interleukin-1 induces c-Fos immunoreactivity in primary afferent neurons of the vagus nerve. *Brain Res* 804, 306–310.

Goehler, L.E., Gaykema, R.P.A., Nguyen, K.T., Lee, J.L, Tilders, F.J.H., Maier, S.F., and Watkins L.R. (1999) Interleukin-1β in immune cells of the abdominal vagus nerve: An immune to nervous system link? *J Neurosci* 17, 2799–2806.

Goehler, L.E., Gaykema, R.P.A., Hansen, M.K., Maier, S.F., and Watkins, L.R. (2000) Vagally mediated fever: A visceral chemosensory modality. *Auton Neurosci* 85, 49–59.

Goehler, L.E., Gaykema, R.P.A., Opitz, N., Reddaway, R., Badr, N.A., and Lyte, M. (2005) Activation in vagal afferents and central autonomic pathways: Early responses to intestinal infection with *Campylobacter jejuni. Brain Behav Immun* 19, 334–344.

Goehler, L.E., Relton, J.K., Dripps, D., Keichle, R., Tartaglia, N., Maier, S.F., and Watkins, L.R. (1997) Vagal paraganglia bind biotinylated interleukin-1 receptor antagonist: A possible mechanism for immune-to-brain communication. *Brain Res Bull* 43, 357–364.

Golanov, E.V., and Reis, D.J. (2001) Neurons of the nucleus of the solitary tract synchronize EEG and elevate cerebral blood flow via a novel medullary area. *Brain Res* 892, 1–12.

Hansen, M.K., and Krueger, J.M. (1997) Subdiaphragmatic vagotomy blocks the sleep- and fever-promoting effects of interleukin-1β. *Am J Physiol* 273, R1246–R1253.

Hansen, M.K., O'Conner, K.A., Goehler, L.E., Watkins, L.R., and Maier, S.F. (2001) The role of the vagus nerve in interleukin-1β-induced fever is dependent on dose. *Am J Physiol* 280, R929–R934.

Hosoi, T., Okuma, Y., Matsuda, T., and Nomura, Y. (2005) Novel pathway for LPS-induced afferent vagus nerve activation: Possible role of nodose ganglion. *Auton Neurosci* 120, 104–107.

Ishizuka, Y., Ishida, Y., Kunitake, T., Kato, K., Hanamori, T., Mitsuyama, Y., and Kannan, H. (1997) Effects of area postrema lesion and abdominal vagotomy on interleukin-1β-induced norepinephrine release in the hypothalamic paraventricular nucleus region in the rat. *Neurosci Lett* 223, 57–60.

Kobierski, L.A., Srivastava, S., and Borsook, D. (2000) Systemic lipopolysaccharide and interleukin-1β activate the interleukin 6: STAT intracellular signaling pathway in neurons of mouse trigeminal ganglion. *Neurosci Lett* 281, 61–64.

Konsman, J.P., Tridon, V., and Dantzer, R. (2000) Diffusion and action of intracerebroventricularly injected interleukin-1 in the CNS. *Neuroscience* 101, 957–967.

Lee, H.Y., Whiteside, M.B., and Herkenham M. (1998) Area postrema removal abolishes stimulatory effects of intravenous interleukin-1β on hypothalamic-pituitary-adrenal axis activity and c-fos mRNA in the hypothalamic paraventricular nucleus. *Brain Res Bull* 46, 495–503.

Marvel, F.A., Chen, C.-C., Badr, N.A., Gaykema, R.P.A., and Goehler L.E. (2004) Reversible inactivation of the dorsal vagal complex blocks lipopolysaccharide-induced social withdrawal and c-Fos expression in central autonomic nuclei. *Brain Behav Immun* 18, 123–143.

Mascarucci, P., Perego, C., Terrazzino, S., and DeSimoni, M.G. (1998) Glutamate release in the nucleus tractus solitarius induced by peripheral lipopolysaccharide and interleukin-1β. *Neuroscience* 86, 1285–1290.

McMenamin, P.G. (1999) Distribution and phenotype of dendritic cells and resident tissue macrophages in the dura mater, leptomeninges and choroid plexus of the rat brain as demonstrated in wholemount preparations. *J Comp Neurol* 405, 553–562.

Nagura, H., Ohtani, H., Masuda, T., Kimura, M., and Nakamura, S. (1991) HLA-DR expression on M cells overlaying Peyer's patches is a common feature of human small intestine. *Acta Pathol Jpn* 41, 818–823.

Nance, D.M., and Burns, J. (1989) Innervation of the spleen in the rat: Evidence for absence of afferent innervation. *Brain Behav Immun* 3, 281–290.

Niijima A. (1996) The afferent discharges from sensors for interleukin-1β in the hepatoportal system in the anesthetized rat. *J Auton Nerv Syst* 61, 287–291.

Opp, M.R. (2005) Cytokines and sleep. *Sleep Med Rev* 9, 355–364.

Opp, M.R., and Toth, L.A. (1998) Somnogenic and pyrogenic effects of interleukin-1 beta and lipopolysaccharide in intact and vagotomized rats. *Life Sci* 62, 923–936.

Popper, P., Mantyh, C.R., Vigna, S.R., Maggio, J.E., and Mantyh, P.W. (1988) The localization of sensory nerve fibers and receptor binding sites for sensory neuropeptides in canine lymph nodes. *Peptides* 9, 257–267.

Quan, N., Whiteside, M., and Herkenham, M. (1998) Time course and localization patterns of interleukin-1β messenger RNA expression in brain and pituitary after peripheral administration of lipopolysaccharide. *Neuroscience* 83, 281–293.

Riche, D., De Pommery, J., and Menetrey, D. (1990) Neuropeptides and catecholamines in efferent projections of the nuclei of the solitary tract in the rat. *J Comp Neurol* 293, 399–424.

Romeo, H., Tio, D.L., Rahman, S.U., Chiappelli, F., and Taylor, A.N. (2001) The glossopharyngeal nerve as a novel pathway in immune-to-brain communication: Relevance to neuroimmune surveillance of the oral cavity. *J Neuroimmunol* 115, 91–100.

Saper, C.B., Scammell, T.E., and Lu, J. (2005) Hypothalamic regulation of sleep and circadian rhythms. *Nature* 437, 1257–1263.

Scafers, M., Svensson, C., Sommer, C., and Sorkin, L.S. (2003) Tumor necrosis factor-a induces mechanical allodynia after spinal nerve ligation by activation of p38 MAPK in primary sensory neurons. *J Neurosci* 23, 2517–2521.

Sykes, R.M., Spyer, K.M., and Izzo, P.N. (1997) Demonstration of glutamate immunoreactivity in vagal sensory afferents in the nucleus tractus solitarius of the rat. *Brain Res* 762, 1–11.

Vernet-der Garabedian, B., Lemaigre-Dubreuil, Y., and Mariani, J. (2000) Central origin of IL-1β produced during peripheral inflammation: Role of meninges. *Mol Brain Res* 75, 259–263.

Wan, W., Janz, L., Vriend, C.Y., Sorensen, C.M., Greenberg, A.H., and Nance, D.M., (1993) Differential induction of c-Fos immunoreactivity in hypothalamus and brainstem nuclei following central and peripheral administration of endotoxin. *Brain Res Bull* 32, 581–587.

Wang, X., Wang, B.-R., Duan, W.-L., Liu, H.L., and Ju, G. (2000) The expression of IL-1 receptor type 1 in nodose ganglion and vagal paraganglia in the rat. *Chin J Neurosci* 16, 90–93.

Wang, X., Wang, B.-R., Duan, X.-L., Zhang, P., Ding, Y.-Q., Jia, Y., Jiao, X.-Y., and Ju, G. (2002) Strong expression of interleukin-1 receptor type 1 in the rat carotid body. *J Histochem Cytochem* 50, 1677–1684.

Zhang, J.-M., Li, H., Liu, B., and Brull, S.J. (2002) Acute topical application of tumor necrosis factor a evokes protein kinase A-dependent responses in rat sensory neurons. *J Neurophysiol* 88, 1387–1392.

7 Aging, Sleep, and Immunity

Laura-Yvette Gorczynski, Ender Terzioglu, Thierry Waelli, and R.M. Gorczynski

7.1 Introduction

It is a truism that sleep remains an important enigma in neurobiology and physiology. It is trite to acknowledge that sleep clearly has an important adaptive value to the organism as a whole, yet defining that function remains elusive (Krueger, Majde, and Obal 2003; Siegel 2003). One prominent thought has been that sleep promotes recovery from infectious challenge. Consistent with this notion are the common observations of lethargy and increased desire to sleep that accompany mild infections such as colds or "the flu." Accompanying these ideas are those suggesting that the lack of sleep increases susceptibility to infectious disease (Opp and Toth 2003). Despite these widespread beliefs there is little empirical evidence to support the hypotheses that increased sleep aids recovery from, and lack of sleep increases susceptibility to, infections. However, there are data now available addressing the possible molecular regulation of the interactions between infection and altered sleep behavior. Amongst the changes observed are evidence for a cytokine cascade within the brain, including interleukin (IL)-1 and tumor necrosis factor (TNF)-α, along with a number of other substances including growth hormone-releasing hormone (GHRH), prolactin (PL), nitric oxide (NO) and nuclear factor kappa B (NF-kB). These substances, as we shall see below, are also implicated in the regulation of normal spontaneous sleep. The review that follows will focus on altered cytokine production as being of primary importance in sleep physiology in health and disease, but will also pay attention to the role of disruptions in these processes during normal human aging, and its effect on human health.

7.2 Altered Sleep and Concomitant Alterations in Immune Function

If sleep is indeed a restorative process that is important for the appropriate functioning of the immune system, it should come as no surprise that a number of researchers have attempted to investigate possible correlations between disordered sleep and disease, both incidence and/or progression, particularly those diseases for which there is already evidence for an immunological component. This explains the interest in infectious disease (see above). In one study attempting to dissect the immune changes which might accompany sleep disruption, immune functioning between patients with chronic insomnia (17 adults meeting criteria for a chronic primary insomnia disorder) and good sleepers (19 adults with a regular sleep schedule and no complaint of sleep disturbances) was compared (Savard, Laroche, Simard, Ivers, and Morin 2003). In this study a detailed analysis of PBL samples at baseline and before a second night of polysomnographic assessment included enumeration of blood cell counts (i.e., white blood cells, monocytes, lymphocytes, and different cell subsets, including $CD3^+$, $CD4^+$, $CD8^+$, $CD16^+$, and $CD56^+$ cells), natural killer cell activity, and cytokine production, i.e., IL-1β, IL-2, and interferon (IFN)γ. Perhaps not surprisingly few differences were found, with some evidence for higher levels of $CD3^+$, $CD4^+$, and $CD8^+$ cells in good sleepers.

However, more subtle interactions have also been reported. Thus, the severity of disordered sleep in depressed and/or alcoholic subjects also shows a correlation with altered innate and cellular immune functions, and furthermore may be related to changes in cytokines/chemokines (Irwin 2002). Alcoholics are recognized to be at increased risk for infectious disease, along with the profound disturbances of sleep and cellular immunity. Accordingly, one study attempted to investigate the possible interrelationships between sleep, nocturnal expression of immunoregulatory cytokines, and natural killer (NK) cell activity in alcoholic patients in comparison to control subjects (Redwine, Dang, Hall, and Irwin 2003). Along with losses of delta sleep and increases of rapid eye movement sleep in comparison to control subjects, the alcoholic patients showed lower levels of IL-6 production, suppression of the IL-6/IL-10 ratio, and a reduction of NK cell activity. In addition, they showed a persistently low ratio of IFNγ:IL-10 with decreased NK cell activity throughout the night, which the authors concluded as implicating sleep in the regulation of immune function, and further that disordered sleep might contribute to immune alterations in patients with chronic alcoholism. Partial night sleep deprivation studies in humans, aimed to replicate the pattern of sleep loss found in clinical samples, was indeed found to produce a pattern of immune alterations similar to that found in depressed and alcoholic patients (Irwin 2002).

Depression itself is well-recognized to be associated with sleep disruption, and with immune impairment (Anisman and Merali 2002) but the nature of the relationship, and indeed any causality, remains obscure. Immune activation, and increases in the activity of several cytokines, including IL-1, IL-2, IL-6, TNF-α, and their receptors, have all been recorded during depressive illness. However, despite successful treatment of depression, altered cytokine patterns apparently remain, suggesting that cytokines may be trait markers of depression, or epigenetic effects of the illness (Anisman et al. 2002). Interestingly, it has been recognized for some time that patients receiving cytokine immunotherapy frequently show depressive symptoms, which may be attenuated by antidepressant medication, supporting a causal role for cytokines in depressive disorders, including in the sleep disruption seen. The mechanisms implicated in these processes (cytokine-induced depression) are extensive, however, with the affective changes potentially stemming from both the neuroendocrine and central neurochemical changes elicited by cytokines. Animal studies also support the view that therapeutic administration of inflammatory cytokines can induce typical major depression, with further support coming from evidence that stimulated cytokine-release during experimental endotoxemia provokes transient deterioration in mood and memory (Pollmacher, Haack, Schuld, Reichenberg, and Yirmiya 2002). However, in most of these studies of animal models of acute infections it is only in the presence of very high amounts of cytokines produced in the periphery that actions within the CNS are observed, while in cases of (arguably) greater interest clinically, such as chronic infection and inflammation, the levels of circulating cytokines is actually quite low. Studies from Pollmacher's group showed that while low levels (and high levels) of circulating cytokines can impinge on CNS responses in general, and sleep in particular, the qualitative effects were quite different. Non-REM sleep was promoted by a slight increase in cytokine levels, but suppressed by greater increases (Pollmacher et al. 2002). These results should be taken into account in building any generalized theory of an effect of cytokines on mood, sleep, and depression. Fatigue and sleep disturbances are also common in cancer patients as well as in those receiving cytokine therapy, but the role of sleep in cancer is still relatively uninvestigated (Krueger et al. 2003).

As has commonly been the case in the research field, analysis of inbred strains of mice has been used to attempt to uncover possible genetic variations in the expression of sleep patterns. It is of interest in this regard that while the genetic influence on many behaviors has been shown to be gender specific, to date, there has been no evidence to suggest variations in sleep patterns in different strains of female mice. Koehl et al. have recently addressed this issue, taking into account a possible confounder of the estrous cycle to eliminate effects solely due to reproductive hormones on sleep (Koehl, Battle, and Turek 2003). These data showed that there was in fact an important impact of the genetic background on the regulation of non-rapid eye movement sleep over a 24-h period, with clear strain differences in rapid eye movement sleep distribution over the light–dark cycle. In contrast, and as had been reported by our group earlier in a human study (Moldofsky, Lue, Davidson, and Gorczynski 1989), the estrous cycle had much less influence on non-rapid eye movement sleep and rapid eye movement sleep. These data, if extrapolated to humans, clearly imply that genetic factors must be accounted for in comparisons of human populations studied in regards to altered sleep behavior, particularly given what we already know about the effects of variations in genetic background on immune function.

7.3 Cytokines and Sleep

As will be discussed in more detail below, sleep is affected by infection, a primary driving force for nonspecific inflammation and cytokine production. It should come as no surprise therefore that a number of groups have examined in detail the evidence for an interaction between modulation of cytokine production and altered sleep behavior. This is reviewed in more detail elsewhere in this volume, but some brief mention is necessary before further discussion of altered sleep in disease.

As a precursor to an introduction of the role of cytokines in altering sleep behavior, consider the changes in cytokines reported in a number of sleep disorders. The latter are quite common in the general population and are associated with significant medical, psychological, and social disturbances. Deep sleep has an inhibitory influence on the hypothalamic-pituitary-adrenal (HPA) axis, in contrast to activation of the HPA axis or administration of glucocorticoids, which has been reported to lead to arousal and sleeplessness. Not surprisingly then, insomnia, the most common sleep disorder, is associated with a 24-h increase of corticotropin and cortisol secretion, consistent with a disorder of central nervous system hyperarousal

(Vgontzas and Chrousos 2002). Sleepiness and generalized fatigue, both prevalent in the general population, are now thought to be associated with elevations in the levels of the proinflammatory cytokines IL-6 and TNF-α, both of which are also elevated in disorders associated with excessive daytime sleepiness, including sleep apnea (Mills and Dimsdale 2004), narcolepsy, and idiopathic hypersomnia. Indeed, sleep deprivation reportedly leads to sleepiness and daytime hypersecretion of IL-6 (Vgontzas et al. 2002). Hu et al. performed a similar investigation in mice, examining serum cytokine and chemokine levels following 36 of sleep deprivation, or after exposure to a known physical stressor (rotational stress) (Hu et al. 2003). Changes in inflammatory cytokines/chemokines (IL-1β, TNF-α, IL-1ra, IL-6, and MIP- 1β, MCP-1) were observed following each manipulation, but the qualitative and quantitative patterns differed in the two scenarios. Only physical stress was associated with measured increases in serum corticosterone levels, and with independent evidence (using in vitro immune allo stimulation) for a generalized immunosuppression secondary to the experimental manipulation. This group concluded that altered cytokine production following sleep perturbation occurred by a different mechanism from that (HPA axis) attributed to stress per se.

Given that independent studies have suggested that IL-1β and TNF-α promote slow-wave sleep (SWS), whereas IL-10 inhibits the synthesis of both cytokines and promotes waking, Toth et al. studied mice in which the gene encoding IL-10 had been deleted by homologous recombination, IL-10KO mice (Toth and Opp 2001). Under basal conditions, IL-10KO mice spent more time in SWS during the dark phase of the light–dark cycle than did their genetically intact counterparts. While the two groups of animals had comparable responses after treatment with IL-1, IL-10, or influenza virus, the response to LPS injection was quite different. In the IL-10 KO mice, LPS induced an initial transient increase and a subsequent prolonged decrease in SWS, along with a marked hypothermic response, observations which were not seen in the wild-type controls. The authors concluded that IL-10 plays a crucial role in the regulation of normal sleep behavior patterns. Kushikata et al. also has reported on another cytokine, IL-4, which, in simple immunological terms, is often thought to act to counter some of the effects of inflammatory and type-1 cytokines. In this study spontaneous sleep was observed in rabbits receiving one of four doses of IL-4 (0.25, 2.5, 25, and 250 ng) injected intracerebroventricularly during the rest (light) period, and another group receiving 25 ng during the active (dark) cycle (Kushikata, Fang, Wang, and Krueger 1998). Appropriate time-matched control injections of saline were performed in the same rabbits on different days. IL-4 administered at dark onset had no effect on sleep. In contrast, the three highest doses of IL-4 inhibited spontaneous non-REM sleep if the IL-4 was given during the light cycle, while the highest dose of IL-4 (250 ng) also decreased REM sleep.

Other inflammatory cytokines, besides IL-1β and TNF-α are now thought to be important in sleep induction. Central administration of rat recombinant IL-6 (100 and 500 ng) increased NREM sleep in rats, with evidence for a subsequent suppression of NREM sleep in the aftermath. Moreover, the effect was not related to a simple pyrogen effect, since IL-6-induced febrile responses at doses lower (50 ng) than those required to alter sleep. REM sleep was not altered by any of the doses of IL-6 tested (Hogan, Morrow, Smith, and Opp 2003). However, when the authors attempted to gain further corollary support for the role for IL-6 using central administration of monoclonal or polyclonal anti-rat IL-6 antibodies, none of the parameters monitored in the study were changed. Thus, while IL-6 may possess sleep modulatory properties, the authors suggest it is unlikely to be involved in the regulation of spontaneous sleep in healthy animals since antagonizing the IL-6 system using antibodies did not alter sleep. A different conclusion was reached by Vgontzas et al. from circadian studies in humans (Vgontzas, Bixler, Lin, Prolo, Trakada, and Chrousos 2005). It has been reported in the human population that IL-6 is elevated in disorders of excessive daytime sleepiness such as narcolepsy and obstructive sleep apnea, correlating positively with body mass index and thus acting potentially as a mediator of sleepiness in obesity. Secretion of this cytokine is also reported to be stimulated by total acute or partial short-term sleep loss, again possibly reflecting the increased sleepiness experienced by sleep-deprived individuals. Measurement of the 24-h secretory pattern of IL-6 in healthy young adults shows that IL-6 is secreted in a biphasic circadian pattern with two nadirs at about 08.00 and 21.00, and two peaks at about 19.00 and 05.00 h. Following sleep deprivation or in disorders with disrupted sleep performance (insomnia) IL-6 peaks during the day and, concomitant with altered cortisol secretion, contributes to sleepiness and deep sleep (when cortisol levels are low) or feelings of tiredness, fatigue and poor sleep (when cortisol levels are high). Furthermore, Vgontzas et al. have reported that IL-6 is somnogenic in rats, with a diurnal (secretion) rhythm that follows the sleep/wake cycle in these animals, supporting an important role for IL-6 as a mediator of sleep behavior.

Other groups have concentrated less on the inflammatory cytokines (or their counter regulatory cytokines) referred to above, and more on their receptors and/or other molecules in sleep behavior. Thus, Haack et al. have characterized concentrations of soluble TNF^rs and $IL\text{-}2^r$ during normal sleep and wakefulness, as well as during a 24-h sleep deprivation (Haack, Pollmacher, and Mullington 2004). Plasma levels of the TNF^r p55, TNF^r p75, and $IL\text{-}2^r$ remained essentially unchanged during nocturnal sleep and nocturnal wakefulness, although they did observe significant diurnal variations for both TNF^r p55 and TNF^r p75, but not $IL\text{-}2^r$. Peak levels for both TNF^rs occurred ~6:00 in the morning, well before that for cortisol, and fluctuated inversely with the diurnal rhythm of temperature, consistent with the hypothesis that diurnal variations in levels of TNF^rs plays a role in normal diurnal temperature regulation. Finally, OConnor et al. recently reported on their investigations into the role for high mobility group box 1 (HMGB1), an abundant, highly conserved cellular protein, widely known as a nuclear DNA-binding protein, in the regulation of sleep. HMGB1 is implicated as a proinflammatory cytokine with a known role as a late mediator of endotoxin lethality, along with a defined ability to stimulate release of proinflammatory cytokines from monocytes (OConnor et al. 2003). Intracerebroventricular administration of HMGB I has been reported to increase TNF-α expression in mouse brain and induce aphagia and taste aversion. OConnor et al. also showed that intracerebroventricular injections of HMGB1-induced fever and hypothalamic IL-1 in rats, while intrathecal administration of HMGB1 lowered the response threshold to calibrated stimuli. While LPS administration elevated IL-1 and TNF-α mRNA levels in various brain regions, HMGB1 mRNA was unchanged. While not completely discounting the possibility that HMGB1 protein is released in brain in response to LPS, their data does imply that HMGB1 has proinflammatory characteristics within the CNS (and thus potentially sleep-promoting properties), which may be triggered independent of LPS signaling.

7.4 Relationship of Infection to Altered Sleep Behavior

Sleep has been proposed as an innate host defense, exerting effects on both specific and nonspecific immunity. One of the most seminal studies dealing with the effects of sleep on immune potential was that from Brown's group showing that depriving influenza virus-immune mice of sleep for 7 h following total respiratory tract viral challenge abrogated antiviral immunity within the lungs and lowered the level of anti-influenza antibody in lung homogenates (Brown 1989). In immune mice nasobronchial immunity to influenza virus reflects the function of secretory IgA (S-IgA) within the mucosal mucocilliary blanket, while serum IgG mediates protection within the lung parenchyma. While provocative, unfortunately a follow-up investigation by Renegar et al., attempting to duplicate these studies in immune mice, was unable to abrogate mucosal anti-influenza viral immunity with a single postviral-challenge sleep-deprivation episode (Renegar, Floyd, and Krueger 1998). Furthermore, even one pre- and two postchallenge sleep-deprivation episodes in young adult or old mice, or two prechallenge sleep-deprivation episodes in old mice had no effect on viral immunity in this follow-up study. Sleep deprivation did not depress the level of serum influenza-specific IgG antibodies, and there seemed actually to be an increased influenza-specific serum IgG compared with normally sleeping mice in aged immune mice boosted 3 weeks before challenge and sleep deprived once before and twice after challenge. An independent assessment of effects in the two age groups studied further revealed no evidence for differences in antiviral respiratory immunity between young and old mice.

At this stage then, despite the intense interest generated by Brown's initial observation, it is evident that there is no categorical evidence for an effect of sleep deprivation in this particular model of viral immunity in animals. That having been acknowledged, we note that sickness per se has been taken to refer to a combination of subjective, behavioral and physiological changes that sick individuals exhibit during the course of an infection. Many of these changes are thought to be associated with, and or causally related to, the effects of IL-1 and other proinflammatory cytokines on brain receptors that are essentially identical to those characterized on immune cells (Dantzer 2004). The expression and action of a number of proinflammatory cytokines in the brain in response to peripheral cytokines, as we have already discussed above, are regulated by a number of molecular intermediates including anti-inflammatory cytokines such as IL-10 and IL-1ra, growth factors such as insulin-like growth factor-1 (IGF-1), hormones such as glucocorticoids and neuropeptides, including vasopressin, alpha-melanotropin, substance P, and somatostatin. Other groups have thus addressed whether sleep disruption may impinge on viral resistance secondary to intervening alterations in some of these mediators, and/or whether cytokines produced in response to viral infection may alter sleep physiology.

In one such study, Hermann, Mullington, Hinzeselch, Schreiber, Galanos, and Pollmacher (1998) used endotoxin (associated with bacterially induced inflammatory processes) as a modifier of sleep. Limited host defense activation by endotoxin does not affect daytime sleepiness and NREM sleep in humans. However, the Hermann study investigated the effects of a more intensive stimulation with Salmonella abortus equi endotoxin (0.8 ng/kg), given 12 h following host response priming by GMCSF (300 mu/g, subcutaneously), on daytime sleep and sleepiness in 10 healthy subjects and placebo controls. Endotoxin induced increases in rectal temperature, and in plasma levels of TNF-α, IL-6, IL-1ra, and cortisol. In a nap occurring 1 h following endotoxin administration, total sleep time and NREM sleep stage 2 were reduced, whereas wakefulness and sleep onset latency were increased. After this first nap sleepiness transiently increased, to a peak occurring prior to a second nap, but this nap and the following ones were not influenced by endotoxin. The authors concluded that one cause of daytime sleepiness during infections *may be prior sleep disruption* and, furthermore, that this kind of sleepiness is not necessarily associated with an increased pressure for further "catch-up" sleep. Mullington et al. also intrigued by data in the animal literature suggesting that in animals at least, activation of host defense mechanisms increases NREM sleep amount and intensity, which is likely to promote the ability of the host immune system to defend itself against challenges from the environment, also explored the effects of various doses of endotoxin on host response, including nocturnal sleep in healthy volunteers (Mullington et al. 2000). If given before nocturnal sleep onset, endotoxin caused a dose-dependent increase in rectal temperature, heart rate, and plasma levels of TNF-α, TNF^{r}s, IL-1ra, IL-6, and cortisol. While the lowest doses increased circulating levels of cytokines and cytokine^{r}s, they did not affect rectal temperature, heart rate, or cortisol, but did augment the amount and intensity (delta power) of SWS (stages 3 and 4). In contrast, the highest dose of endotoxin disrupted sleep, further emphasizing the potential complexity of sleep-related changes associated with/caused by, infection.

Additional studies have explored sleep changes in relation to cytokines associated with viral (not bacterial) infections. The effects of two low doses of IFNα (a cytokine produced early following viral infection) on nocturnal sleep was studied in 18 healthy men by means of polysomnographic sleep recordings (Spathschwalbe, Lange, Perras, Fehm, and Born 2000). Subjects were allowed to sleep from 23:00 h to 07:00 h following subcutaneous administration of 1000 to 10,0000 U/kg of recombinant IFNα or placebo at 19:00 h. The higher dose of IFNα suppressed slow wave sleep while increasing time spent in shallow sleep during the first half of sleep time. In addition, REM latency was postponed and the time spent in REM sleep was decreased after IFNα treatment (versus placebo). It should be noted that in contrast to these data in humans, similar studies in animals suggests IFNα has in fact a sleep-promoting effect! Interestingly, despite these studies on virally induced IFNs, and those discussed earlier on other inflammatory cytokines, there is surprisingly little data available concerning the sleep effects of IFNγ. Intracerebroventricular injection of human IFNγ in rabbits led to dose-dependent increases in NREM sleep, electroencephalographic slow wave activity and brain temperature (Kubota, Majde, Brown, and Krueger 2001). All of these effects were attenuated after heat inactivation of the IFNγ. IFNγ suppressed REM sleep only if given during the light period, but not when given at dark onset. Interestingly, while a TNF^r fragment did not affect sleep parameters when given at dark onset, it attenuated IFNγ-induced NREM sleep and the increased brain temperature, consistent with the notion not only that IFNγ is involved in sleep responses during infection, but that it may have a synergistic interaction with TNF-α in the process.

Based on data from an interesting study precipitated by work with HIV-positive individuals Hogan et al. have suggested that chemokines, and not (just) cytokines, may be important "players" in the regulation of virally induced changes in sleep physiology (Hogan, Hutton, Smith, and Opp 2001). It is known that sleep is altered early in the course of HIV infection, before the onset of AIDS, consistent with effects of the virus on neural processes. Results from earlier studies had indicated that HIV envelope glycoproteins were potential mediators of those responses. Accordingly, since some CC-chemokine receptors are recognized now to be coreceptors for HIV and to bind HIV envelope glycoproteins, Hogan's study investigated in animal models whether selected CC chemokine ligands altered sleep and whether their mRNAs were detectable in brain regions important for sleep. CCL4/MIP-1β beta, but not CCL5/RANTES, injected intraventricularly into rats prior to dark onset increased non-REM sleep, fragmented sleep and induced fever. Furthermore, they found that mRNA for the chemokine receptor CCR3 was constitutively expressed in multiple brain regions.

Before concluding this section, the reader's attention should be drawn to an interesting study by Morrow et al. evaluating sleep–wake behavior and core body temperature of mice with homologous deletion of the gene encoding IL-6 (IL-6KO) and C57BL/6J control mice after intraperitoneal administration of 10 μg lipopolysaccharide (LPS) (Morrow and Opp 2005). To assess the additional possible role for a circadian rhythmicity in the feedback mechanisms which regulate responses to immune challenge responses to LPS were measured after administration of LPS at the beginning of both the light and dark portions of the light:dark cycle. LPS-induced increases in NREM sleep of both IL-6KO and control mice, but the increase was less pronounced in the KO mice than in the wild-type controls. The greatest differences in LPS-induced increases in NREM sleep were observed when LPS was given after light-onset (increase over vehicle for controls: 23±0.5%; for IL-6 KO mice 4.1±0.5%). For both IL-6KO and controls REM sleep was suppressed to the same extent after LPS, regardless of the timing of administration. In addition to the sleep-related changes, LPS-induced a febrile response in wild-type mice regardless of time of administration (~1.6°C), while in contrast, the same doses of LPS-induced hypothermia in the IL-6 KO mice (ranging from declines of 5 to 2.2°C depending upon administration after dark versus light onset respectively. Clearly based on these data, bacterial infection per se can potentially have profound effects on sleep behavior (and core body temperature) in rodents, which may not be seen in man.

7.5 Stress, Hormonal Change, Cytokines, and Sleep Behavior

A number of studies by several groups, including our own, have shown unequivocally that there exists an intimate interrelationship between alterations in inflammatory-type cytokine levels and sleep, especially for the cytokines IL-1 and TNF-α (Dinarello 1994; Floyd and Krueger 1997; Krueger, Obal, and Fang 1999; Krueger et al. 1995; Nakano, Kubo, Sogawa, Teshima, and Nakagawa 1992). The human literature on this subject is reviewed elsewhere (Vgontzas et al. 2002). Typical examples for a role for cytokine mediated effects on sleep behavior include the following: intraventricular injection of TNF-α or IL-1 promotes sleep in sheep/cats, respectively; alterations in serum levels of IL-1 occur during REM and SWS in humans (Moldofsky, Lue, Eisen, Keystone, and Gorczynski 1986); and sleep deprivation itself has been reported to alter serum cytokine levels (Moldofsky et al. 1989).

Several studies have made use of genetically manipulated animals to confirm and/or extend these findings. Thus, sleep patterns in a number of mice bearing gene deletions for cytokines and/or cytokine receptors (e.g., IL-1, TNF-α) support a general role for those cytokines in the regulation of sleep physiology (Dinarello 1994; Fang, Wang, and Krueger 1997; Fang, Wang, and Krueger 1998). Similar animal research, as described earlier, has implicated a role for another cytokine, IL-10 in control of sleep behavior (Toth et al. 2001), and indeed of yet another inflammatory cytokine, IL-6, though these latter data are still of a more controversial nature (Spath-Schwalbe et al. 1998). However, at least in the human literature, a more convincing case can be made for an important role for IL-6 in altered sleep behavior. Thus, both TNF-α-receptor and serum IL-6 have both been reported to vary in humans subjected to sleep deprivation (Shearer et al. 2000) while other groups have explored a role for IL-6 alone (Redwine, Hauger, Gillin, and Irwin 2000; Vgontzas et al. 2002).

"Stress," in some form, whether physical and/or psychological in nature, is proposed to be "read" as a major *danger* signal which can trigger changes in immune responsiveness (Gallucci and Matzinger 2001). Cytokines and chemokines play an important role in providing communication pathways between the CNS and the immune system (Webster, Tonelli, and Sternberg 2002), and stress-related changes in immune functioning are a common tool used to explore that interaction (Elenkov and Chrousos 1999; Kronfol and Remick 2000; Wilder 1995). A number of studies exploring interactions between sleep and cytokines are compounded by a potential interactive effect of "stress per se" on the changes measured (Shephard, Castellani, and Shek 1998). Physical stress produces changes in cellular redistribution and/or migration (Dhabhar and Mcewen 1997), along with concomitant changes in chemokine and/or cytokine levels (Benveniste 1998; Mulla and Buckingham 1999), which may account at least in part for some of the changes in cell migration (Heijnen 2000; Ottaway and Husband 1992). Mental stress has also been reported to alter cytokine production of IL-1, TNF-α, and IL-1ra (Marshall and Born 2002; Steptoe, Willemsen, Owen, Flower, and Mohamedali 2001; Uchakin, Tobin, Cubbage, Marshall, and Sams 2001). In a recent study exploring cytokine changes with infection, it was reported that there was an association between stress levels and increased IL-6 production, beyond that accorded to infection per se (Cohen, Doyle, and Skoner 1999).

We recently compared cytokine changes and stress-hormone changes (e.g., of the HPA axis (Besodovsky and Sorkin 1977) in mice subjected to one of two types of stressor, the one a physical stress (rotation Gorczynski and Kennedy (1984), and Shimizu et al. (2000)) the other that associated with sleep deprivation (Hu et al. 2003). The sleep deprivation model used was based on the "floating platform" test used for rodents by several groups (Fadda and Fratta 1997; Pokk and Vali 2001), although it should be noted that there has been some concern voiced that under these conditions animals show other features of stress, besides the rapid eye movements typical of sleep deprivation; including the "stress" of isolation, (relative) immobilization and falling into the water (Pokk and Vali 2001). The data we reported from this study suggested that it was the level of inflammatory-type cytokines which were most perturbed (IL-1, TNF-α, and IL-1ra) in both instances, but the changes which followed sleep disruption were not easily explained as being simply a "stress" effect, monitoring stress in the two models by following serum corticosterone levels. No further investigations were performed to provide insight into the mechanism(s) by which cytokine changes occur during sleep disruption, independent of activation of (presumably) the HPA axis (as for stress). This issue was investigated further by Andersen, Bignotto, Machado, and Tufik (2004), who studied paradoxical sleep deprivation (PSD) and restraint, electrical foot-shock, cold and forced swimming stress for their ability to alter male reproductive function and levels of steroid hormones in adult Wistar male rats. These workers concluded that different stress modalities resulted in distinct steroid hormone responses, with PSD and footshock being the most similar.

Note that acute stress produces rapid changes in serum cortisol levels, while the effect of sleep deprivation produces changes in a much slower fashion (Weitzman, Zimmermann, Czeisler, and Ronda 1983). Furthermore, it is generally believed that in animal models sleep loss may be more stressful than in humans, and significant alterations in immunity have been reported in sleep deprived mice (Toth and Rehg 1998), a feature which we were unable to replicate in the model described. A dissociation between cytokine changes and cortisol levels was also reported in human subjects following exposure to low levels of endotoxin (Mullington et al. 2000).

Other groups have addressed the relationship between sleep disruption, cortisol levels and cytokines (Rogers 2001; Vgontzas et al. 2002), while the role of IL-1 as a mediator of these interactions, in particular in an elderly human population, has also been studied (Prinz, Bailey, and Woods 2000). One such study concluded that "the (activated) HPA axis stimulates arousal, while IL-6 and TNF-α are possible mediators of excessive daytime sleepiness" (Vgontzas et al. 2002). IFNγ may interact with TNF-α in producing altered sleep physiology in some circumstances (Kubota et al. 2001), while in at least one study it has been suggested that the chemokines CCL4 and MIP-1α, not (or not only) cytokines, are implicated in regulation of sleep physiology, at least in HIV-1-infected individuals (see Hogan et al. (2001)).

It should be noted that altered sleep physiology is not simply a function of altered cytokines and stress hormones. Slow wave steep (SWS) is also characterized by maximum release of growth hormone (GH) and a minimal release of cortisol, likely regulated by GHRH stimulation during a period of relative somatostatin withdrawal, accompanied by elevated levels of circulating ghrelin (VanCauter et al. 2004). Lange et al. recently reported on a study designed to investigate whether this hormonal pattern during SWS leads, both to a generally increased T-cell cytokine production, as well as to a shift toward type-1 cytokine production (Lange, Dimitrov, Fehm, and Bom 2006). Blood was sampled from eight humans during SWS, and cultures were stimulated with ionomycin and phorbol–myristate–acetate (PMA) in the absence and presence of GH neutralizing antibody or with physiological concentrations of cortisol. Anti-GH decreased IFNγ producing CD4$^+$ cells with no effect on other cytokines, while cortisol, both alone and in combination with anti-GH decreased IFNγ, TNF-α, and IL-2 production by CD4$^+$ and CD8$^+$ cells, with an even greater effect on IL-4 producing CD4$^+$ cells, thus overall biasing production toward type-1 cytokines (increased IFNγ:IL-4 ratio). These data were taken to support the hypothesis that release of GH, by increasing production of IFNγ, may contribute to the shift toward type-1 cytokine production seen during SWS.

7.6 Age-Related Changes in Sleep and Relation to Altered Immunity

A focus of our own interest, as indeed for a number of other groups, has been the age-related changes which occur in sleep performance, and their interrelationship with simultaneous changes known to occur with immune functioning with age. Human sleep in old age is characterized by a number of changes, including sleep fragmentation and reductions in sleep efficiency and amounts of visually scored slow-wave and REM sleep, as well as amplitude of the diurnal sleep/wake rhythm. While acknowledging that a large body of evidence suggests that major declines in immunity with age are related to altered functions in T lymphocytes (and or in their numbers/subset distribution) (Globerson and Effros 2000), it is also evident that significant changes in cytokine responses also occur with age, and include changes in production of the very cytokines discussed above which can impact on sleep in humans (experimental animals) (Caruso et al. 1996; Gorczynski et al. 1997; Gorczynski, Dubiski, Munder, Cinader and Westphal, 1993; Hobbs et al. 1993; Sindermann, Kruse, Frercks, Schutz and Kirchner, 1993). One could thus consider the possibility that approaches taken to restore immune functioning in the elderly, if they included addressing similar cytokine changes, might thus also be reflected in the restoration of normal sleep behavior.

A model organism for the study of circadian rhythmicity and the effects of age on the circadian system has been the golden hamster. Surprisingly, nothing is known about the effects of advanced age on sleep in this species. As a first step in determining the effects of aging on sleep in the golden hamster, Naylor et al. recorded sleep over a 24-h period in young (3 months) and old (17 to 18 months) golden hamsters entrained to a 14:10 light:dark (LD) cycle (Naylor, Buxton, Bergmann, Easton, Zee, and Turek 1998). Aged hamsters were found to exhibit small increases in overall NREM sleep time, with no significant differences in REM sleep, median sleep episode length, or the number of arousals. The most striking differences between the sleep of young and old hamsters however was in NREM delta sleep, with older animals showing approximately one-third less delta sleep per NREM epoch than young hamsters, a phenomenon also reported earlier for humans, but not, interestingly, in another animal used popularly in such research, the rat. In these animals a characteristic of older age is a decrease in the mean duration of sleep bouts, an increase in the number of sleep bouts, and a modest reduction of REM sleep. From comparisons of sleep behavior in young adult rats (3 months), middle-aged (12 months) animals, and older (24 months) rats during 24 h under constant dim light, the most significant changes in reductions in high-voltage NREM sleep

("HS2"), the mean length of sleep bouts, and REM-onset duration were already evident by one year of age (the "rat equivalent" of midlife).

A major feature of sleep alteration with age is the altered circadian rhythmicity of the sleep (Weinert 2000). Both in aging animals and humans, all rhythm characters change, with the most prominent changes being a decrease in the amplitude and the diminished ability to synchronize with a periodic environment. Susceptibility to both photic and nonphotic cues is decreased, and in consequence, both internal and external temporal orders are disturbed both under steady-state conditions and, even more dramatically, following changes in the periodic environment. There are potentially multiple causes for these changes, many of which may be related to alterations within the SCN itself. As but one example, the number of functioning neurons decreases with advancing age as indeed likely does the coupling between them. In consequence, the SCN is significantly less able to produce stable rhythms and to transmit timing information to target sites. Interestingly it is thought that this change occurs in stepwise fashion, such that initially only the ability to synchronize with the periodic environment is diminished, while the rhythms themselves continue to be well represented. This being so it should in principle be possible to treat (some) age-dependent sleep disturbances pharmacologically; or by increasing the magnitude of the light–dark (LD) zeitgeber; or by strengthening feedback effects (e.g., by increasing the daily activity). At least in some preliminary studies these predictions seem to hold true (Weinert 2000).

As a further attempt to understand altered sleep with age, Mazzoccoli et al. explored whether age-related changes in neurotransmitters, hormones and cytokines, and in particular age-related changes in the 24-h hormonal and nonhormonal rhythms of their production, might be in some way responsible for altered sleep and/or immune functioning with age (Mazzoccoli et al. 1997). Cortisol, melatonin, thyrotropin-releasing hormone (TRH), thyroid-stimulating hormone (TSH), free thyroxine (T4), growth hormone (GH), insulin-like growth factor I (IGF-I), and IL-2 serum levels were measured, along with a detailed lymphocyte subpopulation analysis, on blood samples collected over a 24-h period from healthy young subjects (aged 36 to 58 years) and healthy older subjects (aged 65 to 78 years). The values of $CD20^+$ (B cells) and $CD25^+$ (activated T cells, expressing the alpha chain of the IL-2^r) were higher in elderly subjects. In contrast, there was no statistically significant difference in the observed values of CD2 (total lymphocytes), CD4 (helper/inducer T cells), CD8 (suppressor/cytotoxic T cells), CD16 (natural killer cells), cortisol, melatonin, TRH, TSH, FT4, GE-I, IGF-I, and IL-2. However, while the group of elderly subjects continued to show a circadian rhythmicity for changes in $CD2^+$, $CD8^+$, $CD16^+$, and $CD25^+$ cells, as well as in cortisol, melatonin, TSH and GH there were significant phase changes in the cycling observed, leading the authors to conclude that not only is aging associated with enhanced responsiveness of the T-cell compartment, but that significant alterations in temporal architecture of the neuro-endocrine-immune system also occur with age. Despite these clear difference however, the response to total sleep deprivation, a powerful stimulus for sleep, does not apparently change significantly with age, despite the aforementioned decreased sleep continuity, slow wave sleep (SWS), growth hormone (GH) release and an increased hypothalamo-pituitary-adrenocortical system activity which is a characteristic of sleep in aging populations.

Independently, Prinz et al. (2000) have addressed in some detail the issue of whether the increased cortisol levels at the circadian nadir which accompany aging, reflecting an increased stress responsivity and a longer-lasting glucocorticoid increase, are responsible for individual differences in age-related sleep impairments. In this particular study they compared sleep, cortisol, and sleep–cortisol correlations under baseline and "stress" conditions in men and women, with the mildly stressful procedure being the insertion of a 24-h indwelling intravenous catheter. Healthy, nonobese subjects (60 women and 28 men) were included in the study, with mean ages in years of the female/male population being 70.6 (±6.2) and 72.3 (±5.7), respectively. 24h urines were assayed for cortisol, serum for IL-1β, and EEGs were also assessed by polysomnography and EEG power spectral analysis. These data showed that healthy older subjects (men and women) with the highest levels of free cortisol (24-h urine level) under a mild stress condition had impaired sleep, as reflected in lower sleep efficiency, less time in stages 2 to 4 sleep, and more EEG beta activity during NREM sleep. Interestingly, men had the highest levels of free urine cortisol under both baseline and mild stressful conditions, and cortisol and sleep correlated most strongly in men (accounting for ~36% of the variance in NREM sleep stress responses). In the female (but not the male population) higher cortisol levels in response to stress were associated with increased circulating levels of IL-1β, explaining ~24% of the variance in a subset of women. These intriguing data suggest that future research in this area must pay attention to gender-related effects in the elderly when we attempt to make correlations between altered neurohormonal/immunological parameters and sleep physiology. A recent study by Hawkley et al. has also made note of the profound effect of cumulative stressors of various types (social, psychological, physical) on immune impairment in the elderly, though no attempt was made to correlate those changes with altered sleep physiology (Hawkley and Cacioppo 2004).

What about studies which have directly assessed whether the sleep changes occurring during immune challenge are themselves affected by age? In one example of this type of investigation, sleep alterations induced by the administration of lipopolysaccharide (LPS) in young and middle-aged rats were examined (Schiffelholz and Lancel 2001). After vehicle challenge, middle-aged rats exhibited less pre-REM sleep as well as REM sleep itself, due to both a smaller number and shorter duration of REM sleep episodes, than young rats (see above). While LPS was observed to elevate body temperature, increase non-REM sleep, and suppress both pre-REM sleep and REM sleep in both young and middle-aged rats, the effects were not identical. Thus, in the younger animals, LPS enhanced slow-wave activity in the EEG within non-REM sleep, presumably reflecting an increase in sleep intensity, while it attenuated the same measures in the older animals. These striking observations are the first to hint at the possibility that the alteration in sleep in response to infection and immune challenge may be different in young versus aged cohorts. This hypothesis was tested directly in a model study by Imeri et al., using intracerebroventricular injections of IL-1β given to young and aged rats whose subsequent sleep–wake behavior was analyzed (Imeri, Ceccarelli, Mariotti, Manfridi, Opp, and Mancia 2004). Under basal conditions and in the absence of an immune challenge, the sleep patterns of young (3 months) and aged rats (25 to 27 months) were not significantly different. However, in young animals, IL-1β (2.5 ng) enhanced non-REM sleep, inhibited REM sleep, and induced fever. In contrast, in the aged animals, IL-1β did not alter non-REM sleep, although changes in REM sleep and brain temperature were equivalent to those seen in young animals. It has been postulated that enhanced non-REM sleep is important in facilitating recovery from microbial infection, which leads to the provocative conclusion that this alteration in non-REM sleep unique to the elderly population may contribute to the increased infection-induced morbidity and mortality of aged organisms. Our own research in sleep and aging has focused around two other of our areas of interest. In the one case we have reported on the role of a novel extract prepared from fetal sheep liver (CLP) in inducing changes in cytokine production from aged mice (Gorczynski et al. 2005; Gorczynski et al. 1997; Gorczynski et al. 1993). While our detailed reports to date only refer to altered cytokine production following CLP injection, it is important to note in the context of this discussion on sleep changes in the elderly that anecdotal records from patients treated with CLP intramuscularly (im) in the clinic clearly suggest a significant sleep-promoting effect also of this material (Waelli, unpublished observations). Exhaustive studies in aged mice showed an increased background production of inflammatory-type cytokines (TNF-α, IL-1β, and IL-6) compared with young mice of the same strain, and mitogen-induced cytokine production from lymphocytes of aged mice showed a bias toward type-2 cytokine production, rather than the type-1 cytokines derived predominantly from cells of young mice (Gorczynski et al. 1997). However, ongoing administration of CLP over a 4-week period resulted in restoration of both of these profiles toward those seen in younger animals.

The active moieties in CLP are now being defined, and apparently represent a complex mixture of (at least) LPS, fetal hemoglobin, MIF, and GSH. In addition to these investigations, we have published at length on the role of a molecule, CD200, in the regulation of nonspecific inflammation, and other acquired immune responses in animal model systems, and in man. Other groups have also shown that in a CD200 KO mouse, increased inflammation is seen within the brain and CNS, consistent with a physiological role of this molecule in regulating inflammatory processes within the CNS. We thus postulated that both the background levels of (inflammatory) cytokine production in aged animals, and the augmentation of those levels following peripheral LPS administration, might be regulated as a reflection of alterations in expression of CD200 which would again be pertinent to observations concerning a sleep-controlling effect of CLP.

In the first series of studies, we analyzed in young and aged groups of C57BL/6 mice, resting levels of expression in the brain both of CD200 and a number of the inflammatory cytokines reported to be key to promoting sleep behavior both in man and in experimental animals. Those levels were then measured in similar groups of mice 48 h following ip administration of LPS as a mimic of exposure to an infectious insult. Typical data for one of three such studies are shown in Fig. 7.1.

Comparison of expression of various mRNAs in brain of young (8wk) or aged (80wk) BL/6 mice at rest and 2d following peripheral (ip) administration of 5μg/mouse LPS

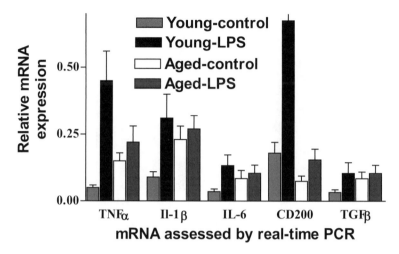

FIGURE 7.1: Real-time mRNA for various cytokines/CD200 in the brain of groups of young/aged BL/6 mice before and after injection of LPS (5 μg/mouse) ip. Data show mean ± SD for five mice per group. All mRNAs are expressed relative to a composite control for three housekeeping genes, GAPDH, RpL13A, ribosomal protein L13a, and CypA, cyclophlinA (Lee, Manuel, and Gorczynski 2006).

It is clear from these data that under basal conditions aged mice express less CD200 in the CNS, and comparatively more of the inflammatory cytokines IL-1β, TNF-α, IL-6, and TGFβ than their young counterparts, as might be predicted from the anti-inflammatory regulatory role postulated for CD200. Interestingly, following peripheral LPS administration, dramatic increases in expression of CD200 and the aforementioned cytokines was observed for young mice (ranging from 4- to 10-fold increases), but not in the aged animals. Both the altered basal levels of cytokines, and the altered response to LPS stimulation, is consistent with other data reported in the literature, and may help explain in part the disrupted sleep patterns in elderly animals. Equally striking however, were the effects observed when we repeated the same studies in groups of mice previously treated chronically (4 weekly injections) with CLP (Fig. 7.2).

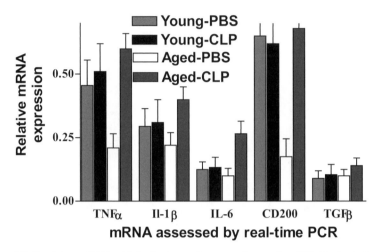

Figure 7.2. Real time mRNA with brain tissue of groups of young/aged BL/6 mice before and after injection of LPS (5 μg/mouse) ip (see also Fig. 7.1). Unlike the data in Fig. 7.1, all mice were pretreated for 4 weeks with either saline (PBS) ip or CLP (100 μg/mouse im). LPS was given 7 days after the last CLP injection.

In this case, CLP produced little change in the LPS stimulated levels of CD200 or cytokine mRNAs in young mice when compared with PBS injected control young mice. However, a clear effect was seen when CLP was given to aged mice in comparison to PBS controls. Following CLP injection the response (to LPS) was restored such that in these animals LPS trigger significantly more marked increases in CD200 and the inflammatory cytokines shown above compared with controls. One model to explain these data suggests that CD200 is an important regulator of inflammation in the CNS, and thus the decline in expression of CD200 with age results in increased endogenous inflammatory cytokine production, with resultant sleep disruption (including fragmented sleep). CLP injections, by reversing the decline in CD200 expression, simultaneously restores the pattern of endogenous cytokine production to that more typical of younger animals, and similarly (we predict) restores normal sleep physiology. This hypothesis implies a potentially important effect for peripheral CLP injections on both sleep and CNS inflammation.

7.7 Summary and Conclusions

The physiological necessity of sleep is readily apparent, but besides saying sleep is "restorative" we have not really improved our understanding of why we sleep. The notion that sleep is important for preserving our innate host defense mechanisms, exerting effects on both specific and nonspecific immunity, is an intriguing one, and there is, as has been discussed, some evidence for this notion, in particular work exploring the correlations behind disrupted sleep and infection, both in young and elderly individuals. There is a large body of evidence that suggests that key cytokines produced within the CNS are important in regulating sleep behavior, including IL-1β, TNF-α, and IL-6. Also consistent with this information is the evidence that alterations in expression of these molecules occur in a number of disease states, and in the elderly, and correlate well with disruption of normal sleep patterns. Considerably less evidence is available to suggest that deliberate attempts to modulate expression of those cytokines can prove beneficial to sleep performance. In addition, there are a number of chronic disease states, including cancer and autoimmune disorders, where sleep disruption, and/or altered expression of inflammatory cytokines is known to occur, yet corollary changes in the alternate variables have yet to be investigated. It is not being overly optimistic to anticipate significant advances, both in basic science understanding and in pharmaceutical intervention, in these areas over the next several years.

Acknowledgments. This study was supported by grants from the Canadian Space Agency (#9F007-001142, to R.M.G.); Trillium Therapeutics Inc. (to R.M.G.); the Turkish Science Foundation (to R.M.G. and E.T.); and from Clinique La Prairie, Montreux, Switzerland (to R.M.G.).

References

Andersen, M.L., Bignotto, M., Machado, R.B., and Tufik, S. (2004) Different stress modalities result in distinct steroid hormone responses by male rats. *Braz J Med Biol Res* 37, 791–797.

Anisman, H., and Merali, Z. (2002) Cytokines, stress, and depressive illness. *Brain Behav Immun* 16, 513–524.

Benveniste, E.N. (1998) Cytokine actions in the central nervous system. *Cytokine Growth Factor Rev* 9, 259–275.

Besodovsky, H., and Sorkin, E. (1977) Network of immune neuroendocrine interactions. *Clin Exp Immunol* 27, 1–12.

Brown, C.M. (1989) *Reg Immunol* 2, 321–325.

Caruso, C., Candore, G., Cigna, D., DiLorenzo, G., Sireci, G., Dieli, F., and Salerno, A. (1996) Cytokine production pathway in the elderly. *Immunol Res* 15, 84–90.

Cohen, S., Doyle, W.J., and Skoner, D.P. (1999) Psychological stress, cytokine production, and severity of upper respiratory illness. *Psychosom Med* 61, 175–180.

Dantzer, R. (2004) Cytokine-induced sickness behaviour: A neuroimmune response to activation of innate immunity. *Eur J Pharmacol* 500, 399–411.

Dhabhar, F.S., and McEwen, B.S. (1997) Acute stress enhances while chronic stress suppresses cell-mediated immunity in vivo: A potential role for leukocyte trafficking. *Brain Behav Immun* 11, 286–306.

Dinarello, C.A. (1994) The interleukin-1 family: 10 years of discovery. *FASEB J* 8, 1314–1325.

Elenkov, I.J., and Chrousos, G.P. (1999) Stress, cytokine patterns and susceptibility to disease. *Best Pract Res Clin Endoc Met* 13, 583–595.

Fadda, P., and Fratta, W. (1997) Stress-induced sleep deprivation modifies corticotropin releasing factor (CRF) levels and CRF binding in rat brain and pituitary. *Pharmacol Res* 35, 443–446.

Fang, J., Wang, Y., and Krueger, J.M. (1997) Mice lacking the TNF 55kDa receptor fail to sleep more after TNFα treatment. *J Neurosci* 17, 5949–5955.

Fang, J.D., Wang, Y., and Krueger, J.M. (1998) Effects of interleukin-1 beta on sleep are mediated by the type I receptor *Am J Physiol Regul Integr* 43, R655–R660.

Floyd, R.A., and Krueger, J.M. (1997) Diurnal variation of TNFα in the rat brain. *NeuroReport* 8, 915–918.

Gallucci, S., and Matzinger, P. (2001) Danger signals: SOS to the immune system. *Curr Opin Immunol* 13, 114–119.

Globerson, A., and Effros, R.B. (2000) Ageing of lymphocytes and lymphocytes in the aged. *Immunol Today* 21, 515–521.

Gorczynski, R., Chen, Z.Q., Kai, Y., Lee, L., Wong, S., and Marsden, P.A. (2004) CD200 is a ligand for all members of the CD200R family of immunoregulatory molecules. *J Immunol* 172, 7744–7749.

Gorczynski, R.M., Alexander, C., Bessler, W., Fournier, K., Hoffmann, P., Mach, J.P., Rietschel, E.T., Song, L., Terzioglu, E., Waelli, T., Westphal, O., Zahringer, U., and Khatri, I. (2006) MIF and GSH as important mediators of cytokine induction by CLP. *Immunol Lett* 102, 001–010.

Gorczynski, R.M., Alexander, C., Bessler, W., Fournier, K., Hoffmann, P., Mach J.P., Rietschel, E.T., Song, L., Waelli, T., Westphal, O., Zahringer, U., and Khatri, I. (2005) Analysis of interaction of cloned human and/or sheep fetal hemoglobin gamma-chain and LPS in augmenting induction of inflammatory cytokine production in vivo and in vitro. *Immunol Lett* 100, 120–129.

Gorczynski, R.M., Dubiski, S., Munder, P.G., Cinader, B., and Westphal, O. (1993) Age-related changes in interleukin production in BALB/cNNia and SJL/J mice and their modification after administration of foreign macromolecules. *Immunol Lett* 38, 243–251.

Gorczynski, R.M., and Kennedy, M. (1984) Associative learning and regulation of immune responses. *Prog Neuro-Psychopharmacol Biol Psychiat* 8, 593–600.

Haack, M., Pollmacher, T., and Mullington, J.M. (2004) Diurnal and sleep-wake dependent variations of soluble TNF- and IL-2 receptors in healthy volunteers. *Brain Behav Immun* 18, 361–367.

Hawkley, L.C., and Cacioppo, J.T. (2004) Stress and the aging immune system. *Brain Behav Immun* 18, 114–119.

Heijnen, C.J. (2000) Who believes in "communication". Norman cousins lecture, 1999. *Brain Behav Immun* 14, 2–9.

Hermann, D.M., Mullington, J., HinzeSelch, D., Schreiber, W., Galanos, C., and Pollmacher, T. (1998) Endotoxin-induced changes in sleep and sleepiness during the day. *Psychoneuroendocrinology* 23, 427–437.

Hobbs, M.V., Weigle, W.O., Noonan, D.J., Torbett, B.E., MvEvilly, R.J., Koch, R.J., Cardenas, G.J., and Ernst, D.N. (1993) Patterns of cytokine gene expression by CD4+ T cells from young and old mice. *J Immunol* 150, 3602–3614.

Hoek, R.M., Ruuls, S.R., Murphy, C.A., Wright, G.J., Goddard, R., Zurawski, S.M., Blom, B., Homola, M.E., Streit, W.J., Brown, M.H., Barclay, A.N., and Sedgwick, J.D. (2000) Down-regulation of the macrophage lineage through interaction with OX2 (CD200). *Science* 290, 1768–1771.

Hogan, D., Hutton, L.A., Smith, E.M., and Opp, M.R. (2001) beta (CC)-chemokines as modulators of sleep: Implications for HIV-induced alterations in arousal state. *J Neuroimmunol* 119, 317–326.

Hogan, D., Morrow, J.D., Smith, E.M., Opp, and M.R. (2003) Interleukin-6 alters sleep of rats. *J Neuroimmunol* 137, 59–66.

Hu, J., Chen, Z., Gorczynski, C.P., Gorczynski, L.Y., Kai, Y., Lee, L., Manuel, J., and Gorczynski, R.M. (2003) Sleep-deprived mice show altered cytokine production manifest by perturbations in serum IL-1ra, TNFα, and IL-6 levels. *Brain Behav Immun* 17, 498–504.

Imeri, L., Ceccarelli, P., Mariotti, M., Manfridi, A., Opp, M.R., and Mancia, M. (2004) Sleep, but not febrile responses of Fisher 344 rats to immune challenge are affected by aging. *Brain Behav Immun* 18, 399–404.

Irwin, M. (2002) Effects of sleep and sleep loss on immunity and cytokines. *Brain Behav Immun* 16, 503–512.

Koehl, M., Battle, S.E., and Turek, F.W. (2003) Sleep in female mice: A strain comparison across the estrous cycle. *Sleep* 26, 267–272.

Kronfol, Z., and Remick, D.G. (2000) Cytokines and the brain: Implications for clinical psychiatry. *Am J Psychiat* 157, 683–694.

Krueger, J.M., Majde, J.A., and Obal, F. (2003) Sleep in host defense. *Brain Behav Immun* 17, S41–S47.

Krueger, J.M., Obal, F., and Fang, J.D. (1999) Humoral regulation of physiological **sleep: cytokines** and GHRH. *J Sleep Res* 8(Suppl 1), 53–59.

Krueger, J.M., Takahashi, S., Kapas, L., Bredow, S., Roky, R., Fang, J.D., Floyd, R., Renegar, K.B., Guhathakurta, N., Novitsky, S., and Obal, F. (1995) Cytokines in sleep regulation. *Adv Neuroimmunol* 5, 171–188.

Kubota, T., Majde, J.A., Brown, R.A., and Krueger, J.M. (2001) Tumor necrosis factor receptor fragment attenuates interferon-gamma-induced non-REM sleep in rabbits. *J Neuroimmunol* 119, 192–198.

Kushikata, T., Fang, J.D., Wang, Y., and Krueger, J.M. (1998) Interleukin-4 inhibits spontaneous sleep in rabbits. *Am J Physiol Regul Integr* 44, R1185–R1191.

Lange, T., Dimitrov, S., Fehm, H.L., and Bom, J. (2006) Sleep-like concentrations of growth hormone and cortisol modulate type1 and type2 in-vitro cytokine production in human T cells. *Int Immunopharmacol* 6, 216-225.

Lee, L., Manuel, J., and Gorczynski, R.M. (2006) Bone and effects of CD200. *Immunol Lett* 102, 010–020.

Lue, F.A., Bail, M., Jepthah-Ochala, J., Carayanniotis, K., Gorczynski, R.M., and Moldofsky, H. (1988) Sleep and cerebrospinal fluid Interleukin-1 like activity in the cat. *Int J Neurosci* 42, 179–184.

Majde, J.A., and Krueger, J.M. (2005) Links between the innate immune system and sleep. *J Allerg Clin Immunol* 116, 1188–1198.

Marshall, L., and Born, J. (2002) Brain-immune interactions in sleep. In: A. Clow, F. Hucklebridge (Eds.), *Neurobiology of the Immune System*. Academic Press, San Diego, pp. 93–131.

Mazzoccoli, G., Correra, M., Bianco, G., DeCata, A., Balzanelli, M., Giuliani, A., and Tarquini, R. (1997) Age-related changes of neuro-endocrine-immune interactions in healthy humans. *J Biol Regul. Homeost Agent* 11, 143–147.

Mendelson, W.B., and Bergmann, B.M. (1999) Age-related changes in sleep in the rat. *Sleep* 22, 145–150.

Mills, P.J., and Dimsdale, J.E. (2004) Sleep apnea: a model for studying cytokines, sleep, and sleep disruption. *Brain Behav Immun* 18, 298–303.

Moldofsky, H., Lue, F.A., Davidson, J.R., and Gorczynski, R.M. (1989) Effects of sleep deprivation on human immune functions. *FASEB J* 3, 1972–1977.

Moldofsky, H., Lue, F.A., Eisen, J., Keystone, E., and Gorczynski, R.M. (1986) The relationship of interleukin-1 and immune functions to sleep in humans. *Psychosom Med* 48, 309–318.

Morrow, J.D., and Opp, M.R. (2005) Diurnal variation of lipopolysaccharide-induced alterations in sleep and body temperature of interleukin-6-deficient mice. *Brain Behav Immun* 19, 40–51.

Mulla, A., and Buckingham, J.C. (1999) Regulation of the hypothalamo-pituitary-adrenal axis by cytokines. *Best Pract Res Clin Endoc Met* 13, 503–521.

Mullington, J., Korth, C., Hermann, D.M., Orth, A., Galanos, C., Holsboer, F., and Pollmacher, T. (2000) Dose-dependent effects of endotoxin on human sleep. *Am J Physiol Regul Integr* 278, R947–R955.

Nakano, H., Kubo, C., Sogawa, H., Teshima, H., and Nakagawa, T. (1992) Effect of REM sleep deprivation on immune function in rats. *Med Sci Res* 20, 65–66.

Naylor, E., Buxton, O.M., Bergmann, B.M., Easton, A., Zee, P.C., and Turek, F.W. (1998) Effects of aging on sleep in the golden hamster. *Sleep* 21, 687–693.

OConnor, K.A., Hansen, M.K., Pugh, C.R., Deak, M.M., Biedenkapp, J.C., Milligan, E.D., Johnson, J.D., Wang, H.C., Maier, S.F., Tracey, K.J., and Watkins, L.R. (2003) Further characterization of high mobility group box 1 (HMGB1) as a proinflammatory cytokine: Central nervous system effects. *Cytokine* 24, 254–265.

Opp, M.R., and Toth, L.A. (2003) Neural-immune interactions in the regulation of sleep. *Front Biosci* 8, D768–D779.

Ottaway, C.A., and Husband, A.J. (1992) Central nervous system influences on lymphocyte migration. *Brain Behav. Immun* 6, 97–116.

Pokk, P., and Vali, M. (2001) The effects of flumazenil, Ro 154513 and beta-CCM on the behaviour of control and stressed mice in the staircase test. *J Psychopharmacol* 15, 155–159.

Pollmacher, T., Haack, M., Schuld, A., Reichenberg, A., and Yirmiya, R. (2002) Low levels of circulating inflammatory cytokines--Do they affect human brain functions? *Brain Behav Immun* 16, 525–532.

Prinz, P.N., Bailey, S.L., and Woods, D.L. (2000) Sleep impairments in healthy seniors: Roles of stress, cortisol, and interleukin-1 beta. *Chronobiol Int* 17, 391–404.

Redwine, L., Dang, J., Hall, M., and Irwin, M. (2003) Disordered sleep, nocturnal cytokines, and immunity in alcoholics. *Psychosom Med* 65, 75–85.

Redwine, L., Hauger, R.L., Gillin, J.C., and Irwin, M. (2000) Effects of sleep and sleep deprivation on interleukin-6, growth hormone, cortisol, and melatonin levels in humans. *J Clin Endocrinol Metab* 85, 3597–3603.

Renegar, K.B., Floyd, R.A., and Krueger, J.M. (1998) Effects of short-term sleep deprivation on murine immunity to influenza virus in young adult and senescent mice. *Sleep* 21, 241–248.

Rogers, N.L., Szuba, M.P., Staab, J.P., Evans, D.L., Dinges, D.F. (2001) Neuroimmunologic aspects of sleep and sleep loss. Semin Clin Neuropsychiatry. 6(4): 295–307.

Savard, J., Laroche, L., Simard, S., Ivers, H., and Morin C.M. (2003) Chronic insomnia and immune functioning. *Psychosom Med* 65, 211–221.

Schiffelholz, T., and Lancel, M. (2001) Sleep changes induced by lipopolysaccharide in the rat are influenced by age. *Am J Physiol Regul Integr* 280, R398–R403.

Shearer, W.T., Reuben, J.M., Mullington, J.M., Price, N.J., Lee, B.N., Smith, E.O., Szuba, M.P., Van Dongen, H.P., and Dinges, D.F. (2000) Soluble TNF-alpha receptor 1 and IL-6 plasma levels in humans subjected to the sleep deprivation model of spaceflight. *J Interferon Cytokin Res* 20, 547–556.

Shephard, R.J., Castellani, J.W., and Shek, P.N. (1998) Immune deficits induced by strenuous exertion under adverse environmental conditions: manifestations and countermeasures. *Crit Rev Immunol* 18, 545–568.

Shimizu, T., Kawamura, T., Miyaji, C., Oya, H., Bannai, M., Yamamoto, S., Weerasinghe, A., Halder, R., Watanabe, H., Hatakeyama, K., and Abo, T. (2000) Altered interleukin-1 production in mice exposed to rotation stress. *Scand J Immunol* 51, 285–292.

Siegel, J.M. (2003) Why we sleep. *Sci Amer* 289, 92–97.

Sindermann, J., Kruse, A., Frercks, H.J., Schutz, R.M., and Kirchner, H. (1993) Investigations of the lymphokine system in elderly individuals. *Mech Age Dev* 70, 149–159.

Spath-Schwalbe, E., Hansen, K., Schmidt, F., Schrezenmeier, H., Marshall, L., Burger, K., Fehm, H.L., and Born, J. (1998) Acute effects of recombinant human interleukin-6 on endocrine and central nervous sleep functions in healthy men. *J Clin Endocrinol Metab* 83, 1573–1579.

SpathSchwalbe, E., Lange, T., Perras, B., Fehm, H.L., and Born, J. (2000) Interferon-alpha acutely impairs sleep in healthy humans. *Cytokine* 12, 518–521.

Steptoe, A., Willemsen, G., Owen, N., Flower, L., and MohamedAli, V. (2001) Acute mental stress elicits delayed increases in circulating inflammatory cytokine levels. *Clin Sci* 101, 185–192.

Toth, L.A., and Opp, M.R. (2001) Cytokine- and microbially induced sleep responses of interleukin-10 deficient mice. *Am J Physiol Regul Integr* 280, R1806–R1814.

Toth, L.A., and Rehg, J.E. (1998) Effects of sleep deprivation and other stressors on the immune and inflammatory responses of influenza-infected mice. *Life Sci* 63, 701–709.

Uchakin, P., Tobin, B., Cubbage, M., Marshall, G.J., and Sams C. (2001) Immune responsiveness following academic stress in first-year medical students. *J Interferon Cytokin Res* 21, 687–694.

VanCauter, E., Latta, F., Nedeltcheva, A., Spiegel, K., Leproult, R., Vandenbril, C., Weiss, R., Mockel, J., Legros, J.J., and Copinschi G. (2004) Reciprocal interactions between the GH axis and sleep *Growth Horm Igf Res* 14, S10–S17.

Vgontzas, A.N., Bixler, E.O., Lin, H.M., Prolo, P., Trakada, G., and Chrousos, G.P. (2005) IL-6 and its circadian secretion in humans. *Neuroimmunomodulation* 12, 131–140.

Vgontzas, A.N., and Chrousos, G.P. (2002) Sleep, the hypothalamic-pituitary-adrenal axis, and cytokines: multiple interactions and disturbances in sleep disorders. *Endocrinol Metab Clin N Am* 31, 15+.

Webster, J.I., Tonelli, L., and Sternberg, E.M. (2002) Neuroendocrine regulation of immunity. *Annu Rev Immunol* 20, 125–163.

Weinert, D. (2000) Age-dependent changes of the circadian system. *Chronobiol Int* 17, 261–283.

Weitzman, E.D., Zimmermann, J.C., Czeisler, C.A., and Ronda, J. (1983) Cortisol secretion is inhibited during sleep in normal men. *J Clin Endocrinol Metab* 56, 352–358.

Wilder, R.L. (1995) Neuroendocrine-immune system interactions and autoimmunity. *Annu Rev Immunol* 13, 307–338.

8 Cytokines and Sleep: Neuro-Immune Interactions and Regulations

Tetsuya Kushikata, Hitoshi Yoshida, and Tadanobu Yasuda

8.1 Introduction

Over past decades, many researches focused on sleep function. These theories proposed that sleep is needed to form some tissue construction. For example, theories of sleep function(s) deal with body functions such as body growth (Schussler et al. 2006) and immune enhancement (Bryant, Trinder, and Curtis 2004; Opp and Toth 2003) or with brain functions such as stimulation of memory consolidation (Takashima et al. 2006; Walker and Stickgold 2006) and maintenance of synaptic superstructure all involve some aspect of construction (Krueger, Obal, and Fang 1999). Much evidence shows that there is a bidirectional communication between immune system and sleep. A complex cytokine network is involved in sleep regulation (Opp 2005; Turrin and Plata-Salaman 2000). Inflammation-related cytokines have somnogenic or antisomnogenic effects depended on type of cytokines. We provide a short review dealing with an interaction between sleep and immune system.

8.2. Effect of Sleep on Immune System

8.2.1 Sleep Pattern and Immune Function

Alteration of sleep pattern has harmful effect on immune function. Sleep deprivation generally depress immune function. Some instance is found particular illness. Depression has sleep disturbance as a major symptom of depression. Such patients often have immune alteration. For example, natural killer activity is negatively correlated with degree of insomnia without any other symptom of depression such as anxiety, weight loss, cognitive disturbance, or diurnal variation (Irwin, Smith, and Gillin 1992). Patients with chronic insomnia are associated with a shift of interleukin-6 (IL-6) and tumor necrosis factor-α (TNF-α) secretion from nighttime to daytime (Vgontzas et al. 2002). Patient with alcoholism frequently has severe sleep loss in term of continuity and depth (Brower 2003). Sleep disorder increases infection and other immune-related diseases. Alcoholic patients showed lower levels of IL-6 production, suppression of the IL-6/IL-10 ratio, and a reduction of NK cell activity (Redwine, Dang, Hall, and Irwin 2003). Indeed, alcoholics tend to have serious infection disease like HIV, tuberculosis and hepatitis C (Irwin 2002). Sleep deprivation also affects immune response to vaccinations in healthy individuals. For example, only one night sleep disturbance impairs antibody response to hepatitis A vaccine (Lange, Perras, Fehm, and Born 2003). In addition, critically ill patients often have sleep abnormalities. In such patient, nearly half of total sleep time developed during daytime thus, circadian rhythm is nearly lost. These sleep disturbance induce cardiovascular consequence such as hypertension or arrhythmia that may contribute to increase of patients morbidity (Parthasarathy and Tobin 2004).

8.2.2 Sleep Deprivation and Immune System

Sleep deprivation impair immune system. In rats, long-term sleep deprivation increases food intake, energy expenditure and produce skin lesion, bowel ulceration then, finally lead to death (Rechtschaffen, Bergmann, Everson, Kushida, and Gilliland 1989). In such situation, animal showed bacterial translocation with mesenteric lymph nodes infection by bacteria that supposed to major cause of death (Everson and Toth 2000). Recently, in rats, sleep deprivation activate immune system to proinflammatory state but fail to prevent bacteria infection (Everson 2005). These results indicate that sleep deprivation itself seriously inhibits self-defense reaction to infections by impairing innate immune system.

8.3 Effect of Modulation of Immune System on Sleep

8.3.1 Overview

It is well known that nearly all infectious diseases and chronic inflammatory disorders affect sleep. Most individual have experienced the almost irresistible desire for sleep with the onset of "flu." Further, altered of sleep were described in infected rabbits (Toth and Krueger 1988) and rats (Kent, Price, and Satinoff 1988). These changes in sleep in response to infectious challenge are facet of the acute phase response (APR) (Krueger and Majde 1990; Krueger et al. 1994; Majde and Krueger 2005). Many kind of infection like viral, bacterial, fungal, or parasitic, have changes in amount of non-rapid eye movement sleep (NREMS) or rapid eye movement sleep (REMS) (Toth 1999). Excess sleep is reported in patients with infectious mononucleosis (Lambore, McSherry, and Kraus 1991), HIV-1 (Norman et al. 1992), rhinovirus-induced common cold (Drake, Roehrs, Royer, Koshorek, Turner, and Roth 2000). The degree of sleep alteration is depending on kind of infection. For example, sleep pattern in patients with infectious mononucleosis is not changed (Guilleminault and Mondini 1986). Patients seropositive for human immunodeficiency virus develop increase slow wave sleep even in the lack of clinical symptoms (Norman et al. 1992). Conversely, infection of rabies virus caused almost total loss of NREMS in mouse or human (Gourmelon, Briet, Clarencon, Court, and Tsiang 1991).

Influenza is one of common model as infection-related sleep alteration. Large dose of influenza virus given intravenously increased NREMS with fever rapidly in rabbits, whereas inactivated virus had not any effect on sleep (Kimura-Takeuchi, Majde, Toth, and Krueger 1992). Pretreatment of influenza virus (Kimura-Takeuchi et al. 1992) or synthetic double-stranded RNA of the virus inhibits these APR. These results suggest that pretreatment of either the virus itself or the dsRNA would products some substance that is able to inhibit the virus-induced APR.

8.3.2 Microbial Products, Cytokines, Sleep

Microbial-induced responses are mediated via enhanced cytokine production in which including somnogenic cytokine. Several substances or conditions have the capacity to enhance IL-1β and/or TNF-α. For example, infectious diseases increase NREMS and up-regulate IL-1β receptor (Alt et al. 2005; Aho, McNulty, and Coussens 2003). Pathological stimuli such as lipopolysaccharide or neurotropic virus enhance brain TNF-α production (Lang, Silvis, Deshpande, Nystrom, and Frost 2003). Several cytokines are involved in sleep regulation (Table 8.1).

TABLE 8.1. Cytokines involved in sleep regulation.

A. Somnogenic cytokines

IL-1β
IL-1$^\alpha$
TNF-α
IL-2
IL-15
IL-18
Epidermal growth factor
Acidic fibroblast growth factor
Neurotrophin 1 (nerve growth factor)
Neurotrophin 2 (brain-derived neurotropic factor)
Neurotrophin 3
Neurotrophin 4
Glia-derived neurotrophic factor
Platelet Activating Factor
Interferon-α
Interferon-γ
Granulocyte-macrophage colony stimulating factor
Granulocyte stimulating factor

B. Anti-Somnogenic cytokines

IL-4
IL-10
IL-13
Transforming growth factor β
Insulin-like growth factor
Soluble TNF receptor
Soluble IL-1 receptor

8.3.3 Proinflammatory Cytokines on Sleep

The role of humoral mechanisms on sleep regulation has been considered across centuries. Substance-based theories on the humoral mechanism were reported at early last century using sleep-deprived dogs by two independent laboratories,

Ishimori in Japan (Ishimori 1909) and Pièron in France. At 1960s, many studies have focused again on the humoral mechanism.

Pappenheimer, Koski, Fencl, Karnovsky, and Krueger (1975) reported that if cerebrospinal fluid obtained from sleep-deprived goat administered into rats, the rats dramatically increased sleep. The unidentified sleep-inducible substance was referred as factor S. Later, factor S was a muramyl peptide derived from bacterial peptideglycan that chemically unique cell wall component of all bacteria. Muramyl dipeptide and the factor S related peptideglucans were all induce IL-1β, a key immunoregulatory cytokine. IL-1β was shown as a potent somnogen and pyrogen (Majde et al. 2005; Opp 2004).

Since then, much evidence showed that proinflammatory cytokines such as IL-1β or TNF-α are key elements in sleep regulation. Administration of exogenous these substances induce increased in NREMS in various species (reviewed in Krueger, Majde, and Obal (2003)). Administration of exogenous TNF-α (Fang, Wang, and Krueger 1997; Kapas et al. 1992) or IL-1 (Fang, Wang, and Krueger 1998; Krueger, Walter, Dinarello, Wolff, and Chedid 1984) enhances NREMS in a variety of species. Normal sleep pattern was maintained following these substance administrations. Animals have a normal circadian cycle of wakefulness, NREMS, and REMS and easily aroused if disturbed.

Also, changes in brain temperature accompanied with sleep was maintained as normal, thus, autonomic changes with sleep stage was persist (Walter, Davenne, Shoham, Dinarello, and Krueger 1986). Both IL-1β and TNF-α induced an enhancement of electroencephalographic (EEG) slow wave activity (SWA), a parameter of sleep intensity (Pappenheimer et al. 1975). If either IL-1β (Yasuda, Yoshida, Garcia-Garcia, Kay, and Krueger 2005) or TNF-α (Yoshida, Peterfi, Garcia-Garcia, Kirkpatrick, Yasuda, and Krueger 2004) applied to rat cerebral cortex locally, SWA was enhanced only at the ipsilaterally. In addition, locally application of soluble receptor fragment, inhibitor of both IL-1β (Yasuda et al. 2005) and TNF-α (Yoshida et al. 2004), inhibited the enhancement following sleep deprivation. These results showed IL-1β and TNF-α have a capability of physiological sleep.

IL-1 and TNF and their receptor are present in normal brain (Krueger et al. 1999). Both IL-1 and TNF mRNA have diurnal rhythm. For example, in rats, IL-1β mRNA levels in the hypothalamus, cerebral cortex, brainstem, and hippocampus are highest during peak sleep periods (daytime) (Taishi, Bredow, Guha-Thakurta, Obal, and Krueger 1997) and increase in the brain after sleep deprivation (Mackiewicz, Sollars, Ogilvie, and Pack 1996). In cats, IL-1 cerebrospinal fluid levels varied in phase with the sleep–wake cycle (Lue, Bail, Jephthah-Ochola, Carayanniotis, Gorczynski, and Moldofsky 1988). TNF bioactivity levels in rat hypothalamus and cortex are about 10-fold greater during peak sleep periods than during waking hours (Floyd and Krueger 1997). After sleep deprivation, brain IL-1 and TNF mRNA were increased (Taishi et al. 1998). Plasma levels of IL-1 are highest at onset of sleep in human (Moldofsky, Lue, Eisen, Keystone, and Gorczynski 1986). Circulating levels of IL-1 and TNF are also affected by the sleep–wake cycle and sleep deprivation (Entzian, Linnemann, Schlaak, and Zabel 1996; Gudewill, Pollmacher, Vedder, Schreiber, Fassbender, and Holsboer 1992; Hohagen et al. 1993; Uthgenannt, Schoolmann, Pietrowsky, Fehm, and Born 1995), and their highest levels are associated with enhanced sleep or sleepiness. These data strongly suggest that both IL-1 and TNF are involved in sleep regulation.

Some condition that induces of IL-1 or TNF also increases excess NREMS. Therefore, microbial product such as muramyl peptides, viral double-stranded RNA, and lipopolysuccharide induce IL-1 and TNF, and sleep (Krueger and Majde 1994). Some inflammatory mediator involved in sleep regulation. Platelet activating factor (PAF) is one of a key inflammatory mediator (Mathiak, Szewczyk, Abdullah, Ovadia, and Rabinovici 1997). PAF and its receptor are found in brain (Bito, Kudo, and Shimizu 1993; Dray et al. 1989; Aihara, Ishii, Kume, and Shimizu 2000; Brodie 1994), and it affects or is affected of the production IL-1β (Fernandes et al. 2005), TNF$^-\alpha$ (Fernandes et al. 2005; Rola-Pleszczynski and Stankova 1992). PAF interacts production of several other sleep-regulatory substances such as nerve growth factor (Brodie, 1995), brain-derived neurotrophic factor and neurotrophin-3 (Noga, Englmann, Hanf, Grutzkau, Seybold, and Kunkel 2003), nitric oxide (Zhu and He 2005; Mariano, Bussolati, Migliori, Russo, Triolo, and Camussi 2003), prostaglandins (Teather, Lee, and Wurtman 2002; Teather and Wurtman 2003), and prolactin (Camoratto and Grandison 1989). Indeed, PAF enhances NREMS in rabbits (Kushikata, Fang, and Krueger 2006). These results are consistent with the hypothesis that the brain cytokine network is involved in physiological sleep regulation.

It is a common experience and an established experimental finding that mild increases in ambient temperature enhance sleep (Obal, Alfoldi, and Rubicsek 1995; Shoham and Krueger 1988; Szymusiak, Danowski, and McGinty 1991). Although the exact mechanisms of this effect remain unknown, many data suggest that sleep regulation and thermoregulation are closely linked (Krueger and Takahashi 1997). Many sleep-regulatory substances, including IL-1β and TNF-α, also affect thermoregulation (Krueger et al. 1995). A tumor necrosis factor receptor fragment (TNF-RF) inhibits warm-induced sleep responses in rabbits (Takahashi and Krueger 1997). However, somnogenic and pyrogenic capacity of both IL-1β and TNF-α are separable. For example, doses of both substances that increase of NREMS are lower than their pyrogenic doses. Coadministration of an antipyretic with IL-1 blocks IL-1-induced fever but not IL-1-induced sleep (Krueger et al. 1984). Inhibition of NO synthase with arginine analogs blocks IL-1-induced sleep but not IL-1-induced fever (Kapas, Fang, and Krueger 1994).

Inhibition of either IL-1β or TNF$^-\alpha$ decrease amount of NREMS. The regulation of proinflammatory cytokines is complex and, in brain, not very well understood. Nevertheless, some substances associated with specific cytokines such as the IL-1RA or the TNF and IL-1 soluble receptors seem to act as endogenous antagonists and indeed, these substances inhibit spontaneous sleep (Opp, Postlethwaite, Seyer, and Krueger 1992; Takahashi, Kapas, Fang, Seyer, Wang, and Krueger 1996; Takahashi, Kapas, and Krueger 1996). The blockade of TNF or IL-1 using antibodies (Opp and Krueger 1994; Takahashi et al. 1997) inhibits spontaneous sleep. These inhibitors also inhibit sleep rebound after sleep deprivation (Opp et al. 1994; Takahashi, Fang, Kapas, Wang, and Krueger 1997), the excess NREMS associated with acute mild increases in ambient temperature (Takahashi et al. 1997; Kushikata, Takahashi, Wang, Fang, and Krueger 1998) and the excess NREMS associated with administration of bacterial products (Takahashi et al. 1996). Mutant mice lacking the

TNF 55-kDa receptor or mice lacking the IL-1 type 1 receptor sleep less than their respective strain controls (Fang et al. 1997, 1998).

8.3.4 Anti-Inflammatory Cytokines on Sleep

There is a class of anti-inflammatory cytokines, which includes IL-4, IL-10, and IL-13, transforming growth factor-β-1 (TGF β). Each of these cytokines has a unique set of biological activities though both, in one manner or another, inhibit proinflammatory cytokines thus they could inhibit sleep. Indeed, IL-4 inhibits rabbit spontaneous sleep (Kushikata, Fang, Wang, and Krueger 1998), IL-10 inhibits spontaneous NREMS in rats (Opp, Smith, and Hughes 1995) and rabbits (Kushikata, Fang, and Krueger 1999), and IL-13 and TGF β inhibit sleep (Kubota, Fang, Kushikata, and Krueger 2000), thereby providing further evidence that proinflammatory cytokines are involved in physiological sleep regulation. IL-10 inhibits IL-1β and TNF-α (Kanaan, Poole, Saade, Jabbur, and Safieh-Garabedian, 1998; Fiorentino, Zlotnik, Mosmann, Howard, and O'Garra, 1991). IL-10 inhibits IL-1β and TNF-α production (Thomassen, Divis, and Fisher, 1996) and increases the production of the IL-1RA (Joyce, Steer, and Kloda 1996). IL-10 inhibits induction of IL-1 receptor type I and II gene expression (Dickensheets and Donnelly 1997). Further, exogenous IL-10 inhibits production or release of other substances implicated in sleep regulation, e.g., nitric oxide (Dugas, Palacios-Calender, Dugas, Riveros-Moreno, Delfraissy, Kolb, and Moncada 1998; Laffranchi, and Spinas, 1996) and insulin (Laffranchi et al. 1996). In addition, IL-10 increases sleep-inhibitory substances production, e.g., corticotrophin releasing factor (Stefano, Prevot, Beauvillain, and Hughes, 1998).

8.3.5 Molecular Level Modulation of Cytokines on Sleep

Sleep regulation is dependent, in part, on changes in gene expression and production of sleep-regulatory substances (Krueger et al. 1994). Nuclear factor kappa B (NF-κB) is a heterodimeric transcription factor and is a central regulator of proinflammatory cytokine induction. NF-κB is also activated by the same cytokines that promote sleep. For example, IL-1β, TNF-α (Grilli, Chiu, and Lenardo, 1993), nerve growth factor (Carter, Kaltschmidt, Kaltschmidt, Offenhauser, Bohm-Matthaei, Baeuerle, and Barde, 1996), epidermal growth factor (Obata, Biro, Arima, Kaieda, Kihara, Eto, Miyata, and Tanaka, 1996), and interferon α (Chaturvedi, Higuchi, and Aggarwal, 1994) all activate NF-κB. In contrast, several substances that inhibit sleep, for example, IL-4 (Clarke, Taylor-Fishwick, Hales, Chernajovsky, Sugamura, Feldmann, and Foxwell, 1995), IL-10 (Wang, Wu, Siegel, Egan, and Billah, 1995), glucocorticoids (Unlap and Jope, 1995), directly or indirectly, inhibit NF-κB activation. NF-κB activation promotes production of several additional substances thought to be involved in sleep regulation such as nitric oxide (Oddis and Finkel, 1996). Activation of NF-κB is enhanced by sleep deprivation (Chen, Gardi, Kushikata, Fang, and Krueger 1999). A NF-κB cell-permeable inhibitor peptide injected intracerebroventricularly in rats and rabbits significantly inhibited NREMS and REMS if administered during the light period. Moreover, pretreatment of rabbits with IL-1β 12 h before intracerebroventricular injection of the inhibitor peptide, significantly attenuated IL-1β-induced sleep and febrile responses (Kubota, Kushikata, Fang, and Krueger, 2000). These results suggest that NF-κB activation could be involved in sleep regulation interacted with many sleep-related cytokines.

8.4 Conclusion

The fundamental function of the immune system is to distinguish entities within the body as "self" or "nonself" and to eliminate those that are nonselfs. Once immune system is activated, several cytokines will be secreted from lymphocytes that are in proliferation and differentiation processes. On search of mechanism of sleep, links between the innate immune system and sleep regulation have been found out through the cytokine network. The link was unexpected initially, however, as much knowledge of sleep mechanism has been accumulated, the links seemed one of key element of significance of sleep.

References

Aho, A.D., McNulty, A.M., and Coussens, P.M. (2003) Enhanced expression of interleukin-1alpha and tumor necrosis factor receptor-associated protein 1 in ileal tissues of cattle infected with Mycobacterium avium subsp. paratuberculosis. *Infect Immun* 71, 6479–6486.

Aihara, M., Ishii, S., Kume, K., and Shimizu, T. (2000) Interaction between neurone and microglia mediated by platelet-activating factor. *Genes Cells* 5, 397–406.

Alt, J.A., Bohnet, S., Taishi, P., Duricka, D., Obal, F., Jr., Traynor, T., Majde, J.A., and Krueger, J.M. (2005) Influenza virus-induced glucocorticoid and hypothalamic and lung cytokine mRNA responses in dwarf lit/lit mice. *Brain Behav Immun*.

Bito, H., Kudo, Y., and Shimizu, T. (1993) Characterization of platelet-activating factor (PAF) receptor in the rat brain. *J Lipid Mediat* 6, 169–174.

Brodie, C. (1995) Platelet activating factor induces nerve growth factor production by rat astrocytes. *Neurosci Lett* 186, 5–8.

Brodie, C. (1994) Functional PAF receptors in glia cells: Binding parameters and regulation of expression. *Int J Dev Neurosci* 12, 631–640.

Brower, K.J. (2003) Insomnia, alcoholism and relapse. *Sleep Med Rev* 7, 523–539.

Bryant, P.A., Trinder, J., and Curtis, N. (2004) Sick and tired: Does sleep have a vital role in the immune system? *Nat Rev Immunol* 4, 457–467.

Camoratto, A.M., and Grandison, L. (1989) Platelet-activating factor stimulates prolactin release from dispersed rat anterior pituitary cells in vitro. *Endocrinology* 124, 1502–1506.

Carter, B.D., Kaltschmidt, C., Kaltschmidt, B., Offenhauser, N., Bohm-Matthaei, R., Baeuerle, P.A., and Barde, Y.A. (1996) Selective activation of NF-kappa B by nerve growth factor through the neurotrophin receptor p75. *Science* 272, 542–545.

Chaturvedi, M.M., Higuchi, M., and Aggarwal, B.B. (1994) Effect of tumor necrosis factors, interferons, interleukins, and growth factors on the activation of NF-kappa B: Evidence for lack of correlation with cell proliferation. *Lymphokine Cytokine Res* 13, 309–313.

Chen, Z., Gardi, J., Kushikata, T., Fang, J., and Krueger, J.M. (1999) Nuclear factor-kappaB-like activity increases in murine cerebral cortex after sleep deprivation. *Am J Physiol* 276, R1812–R1818.

Clarke, C.J., Taylor-Fishwick, D.A., Hales, A., Chernajovsky, Y., Sugamura, K., Feldmann, M., and Foxwell, B.M. (1995) Interleukin-4 inhibits kappa light chain expression and NF kappa B activation but not I kappa B alpha degradation in 70Z/3 murine pre-B cells. *Eur J Immunol* 25, 2961–2966.

Dickensheets, H.L., and Donnelly, R.P. (1997) IFN-gamma and IL-10 inhibit induction of IL-1 receptor type I and type II gene expression by IL-4 and IL-13 in human monocytes. *J Immunol* 159, 6226–6233.

Drake, C.L., Roehrs, T.A., Royer, H., Koshorek, G., Turner, R.B., and Roth, T. (2000) Effects of an experimentally induced rhinovirus cold on sleep, performance, and daytime alertness. *Physiol Behav* 71, 75–81.

Dray, F., Wisner, A., Bommelaer-Bayet, M.C., Tiberghien, C., Gerozissis, K., Saadi, M., Junier, M.P., and Rougeot, C. (1989) Prostaglandin E2, leukotriene C4, and platelet-activating factor receptor sites in the brain. Binding parameters and pharmacological studies. *Ann N Y Acad Sci* 559, 100–111.

Dugas, N., Palacios-Calender, M., Dugas, B., Riveros-Moreno, V., Delfraissy, J.F., Kolb, J.P., and Moncada, S. (1998) Regulation by endogenous interleukin-10 of the expression of nitric oxide synthase induced after ligation of CD23 in human macrophages. *Cytokine* 10, 680–689.

Entzian, P., Linnemann, K., Schlaak, M., and Zabel, P. (1996) Obstructive sleep apnea syndrome and circadian rhythms of hormones and cytokines. *Am J Respir Crit Care Med* 153, 1080–1086.

Everson, C.A. (2005) Clinical assessment of blood leukocytes, serum cytokines, and serum immunoglobulins as responses to sleep deprivation in laboratory rats. *Am J Physiol Regul Integr Comp Physiol* 289, R1054-63.

Everson, C.A., and Toth, L.A. (2000) Systemic bacterial invasion induced by sleep deprivation. *Am J Physiol Regul Integr Comp Physiol* 278, R905– R916.

Fang, J., Wang, Y., and Krueger, J.M. (1998) Effects of interleukin-1 beta on sleep are mediated by the type I receptor. *Am J Physiol* 274, R655– R660.

Fang, J., Wang, Y., and Krueger, J.M. (1997) Mice lacking the TNF 55 kDa receptor fail to sleep more after TNFalpha treatment. *J Neurosci* 17, 5949–5955.

Fernandes, E.S., Passos, G.F., Campos, M.M., de Souza, G.E., Fittipaldi, J.F., Pesquero, J.L., Teixeira, M.M., and Calixto, J.B. (2005) Cytokines and neutrophils as important mediators of platelet-activating factor-induced kinin B(1) receptor expression. *Br J Pharmacol.*

Fiorentino, D.F., Zlotnik, A., Mosmann, T.R., Howard, M., and O'Garra, A. (1991) IL-10 inhibits cytokine production by activated macrophages. *J Immunol* 147, 3815–3822.

Floyd, R.A., and Krueger, J.M. (1997) Diurnal variation of TNF alpha in the rat brain. *Neuroreport* 8, 915–918.

Gourmelon, P., Briet, D., Clarencon, D., Court, L., and Tsiang, H. (1991) Sleep alterations in experimental street rabies virus infection occur in the absence of major EEG abnormalities. *Brain Res* 554, 159–165.

Grilli, M., Chiu, J.J., and Lenardo, M.J. (1993) NF-kappa B and Rel: Participants in a multiform transcriptional regulatory system. *Int Rev Cytol* 143, 1–62.

Gudewill, S., Pollmacher, T., Vedder, H., Schreiber, W., Fassbender, K., and Holsboer, F. (1992) Nocturnal plasma levels of cytokines in healthy men. Eur *Arch Psychiatry Clin Neurosci* 242, 53–56.

Guilleminault, C., and Mondini, S. (1986) Mononucleosis and chronic daytime sleepiness. A long-term follow-up study. *Arch Intern Med* 146, 1333–1335.

Hohagen, F., Timmer, J., Weyerbrock, A., Fritsch-Montero, R., Ganter, U., Krieger, S., Berger, M., and Bauer, J. (1993) Cytokine production during sleep and wakeful-ness and its relationship to cortisol in healthy humans. *Neuropsychobiology* 28, 9–16.

Irwin, M. (2002) Effects of sleep and sleep loss on immunity and cytokines. *Brain Behav Immun* 16, 503–512.

Irwin, M., Smith, T.L., and Gillin, J.C. (1992) Electroencephalographic sleep and natural killer activity in depressed patients and control subjects. *Psychosom Med* 54, 10–21.

Ishimori, K. (1909) True cause of sleep: A hypnogenic substance as evidenced in the brain of sleep-deprived animals. *Tokyo Igakkai Zassi* 23, 429–459.

Joyce, D.A., Steer, J.H., and Kloda, A. (1996) Dexamethasone antagonizes IL-4 and IL-10-induced release of IL-1RA by monocytes but augments IL-4-, IL-10-, and TGF-beta-induced suppression of TNF-alpha release. *J Interferon Cytokine Res* 16, 511–517.

Kanaan, S.A., Poole, S., Saade, N.E., Jabbur, S., and Safieh-Garabedian, B. (1998) Interleukin-10 reduces the endotoxin-induced hyperalgesia in mice. *J Neuro-immunol* 86, 142–150.

Kapas, L., Fang, J., and Krueger, J.M. (1994) Inhibition of nitric oxide synthesis inhibits rat sleep. *Brain Res* 664, 189–196.

Kapas, L., Hong, L., Cady, A.B., Opp, M.R., Postlethwaite, A.E., Seyer, J.M., and Krueger, J.M. (1992) Somnogenic, pyrogenic, and anorectic activities of tumor necrosis factor-alpha and TNF-alpha fragments. *Am J Physiol* 263, R708–R715.

Kent, S., Price, M., and Satinoff, E. (1988) Fever alters characteristics of sleep in rats. *Physiol Behav* 44, 709–715.

Kimura-Takeuchi, M., Majde, J.A., Toth, L.A., and Krueger, J.M. (1992) Influenza virus-induced changes in rabbit sleep and acute phase responses. *Am J Physiol* 263, R1115–R1121.

Krueger, J.M., and Majde, J.A. (1990) Sleep as a host defense: its regulation by microbial products and cytokines. *Clin Immunol Immunopathol* 57, 188–199.

Krueger, J.M., and Majde, J.A. (1994) Microbial products and cytokines in sleep and fever regulation. *Crit Rev Immunol* 14, 355–379.

Krueger, J.M., Majde, J.A., and Obal, F. (2003) Sleep in host defense. *Brain Behav Immun* 17(Suppl 1), S41–S47.

Krueger, J.M., Obal, F., Jr., and Fang, J. (1999) Why we sleep: A theoretical view of sleep function. *Sleep Med Rev* 3, 119–129.

Krueger, J.M., Obal, F., Jr., and Fang, J. (1999) Humoral regulation of physiological sleep: cytokines and GHRH. *J Sleep Res* 8(Suppl 1), 53–59.

Krueger, J.M., and Takahashi, S. (1997) Thermoregulation and sleep. Closely linked but separable. *Ann N Y Acad Sci* 813, 281–286.

Krueger, J.M., Takahashi, S., Kapas, L., Bredow, S., Roky, R., Fang, J., Floyd, R., Renegar, K.B., Guha-Thakurta, N., and Novitsky, S. (1995) Cytokines in sleep regulation. *Adv Neuroimmunol* 5, 171–188.

Krueger, J.M., Toth, L.A., Floyd, R., Fang, J., Kapas, L., Bredow, S., and Obal, F., Jr. (1994) Sleep, microbes and cytokines. *Neuroimmunomodulation* 1, 100-109.

Krueger, J.M., Walter, J., Dinarello, C.A., Wolff, S.M., and Chedid, L. (1984) Sleep-promoting effects of endogenous pyrogen (interleukin-1). *Am J Physiol* 246, R994–R999.

Kubota, T., Fang, J., Kushikata, T., and Krueger, J.M. (2000) Interleukin-13 and transforming growth factor-beta1 inhibit spontaneous sleep in rabbits. *Am J Physiol Regul Integr Comp Physiol* 279, R786–R792.

Kubota, T., Kushikata, T., Fang, J., and Krueger, J.M. (2000) Nuclear factor-kappaB inhibitor peptide inhibits spontaneous and interleukin-1beta-induced sleep. *Am J Physiol Regul Integr Comp Physiol* 279, R404–R413.

Kushikata, T., Fang, J., and Krueger, J.M. (2006) Platelet activating factor and its metabolite promote sleep in rabbits. *Neurosci Lett* 394, 233–238.

Kushikata, T., Fang, J., and Krueger, J.M. (1999) Interleukin-10 inhibits spontaneous sleep in rabbits. *J Interferon Cytokine Res* 19, 1025–1030.

Kushikata, T., Fang, J., Wang, Y., and Krueger, J.M. (1998) Interleukin-4 inhibits spontaneous sleep in rabbits. *Am J Physiol* 275, R1185–R1191.

Kushikata, T., Takahashi, S., Wang, Y., Fang, J., and Krueger, J.M. (1998) An interleukin-1 receptor fragment blocks ambient temperature-induced increases in brain temperature but not sleep in rabbits. *Neurosci Lett* 244, 125–128.

Laffranchi, R., and Spinas, G.A. (1996) Interleukin 10 inhibits insulin release from and nitric oxide production in rat pancreatic islets. *Eur J Endocrinol* 135, 374–378.

Lambore, S., McSherry, J., and Kraus, A.S. (1991) Acute and chronic symptoms of mononucleosis. *J Fam Pract* 33, 33–37.

Lang, C.H., Silvis, C., Deshpande, N., Nystrom, G., and Frost, R.A. (2003) Endotoxin stimulates in vivo expression of inflammatory cytokines tumor necrosis factor alpha, interleukin-1beta, -6, and high-mobility-group protein-1 in skeletal mus-cle. *Shock* 19, 538–546.

Lange, T., Perras, B., Fehm, H.L., and Born, J. (2003) Sleep enhances the human antibody response to hepatitis A vaccination. *Psychosom Med* 65, 831–835.

Lue, F.A., Bail, M., Jephthah-Ochola, J., Carayanniotis, K., Gorczynski, R., and Moldofsky, H. (1988) Sleep and cerebrospinal fluid interleukin-1-like activity in the cat. *Int J Neurosci* 42, 179–183.

Mackiewicz, M., Sollars, P.J., Ogilvie, M.D., and Pack, A.I. (1996) Modulation of IL-1 beta gene expression in the rat CNS during sleep deprivation. *Neuroreport* 7, 529–533.

Majde, J.A., and Krueger, J.M. (2005) Links between the innate immune system and sleep. *J Allergy Clin Immunol* 116, 1188–1198.

Mariano, F., Bussolati, B., Migliori, M., Russo, S., Triolo, G., and Camussi, G. (2003) Platelet-activating factor synthesis by neutrophils, monocytes, and endothelial cells is modulated by nitric oxide production. *Shock* 19, 339–344.

Mathiak, G., Szewczyk, D., Abdullah, F., Ovadia, P., and Rabinovici, R. (1997) Platelet-activating factor (PAF) in experimental and clinical sepsis. *Shock* 7, 391–404.

Moldofsky, H., Lue, F.A., Eisen, J., Keystone, E., and Gorczynski, R.M. (1986) The relationship of interleukin-1 and immune functions to sleep in humans. *Psycho-som Med* 48, 309–318.

Noga, O., Englmann, C., Hanf, G., Grutzkau, A., Seybold, J., and Kunkel, G. (2003) The production, storage and release of the neurotrophins nerve growth factor, brain-derived neurotrophic factor and neurotrophin-3 by human peripheral eosinophils in allergics and non-allergics. *Clin Exp Allergy* 33, 649–654.

Norman, S.E., Chediak, A.D., Freeman, C., Kiel, M., Mendez, A., Duncan, R., Simoneau, J., and Nolan, B. (1992) Sleep disturbances in men with asymptomatic human immunodeficiency (HIV) infection. *Sleep* 15, 150–155.

Obal, F., Jr., Alfoldi, P., and Rubicsek, G. (1995) Promotion of sleep by heat in young rats. *Pflugers Arch* 430, 729–738.

Obata, H., Biro, S., Arima, N., Kaieda, H., Kihara, T., Eto, H., Miyata, M., and Tanaka, H. (1996) NF-kappa B is induced in the nuclei of cultured rat aortic smooth muscle cells by stimulation of various growth factors. *Biochem Biophys Res Commun* 224, 27–32.

Oddis, C.V., and Finkel, M.S. (1996) NF-kappa B and GTP cyclohydrolase regulate cytokine-induced nitric oxide production by cardiac myocytes. *Am J Physiol* 270, H1864–H1868.

Opp, M.R. (2004) Cytokines and sleep: The first hundred years. *Brain Behav Immun* 18, 295–297.

Opp, M.R. (2005) Cytokines and sleep. *Sleep Med Rev* 9, 355–364.

Opp, M.R., and Krueger, J.M. (1994) Interleukin-1 is involved in responses to sleep deprivation in the rabbit. *Brain Res* 639, 57–65.

Opp, M.R., Postlethwaite, A.E., Seyer, J.M., and Krueger, J.M. (1992) Interleukin 1 receptor antagonist blocks somnogenic and pyrogenic responses to an interleukin 1 fragment. *Proc Natl Acad Sci USA* 89, 3726–3730.

Opp, M.R., Smith, E.M., and Hughes, T.K., Jr. (1995) Interleukin-10 (cytokine synthesis inhibitory factor) acts in the central nervous system of rats to reduce sleep. *J Neuroimmunol* 60, 165–168.

Opp, M.R., and Toth, L.A. (2003) Neural-immune interactions in the regulation of sleep. *Front Biosci* 8, d768–d779.

Pappenheimer, J.R., Koski, G., Fencl, V., Karnovsky, M.L., and Krueger, J. (1975) Extraction of sleep-promoting factor S from cerebrospinal fluid and from brains of sleep-deprived animals. *J Neurophysiol* 38, 1299–1311.

Parthasarathy, S., and Tobin, M.J. (2004) Sleep in the intensive care unit. *Intensive Care Med* 30, 197–206.

Rechtschaffen, A., Bergmann, B.M., Everson, C.A., Kushida, C.A., and Gilliland, M.A. (1989) Sleep deprivation in the rat: X. Integration and discussion of the find-ings. *Sleep* 12, 68–87.

Redwine, L., Dang, J., Hall, M., and Irwin, M. (2003) Disordered sleep, nocturnal cytokines, and immunity in alcoholics. *Psychosom Med* 65, 75–85.

Rola-Pleszczynski, M., and Stankova, J. (1992) Differentiation-dependent modulation of TNF production by PAF in human HL-60 myeloid leukemia cells. *J Leukoc Biol* 51, 609–616.

Schussler, P., Uhr, M., Ising, M., Weikel, J.C., Schmid, D.A., Held, K., Mathias, S., and Steiger, A. (2006) Nocturnal ghrelin, ACTH, GH and cortisol secretion after sleep deprivation in humans. *Psychoneuroendocrinology* 31, 915–923.

Shoham, S., and Krueger, J.M. (1988) Muramyl dipeptide-induced sleep and fever: effects of ambient temperature and time of injections. *Am J Physiol* 255, R157–R165.

Stefano, G.B., Prevot, V., Beauvillain, J.C., and Hughes, T.K. (1998) Interleukin-10 stimulation of corticotrophin releasing factor median eminence in rats: Evidence for dependence upon nitric oxide production. *Neurosci Lett* 256, 167–170.

Szymusiak, R., Danowski, J., and McGinty, D. (1991) Exposure to heat restores sleep in cats with preoptic/anterior hypothalamic cell loss. *Brain Res* 541, 134–138.

Taishi, P., Bredow, S., Guha-Thakurta, N., Obal, F., Jr., and Krueger, J.M. (1997) Diurnal variations of interleukin-1 beta mRNA and beta-actin mRNA in rat brain. *J Neuroimmunol* 75, 69–74.

Taishi, P., Chen, Z., Obal, F., Jr., Hansen, M.K., Zhang, J., Fang, J., and Krueger, J.M. (1998) Sleep-associated changes in interleukin-1beta mRNA in the brain. *J Interferon Cytokine Res* 18, 793–798.

Takahashi, S., Fang, J., Kapas, L., Wang, Y., and Krueger, J.M. (1997) Inhibition of brain interleukin-1 attenuates sleep rebound after sleep deprivation in rabbits. *Am J Physiol* 273, R677–R682.

Takahashi, S., Kapas, L., Fang, J., Seyer, J.M., Wang, Y., and Krueger, J.M. (1996) An interleukin-1 receptor fragment inhibits spontaneous sleep and muramyl dipep-tide-induced sleep in rabbits. *Am J Physiol* 271, R101–R108.

Takahashi, S., Kapas, L., Krueger, J.M. (1996) A tumor necrosis factor (TNF) receptor fragment attenuates TNF-alpha- and muramyl dipeptide-induced sleep and fever in rabbits. *J Sleep Res* 5, 106–114.

Takahashi, S., and Krueger, J.M. (1997) Inhibition of tumor necrosis factor prevents warming-induced sleep responses in rabbits. *Am J Physiol* 272, R1325–R1329.

Takashima, A., Petersson, K.M., Rutters, F., Tendolkar, I., Jensen, O., Zwarts, M.J., McNaughton, B.L., and Fernandez, G. (2006) Declarative memory consolidation in humans: A prospective functional magnetic resonance imaging study. *Proc Natl Acad Sci USA* 103, 756–761.

Teather, L.A., Lee, R.K., and Wurtman, R.J. (2002) Platelet-activating factor increases prostaglandin E(2) release from astrocyte-enriched cortical cell cultures. *Brain Res* 946, 87–95.

Teather, L.A., and Wurtman, R.J. (2003) Cyclooxygenase-2 mediates platelet-activating factor-induced prostaglandin E2 release from rat primary astrocytes. *Neurosci Lett* 340, 177–80.

Thomassen, M.J., Divis, L.T., and Fisher, C.J. (1996) Regulation of human alveolar macrophage inflammatory cytokine production by interleukin-10. *Clin Immunol Immunopathol* 80, 321–324.

Toth, L. (1999) *Microbial Modulation of Sleep.* CDC Press, Boca Raton, FL.

Toth, L.A., and Krueger, J.M. (1988) Alteration of sleep in rabbits by Staphylococcus aureus infection. *Infect Immun* 56, 1785–1791.

Turrin, N.P., and Plata-Salaman, C.R. (2000) Cytokine-cytokine interactions and the brain. *Brain Res Bull* 51, 3–9.

Unlap, T., and Jope, R.S. (1995) Inhibition of NFkB DNA binding activity by gluco-corticoids in rat brain. *Neurosci Lett* 198, 41–44.

Uthgenannt, D., Schoolmann, D., Pietrowsky, R., Fehm, H.L., and Born, J. (1995) Effects of sleep on the production of cytokines in humans. *Psychosom Med* 57, 97–104.

Vgontzas, A.N., Zoumakis, M., Papanicolaou, D.A., Bixler, E.O., Prolo, P., Lin, H.M., Vela-Bueno, A., Kales, A., and Chrousos, G.P. (2002) Chronic insomnia is associated with a shift of interleukin-6 and tumor necrosis factor secretion from nighttime to daytime. *Metabolism* 51, 887–892.

Walker, M.P., and Stickgold, R. (2006) Sleep, memory, and plasticity. *Annu Rev Psychol* 57, 139–166.

Walter, J., Davenne, D., Shoham, S., Dinarello, C.A., and Krueger, J.M. (1986) Brain temperature changes coupled to sleep states persist during interleukin 1-enhanced sleep. *Am J Physiol* 250, R96–R103.

Wang, P., Wu, P., Siegel, M.I., Egan, R.W., and Billah, M.M. (1995) Interleukin (IL)-10 inhibits nuclear factor kappa B (NF kappa B) activation in human mono-cytes. IL-10 and IL-4 suppress cytokine synthesis by different mechanisms. *J Biol Chem* 270, 9558–9563.

Yasuda, T., Yoshida, H., Garcia-Garcia, F., Kay, D., and Krueger, J.M. (2005) Interleukin-1beta has a role in cerebral cortical state-dependent electroencepha-lographic slow-wave activity. *Sleep* 28, 177–184.

Yoshida, H., Peterfi, Z., Garcia-Garcia, F., Kirkpatrick, R., Yasuda, T., and Krueger, J.M. (2004) State-specific asymmetries in EEG slow wave activity induced by local application of TNFalpha. *Brain Res* 1009, 129–136.

Zhu, L., and He, P. (2005) Platelet-activating factor increases endothelial [Ca2+]i and NO production in individually perfused intact microvessels. *Am J Physiol Heart Circ Physiol* 288, H2869–H2877.

9 Selective REM Sleep Deprivation and Its Impact on the Immune Response

Javier Velazquez Moctezuma, José Ángel Rojas Zamorano, Enrique Esqueda León, Andrés Quintanar Stephano, and Anabel Jiménez Anguiano

9.1 Introduction

Shortly after the seminal paper of Aserinsky and Kleitman (1953) in which they reported the existence of the rapid eye movements sleep stage, the French school lead by Jouvet called this sleep stage as Sommeil Paradoxale, i.e., paradoxical sleep, mainly because some of its features were difficult to explain at that time (Jouvet and Michel 1960a, 1960b). After almost 50 years of intensive experimental analysis concerning this stage, the full understanding of the biological role of REM sleep remains elusive.

Based on one of the main features of REM sleep, i.e., muscular atonia, Jouvet and his group designed the so-called "island technique" for selective REM sleep deprivation in animals. In brief, this technique consists in placing the animal (mainly cats and rats) on a small platform surrounded by water. The diameter of the platform is such that the animal must remain standing or sit to remain out of the water.

Thus, when atonia characteristic of REM sleep occurs, the animal awakes as it contacts the water and is therefore prevented from falling into this sleep phase. In this way it has been easy to selectively deprive an animal of REM sleep without greatly affecting slow wave sleep (Jouvet, Vimont, Delorme, and Jouvet 1964; Jouvet-Mounier, Vimont, and Delorme 1965). For years this was the technique used by those who work with animals and REM deprivation and currently, some improvements have made it more suitable.

For a number of years a great volume of published information dealing with REM sleep function has been obtained from experiments using this technique of selective deprivation (Vogel 1975). Unfortunately, the technique has not been free of criticism. From the beginning, it was clear that the technique has a stress component (Mark, Heiner, and Godin 1970). Most of the authors involved in this kind of research have recognized that the particular results that each one of them has obtained are contaminated with a stress component (Kovalzon and Tsibulsky 1984; Coenen and Van Luijtelaar 1985).

Several attempts have been made to generate a suitable control of stress, however, most have been unsuccessful (Coenen and Van Hulzen 1980).

Nevertheless, the island technique has been used to analyze the effect of prolonged periods (i.e., several days) of selective REM deprivation and despite the almost unavoidable stress component, it has been possible to elucidate some of the functions that this sleep stage exerts on an organism. Moreover, some other techniques have been tested for selective REM sleep deprivation and none of them has been adopted by the scientific community involved in this issue (Coenen and Van Luijtelaar 1985; Elomaa 1979; Van Hulzen and Coenen 1980; Van Luijtelaar and Coenen 1982, 1986; Bergmann, Kushida, Everson, Gilliland, Obermeyer, and Rechtschaffen 1989).

The stress component of the island technique becomes an essential issue when someone is trying to study the effect of REM deprivation on the immune response.

This is relevant because it is well known that stressful situations have an impact on the immune system. Once the hypothalamus-hypophysis-adrenal axis is activated by a particular stressful situation, one of the consequences will be the rise of circulating levels of corticosterone from the adrenal glands. This steroid hormone has multiple actions on the organism. One of the most conspicuous actions of corticosterone is the suppression of the immune response (Mendelson, Guthrie, Guynn, Harris, and Wyatt 1974).

9.2 Human Studies

The partial loss of sleep is commonly observed in modern societies and surprisingly, there are only few studies concerning partial loss of sleep and its impact on the immune system.

Concerning human studies, the selective deprivation of REM sleep has been extremely difficult to perform. A fully trained staff is needed to watch the polygraphic recording continuously and awake the subject as soon as it falls into REM sleep. As the sleep recording progresses, the pressure of the subject to enter in REM sleep increases and the attempts become more frequent, making it very difficult to reliably prevent REM sleep (Endo et al. 1988).

Mainly due to this technical difficulty, most of the research has shifted toward total sleep deprivation or partial sleep deprivation, the latter by suppressing the second half of the night where it has been shown that the highest percentage of REM sleep occurs (Irwin, Mascovich, Gillin, Willoughby, Pike, and Smith 1994). The effects on the immune system of this manipulation have been attributed to the loss of REM sleep.

The pioneering study on the relationship between sleep loss and the immune system in humans was performed by Palmblad et al. three decades ago. In this study, the group of Palmblad kept a group of healthy women awake for 77 h by submitting them to a simulated stressful situation of a battlefield. They did not find differences before, during or after sleep deprivation in the number of polymorphonuclear leukocytes, monocytes, or lymphocytes-B.

However, interferon production was increased, while leukocyte phagocytic activity decreased (Palmblad 1976). In a second study, the group assessed the effect of 64 h of sleep deprivation in healthy men. Compared to control values, sleep deprived subjects displayed a decrease in lymphocyte blastogenesis induced by phytohemaglutinin (PHA) as well as DNA synthesis of blood lymphocytes

(Palmblad 1979). Seeking for possible mediators of these effects the same group assessed the effect of sleep deprivation on thyroid hormones and catecholamines, finding an increase in thyroid hormones while catecholamines remained unchanged (Palmblad 1979).

Nevertheless, the effect of sleep deprivation on lymphocyte blastogenesis seems to depend on the factor used to induce it. Moldofsky et al. reported that when blastogenesis was induced in blood samples from sleep deprived males, an increase can be observed if the stimulation is done with pokeweed mitogen (PWM).

This group also reported that both PHA and PWM induced an increase of blastogenesis during the control observation before REM deprivation and a decrease when the observations were done during the recovery from sleep deprivation (Moldofsky, Lue, Davidson, and Gorczynski 1989a; Moldofsky, Lue, Davidson, Jephthah-Ochoa, Carayanniotis, and Gorczynski 1989b).

Dinges and his group perform a detailed study in 20 healthy young adults of both sexes. After 64 h of total sleep deprivation, they reported an increase total count of white blood cells, granulocytes, monocytes, and natural killer (NK) cells. Discordant to the abovementioned studies, Dinges et al. did not find any change in blastogenesis of lymphocytes using PHA, PWA and also Concanavalin A. The assessment of interleukins 2, 6, 12, and TNF-α did not show significant differences (Dinges et al. 1994).

In the same year, Irwin and colleagues studied the relationship between sleep and the immune system using partial sleep deprivation. Healthy male subjects were forced to awaken one night at 03:00 h and their blood was assayed for NK cell activity between 07:00 and 09:00 h. They found a reduction of NK activity after the partial loss of sleep that was directly correlated with the activity values observed during the basal assessment, i.e., the higher the activity at baseline, the higher the decrease after sleep deprivation.

Similar results were found years later in a study perform in healthy volunteers with 48 h of total sleep deprivation. The results showed that the proportion of NK cells was decreased during sleep deprivation and returned to control values during the recovery period (Ozturk, Pelin, Zarandeniz, Kaynak, Cakar, and Gozukirmizi 1999).

In a splendid review, Michael Irwin pointed out that only small amounts of sleep lost is needed to experience an impact on both the cellular and the humoral immunity in which interleukins 1 and 6 as well as the natural killer cells are particularly involved (Irwin 2002). In addition, new sources of information concerning the relationship between sleep loss and the immune response are currently represented by several psychiatric disorders in which sleep is impaired and the immune response is altered, as in depression and alcoholism (Redwine, Dang, Hall, and Irwin 2002).

9.3 Animal Studies

As it was mentioned before, the main technique used to achieve selective deprivation of REM sleep has been the so-called island technique. The stress component inherent to this technique has also been recognized by all the sleep researchers.

As stress is a well known factor that alters the immune system, it is critical to avoid any stress component in the analysis of the effect of selective REM deprivation on the immune response. Several attempts have been made to diminish the stress component of REM deprivation.

In Brasil, Sergio Tufik's group has worked on successive modifications of the island technique. They have reported that a maximum of stress elimination is achieved using multiple platforms and using animals that were previously kept in the same cage and conform what they call, a socially stable group (Suchecki and Tufik 2000; Suchecki, Duarte Palma, and Tufik 2000)

It has been reported that REM sleep deprivation induces a reduction of primary antibody response to sheep erythrocytes and a decrease antigen uptake in the spleen and liver (Moldofsky 1994). Conversely, when Allan Rechstshaffen and his group reported a new method for selective REM sleep deprivation, they included a study in which they analyzed the effect of this manipulation on some immune functional parameters. They reported that REM deprivation had no effect on the number and the functional activity of spleen cells (Benca 1989).

The stress component in the island technique for REM deprivation is recognized. Despite of this recognition, most of the studies perform to elucidate the functions of REM sleep, have been done using this technique. This situation has elicited the need to separate the effects of REM sleep deprivation from the effects of stress. Furthermore, it is common to find that some authors use the terms stress and REM deprivation as the same phenomenon.

In an effort to define the effects of REM deprivation and the effects of a common stressor on the immune system, we have performed a study comparing REM deprivation with a frequently used stressor in rats, immobilization.

This immobilization was achieved by placing the rats inside a plexiglass cylinder (10 cm long; 4 cm diameter). REM sleep deprivation was achieved by placing the animal on a small platform, surrounded by water up to 1 cm above the platform. Rats were kept in individual chambers.

In the long-term deprived group, water was daily changed. In addition, as the stress response has a habituation component, the observations were made both acutely and chronically.

Independent groups of male adult rats were submitted to short- (24 h) or long-term (240 h) selective REM sleep deprivation as well as to short (6 h) or long-term immobilization (6 h daily for 10 days). In addition, a control group of intact animals was used. Once the procedure ended, blood was obtained from cardiac puncture and lymphocyte population bearing surface markers were determined using selective antibodies that were analyzed through a flow cytometer.

These results indicate that the immune system responds selectively to specific demands that the environment imposes on the organism. In addition, it seems that REM sleep deprivation has a direct effect on the immune system in which a stress component inherent to the deprivation technique is not influencing such effects.

Moreover, it seems that the nature of the stressor will determine the characteristics of the immune response. It is relevant to remember that decades ago, it was shown that during REM sleep, there is an increase of protein synthesis in the brain.

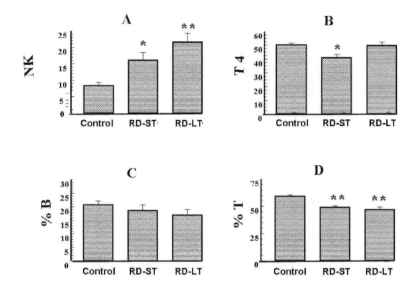

FIGURE 9.1. Effect fo short-term (RD-ST = 24 hours) and long-term (RD-LT = 240 hours) REM sleep deprivation on different subsets of lymphocytes. *=p<0.05; **=<0.01 Dunnet test compared to control.

As it can been seen in Fig. 9.1, REM sleep deprivation induced an increase in the number of NK cells, significant for the short term and augmenting even more after the long-term deprivation.

It seems that this effect is not limited by a habituation process. It would be expected that after constant submission to a stressful situation, the organism would trigger adaptation mechanisms tending to reverse the alterations due to the basic impact of the stressor. On the other hand, concerning REM sleep deprivation and the NK cells, there is an increase of the effect after 10 days of REM deprivation.

Conversely, T4 lymphocytes showed an initial decrease that seemed to be compensated in the long term and the values return to control levels.

The percentage of B lymphocytes showed a trend to decrease that did not reach significant levels. Concerning the percentage of T lymphocytes, 24 h of REM deprivation induced a significant reduction, which remained similar after 10 days of REM deprivation.

Concerning the effects of immobilization, the picture is quite different from that observed after REM deprivation. Both short- and long-term immobilization stresses did not induce any change in the NK cells, nor the T4 lymphocytes (Fig. 9.2).

The percentage of B lymphocytes showed an increase after short-term REM deprivation that is not longer observed after 10 days of immobilization.

On the other hand, the percentage of T lymphocytes showed an initial decrease that is reversed after constant application of the stressor.

Thus, it seems that the initial response to the stressor is efficiently compensated by the habituation mechanisms triggered by the organism.

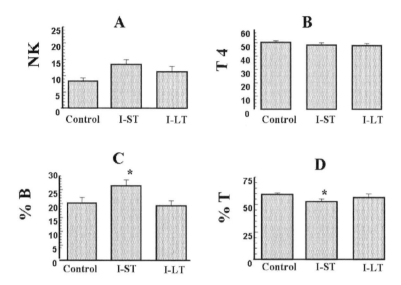

FIGURE 9.2. Effect of short-term (I-ST = 6 hours/one day) and long-term (I-Lt = 6 hours/10 days) immobilization stress on different subsets of lymphocytes. *=p<0.05 Dunnet test compared to control.

Therefore, it is possible that during this stage, chemical signals from the brain are synthesized in order to modulate the activity of the immune system to respond to external challenges. The characteristics of these chemical signals and their effects on the immune system will depend on the nature of the method and the duration of the deprivation.

Acknowledgment. The authors would like to express their gratitude to Dr. Aldebarán Prospero Garcia for her helpful advice on manuscript language.

References

Aserinsky, E., and Kleitman, N. (1953) Regularly occurring periods of eye motility, and concomitant phenomena, during sleep. *Science* 118(3062), 273–274.

Benca, R.M., Kushida, C.A., Everson, C.A., Kalski, R., Bergmann, B.M., Rechtschaffen A. (1989) Sleep deprivation in the rat: VII. Immune function. *Sleep* 12(1): 47–52.

Bergmann, B.M., Kushida, C.A., Everson, C.A., Gilliland, M.A., Obermeyer, W., and Rechtschaffen, A. (1989) Sleep deprivation in the rat: II. Methodology. *Sleep* 12, 5–12.

Coenen, A.M.L., and Van Luijtelaar, E.L.J.M. (1985) Stress induced by three procedures of deprivation of paradoxical sleep. *Physiol Behav* 35, 501–504.

Coenen, A.M., and Van Hulzen, Z.J. (1980) Paradoxical sleep deprivation in animal studies: Some methodological considerations. *Prog Brain Res* 53, 325–330.

Elomaa, E. (1979) The cuff pedestal: An alternative to flowerpots? *Physiol Behav* 23(4), 669–672.

Dinges, D., Douglas, S., Zaugg, L., Campbell, D., McMann, J., Whitehouse, W., Orne, E., Kapoor, S., Icaza, E., and Orne, M. (1994) Leukocytosis and natural killer cell function parallel neurobehavioral fatigue induced by 64 hours of sleep deprivation. *J Clin Invest* 93 1930–1939.

Dinges, D.F., Douglas, S.D., Hamarman, S., Zaugg, L., Kapoor, S. (1995) Sleep deprivation and human immune function. Adv Neuroimmunol. 5(2): 97–110.

Endo, T., Roth, C., Landolt, H.P., Werth, E., Aeschbach, D., Achermann, P., and Borbély, A.A. (1998) Selective REM sleep deprivation in humans: Effects on sleep and sleep EEG. *Am J Physiol Regul Integr Comp Physiol* 274, 1186–1194.

Horne, J.A., McGrath, M.J. (1984) The consolidation hypothesis for REM sleep function: stress and other confounding factors–a review. *Biol Psychol* 18(3): 165–184.

Irwin, M., Mascovich, A., Gillin, J.C., Willoughby, R., Pike, J., and Smith, T. (1994) Partial sleep deprivation reduces natural killer cell activity in humans. *Psychosom Med* 56, 493–498.

Irwin, M. (2002) Effects of sleep and sleep loss on immunity and cytokines. *Brain Behav Immun* 16(5), 503–512.

Jouvet, M., and Michel, F. (1960a) New research on the structures responsible for the "paradoxical phase" of sleep. *J Physiol (Paris)* 52, 130–131.

Jouvet, M., and Michel, F. (1960b) Release of the "paradoxal phase" of sleep by stimulation of the brain stem in the intact and chronic mesencephalic cat. *CR Seances Soc Biol Fil* 154, 636–641.

Jouvet, D., Vimont, R., Delorme, F., and Jouvet, M. (1964) Etude de la privation selective de la phase paradoxale de sommeil chez le chat. *CR Soc Biol* 158, 756–759.

Jouvet-Mounier, D., Vimont, P., and Delorme, F. (1965) Study of the effects of sleep deprivation in the adult cat. *J Physiol (Paris)* 57(5), 636–637.

Kovalzon, V.M., and Tsibulsky, V.L. (1984) REM sleep deprivation, stress and emotional behavior in rats. *Behav Brain Res* 14, 235–245.

Mark, J., Heiner, L., and Godin, Y. (1970) Stress during experimental deprivation of paradoxical sleep. *J Physiol (Paris)* 62(Suppl 1), 185.

Mendelson, W., Guthrie, R.D., Guynn, R., Harris, R.L., and Wyatt, R.J. (1974) Rapid eye movement (REM) sleep deprivation, stress and intermediary metabolism. *J Neurochem* 22(6), 1157–1159.

Moldofsky, H. (1994) Central nervous system and peripheral immune functions and the sleep-wake system. *J Psychiatry Neurosci.* 19(5): 368–374.

Moldofsky, H., Lue, F., Davidson, J., and Gorczynski, R. (1989a) Effects of sleep deprivation in human immune functions. *FASEB J* 3, 1972–1977.

Moldofsky, H., Lue, F., Davidson, J., Jephthah-Ochoa, J., Carayanniotis, K., and Gorczynski, R. (1989b) The effect of 64 hours of wakefulness on immune functions and plasma cortisol in humans. In: J. Horne (Ed.). *Sleep'88.* Gustav Fisher Verlag, Stuttgart, pp. 185–187.

Ozturk, L., Pelin, Z., Karadeniz, D., Kaynak, H., Cakar, L., and Gozukirmizi, E. (1999) Effects of 48 hours sleep deprivation on human immune profile. *Sleep Res Online* 2(4), 107–111.

Palmblad, J., Cantell, K., Strander, H., Froberg, J., Karlsson, C.-G., Levi, L., Granstrom, M., Unger, P. (1976) Stressor exposure and immunologic response in man: interferon producing capacity and phagocytosis. *Psychosom Res* 20: 193–199.

Palmblad, J., Petrini, B., Wasserman, J., Akerstedt, T. (1979) Lymphocyte and Granulocyte reactions during sleep deprivation. *Psychosomatic Medicine* 41(4): 273–278.

Redwine, L. Dang, J., Hall, M., Irwin, M. (2003) Disordered sleep, nocturnal cytokines, and immunity in alcoholics. *Psychosom Med* 65(1), 75–85.

Tobler, I., Murison, R., Ursin, R., Ursin, H., and Borbely, A.A. (1983) The effect of sleep deprivation and recovery sleep on plasma corticosterone in the rat. *Neurosci Lett* 35(3): 297–300.

Suchecki, D., and Tufik, S. (2000) Social stability attenuates the stress in the modified multiple platform method for paradoxical sleep deprivation in the rat. *Physiol Behav* 68(3), 309–316.

Suchecki, D., Duarte Palma, B., and Tufik, S. (2000) Sleep rebound in animals deprived of paradoxical sleep by the modified multiple platform method. *Brain Res* 875(1/2), 14–22.

Van Hulzen, Z.J., and Coenen, A.M. (1980) The pendulum technique for paradoxical sleep deprivation in rats. *Physiol Behav* 25(6), 807–811.

Van Luijtelaar, E.L., and Coenen, A.M. (1982) Differential behavioural effects of two instrumental paradoxical sleep deprivation techniques in rats. *Biol Psychol* 15(1/2), 85–93.

Van Luijtelaar, E.L., and Coenen, A.M. (1986) Electrophysiological evaluation of three paradoxical sleep deprivation techniques in rats. *Physiol Behav* 36(4), 603–609.

Velazquez-Moctezuma, J., Dominguez-Salazar, E., Cortes-Barberena, E., Najera-Medina, O., Retana-Marquez, S., Rodriguez-Aguilera, E., Jimenez-Anguiano, A., Cortes-Martinez, L., Ortiz-Muniz, R. (2004) Differential effects of rapid eye movement sleep deprivation and immobilization stress on blood lymphocyte subsets in rats. Neuroimmunomodulation. 11(4): 261–267.

Vogel, G.W. (1975) A review of REM sleep deprivation. *Arch Gen Psychiat* 32, 749–761.

PART III
Clinical Research

10 Sleep and Immune Correlates: Translational Research in Clinical Populations

Sarosh J. Motivala and Michael Irwin

10.1 The Rationale for Studying Sleep and Immunity in Clinical Populations

Substantial evidence demonstrates that sleep and the immune system interact via bidirectional pathways (Krueger et al. 1995; Opp 2005). Experimental strategies that target sleep (e.g., sleep deprivation) or modify the immune system (via the cytokine network) are beginning to identify the mechanisms that link sleep and immunity. Basic studies show that inflammatory cytokine administration alters sleep macroarchitecture and continuity (Spath-Schwalbe et al. 1998; Späth-Schwalbe, Lange, Perras, Fehm, and Born 2000; Hogan, Morrow, Smith, and Opp 2003). Conversely, sleep behavior influences immunity; with as few as 4 h of sleep loss altering lymphocyte trafficking, cytokine expression, and effector cell activity (Motivala and Irwin 2005); furthermore, prolonged sleep deprivation in rats (>2 weeks) results in death due to poor host defense of normally controlled bacterial pathogens that progressively penetrate into multiple organ systems (Everson and Toth 2000). These studies suggest that chronic sleep impairments show a reciprocal relationship with immune dysregulation. There is growing interest in testing the impact of chronic loss of sleep on the immune system in clinical disorders in which there are marked sleep impairments. Such studies have the potential to provide insights into the reciprocal influence of cytokines on sleep, and whether increases in inflammatory markers in many of these disorders might contribute to abnormalities in sleep continuity or sleep depth in these populations. In addition, immune dysregulation due to sleep impairments may impact the incidence of infectious diseases and the development of inflammatory disorders in these vulnerable populations. This chapter focuses on these clinical populations who have disordered sleep and/or inflammation and examines the hypothesis that sleep impairments initiate increased expression of inflammatory cytokines that are part of a positive feedback loop that promotes and perpetuates sleep impairments and low-grade inflammatory processes.

10.2 Sleep and Immunity in Primary Insomnia

Insomnia is the most prevalent type of sleep disorder; rates are as high as 10% in the general population and the prevalence increases with age (NIH 2005). Rates of chronic insomnia can be as high as 30% in older adults, greater in frequency and severity than any other age group (Petit, Azad, Byszewski, Sarazan, and Power 2003). Insomnia typically involves difficulties in initiating sleep, maintaining sleep, and/or waking too early. There can be substantial daytime dysfunction with the disorder, including fatigue and lethargy. Sometimes insomnia is secondary to other psychiatric disorders such as substance dependence or is subsumed as a core symptom, such as with major depressive disorder. Also, insomnia often manifests as a result of an underlying medical condition such as rheumatoid arthritis, cardiopulmonary disorders, or low back pain (NIH 2005). However, insomnia can be present in the absence of psychiatric or medical conditions, and in such cases is termed primary insomnia. The chronicity of primary insomnia can be long lasting, although few studies have characterized its natural course (NIH 2005). According to the Diagnostic and Statistical Manual—IV, Revised (Association 2000), primary insomnia diagnosis requires a duration of at least one month, but a number of clinical trials have extended entrance criteria to 6 months (Morin, Bastien, Guay, Radouco-Thomas, Leblanc, and Vallieres 2004).

10.2.1 Effects of Experimental Sleep Loss

Few clinical studies have studied sleep and immunity in persons with primary insomnia. However, sleep deprivation studies, in which healthy participants are required to undergo extended periods of wakefulness, provide some clues. Partial sleep deprivation studies typically involve 4 h of sleep deprivation and as such, are more closely related to sleep loss in insomnia patients. In studies with healthy participants, partial sleep deprivation results in decreased natural killer (NK) cell activity, independent of changes in actual numbers of natural killer cells, which also decrease (Irwin, McClintick, Costlow, Fortner, White, and Gillin 1996). This subset of cells represent an important line of defense against viral infections and in tumor surveillance (Ben-Eliyahu, Shakhar, Page, Stefanski, and Shakhar 2000). Although the implications of decreased NK activity in healthy participants are unclear, some studies show that decreases NK activity in cancer patients is associated with poorer prognosis (Pross and Lotzova 1993). Cytokine levels also fluctuate in response to partial sleep deprivation. Circulating levels of interleukin 6 (IL-6) are readily detected systemically and this cytokine is thought to exhibit potent systemic inflammatory effects including activation of acute phase proteins (Isomaki and Punnonen 1997). IL-6 is produced by multiple sources including monocytes, fibroblasts, endothelial cells, smooth muscle cells, and adipose tissue. IL-6 is thought to regulate systemic inflammation by stimulating production of acute phase reactants by the liver. It also stimulates both B-cell maturation into plasma cells as well as T-cell differentiation to cytotoxic T cells. However, elevated levels of plasma IL-6 are associated with polyarthritis and rheumatoid arthritis. IL-6 has a circadian rhythm, peaking at night, with lower levels during the day (Bauer et al. 1994). Redwine, Hauger, Gillin, and Irwin (2000) found that in healthy men, levels of IL-6 increased with peak values occurring 2.5 h after sleep onset; however, during partial sleep deprivation, the nocturnal increase of IL-6 was delayed and did not occur until after sleep onset at 3 AM Sleep deprivation did not influence the nocturnal secretion of cortisol or melatonin, which taken together suggest that sleep, rather than a circadian pacemaker, influences nocturnal IL-6 and growth hormone secretion (Redwine et al. 2000). A sleep induced increase in IL-6 is in contrast to findings by Born, Lange, Hansen, Molle, and Fehm (1997), who found that IL-6 concentrations were flat during sleep and during sleep deprivation. However, blood sampling was limited to 3-h intervals. Thus, the frequency of blood sampling may not have been adequate to ascertain nocturnal increases in IL-6 during normal sleep or the effects of sleep deprivation on this cytokine. In contrast, other researchers have found that partial sleep deprivation each night for a week results in increased circulating levels of IL-6 (Vgontzas et al. 2004a).

10.2.2 Primary Insomnia

Clinical studies with insomnia patients corroborate the sleep–immune findings of partial sleep deprivation, with studies indicating evidence of immune dysregulation. In general, insomniacs show increased levels of inflammatory cytokines (Vgontzas et al. 2002; Burgos et al. 2005) and decreased T-helper (CD3+, CD4+), T-cytotoxic (CD8+) cell numbers (Savard, Laroche, Simard, Ivers, and Morin 2003), and decreased NK cell activity (Irwin, Clark, Kennedy, and Ziegler 2003). Irwin and colleagues (2003) discovered that nondepressed insomniacs had diminished NK cell activity and a trend for less lymphokine activated killer (LAK) cell activity (Fig. 10.1). In this study, insomnia patients also showed increased levels of catecholamines across the night. Prolonged elevations in sympathetic nervous system activity can suppress NK activity (Friedman and Irwin 1997). Thus, heightened autonomic arousal during the night may promote a suppression in innate immunity the following morning (Irwin, Mascovich, Gillin, Willoughby, Pike, and Smith 1994; Irwin et al. 1996).

Insomnia patients also have increased circulating levels of inflammatory cytokines, although the timing of the expression is less clear. Burgos and colleagues (2005) found elevated levels of IL-6 in insomnia patients versus control subjects, from 3 AM to 9 AM, but not before sleep onset. In contrast, other researchers have found increased levels of IL-6 in insomniacs during late afternoon to evening (2 PM to 9 PM) (Vgontzas et al. 2002). Unfortunately, in both studies, samples were small with 11 insomniacs in each study. Overall, clinical studies with this patient population demonstrates altered cytokine expression and NK cell activity that are similar to experimental sleep deprivation studies, suggesting that poor sleep dysregulates aspects of the immune system in this population.

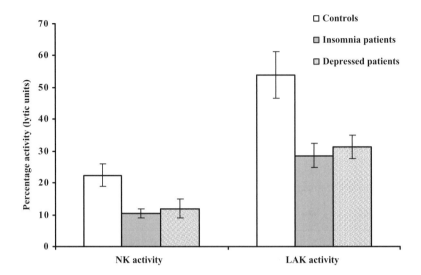

FIGURE 10.1. NK activity is significantly decreased in nondepressed insomniacs and depressed patients, $p < 0.02$. Findings are similar for LAK activity, $p = 0.07$.

These studies also suggest that poor sleep promotes increased daytime levels of inflammatory cytokines. As the links between low-grade inflammation and atherosclerosis are becoming increasingly clear (Ridker, Rifai, Stampfer, and Hennekens 2000), sleep–immune studies with insomnia patients have implications for the understanding the pathways linking insomnia and cardiovascular disease mortality.

10.2.3 Insomnia in Aging: Influence on Immunity

Normal healthy aging is associated with increasing sleep fragmentation, increasingly light sleep (stages 1 and 2), and less slow wave sleep (SWS, stages 3 and 4) (Benca, Obermeyer, Thisted, and Gillin 1992; Ehlers and Kupfer 1997; Van Cauter, Leproult, and Plat 2000). These changes are thought to contribute to daytime fatigue, depression, and impairments in health functioning in older adults. Such changes in sleep parameters may explain why self reported poor sleep is one of the most common complaints in elderly adults. Furthermore, the elderly are at higher risk of developing insomnia than any other age group, and this sleep impairment is independent of medical and psychiatric illness, medication use, circadian rhythm changes, and psychosocial factors which can all contribute to sleep complaints (Ancoli-Israel 2000).

As of yet, there are no studies integrating cytokine and sleep assessment in older adults with insomnia. Considering that older adults are particularly vulnerable to insomnia, this would be an important population to study. Elderly persons exhibit many of the prominent immune changes observed after acute or prolonged sleep deprivation described earlier. Regarding innate immune functions, aging, like sleep deprivation, leads to diminished NK cell activity when considered on a per cell basis (Irwin et al. 1994). There is also evidence that aging is associated with progressive increases in circulating levels of inflammatory cytokines such as IL-6 (Glaser and Kiecolt-Glaser 2005). For example, as compared to young subjects, healthy elderly subjects showed an increased production of interleukin 1 (IL-1) and tumor necrosis factor (TNF) that was particularly pronounced during nocturnal sleep (Born et al. 1997). In younger adults, levels of IL-6 are lower during SWS as compared to stages 1 and 2 sleep; however, aging is associated with progressive decreases in SWS, and this change may be important in explaining the increases in IL-6 seen in older adults. Although these findings are promising, there are some important caveats. Aging-related thymic involution accounts for reduced numbers of circulating lymphocytes in older adults, and this decrease is opposite to what is observed following sleep deprivation. Accordingly, it should be cautioned against considering alterations in immune function in the aged as a consequence of a "chronic sleep deprivation," since the effects of poor sleep in elderly persons interact with a number of other physiological conditions not related to sleep. However, there appears to be a convergence of declines in host defense, increases in systemic low-grade inflammation and poorer sleep in older adults, which suggests that the health consequences of sleep–immune interactions are particularly salient in this population.

10.2.4 Cytokines: Role in the Regulation of Sleep

Although sleep deprivation studies and insomnia studies both indicate that inflammatory cytokines are elevated subsequent to sleep loss, cytokines themselves appear to impact sleep. Animal studies indicate that administration of inflammatory cytokines have varying effects on sleep, depending on the nature, timing, and dose of cytokine administered. IL-6 has been described as a sleep modulator and its administration produces sleep fragmentation in rats (Hogan et al. 2003). IL-1 and TNF produce increases in nonrapid eye movement (NREM) sleep in mice, rats, and rabbits. However, the effects of proinflammatory cytokines on sleep continuity and fragmentation in humans is far less clear, partially because cytokine administration studies are rare. In healthy subjects, IL-6 and interferon-gamma (IFN-) administration decreases amounts of SWS in the first half of the night and substantially increases SWS in the second half of the night (Spath-Schwalbe et al. 1998; Späth-Schwalbe et al. 2000). Along with these changes in sleep, administration of inflammatory cytokines produces feelings of fatigue and lethargy. In contrast, medications that block the effects of inflammatory cytokines appear to improve sleep and fatigue during the day. In rheumatoid arthritis (RA) patients with markedly poor sleep, a single dose administration of infliximab, a TNF antagonist approved for the treatment of RA, produced rapid, marked improvements in objective measurement of sleep efficiency and sleep latency, before any improvement in clinical markers such as joint pain and swelling (Zamarron, Maceiras, Mera, and Gomez-Reino 2004). The study was an open label trial with only six participants. Similarly, in a small sample of obese sleep apnea patients ($n = 8$), administration of etanercept, another TNF-antagonist medication, produced decreases in self-reported lethargy and less sleepiness, measured via increased sleep onset times in response to the multiple sleep latency test, a polysomnography-based measure of sleepiness (Vgontzas, Zoumakis, Lin, Bixler, Trakada, and Chrousos 2004b). These studies suggest that blockage of inflammatory cytokines lead to improved sleep.

Overall, translational research, using the framework and models generated by research from animal and human experimental studies, suggest that insomnia is associated with altered immune function, which includes increased inflammatory cytokine expression and decreased NK cell activity. These effects on immunity may be particularly important in clinical populations. NK cells play a role in tumor surveillance and decreased NK activity predicts poor prognosis in cancer patients. The prevalence of insomnia is as high as 50% in cancer patients (O'Donnell 2004), and fatigue and sleep complaints are increasingly being recognized as important detractors of quality of life in these patients (Carr et al. 2002). Given that clinical studies with insomnia patients and experimental studies on sleep loss both indicate decreases in NK activity, the effects of sleep impairment on immunity may have implications for the link between insomnia and cancer morbidity (Savard et al. 1999) and mortality (Kripke, Simons, Garfinkel, and Hammond 1979). The other primary immunologic finding in insomnia is inflammatory cytokine expression. In a 12-year follow-up of 1870 middle-aged men and women, sleep complaints (e.g., difficulty maintaining sleep) predicted coronary artery disease mortality (Mallon, Broman, and Hetta 2002). Much evidence documents that in cardiovascular disease, atherosclerotic plaque formation is driven by inflammatory processes (Ridker et al. 2000), and that increased levels of IL-6 are also implicated in the development of type 2 diabetes and hypertension (Boos and Lip 2006). Insomnia patients have elevated levels of the cytokines that promote low-grade inflammation and furthermore, sleep deprivation studies indicate that sleep loss produces increases in these cytokines as well. Together, these experimental and clinical studies are not only demonstrating links between sleep and immune processes, but are also identifying pathways through which chronic insomnia impacts health.

10.3 Sleep and Immunity in Depression and Substance Dependence

Among the most prevalent psychiatric disorders, depression and substance dependence both involve prominent sleep dysregulation. The lifetime risk of developing a major depressive disorder ranges from 5 to 25%, with up to 40% of depressed outpatients and 90% of depressed inpatients having sleep impairments as documented via polysomnography (American Psychiatric Association and American Psychiatric Association, Task Force on DSM-IV, 1994). Similarly, the lifetime prevalence for alcohol dependence is approximately 14% (American Psychiatric Association and American Psychiatric Association, Task Force on DSM-IV 1994) and approximately 30 to 70% of alcohol dependent patients report chronic difficulty sleeping (Brower 2001).

10.3.1 Depression and Immunity

In depressed patients, the most common sleep difficulty is inability to initiate or maintain sleep, although less frequently, hypersomnia is present. Of all depressive symptoms, depressed patients report sleep complaints as the primary presenting problem to health care practitioners (American Psychiatric Association and American Psychiatric Association, Task Force on DSM-IV 1994). Depressed patients also have well-documented alterations in immunity, including decreased NK activity, decreased proliferative response to antigenic challenge, and increased levels of inflammatory cytokines. Emerging evidence indicates that poor sleep in depression is related to, and may mediate immune alterations in these patients. Similar to primary insomnia patients, self-reported sleep quality negatively correlates with NK activity in depression, whereas other depression-related symptoms such as weight loss, poor concentration, or diurnal variation in moods are not related to NK activity (Irwin et al. 2003). Likewise, EEG studies reveal that disturbances of sleep continuity (e.g., prolonged sleep latency, declines of total sleep time) correlate with alterations of innate and cellular immune function among depressed patients (Irwin, Smith, and Gillin 1992; Cover and Irwin 1994). Moreover in bereaved subjects, causal statistical analyses have shown that disordered sleep mediates the relationship between severe life stress and decreased NK responses (Hall, Baum, Buysse, Prigerson, Kupfer, and Reynolds 1998). Depressed patients have increased levels of inflammatory markers such as IL-6, soluble intercellular adhesion molecule (sICAM) and C-reactive protein (Liukkonen et al. in press; Miller, Stetler, Carney, Freedland, and Banks 2002; Motivala, Sarfatti, Olmos, and Irwin 2005). As mentioned earlier, low-grade inflammation, as reflected by systemic elevations in these measures, promotes atherosclerosis and cardiovascular disease.

The elevation of inflammatory markers in depressed patients is consistent both in depressed patients recovering from an acute coronary artery event (Lesperance, Frasure-Smith, Theroux, and Irwin 2004) as well as with depressed patients who are otherwise healthy (Miller et al. 2002; Motivala et al. 2005). The relation between depression and inflammatory markers do not appear to be due to factors such as smoking or obesity (Lesperance et al. 2004; Motivala et al. 2005), but may be related to poor sleep. Motivala and colleagues (2005) found that in depressed patients, polysomnographic assessment of prolonged sleep latency and increases of rapid eye movement (REM) density were associated with elevated levels of IL-6 and sICAM (Fig. 10.2). Moreover, these sleep measures were better predictors of IL-6 and sICAM than depression or depressive symptoms, providing initial evidence that sleep disturbance has a key role in alterations of inflammatory markers in depression. To our knowledge, this study is the only one to date that examined the relationship between sleep and cytokines in depression, although corroborating results can be found in other populations. For example, caregivers of patients with Alzheimer's disease face heavy demands, and this population is chronically stressed with significant depressive symptoms (Sanders and Adams 2005). Like patients with major depression, caregivers of dementia patients sleep less and have more fragmented sleep than noncaregiver controls.

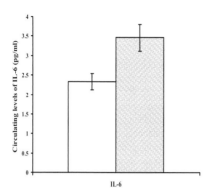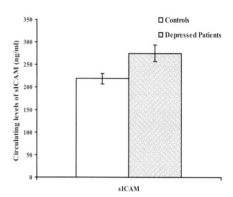

FIGURE 10.2. Depressed subjects ($n = 22$) have significantly higher circulating levels of IL-6 ($p < 0.01$) and sICAM ($p < 0.02$) than nondepressed controls ($n = 18$).

Furthermore, longer sleep onset time correlates with higher IL-6 in this sample (Kanel et al. 2006). Links between sleep and immunity are only beginning to be examined in depressed patients; thus, diurnal variation in cytokine expression in this population is unclear and the impact of experimental sleep deprivation paradigms on cytokines are also unknown in this population.

10.3.2 Alcohol Dependence, Sleep, and Immunity

Diagnostic criteria for alcohol dependence include at least three of seven patterns of substance use causing significant impairment over a 12-month period. Criteria can include tolerance to alcohol, presence of withdrawal symptoms, intake of larger amounts than intended, persistent desire to drink, much time spent in activities related to drinking, abandonment of occupational, social or recreational activities because of alcohol use, and continued intake despite recurrent problems related to drinking (American Psychiatric Association and American Psychiatric Association, Task Force on DSM-IV 1994).

Alcohol dependent patients have profound disturbances of sleep continuity and sleep architecture including decreases of total sleep time, declines of delta sleep, and increases of REM sleep (Irwin, Miller, Gillin, Demodena, and Ehlers 2000; Irwin, Gillin, Dang, Weissman, Phillips, and Ehlers 2002).

In addition to poor sleep, alcohol dependent patients are also more prone to infectious diseases such as pneumonia (Nelson and Kolls 2002), tuberculosis (Buskin, Gale, Weiss, and Nolan 1994), hepatitis C (Balasekaran et al. 1999) and possibly HIV infection (Crum, Galai, Cohn, Celentano, and Vlahov 1996), suggesting broad impairments in host defense.

When immune parameters are examined more specifically in alcohol dependence, studies show that these patients have decreases in NK activity, interleukin-2 (IL-2) stimulated killer cell activity (LAK activity), and ex vivo production of IL-6 (Irwin and Miller 2000). There is also a shift in cytokine expression reflecting greater T-helper 2 cytokine expression that persists after weeks of abstinence (Redwine, Dang, Hall, and Irwin 2003).

Throughout the night, alcohol dependent persons show a shift toward a Th2 cytokine expression with a predominance of this Th2 response, evidenced by a decreased IFN-γ/IL-10 ratio. In contrast, healthy control subjects show an increase in Th1 cytokine expression during the later half of the nocturnal period. T helper cell differentiation into Th1 or Th2 subtypes allow for greater flexibility in coordinating cell-mediated immunity (Th1 cells) or coordinating antibody production (Th2 cells).

Th1 cytokines such as IFN, IL-2, and granulocyte colony stimulating factor can stimulate monocyte/macrophage and NK cell activity, in turn leading to heightened defense against intracellular pathogens such as *Pneumocystis carinii* and *M. tuberculosis* (Curtis 2005). The preferential nocturnal shift toward Th2 cytokines in alcoholic patients may reflect a primary defect in the coordination of T-cell differentiation into Th1 versus Th2 cell subtypes and this may underlie their greater susceptibility to pneumonia and tuberculosis.

Inflammatory cytokines are also dysregulated in alcoholic patients. Daytime circulating levels of inflammatory cytokines such as TNF and IL-6 are elevated in alcohol dependent patients, but these levels may decrease during periods of abstinence (Gonzalez-Quintela, Dominguez-Santalla, Perez, Vidal, Lojo, and Barrio 2000).

In contrast, in vitro production assays, in which whole blood is stimulated with an antigen such as liposaccharide (LPS), indicate that nocturnal production of IL-6 is reduced in alcoholic patients in the first half of the night as compared with controls, with subsequent increases in the second half of the night yielding levels that are comparable with controls (Redwine et al. 2003). Sleep macroarchitecture also appears related to IL-6 in alcohol dependence; increased REM sleep predicts morning higher stimulated IL-6 production. These data along with evidence that IL-6 is elevated during REM sleep (Redwine et al. 2000) raise the possibility that greater amounts of REM sleep during the late part of the night contribute to the nocturnal rise of IL-6 in alcohol dependent persons. Deficits in SWS in alcohol dependence may also lead to abnormal elevations of IL-6 and TNF. For example, experimental sleep deprivation typically induces a rebound increase in SWS as described above. However, alcoholics evidence a defect in SWS recovery, which is coupled with an exaggerated recovery night increase in circulating levels of IL-6 and TNF in alcoholics as compared to controls (Irwin, Rinetti, Redwine, Motivala, Dang, and Ehlers 2004) (see Fig. 10.3). Finally, given that daytime elevations of IL-6 correlate with fatigue (Vgontzas, Papanicolaou, Bixler, Kales, Tyson, and Chrousos 1997), such increases in the production or circulating levels of IL-6 may have implications for daytime fatigue in alcohol dependent persons.

Just as sleep can influence cytokine expression, cytokines can affect sleep in alcohol dependence. In clinical studies with alcoholic patients, higher levels of circulating IL-6 before sleep (11 PM) were associated with prolonged sleep latency (Irwin and Rinetti 2004) (Fig. 10.4). In addition, IL-6 is associated with increases in REM sleep during the later half of the night. In contrast, expression of the anti-inflammatory cytokine, IL-10 before sleep predicted increases in delta sleep, accounting for over 23% of the variance in delta sleep independent of age, and alcohol consumption (Redwine et al. 2003).

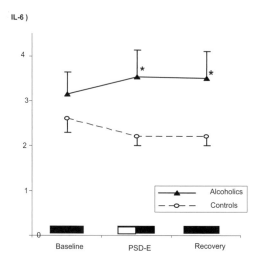

FIGURE 10.3. Alcoholics ($n = 15$) have higher circulating levels of IL-6 as compared to controls ($n = 16$) across consecutive nights of baseline sleep, early partial sleep deprivation (4 h) and recovery sleep.

In alcohol dependent patients, there is evidence that the bidirectional effects between sleep parameters and cytokine expression are altered and that dysregulation persists even during periods of abstinence. Peripheral proinflammatory cytokines are capable of exerting direct effects on central nervous system function, promoting behavioral consequences that include increased fatigue, anxiety, and depressed mood. Like insomnia, alcohol dependence is associated with increased circulating levels of IL-6 during the day.

High levels of this cytokine are associated with severity of fatigue symptoms in women recovering from breast cancer (Collado-Hidalgo, Bower, Ganz, Cole, and Irwin 2006), and experimental immune activation results in increases of anxiety and depressed mood, which are coupled with declines of memory functions (Reichenberg et al. 2001).

No studies to date have examined whether disordered sleep and daytime elevations of proinflammatory cytokines contribute to fatigue or depressive symptoms in recovering alcoholics. However, in sleep apnea patients, administration of a TNF antagonist medication (etanercept) reduced IL-6 as well as daytime sleepiness (Vgontzas et al. 2004a, 2004b). Importantly, the contribution of cytokines in the regulation of sleep suggests that cytokine mechanisms are a possible target for interventions to optimize sleep initiation and sleep quality in alcoholic and other at risk populations.

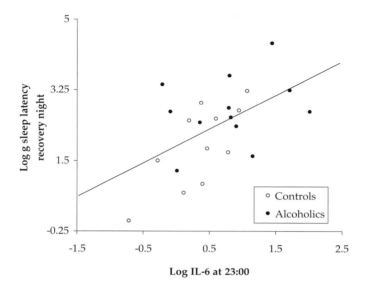

FIGURE 10.4. Scatterplot of IL-6 before sleep onset and sleep latency (both log transformed to normalize distributions) during a night of recovery sleep in both alcoholics and controls ($r = 0.53$, $p < 0.01$). The relationship between these variables was similar for initial baseline sleep ($r = 0.37$, $p = 0.08$) and during partial sleep deprivation ($r = 0.56$, $p < 0.01$).

10.4 Sleep and Cytokines in Rheumatoid Arthritis

The prevalence of sleep disturbance can be quite high in autoimmune disorders; for example, 50% of multiple sclerosis patients report clinically relevant sleep complaints (Fleming and Pollak 2005) as do 61% of systemic lupus patients (Tench, McCurdie, White, and D'Cruz 2000). Perhaps the highest prevalence is in rheumatoid arthritis (RA) patients, between 50 and 75% of patients complain of difficulties initiating or maintaining sleep (Drewes 1999). Fatigue is a debilitating symptom in each of these conditions, and studies show that poor sleep is significant predictor of fatigue, often more so than other disease-specific variables (Nicassio, Moxham, Schuman, and Gevirtz 2002). Although the underlying causes of sleep difficulties can include medication effects as well as chronic pain, studies with RA patients are indicating that inflammatory cytokines might be important promoters of sleep dysregulation.

RA is a chronic autoimmune disorder in which patients develop bilateral polyarticular pain. This disorder can be highly debilitating, with significant pain and fatigue early in the disease; subsequent bone resorption and possible deformation of joints, typically in the hands occur later in the disease. The etiology of the disease is unknown, but RA disease progression is driven primarily by inflammatory cytokines, including TNF, IL-1, and IL-6. Biologic agents, such as etanercept, infliximab, and humira, work by specifically neutralizing the activity of TNF, which in turn down-regulates expression of other inflammatory cytokines (Feldmann and Maini 2001). Very high levels of inflammatory cytokines are found in synovial fluid in the joints of RA patients, and systemic levels are also increased. RA patients typically show high plasma levels of IL-6, the magnitude of which correlates with subsequent disease progression; and plasma IL-6 levels decrease drastically following treatment with biologic agents (Charles et al. 1999). Thus, inflammatory cytokines that are involved in disease progression are the same cytokines that interact with sleep processes.

RA patients report frequent difficulties with sleep; polysomnography studies show that RA patients have trouble falling asleep and have poorer sleep efficiency with multiple awakenings during the night as compared to healthy controls (Hirsch et al. 1994). The sleep of RA patients also shows frequent alpha-EEG arousals during SWS (Mahowald, Mahowald, Bundlie, and Ytterberg 1989; Drewes, Svendsen, Taagholt, Bjerregard, Nielsen, and Hansen 1998). Although poor sleep is often speculated as due to nocturnal pain, the latter may not be the primary causative agent. Sleep impairments, measured via polysomnography, are found not only in RA patients experiencing a pain flare, but also in those who are not experiencing pain (Drewes et al. 1998). Furthermore, treatment with nonsteroidal anti-inflammatory medications reduced pain severity but did not improve sleep in a sample of RA patients (Lavie, Nahir, Lorber, and Scharf 1991). In turn, poor sleep may contribute to morning levels of fatigue and pain in RA (Nicassio and Wallston 1992; Drewes et al. 1998). No study to date has studied the roles of cytokines on sleep, fatigue, and pain in this population. There is a striking absence of studies examining the role of inflammatory processes in mediating these links between sleep and disease symptoms in RA patients. Such studies are needed to advance our understanding of sleep and its association with other RA symptoms.

It is known that a poor night's sleep can increase pain sensitivity the following morning. In a sample of healthy men, experimental disruption of NREM sleep over three nights induced increased pain sensitivity and muscle tenderness (Moldofsky and Scarisbrick 1976), a finding more recently replicated in healthy middle-aged women (Lentz, Landis, Rothermel, and Shaver 1999). To date, there is no study examining experimental sleep deprivation, even for brief periods of time in RA patients. However, poor sleep is associated with increased morning tenderness in the peripheral joints of RA patients who are experiencing an acute pain flare (Moldofsky, Lue, and Smythe 1983). The biological mechanisms that underlie the link between poor sleep and increases of pain and fatigue are as of yet unexplored in RA, although elevations in proinflammatory cytokine activity are implicated (Moldofsky 2001). As noted above, RA patients with poor sleep who were given a single dose of infliximab showed dramatic improvements in sleep efficiency and sleep latency immediately following administration. Furthermore, these improvements occurred *before* these subjects improved on clinical markers such as joint pain and swelling (Zamarron et al. 2004). This is consistent with the hypothesis of cytokine dysregulation of sleep, since the medication produces rapid decreases in inflammatory cytokines and blocks the activity of TNF.

Sleep research with RA patients provides a valuable opportunity to test relationships between sleep and cytokines, particularly because cytokines play such a major role in the severity of the disease. A variety of factors such as medication types, degree of disease progression, and the presence of depressed mood and fatigue in this population make these studies difficult. However, inclusion of cytokine assessment in these studies would provide valuable corroboration as to whether inflammatory cytokines contribute to sleep deficits seen in RA and by extension, impact pain sensitivity, fatigue, and lethargy during the day.

10.5 Future Directions

In this chapter, we have described sleep impairments and immune dysregulation across multiple clinical samples including insomnia, major depression, substance dependence, and RA. The possible effect of sleep disruption has tremendous importance because of the pervasiveness of insomnia complaints with up to 69% of adults report having sleep problems on a few nights a week or more. Moreover in recent decades, decreases in the mean duration of sleep and an increase in shift work have occurred, which might pose substantial public health burden that has not yet been evaluated. Together, these data indicate that sleep is a new "vital sign" that should be more carefully characterized in the management of patients, particularly in vulnerable populations such as the elderly or the immunocompromised. Ongoing basic and clinical research aims to provide insight into the molecular and cellular pathways by which sleep and the immune system are interrelated, which will ultimately lead to novel interventions and therapies for insomnia which have the potential to promote health. We have hypothesized that cytokines and sleep have a bidirectional relationship, which promotes mutual dysregulation, promoting a vicious cycle of greater and greater impairment. Basic studies show that cytokines impact sleep depending on the dose and time of administration (Opp 2005) and sleep deprivation studies indicate that immunity is affected by extended bouts of wakefulness (Irwin 2002). Translational research, reported here in this chapter, is extending these findings to clinical populations. We hypothesize that poor sleep promotes elevated inflammatory cytokine expression during the day and this in turn promotes greater sleep impairments the following night. Continued work utilizing sleep deprivation protocols, as well as administration of inflammatory cytokine antagonists in clinical populations, are important to test this hypothesis and determine the effects of this bidirectional relationship on health outcomes.

References

American Psychiatric Association and American Psychiatric Association Task Force on DSM-IV (1994) *Diagnostic and Statistical Manual of Mental Disorders: DSM-IV*. American Psychiatric Association, Washington, DC.

Ancoli-Israel, S. (2000) Insomnia in the elderly: A review for the primary care practitioner. *Sleep* 23(Suppl 1), S23–S30; discussion S36–S38.

Association, A.P. (2000) *Diagnostic and Statistical Manual of Mental Disorders*. American Psychiatric Association, Washington, DC.

Balasekaran, R., Bulterys, M., Jamal, M.M., Quinn, P.G., Johnston, D.E., Skipper, B., Chaturvedi, S., and Arora, S. (1999) A case-control study of risk factors for sporadic hepatitis C virus infection in the southwestern United States. *Am J Gastroenterol* 94(5), 1341–1346.

Bauer, J., Hohagen, F., Ebert, T., Timmer, J., Ganter, U., Krieger, S., Lis, S., Postler, E., Voderholzer, U., and Berger, M. (1994) Interleukin-6 serum levels in healthy persons correspond to the sleep-wake cycle. *Clin Invest* 72(4), 315.

Benca, R.M., Obermeyer, W.H., Thisted, R.A., and Gillin, J.C., (1992) Sleep and psychiatric disorders. A meta-analysis. *Arch Gen Psychiatry* 49(8), 651–668; discussion 669–670.

Ben-Eliyahu, S., Shakhar, G., Page, G.G., Stefanski, V., and Shakhar, K. (2000) Suppression of NK cell activity and of resistance to metastasis by stress: A role for adrenal catecholamines and beta-adrenoceptors. *Neuroimmunomodulation* 8(3), 154–164.

Boos, C.J., and Lip, G.Y. (2006) Is hypertension an inflammatory process? *Curr Pharm Des* 12(13), 1623–1635.

Born, J., Lange, T., Hansen, K., Molle, M., and Fehm, H.L. (1997) Effects of sleep and circadian rhythm on human circulating immune cells. *J Immunol* 158(9), 4454–4464.

Brower, K.J. (2001) Alcohol's effects on sleep in alcoholics. *Alcohol Res Health* 25(2), 110–125.

Burgos, I., Richter, L., Klein, T., Fiebich, B., Feige, B., Lieb, K., Voderholzer, U., and Riemann, D. (2006) Increased nocturnal interleukin-6 excretion in patients with primary insomnia: a pilot study. *Brain Behav Immun* 20(3): 246–253.

Buskin, S.E., Gale, J.L., Weiss, N.S., and Nolan, C.M. (1994) Tuberculosis risk factors in adults in King County, Washington, 1988 through 1990. *Am J Pub Health* 84(11), 1750–1756.

Carr, D., Goudas, L., Lawrence, D., Pirl, W., Lau, J., DeVine, D., Kupelnick, B., and Miller, K. (2002) Management of cancer symptoms: Pain, depression, and fatigue. *Evid Rep Technol Assess* (Summ)(61), 1–5.

Charles, P., Elliott, M.J., Davis, D., Potter, A., Kalden, J.R., Antoni, C., Breedveld, F.C., Smolen, J.S., Eberl, G., deWoody, K., Feldmann, M., and Maini, R.N. (1999) Regulation of cytokines, cytokine inhibitors, and acute-phase proteins following anti-TNF-alpha therapy in rheumatoid arthritis. *J Immunol* 163(3), 1521–1528.

Collado-Hidalgo, A., Bower, J.E., Ganz, P.A., Cole, S.W., and Irwin, M.R. (2006) Inflammatory biomarkers for persistent fatigue in breast cancer survivors. *Clin Cancer Res* 12(9), 2759–2766.

Cover, H., and Irwin, M. (1994) Immunity and depression: Insomnia, retardation, and reduction of natural killer cell activity. *J Behav Med* 17(2), 217–223.

Crum, R.M., Galai, N., Cohn, S., Celentano, D.D., and Vlahov, D. (1996) Alcohol use and T-lymphocyte subsets among injection drug users with HIV-1 infection: A prospective analysis. *Alcohol Clin Exp Res* 20(2), 364–371.

Curtis, J.L. (2005) Cell-mediated adaptive immune defense of the lungs. *Proc Am Thorac Soc* 2(5), 412–416.

Drewes, A.M. (1999) Pain and sleep disturbances with special reference to fibromyalgia and rheumatoid arthritis. *Rheumatology (Oxford)* 38(11), 1035–1038.

Drewes, A.M., Svendsen, L., Taagholt, S.J., Bjerregard, K., Nielsen, K.D., and Hansen, B. (1998) Sleep in rheumatoid arthritis: A comparison with healthy subjects and studies of sleep/wake interactions. *Br J Rheumatol* 37, 71–81.

Ehlers, C.L., and Kupfer, D.J. (1997) Slow-wave sleep: Do young adult men and women age differently? *J Sleep Res* 6(3), 211–215.

Everson, C.A., and Toth, L.A. (2000) Systemic bacterial invasion induced by sleep deprivation. *Am J Physiol Regul Integr Comp Physiol* 278(4), R905–R916.

Feldmann, M., and Maini, R.N. (2001) Anti-TNF alpha therapy of rheumatoid arthritis: What have we learned? *Annu Rev Immunol* 19, 163–196.

Fleming, W.E., and Pollak, C.P. (2005) Sleep disorders in multiple sclerosis. *Semin Neurol* 25(1), 64–68.

Friedman, E.M., and Irwin, M. (1997) Modulation of immune cell function by the autonomic nervous system. *Pharmacol Ther* 74, 27–38.

Glaser, R., and Kiecolt-Glaser, J.K. (2005) Stress-induced immune dysfunction: Implications for health. *Nat Rev Immunol* 5(3), 243–251.

Gonzalez-Quintela, A., Dominguez-Santalla, M.J., Perez, L.F., Vidal, C., Lojo, S., and Barrio, E. (2000) Influence of acute alcohol intake and alcohol withdrawal on circulating levels of IL-6, IL-8, IL-10 and IL-12. *Cytokine* 12(9), 1437–1440.

Hall, M., Baum, A., Buysse, D.J., Prigerson, H.G., Kupfer, D.J., and Reynolds, C.F. (1998) Sleep as a mediator of the stress-immune relationship. *Psychosom Med* 60, 48–51.

Hirsch, M., Carlander, B., Verge, M., Tafti, M., Anaya, J.M., Billiard, M., and Sany, J. (1994) Objective and subjective sleep disturbances in patients with rheumatoid arthritis. A reappraisal. *Arthritis Rheum* 37(1), 41–49.

Hogan, D., Morrow, J.D., Smith, E.M., and Opp, M.R. (2003) Interleukin-6 alters sleep of rats. *J Neuroimmunol* 137(1/2), 59–66.

Irwin, M. (2002) Effects of sleep and sleep loss on immunity and cytokines. *Brain Behav Immun* 16(5), 503–512.

Irwin, M., and Miller, C. (2000) Decreased natural killer cell responses and altered interleukin-10 production in alcoholism: An interaction between alcohol dependence and African-American ethnicity. *Alcohol Clin Exp Res* 24(4), 560–569.

Irwin, M.R., and Rinetti, G. (2004) Disordered sleep, nocturnal cytokines, and immunity: Interactions between alcohol dependence and African-American ethnicity. *Alcohol* 32(1), 53–61.

Irwin, M., Mascovich, A., Gillin, J.C., Willoughby, R., Pike, J., and Smith, T.L. (1994) Partial sleep deprivation reduces natural killer cell activity in humans. *Psychosom Med* 56, 493–498.

Irwin, M., Clark, C., Kennedy, B., JC, G., and Ziegler, M. (2003) Nocturnal catehcolamines and immune function in insomniacs, depressed patients, and control subjects. *Brain Beh Immun* 17, 365–372.

Irwin, M., Miller, C., Gillin, J.C., Demodena, A., and Ehlers, C.L. (2000) Polysomnographic and spectral sleep EEG in primary alcoholics: An interaction between alcohol dependence and African-American ethnicity. *Alcohol Clin Exp Res* 24(9), 1376–1384.

Irwin, M., Gillin, J.C., Dang, J., Weissman, J., Phillips, E., and Ehlers, C. (2002) Sleep deprivation as a probe of homeostatic sleep regulation in primary alcoholics. *Biol Psychiatry* 51(8), 632–641.

Irwin, M., McClintick, J., Costlow, C., Fortner, M., White, J., and Gillin, J.C. (1996) Partial night sleep deprivation reduces natural killer and cellular immune responses in humans. *FASEB J* 10, 643–653.

Irwin, M., Rinetti, G., Redwine, L., Motivala, S., Dang, J., and Ehlers, C. (2004) Nocturnal proinflammatory cytokine-associated sleep disturbances in abstinent African American alcoholics. *Brain Behav Immun* 18(4), 349–360.

Irwin, M., Smith, T.L., and Gillin, J.C. (1992) Electroencephalographic sleep and natural killer activity in depressed patients and control subjects. *Psychosom Med* 54, 107–126.

Isomaki, P., and Punnonen, J. (1997) Pro- and anti-inflammatory cytokines in rheumatoid arthritis. *Ann Med* 29(6), 499–507.

Kanel, R., Dimsdale, J.E., Ancoli-Israel, S., Mills, P.J., Patterson, T.L., McKibbin, C.L., Archuleta, C., and Grant, I. (2006) Poor sleep is associated with higher plasma proinflammatory cytokine interleukin-6 and procoagulant marker fibrin D-Dimer in older caregivers of people with Alzheimer's disease. *J Am Geriatr Soc* 54(3), 431–437.

Kripke, D.F., Simons, R.N., Garfinkel, L., and Hammond, E.C. (1979) Short and long sleep and sleeping pills. Is increased mortality associated? *Arch. Gen. Psychiatry* 36, 103–116.

Krueger, J.M., Takahashi, S., Kapás, L., Bredow, S., Roky, R., Fang, J., Floyd, R., Renegar, K.B., Guha-Thakurta, N., and Novitsky, S. (1995) Cytokines in sleep regulation. *Adv Neuroimmunol* 5, 171–188.

Lavie, P., Nahir, M., Lorber, M., and Scharf, Y. (1991) Non-steroid anti-inflammatory drug therapy in RA patients. Lack of association between clinical improvement and effects on sleep. *Arthritis Rheum* 34, 655–659.

Lentz, M.J., Landis, C.A., Rothermel, J., and Shaver, J.L. (1999) Effects of selective slow wave sleep disruption on musculoskeletal pain and fatigue in middle-aged women. *J Rheumatol* 26(7), 1586–1592.

Lesperance, F., Frasure-Smith, N., Theroux, P., and Irwin, M. (2004) The association between major depression and levels of soluble intercellular adhesion molecule 1, interleukin-6, and C-reactive protein in patients with recent acute coronary syndromes. *Am J Psychiatry* 161(2), 271–277.

Liukkonen, T., Silvennoinen-Kassinen, S., Jokelainen, J., Rasanen, P., Leinonen, M., Meyer-Rochow, V.B., and Timonen, M. (in press) The association between C-reactive protein levels and depression: Results from the Northern Finland 1966 Birth Cohort Study. *Biol Psychiatry* (Corrected Proof).

Mahowald, M.W., Mahowald, M.L., Bundlie, S.R., and Ytterberg, S.R. (1989) Sleep fragmentation in rheumatoid arthritis. *Arthritis Rheum* 32(8), 974–983.

Mallon, L., Broman, J.E., and Hetta, J. (2002) Sleep complaints predict coronary artery disease mortality in males: A 12-year follow-up study of a middle-aged Swedish population. *J Intern Med* 251(3), 207–216.

Miller, G.E., Stetler, C.A., Carney, R.M., Freedland, K.E., and Banks, W.A. (2002) Clinical depression and inflammatory risk markers for coronary heart disease. *Am J Cardiol* 90(12), 1279–1283.

Moldofsky, H. (2001) Sleep and pain. *Sleep Med Rev* 5(5), 385–396.

Moldofsky, H., and Scarisbrick, P. (1976) Induction of neurasthenic musculoskeletal pain syndrome by selective sleep stage deprivation. *Psychosom Med* 38(1), 35–44.

Moldofsky, H., Lue, F.A., and Smythe, H.A. (1983) Alpha EEG sleep and morning symptoms in rheumatoid arthritis. *J Rheumatol* 10(3), 373–379.

Morin, C.M., Bastien, C., Guay, B., Radouco-Thomas, M., Leblanc, J., and Vallieres, A. (2004) Randomized clinical trial of supervised tapering and cognitive behavior therapy to facilitate benzodiazepine discontinuation in older adults with chronic insomnia. *Am J Psychiatry* 161(2), 332–342.

Motivala, S., and Irwin, M. (2005) Intrinsic changes affecting sleep loss/deprivation: Immunologic changes. In: C. Kushida (Ed.), *Sleep Deprivation: Basic Science, Physiology and Behavior*. Marcel Dekker, New York, pp. 192, 359–386.

Motivala, S.J., Sarfatti, A., Olmos, L., and Irwin, M.R. (2005) Inflammatory markers and sleep disturbance in major depression. *Psychosom Med* 67(2), 187–194.

Nelson, S., and Kolls, J.K. (2002) Alcohol, host defence and society. *Nat Rev Immunol* 2(3), 205–209.

Nicassio, P.M., Moxham, E.G., Schuman, C.E., and Gevirtz, R.N. (2002) The contribution of pain, reported sleep quality, and depressive symptoms to fatigue in fibromyalgia. *Pain* 100(3), 271–279.

Nicassio, P.M., and Wallston, K.A. (1992) Longitudinal relationships among pain, sleep problems, and depression in rheumatoid arthritis. *J Abnorm Psychol* 101(3), 514–520.

NIH (2005) National Institutes of Health State of the Science Conference statement on manifestations and management of chronic insomnia in adults, June 13–15, 2005. *Sleep* 28(9), 1049–1057.

O'Donnell, J.F. (2004) Insomnia in cancer patients. *Clin Cornerstone* 6(Suppl 1D), S6–S14.

Opp, M.R. (2005) Cytokines and sleep. *Sleep Med Rev* 9(5), 355–364.

Petit, L., Azad, N., Byszewski, A., Sarazan, F.F., and Power, B. (2003) Non-pharmacological management of primary and secondary insomnia among older people: Review of assessment tools and treatments. *Age Ageing* 32(1), 19–25.

Pross, H.F., and Lotzova, E. (1993) Role of natural killer cells in cancer. *Nat Immun* 12(4–5), 279–292.

Redwine, L., Dang, J., Hall, M., and Irwin, M. (2003) Disordered sleep, nocturnal cytokines, and immunity in alcoholics. *Psychosom Med* 65(1), 75–85.

Redwine, L., Hauger, R.L., Gillin, J.C., and Irwin, M. (2000) Effects of sleep and sleep deprivation on interleukin-6, growth hormone, cortisol, and melatonin levels in humans. *J Clin Endocrinol Metab* 85(10), 3597–603.

Reichenberg, A., Yirmiya, R., Schuld, A., Kraus, T., Haack, M., Morag, A., and Pollmächer, T. (2001) Cytokine-associated emotional and cognitive disturbances in humans. *Arch Gen Psychiatry* 58(5), 445–452.

Ridker, P.M., Rifai, N., Stampfer, M.J., and Hennekens, C.H. (2000) Plasma concentration of interleukin-6 and the risk of future myocardial infarction among apparently healthy men. *Circulation* 101(15), 1767–1772.

Sanders, S., and Adams, K.B. (2005) Grief reactions and depression in caregivers of individuals with Alzheimer's disease: Results from a pilot study in an urban setting. *Health Soc Work* 30(4), 287–295.

Savard, J., Laroche, L., Simard, S., Ivers, H., and Morin, C.M. (2003) Chronic insomnia and immune functioning. *Psychosom Med* 65(2), 211–221.

Savard, J., Miller, S.M., Mills, M., O'Leary, A., Harding, H., Douglas, S.D., Mangan, C.E., Belch, R., and Winokur, A. (1999) Association between subjective sleep quality and depression on immunocompetence in low-income women at risk for cervical cancer. *Psychosom Med* 61(4), 496–507.

Spath-Schwalbe, E., Hansen, K., Schmidt, F., Schrezenmeier, H., Marshall, L., Burger, K., Fehm, H.L., and Born, J. (1998) Acute effects of recombinant human interleukin-6 on endocrine and central nervous sleep functions in healthy men. *J Clin Endocrinol Metab* 83(5), 1573–1579.

Späth-Schwalbe, E., Lange, T., Perras, B., Fehm, H.L., and Born, J. (2000) Interferon-alpha acutely impairs sleep in healthy humans. *Cytokine* 12(5), 518–521.

Tench, C.M., McCurdie, I., White, P.D., and D'Cruz, D.P. (2000) The prevalence and associations of fatigue in systemic lupus erythematosus. *Rheumatology* 39(11), 1249–1254.

Van Cauter, E., Leproult, R., and Plat, L. (2000) Age-related changes in slow wave sleep and REM sleep and relationship with growth hormone and cortisol levels in healthy men. *JAMA* 284(7), 861–868.

Vgontzas, A.N., Papanicolaou, D.A., Bixler, E.O., Kales, A., Tyson, K., and Chrousos, G.P. (1997) Elevation of plasma cytokines in disorders of excessive daytime sleepiness: Role of sleep disturbance and obesity [see comments]. *J Clin Endocrinol Metab* 82(5), 1313–1316.

Vgontzas, A.N., Zoumakis, E., Bixler, E.O., Lin, H.M., Follett, H., Kales, A., and Chrousos, G.P. (2004a) Adverse effects of modest sleep restriction on sleepiness, performance, and inflammatory cytokines. *J Clin Endocrinol Metab* 89(5), 2119–2126.

Vgontzas, A.N., Zoumakis, E., Lin, H.M., Bixler, E.O., Trakada, G., and Chrousos, G.P. (2004b) Marked decrease in sleepiness in patients with sleep apnea by Etanercept, a tumor necrosis factor-{alpha} antagonist. *J Clin Endocrinol Metab* 89(9), 4409–4413.

Vgontzas, A.N., Zoumakis, M., Papanicolaou, D.A., Bixler, E.O., Prolo, P., Lin, H.M., Vela-Bueno, A., Kales, A., and Chrousos, G.P. (2002) Chronic insomnia is associated with a shift of interleukin-6 and tumor necrosis factor secretion from nighttime to daytime. *Metabolism* 51(7), 887–892.

Zamarron, F., Maceiras, F., Mera, A., and Gomez-Reino, J.J. (2004) Effects of the first infliximab infusion on sleep and alertness in patients with active rheumatoid arthritis. *Ann Rheum Dis* 63, 88–90.

11 The Stress of Inadequate Sleep and Immune Consequences

Beatriz Duarte Palma, Sergio Tufik, and Deborah Suchecki

11.1 Introduction

The interaction between sleep and immune system has been under study for some years. Although such studies have began many years ago, only in the last decade the investigation had taken place systematically. Two major phenomena have guided the research on this interaction. The first one corresponds to the lethargy and strong sleepiness that occur during infections and inflammatory conditions. The second one is related to data showing that one of the consequences of sleep deprivation includes increased susceptibility to infection. Thus, the effects of sleep deprivation have been examined to improve our comprehension of the physiological function of specific endogenous immune-regulators during sleep as well as the importance of sleep for the immune system.

During the last decades some authors have claimed that the general population is being subjected to a reduction in hours of sleep, as a consequence of social demands and working pressures, including shift-work. Therefore, sleep reduction or deprivation represents a risk factor for immunological insult. Sleep deprivation can, thus, be considered a biological stress given that sleep is essential to life and to health.

Numerous studies carried out both in humans and in animals suggest the existence of a close relationship between sleep deprivation and the activation of the hypothalamic-pituitary-adrenal (HPA) axis (Horne 2000; Van Reeth, Weibel, Spiegel, Leproult, Dugovic, and Maccari 2000; Vgontzas and Chrousos 2002).

Most of the times sleep deprivation leads to typical signs of stress, including elevated glucocorticoid and catecholamine plasma levels. In humans, sleep deprivation resulted, in fact, in elevated levels of plasma cortisol when compared to a normal night of sleep, although these levels did not reach waking or stress values, when the activity of the HPA axis is maximum (Lange, Perras, Fehm, and Born 2003). Curiously, it has been shown that elevated plasma levels are not detected in the morning following the night of sleep deprivation, the elevation is rather manifested in the evening following sleep deprivation, when cortisol levels should be lower (Leproult, Copinschi, Buxton, and Van Cauter 1997). In the end, the cortisol circadian rhythm becomes flattened and levels are continually elevated.

In animal studies, sleep deprivation-induced elevation of corticosterone and catecholamine levels is evident (Tobler, Murison, Ursin, Ursin, and Borbely 1983;

Suchecki, Lobo, Hipólide, and Tufik 1998; Sgoifo, Buwalda, Roos, Costoli, Merati, and Meerlo 2006). It is, therefore, natural to assume that these hormones play a major role on the immune outcomes of sleep deprivation, given the well-documented impact of stress, and especially of glucocorticoids on the immune system (Biondi 2001).

As a general rule, glucocorticoids inhibit the activity of the lymphoreticular system, with inhibition of lymphocytes and macrophages activity; they reduce the production of many cytokines and some inflammatory mediators, in addition to modulate molecular adhesion, cellular migration, differentiation and modulation (for review, see Adcock, 2000). Nonetheless, immunosuppressive effects are not exclusive of glucocorticoids and catecholamines. In a classical study, Dantzer and Kelley (1989) showed that stressed adrenalectomized rats (ADX) exhibit some, but not all signs of immunosuppression. Such finding indicates the existence of other substances that result in a negative impact on the immune system, including opioids (Black 1994).

Numerous evidence point out to a close relationship between HPA axis activity and immune function. In fact this is a bimodal relationship in which glucocorticoids may function as an inhibitory or permissive factor, depending on its circulating levels. Disturbances at any level of the HPA axis or on the action of glucocorticoids results in immune system imbalance, which, consequently is harmful to the organism. Glucocorticoids act as modulators of the immune system, for either excess or deficient release are associated with illness. On one hand, hyperstimulation of the HPA axis, with excessive secretion of glucocorticoids results in intense immunosuppression which ultimate result being increased susceptibility to infection; on the other hand, insufficient secretion of glucocorticoids leads to inflammatory and autoimmune diseases (for a detailed review, see Webster, Tonelli, and Sternberg 2002).

The use of animal models has led to many important findings highlighting how essential the stress response is for both the physiological regulation of the immune system and for the development and manifestation of inflammatory and autoimmune diseases. The rat strains Lewis (LEW/N) and Fischer (F344/N) are widely used to assess neuroendocrine regulation of many aspects of auto-immunity. LEW/N rats are more susceptible to the development of autoimmune/inflammatory disorders in response to antigenic stimuli, since this strain exhibits a hyporesponsiveness of the HPA axis. The F344/N strain, on the contrary, is relatively resistant to these disorders due to its augmented activity of the HPA axis (Moncek, Kvetnansky, and Jezova 2001).

Some authors consider sleep as a major association pathway between stress and the immune system. For instance, some behavioral features of a stressful situation such as intrusive and undesired thoughts, avoidance of reality and fear, are associated with sleep disorders (Hall, Baum, Buysse, Prigerson, Kupfer, and Reynolds 1997; Hall, Buysse, Dew, Prigerson, Kupfer, and Reynolds 1998), nonrestorative, bad quality sleep, in turn, can lead to immune disorders. According to this hypothesis, is the sleep impairment started out by stress that triggers the impairment of the immune system, and not the direct action of stress itself. Sleep, therefore, can be considered as an integral and functional part of the immune system.

The literature on sleep deprivation is very prolific, both in human beings and in animals, and indicates that one of the outcomes is the alteration of several immune parameters. The purpose of the present chapter is not to review different results and methods of forced sleep deprivation for the reader can consult numerous comprehensive reviews on this topic (Moldofsky 1993; Toth 1995; Benca and Quintas, 1997; Rogers, Szuba, Staab, Evans, and Dinges 2001; Irwin 2002; Bryant, Trinder, and Curtis 2004). Our goal is to discuss the impact of stress of inadequate sleep in naturally occurring conditions, such as sleep disorders and shift-work, on the immune system.

11.2 Inadequate Sleep and Disease

An ever-growing number of studies have been performed in order to assess the impact of inadequate sleep on the quality of life, morbidity and mortality. For instance, a retrospective study conducted in Denmark demonstrated that nighttime-working women showed augmented risk for breast cancer than daytime-working women (Hansen 2001). Likewise, increased risk for infections was reported by workers whose working hours were comprised to the night, when compared to daytime workers (Mohren, Jansen, Kant, Galama, van den Brandt, and Swaen 2002). This is somehow an expected outcome, given that shift workers experience a complete alteration of circadian rhythm of some hormones, such as cortisol and melatonin, in addition to temperature (Fujiwara, Shinkai, Kurokawa, and Watanabe 1992; Hennig, Kieferdorf, Moritz, Huwe, and Netter 1998; Vangelova 2000). In an elegant in vitro study, Rogers, van den Heuvel, and Dawson (1997) evaluated the influence of cortisol and melatonin on human lymphocyte proliferation after stimulation with concanavalin A (CON A) and incubation with either cortisol alone (such as during the day), melatonin alone (such as during the night), or with the combination of cortisol and melatonin (such as in shift workers). Cortisol alone produced immunosuppression as expected. The combination of cortisol and melatonin inhibit even further the lymphocyte proliferation, which is rather surprising, since melatonin has been claimed to be an immunoestimulant agent (Carrillo-Vico, Guerrero, Lardone, and Reiter 2005). These findings, therefore, suggest that concomitant increase of both hormones disrupts the correct functioning of the immune system.

The correlation between life expectancy and time of sleep has been demonstrated by some (Pollack, Perlick, Linsner, Wenston, and Hsieh 1990; Kripke, Simons, Garfinkel, and Hammond 1979), but not all authors (Wingard and Berkman, 1983; Nilsson, Hedblad, and Berglund 2001). Two recent studies are noteworthy mentioning and deserve to be divulged due to their importance for public health and because of the growing amount of people working at night or whose sleep is seriously impaired. The first study reveals that partial sleep deprivation following immunization against hepatitis A reduces the production of antibody titers, in addition to changes in hormones secretion, including prolactin and cortisol (Lange et al. 2003). The second study reports that, compared to individuals who sleep, partially sleep-deprived individuals also exhibit impairment of antibody production against Influenza virus. The antibody titers were determined in the fourth week postimmunization, at a time when production reaches a peak (Spiegel, Sheridan, and

Van Cauter 2002). These findings highlight the importance of ideal sleep, given that the effects of sleep deprivation can be manifested much later.

11.3 Immune Function in Sleep Disorders

Insomnia and obstructive sleep apnea are the most prevalent sleep disorders in the general population (Ancoli-Israel 1993). Such disturbances result in major alterations in the immune system and may be mediated by the augmented activity of the HPA axis and/or sympathetic nervous system, as we will soon describe.

There is evidence of the association of insomnia and elevated HPA axis activity. Primary insomnia patients connect the onset of the disorder to some stressful event. Chronic internalization of feelings may cause a psychological and physiological arousal, leading, ultimately to insomnia. Perlis, Giles, Mendelson, Bootzin, and Wyatt (1997), for instance, propose the existence of a mechanism of cortical hyperarousal in insomniac patients triggered by emotional, cognitive, and physiological components. Therefore, the hyperactivity exacerbates vigilance and impact negatively on sleep, thus forming a vicious circle in which difficulty to sleep becomes the stressful factor itself. This phenomenon had been proposed by Spath-Schwalbe and co-workers, in 1991, who demonstrated that increased HPA axis activity produces sleep fragmentation, which in turn elevates cortisol-circulating levels. One of the first studies on the relationship between the HPA axis activity and sleep showed that poor sleepers exhibited augmented 24 h urinary cortisol levels (Johns, Gay, Masterton, and Bruce 1971). This association was later confirmed by Vgontzas et al. (1998) who showed a positive correlation between elevated levels of cortisol, increased activity of the sympathetic system and time of waking in chronic insomniacs. In addition, plasma levels of ACTH and cortisol measured throughout a 24-h period are elevated in chronic insomniacs exactly during the nadir of the circadian rhythm, i.e., between approximately 22:00 and 02:00 h, although no change in the circadian pattern of secretion has been observed besides a flattening of the amplitude (Vgontzas et al. 2001). Similar results were reported by Rodenbeck, Huether, Ruther, and Hajak (2002) who showed that patients with severe primary insomnia exhibit high cortisol levels at the beginning of the night.

Convincing evidence confirm the correlation between immune alterations and insomnia, as, for instance, is shown in chronic insomnia patients, whose CD3+, CD4+, and CD8+ cells are reduced (Savard, Laroche, Simard, Ivers, and Morin 2003). Moreover, a decrease in the number of natural killer (NK) cells is also associated with augmented sympathetic tonus (Irwin, Clark, Kennedy, Christian Gillin, and Ziegler 2003), and the balance between Th1 and Th2 response is impaired in these patients with predominance of the Th2 response (Sakami et al. 2002). Th2-type cytokines enhance the humoral immune response. In addition, alterations in Th1/Th2 balance are characteristic of autoimmune diseases. It is known that the major effects of glucocorticoids on the immune system are suppression of cellular immunity and enhancement of humoral immunity, a process that also is driven by the Th1 to Th2 shift in cytokine profile production. Thus, these results indicate that the relationship between sleep and the etiology of immune-related diseases should be considered.

Recurrent nocturnal hypoxia and sleep fragmentation are related to obstructive sleep apnea, and these are also associated with increased activation of the HPA and the sympathetic systems. Ultimately, the latter alteration is regarded as one of the mechanisms responsible for hypertension that are commonly associated with to obstructive sleep apnea (Waradekar, Sinoway, Zwillich, and Leuenberger 1996). Patients with sleep apnea show higher nocturnal urinary and plasma catecholamine levels than healthy individuals (Fletcher, Miller, Schaaf, and Fletcher 1987). Although the assessment of the HPA axis activity in these patients is scarce, there is a study that shows a greater tryptophan-induced cortisol response, indicating augmented sensitivity of the axis (Hudgel and Gordon 1997). Despite this scarcity, increased sympathetic tonus and especially, augmented sleep fragmentation are connected to increased stress response and it seems unquestionable that such poor sleep quality leads to higher HPA axis activity. In a recent review, the HPA axis hyperactivity was attributed as being responsible for some morbid conditions frequently associated to sleep apnea, such as metabolic syndrome and increased aldosterone plasma levels (Buckley and Schatzberg 2005). In this pathological condition, high levels of proinflammatory cytokines, including TNF and IL-6 (Alberti et al. 2003) and of C-reactive protein (CRP) are observed (Shamsuzzaman, Winnicki, Lanfranchi, Wolk, Kara, and Accurso 2002). This myriad of altered inflammatory mediators, together with apnea pathophysiology is extremely harmful due to their contribution to the development of severe complications, which include increased risk to cardiovascular problems, metabolic syndrome and neurocognitive impairment. The regular use of CPAP results in restoration of adequate sleep and reduction of inflammatory markers (Yokoe et al. 2003), as well as of catecholamines and of the sympathetic tonus (Hedner, Darpo, Ejnell, Carlson, and Caidahl 1995; Ziegler, Mills, Loredo, Ancoli-Israel, and Dimsdale 2001).

11.4 Hospitalized Patients

The harmful effect of sleep deprivation is particularly important for people with high risk of infection, such as hospitalized, burn victims and elder patients. In addition, it is known that surgical patients suffer from severe sleep disruptions during postoperative period, such that major surgical procedures result in approximately 20% reduction in slow wave sleep and a complete abolition of REM sleep (Rosenberg-Adamsen, Kehlet, Dodds, and Rosenberg 1996; Cronin, Keifer, Davies, King, and Bixler 2001), even with nonopioid pain control, since opioids are known to alter sleep pattern (Kay, Eisenstein, and Jasinski 1969; Knill, Moote, Skinner, and Rose 1990).

Despite the paucity of data, burned victims, especially those with extensive burning, show an intense change in sleep pattern. A single study assessed the sleep of burn victims by polysomnography and reports reduction of slow wave sleep and of REM sleep, in addition to increased awakenings (Gottschlich et al. 1994). A questionnaire-based study reveals that burned children complain of sleep disturbances, especially nightmares, for as long as one year following hospital discharge (Kravitz et al. 1993). An extensive review (Rose, Sanford, Thomas, and Opp 2001) explores in detail the factors that may alter the sleep in burned children, which include metabolic and hormone alterations, medicine (opioids), pain, and hospital environment are some of these factors. To the best of our knowledge there is no study in which the correlation between sleep disorder-induced immune deficiencies in burned victims and the patient's evolution. Nonetheless, chronic disrupted sleep is known to delay some recovery processes, such as tissue wound healing. Indirect evidence indicate that sleep is crucial for tissue repair, insofar as major repairing factors, such as protein synthesis, cell division and growth hormone secretion, are augmented during sleep. Empirical data show that women exposed to short period of sleep deprivation exhibit an impairment of skin barrier homeostasis. The authors attribute this result to increased production of proinflammatory cytokines induced by sleep deprivation (Altemus, Rao, Dhabhar, Ding, and Granstein 2001), although in rats, sleep deprivation failed to impair the wound healing in rats submitted to tissue biopsy (Mostaghimi, Obermeyer, Ballamudi, Martinez-Gonzalez, and Benca 2005). This discrepancy is very likely attributable to the species-specific factors and to differences in the type of wound healing process. Despite this controversy, it is undisputable that stress is a harmful factor for tissue repair (Padgett, Marucha, and Sheridan 1998; Marucha, Kiecolt-Glaser, and Favagehi 1998).

Finally, there is no doubt that sleep deprivation makes the organism more vulnerable to opportunistic infections, which is a common condition in burned victims. The extensive literature points out to the impair of major immune parameters, including increased size of lymph nodes and of proliferation of several bacteria species in sleep-deprived rats (Everson and Toth 2000). Bacterial infection in a healthy, undamaged tissue, suggests that sleep deprivation produces an insufficient immune response that results in a disease susceptible organism.

11.5 Chronic Inflammatory Diseases and Sleep

Sleep disorders are frequently observed in chronic inflammatory diseases. Such disorders are believed to be an intrinsic component of the disease and not simply a consequence of the pain. Nonetheless, fatigue, pain, and muscle rigidity are associated to nonrestorative sleep (Nicassio, Moxham, Schuman, and Gevirtz 2002). What seems to take place is a vicious cycle in which pain and chronic fatigue leads to nonrestorative sleep, which in turn, worsens painful conditions.

Among the chronic inflammatory diseases, fibromyalgia, systemic lupus erythematosus, rheumatoid arthritis, and chronic fatigue syndrome are most frequently associated with impaired sleep (Lashley 2003). This is quite interesting, especially when the HPA axis is included in the equation. The association between HPA axis activity and immune functions is relevant to humans with illnesses including rheumatoid arthritis (Gutierrez, Garcia, Rodriguez, Mardonez, Jacobelli, and Rivero 1999), systemic lupus erythematosus (Gutierrez, Garcia, Rodriguez, Rivero, and Jacobelli 1998), fibromyalgia (Crofford et al. 1994), and chronic fatigue syndrome (Demitrack et al. 1991).

Chrousos and Gold (1992) have proposed the hypothesis that sustained stress system dysfunction, characterized by either hyper or hypoactivity of the HPA axis, play a role in various pathologic states, including a range of autoimmune/-inflammatory, psychiatric, and endocrine diseases (Fig. 11.1). Although the concept of hyperactivity leading to sleep impairment (Kupfer 1995; Steiger 2002) is what comes more easily to min, it is necessary to acknowledge that, once again, glucocorticoids play a bimodal role on sleep regulation, insofar as insufficient cortisol production, as seen in Addison's disease, also results in sleep disruption, which is corrected by administration of hydrocortisone (Garcia-Borreguero et al. 2000; Lovas, Husebye, Holsten, and Bjorvatn 2003).

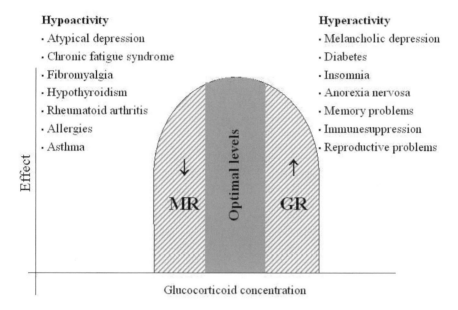

FIGURE 11.1. Theoretical model of the inverted U shaped curve, in which optimal levels of glucocorticoids are essential for the perfect organic functioning. Either hypo (left-hand side) or hyperactivity (right-hand side) of the hypothalamic-pituitary-adrenal axis will lead to unbalanced occupation of MR/GR corticosteroid receptors, likely resulting in many unfavorable conditions (based on Chrousos and Gold 1992).

Lupus patients often report to experience a poor sleep quality. Most of the studies which explore the relationship between lupus and sleep quality use questionnaires as a tool to assess sleep. Recent findings attribute the poor quality of sleep to associated features, such as depression, chronic use of corticoids, lack of physical activity (Costa et al. 2005), and painful condition (Gudbjornsson and Hetta 2001). In addition, a negative correlation between sleep quality and disease activity is observed (Tench, McCurdie, White, and D'Cruz 2000) so that intense production of cytokines and other inflammatory agents may exacerbate the already altered sleep pattern in these patients. The polysomnographic evaluation confirms the patients' complains of poor sleep quality, revealing low sleep efficiency, sleep fragmentation, breathing disorders, and periodic leg movements. The net result, and one of the major findings of the study, is the extreme daytime somnolence, pointing out to nighttime sleep impairment (Valencia-Flores et al. 1999).

In this regard, stress is also a major player in this relationship. Several studies have pointed out stress, either physical or psychological, as a triggering factor for the onset or worsening of lupus (Blumenfield 1978; Wallace 1987; Peralta-Ramirez, Jimenez-Alonso, Godoy-Garcia, and Perez-Garcia 2004). Besides anxiety and depression, lupus patients report that stress is related to aggravation of their physical and mental well-being (Wekking, Vingerhoets, van Dam, Nossent, and Swaak 1991; Da Costa et al., 1999).

In a recent study, Palma, Gabriel, Colugnati, and Tufik (2006) showed that sleep deprivation triggered an early onset of disease manifestation in an animal model of systemic erythematosus lupus (first generation of the New Zealand Black × New Zealand White breeding), without altering either the course or severity of the disease. Interestingly, sleep deprivation induced a potent secretion of corticosterone in these animals, but contrary to a strain that does not develop lupus, whose corticosterone levels returned to basal on the week following sleep deprivation, hormone levels in these animals never returned back to basal throughout the mice entire life span (unpublished data).

11.4 Concluding Remarks

The mutual influence of sleep on the immune system and viceversa is likely mediated in part by the HPA axis. Over or underproduction of cortisol results in disrupted immune function, leading to immunosuppression or immunostimulation, respectively. Glucocorticoids, in turn, also interfere with sleep in a bimodal fashion, and cytokines can also interfere with sleep quality. Therefore, this relationship among sleep, glucocorticoids and cytokines poses as a perfect model of psychoneuroimmunology, in which elements of different systems interact as a manner to maintain a perfect balance. The importance of this closed system comes exactly from its fraily, for any disruption can lead to disease.

References

Adcock, I.M. (2000) Molecular mechanisms of glucocorticosteroid actions. *Pulm Pharmacol Ther* 13, 115–126.

Alberti, A., Sarchielli, P., Gallinella, E., Floridi, A., Floridi, A., Mazzotta, G., and Gallai V. (2003) Plasma cytokine levels in patients with obstructive sleep apnea syndrome: A preliminary study. *J Sleep Res* 12, 305–311.

Altemus, M., Rao, B., Dhabhar, F.S., Ding, W., and Granstein, R.D. (2001) Stress-induced changes in skin barrier function in healthy women. *J Invest Dermatol* 117, 309–317.

Ancoli-Israel, S. (1993) Epidemiology of sleep complaints and disorders. In: M.A. Carskadon (Ed.), *Encyclopedia of Sleep and Dreaming*. Macmillan, New York, pp. 223–225.

Benca, R.M., and Quintas, J. (1997) Sleep and host defenses: A review. *Sleep* 20, 1027–1037.

Biondi, M. (2001) Effects of stress on immune function: An overview. In: R. Ader, D.L. Felten, and N. Cohen (Eds.), *Psychoneuroimmunology*. Wiley, New York, pp. 189–226.

Black, P.H. (1994) Immune system-central nervous system interactions: Effect and immunomodulatory consequences of immune system mediators on the brain. *Antimicrob Agents Chemother* 38, 7–12.

Blumenfield, M. (1978). Psychological aspects of systemic lupus erythematosus. *Prim Care* 5, 159–171.

Bryant, P.A., Trinder, J., and Curtis, N. (2004) Sick and tired: Does sleep have a vital role in the immune system? *Nat Rev Immunol* 4, 457–467.

Buckley, T.M., and Schatzberg, A.F. (2005) On the interactions of the hypothalamic-pituitary-adrenal (HPA) axis and sleep: Normal HPA axis activity and circadian rhythm, exemplary sleep disorders. *J Clin Endocrinol Metab* 90, 3106–3114.

Carrillo-Vico, A., Guerrero, J.M., Lardone, P.J., and Reiter, R.J. (2005) A review of the multiple actions of melatonin on the immune system. *Endocrine* 27, 189–200.

Chrousos, G.P., and Gold, P.W. (1992) The concepts of stress and stress system disorders. Overview of physical and behavioral homeostasis. *JAMA* 267, 1244–1252.

Crofford, L.J., Pillemer, S.R., Kalogeras, K.T., Cash, J.M., Michelson, D., Kling, M.A., Sternberg, E.M., Gold, P.W., Chrousos, G.P., and Wilder, R.L. (1994) Hypothalamic-pituitary-adrenal axis perturbations in patients with fibromyalgia. *Arthritis Rheum* 37, 1583–1592.

Cronin, A.J., Keifer, J.C., Davies, M.F., King, T.S., and Bixler, E.O. (2001) Postoperative sleep disturbance: Influences of opioids and pain in humans. *Sleep* 24, 39–44.

Da Costa, D., Bernatsky, S., Dritsa, M., Clarke, A.E., Dasgupta, K., Keshani, A., and Pineau, C. (2005) Determinants of sleep quality in women with systemic lupus erythematosus. *Arthritis Rheum* 53, 272–278.

Da Costa, D., Dobkin, P.L., Pinard, L., Fortin, P.R., Danoff, D.S., Esdaile, J.M., Clarke, A.E. (1999) The role of stress in functional disability among women with systemic lupus erythematosus: a prospective study. Arthritis Care Res. 12(2):112–9.

Dantzer, R., and Kelley, K.W. (1989) Stress and immunity: An integrated view of relationships between the brain and the immune system. *Life Sci* 44, 1995–2008.

Demitrack, M.A., Dale, J.K., Straus, S.E., Laue, L., Listwak, S.J., Kruesi, M.J., Chrousos, G.P., and Gold, P.W. (1991) Evidence for impaired activation of the hypothalamic-pituitary-adrenal axis in patients with chronic fatigue syndrome. *J Clin Endocrinol Metab* 73, 1224–1234.

Everson, C.A., and Toth, L.A. (2000) Systemic bacterial invasion induced by sleep deprivation. *Am J Physiol Regul Integr Comp Physiol* 278, R905–R916.

Fletcher, E.C., Miller, J., Schaaf, J.W., and Fletcher, J.G. (1987). Urinary catecholamines before and after tracheostomy in patients with obstructive sleep apnea and hypertension. *Sleep* 10, 35–44.

Fujiwara, S., Shinkai, S., Kurokawa, Y., and Watanabe T. (1992) The acute effects of experimental short-term evening and night shifts on human circadian rhythm: The oral temperature, heart rate, serum cortisol and urinary catecholamines levels. *Int Arch Occup Environ Health* 63, 409–418.

Garcia-Borreguero, D., Wehr, T.A., Larrosa, O., Granizo, J.J., Hardwick, D., Chrousos, G.P., and Friedman, T.C. (2000) Glucocorticoid replacement is permissive for rapid eye movement sleep and sleep consolidation in patients with adrenal insufficiency. *J Clin Endocrinol Metab* 85, 4201–4206.

Gottschlich, M.M., Jenkins, M.E., Mayes, T., Khoury, J., Kramer, M., Warden, G.D., and Kagan R.J. (1994) The 1994 Clinical Research Award. A prospective clinical study of the polysomnographic stages of sleep after burn injury. *J Burn Care Rehabil* 15, 486–492.

Gudbjornsson, B., and Hetta, J. (2001). Sleep disturbances in patients with systemic lupus erythematosus: A questionnaire-based study. *Clin Exp Rheumatol* 19, 509–514.

Gutierrez, M.A., Garcia, M.E., Rodriguez, J.A., Mardonez, G., Jacobelli, S., and Rivero, S. (1999) Hypothalamic-pituitary-adrenal axis function in patients with active rheumatoid arthritis: A controlled study using insulin hypoglycemia stress test and prolactin stimulation. *J Rheumatol* 26, 277–281.

Gutierrez, M.A., Garcia, M.E., Rodriguez, J.A., Rivero, S., and Jacobelli, S. (1998) Hypothalamic-pituitary-adrenal axis function and prolactin secretion in systemic lupus erythematosus. *Lupus* 7, 404–408.

Hall, M., Baum, A., Buysse, D.J., Prigerson, H.G., Kupfer, D.J., and Reynolds, C.F., 3rd. (1998) Sleep as a mediator of the stress-immune relationship. *Psychosom Med* 60, 48–51.

Hall, M., Buysse, D.J., Dew, M.A., Prigerson, H.G., Kupfer, D.J., and Reynolds, C.F., 3rd. (1997) Intrusive thoughts and avoidance behaviors are associated with sleep disturbances in bereavement-related depression. *Depress Anxiety* 6, 106–112.

Hansen, J. (2001). Increased breast cancer risk among women who work predominantly at night. *Epidemiology* 12, 74–77.

Hedner, J., Darpo, B., Ejnell, H., Carlson, J., and Caidahl, K. (1995) Reduction in sympathetic activity after long-term CPAP treatment in sleep apnoea: Cardiovascular implications. *Eur Respir J* 8, 222–229.

Hennig, J., Kieferdorf, P., Moritz, C., Huwe, S., and Netter, P. (1998) Changes in cortisol secretion during shiftwork: Implications for tolerance to shiftwork? *Ergonomics* 41, 610–621.

Horne, J.A. (2000) REM sleep–By default? *Neurosci Biobehav Rev* 24, 777–797.

Hudgel, D.W., and Gordon, E.A. (1997) Serotonin-induced cortisol release in CPAP-treated obstructive sleep apnea patients. *Chest* 111, 632–638.

Irwin, M. (2002) Effects of sleep and sleep loss on immunity and cytokines. *Brain Behav Immun* 16, 503–512.

Irwin, M., Clark, C., Kennedy, B., Christian Gillin, J., and Ziegler, M. (2003) Nocturnal catecholamines and immune function in insomniacs, depressed patients, and control subjects. *Brain Behav Immun* 17, 365–372.

Johns, M.W., Gay, T.J., Masterton, J.P., and Bruce, D.W. (1971) Relationship between sleep habits, adrenocortical activity and personality. *Psychosom Med* 33, 499–508.

Kay, D.C., Eisenstein, R.B., and Jasinski, D.R. (1969) Morphine effects on human REM state, waking state and NREM sleep. *Psychopharmacologia* 14, 404–416.

Knill, R.L., Moote, C.A., Skinner, M.I., and Rose, E.A. (1990) Anesthesia with abdominal surgery leads to intense REM sleep during the first postoperative week. *Anesthesiology* 73, 52–61.

Kravitz, M., McCoy, B.J., Tompkins, D.M., Daly, W., Mulligan, J., McCauley, R.L., Robson, M.C., and Herndon, D.N. (1993). Sleep disorders in children after burn injury. *J Burn Care Rehabil* 14, 83–90.

Kripke, D.F., Simons, R.N., Garfinkel, L., and Hammond, E.C. (1979) Short and long sleep and sleeping pills. Is increased mortality associated? *Arch Gen Psychiatry* 36, 103–116.

Kupfer, D.J. (1995). Sleep research in depressive illness: Clinical implications—a tasting menu. *Biol Psychiatry* 38, 391–403.

Lange, T., Perras, B., Fehm, H.L., and Born, J. (2003) Sleep enhances the human antibody responses to hepatitis A vaccination. *Psychosom Med* 65, 831–835.

Lashley, F.R. (2003) A review of sleep in selected immune and autoimmune disorders. *Holist Nurs Pract* 17, 65–80.

Leproult, R., Copinschi, G., Buxton, O., and Van Cauter, E. (1997) Sleep loss results in an elevation of cortisol levels the next evening. *Sleep* 20, 865–870.

Lovas, K., Husebye, E.S., Holsten, F., and Bjorvatn, B. (2003) Sleep disturbances in patients with Addison's disease. *Eur J Endocrinol* 148, 449–456.

Marucha, P.T., Kiecolt-Glaser, J.K., and Favagehi, M. (1998) Mucosal wound healing is impaired by examination stress. *Psychosom Med* 60, 362–365.

Mohren, D.C., Jansen, N.W., Kant, I.J., Galama, J., van den Brandt, P.A., and Swaen, G.M. (2002) Prevalence of common infections among employees in different work schedules. *J Occup Environ Med* 44, 1003–1011.

Moldofsky, H. (1993) Fibromyalgia, sleep disorder and chronic fatigue syndrome. *Ciba Found Symp* 173, 262–271.

Moncek, F., Kvetnansky, R., and Jezova, D. (2001) Differential responses to stress stimuli of Lewis and Fischer rats at the pituitary and adrenocortical level. *Endocr Regul* 35, 35–41.

Mostaghimi, L., Obermeyer, W.H., Ballamudi, B., Martinez-Gonzalez, D., and Benca, R.M. (2005) Effects of sleep deprivation on wound healing. *J Sleep Res* 14, 213–219.

Nicassio, P.M., Moxham, E.G., Schuman, C.E., and Gevirtz, R.N. (2002) The contribution of pain, reported sleep quality, and depressive symptoms to fatigue in fibromyalgia. *Pain* 100, 271–279.

Nilsson, P.M., Nilsson, J.A., Hedblad, B., and Berglund, G. (2001) Sleep disturbances in association with elevated pulse rate for prediction of mortality—Consequences of mental strain? *J Intern Med* 250, 521–529.

Padgett, D.A., Marucha, P.T., and Sheridan, J.F. (1998) Restraint stress slows cutaneous wound healing in mice. *Brain Behav Immun* 12, 64–73.

Palma, B.D., Gabriel, A. Jr., Colugnati, F.A., and Tufik, S. (2006) Effects of sleep deprivation on the development of autoimmune disease in an experimental model of Systemic lupus erythematosus. *Am J Physiol Regul Integr Comp Physiol* 291(5): R1527–1532.

Peralta-Ramirez, M.I., Jimenez-Alonso, J., Godoy-Garcia, J.F., and Perez-Garcia, M. (2004). The effects of daily stress and stressful life events on the clinical symptomatology of patients with lupus erythematosus. *Psychosom Med* 66, 788–794.

Perlis, M.L., Giles, D.E., Mendelson, W.B., Bootzin, R.R., and Wyatt, J.K. (1997) Psychophysiological insomnia: The behavioural model and a neurocognitive perspective. *J Sleep Res* 6, 179–188.

Pollack, C.P., Perlick, D., Linsner, J.P., Wenston, J., and Hsieh, F. (1990) Sleep problems in the community elderly as predictors of death and nursing home placement. *J Community Health* 15, 123–135.

Rodenbeck, A., Huether, G., Ruther, E., and Hajak, G. (2002) Interactions between evening and nocturnal cortisol secretion and sleep parameters in patients with severe chronic primary insomnia. *Neurosci Lett* 324, 159–163.

Rogers, N., van den Heuvel, C., and Dawson, D. (1997) Effect of melatonin and corticosteroid on in vitro cellular immune function in humans. *J Pineal Res* 22, 75–80.

Rogers, N.L., Szuba, M.P., Staab, J.P., Evans, D.L., and Dinges, D.F. (2001) Neuroimmunologic aspects of sleep and sleep loss. *Semin Clin Neuropsychiatry* 6, 295–307.

Rose, M., Sanford, A., Thomas, C., and Opp, M.R. (2001) Factors altering the sleep of burned children. *Sleep* 24, 45–51.

Rosenberg-Adamsen, S., Kehlet, H., Dodds, C., and Rosenberg, J. (1996) Postoperative sleep disturbances: Mechanisms and clinical implications. *Br J Anaesth* 76, 552–559.

Sakami, S., Ishikawa, T., Kawakami, N., Haratani, T., Fukui, A., Kobayashi, F., Fujita, O., Araki, S., and Kawamura, N. (2002–2003) Coemergence of insomnia and a shift in the Th1/Th2 balance toward Th2 dominance. *Neuroimmunomodulation* 10, 337–343.

Savard, J., Laroche, L., Simard, S., Ivers, H., and Morin, C.M. (2003) Chronic insomnia and immune functioning. *Psychosom Med* 65, 211–221.

Sgoifo, A., Buwalda, B., Roos, M., Costoli, T., Merati, G., and Meerlo, P. (2006) Effects of sleep deprivation on cardiac autonomic and pituitary-adrenocortical stress reactivity in rats. *Psychoneuroendocrinology* 31, 197–208.

Shamsuzzaman, A.S., Winnicki, M., Lanfranchi, P., Wolk, R., Kara, T., Accurso, V., and Somers, V.K. (2002) Elevated C-reactive protein in patients with obstructive sleep apnea. *Circulation* 105, 2462–2464.

Spath-Schwalbe, E., Gofferje, M., Kern, W., Born, J., and Fehm, H.L. (1991) Sleep disruption alters nocturnal ACTH and cortisol secretory patterns. *Biol Psychiatry* 29, 575–584.

Spiegel, K., Sheridan, J.F., and Van Cauter, E. (2002) Effect of sleep deprivation on response to immunization. *JAMA* 288, 1471–1472.

Steiger, A. (2002) Sleep and the hypothalamo-pituitary-adrenocortical system. *Sleep Med Rev* 6, 125–138.

Suchecki, D., Lobo, L.L., Hipólide, D.C., and Tufik, S. (1998) Increased ACTH and corticosterone secretion induced by different methods of paradoxical sleep deprivation. *J Sleep Res* 7, 276–281.

Tench, C.M., McCurdie, I., White, P.D., and D'Cruz, D.P. (2000) The prevalence and associations of fatigue in systemic lupus erythematosus. *Rheumatology* 39, 1249–1254.

Tobler, I., Murison, R., Ursin, R., Ursin, H., and Borbely, A.A. (1983) The effect of sleep deprivation and recovery sleep on plasma corticosterone in the rat. *Neurosci Lett* 35, 297–300.

Toth, L.A. (1995) Sleep, sleep deprivation and infectious disease: Studies in animals. *Adv Neuroimmunol* 5, 79–92.

Valencia-Flores, M., Resendiz, M., Castano, V.A., Santiago, V. Campos, R.M., Sandino, S., Valencia, X., Alcocer, J., Ramos, G.G., and Bliwise, D.L. (1999) Objective and subjective sleep disturbances in patients with systemic lupus erythematosus. *Arthritis Rheum* 42, 2189–2193.

Van Reeth, O., Weibel, L., Spiegel, K., Leproult, R., Dugovic, C., and Maccari, S. (2000) Interactions between stress and sleep: From basic research to clinical situations. *Sleep Med Rev* 4, 201–219.

Vangelova, K. (2000) Circadian adjustment and stress in operators under fast rotating 12-hour shifts. *Cent Eur J Public Health* 8, 229–232.

Vgontzas, A.N., Bixler, E.O., Lin, H.M., Prolo, P., Mastorakos, G., Vela-Bueno, A., Kales, A., and Chrousos, G.P. (2001) Chronic insomnia is associated with nyctohemeral activation of the hypothalamic-pituitary-adrenal axis: Clinical implications. *J Clin Endocrinol Metab* 86, 3787–3794.

Vgontzas, A.N., and Chrousos, G.P. (2002) Sleep, the hypothalamic-pituitary-adrenal axis, and cytokines: Multiple interactions and disturbances in sleep disorders. *Endocrinol Metab Clin North Am* 31, 15–36.

Vgontzas, A.N., Tsigos, C., Bixler, E.O., Stratakis, C.A., Zachman, K., Kales, A., Vela-Bueno, A., and Chrousos G.P. (1998) Chronic insomnia and activity of the stress system: A preliminary study. *J Psychosom Res* 45, 21–31.

Wallace, D.J. (1987) The role of stress and trauma in rheumatoid arthritis and systemic lupus erythematosus. *Semin Arthritis Rheum.* 16, 153–157.

Waradekar, N.V., Sinoway, L.I., Zwillich, C.W., and Leuenberger, U.A. (1996) Influence of treatment on muscle sympathetic nerve activity in sleep apnea. *Am J Respir Crit Care Med* 153, 1333–1338.

Webster, J.I., Tonelli, L., and Sternberg, E.M. (2002) Neuroendocrine regulation of immunity. *Annu Rev Immunol* 20, 125–163.

Wekking, E.M., Vingerhoets, A.J., van Dam, A.P., Nossent, J.C., and Swaak, A.J. (1991) Daily stressors and systemic lupus erythematosus: A longitudinal analysis—First findings. *Psychother Psychosom* 55, 108–113.

Wingard, D.L., and Berkman, L.F. (1983) Mortality risk associated with sleeping patterns among adults. *Sleep* 6, 102–107.

Yokoe, T., Minoguchi, K., Matsuo, H., Oda, N., Minoguchi, H., Yoshino, G., Hirano, T., and Adachi, M. (2003) Elevated levels of C-reactive protein and interleukin-6 in patients with obstructive sleep apnea syndrome are decreased by nasal continuous positive airway pressure. *Circulation* 107, 1129–1134.

Ziegler, M.G., Mills, P.J., Loredo, J.S., Ancoli-Israel, S., and Dimsdale, J.E. (2001) Effect of continuous positive airway pressure and placebo treatment on sympathetic nervous activity in patients with obstructive sleep apnea. *Chest* 120, 887–893.

12 Neuroimmunology of Pregnancy-Related Sleep Disturbances

Michele L. Okun and Mary E. Coussons-Read

12.1 Introduction

Sleep is undoubtedly disturbed during pregnancy. The disturbances include an increase in nocturnal awakenings and greater periods of time spent awake during the night, as well as complaints of fatigue during the day. Although nausea (i.e., morning sickness) is most commonly expected in the first trimester, an increase in fatigue, which is often a consequence of disturbed sleep, is actually the first symptom of pregnancy (Lee 2006). Various contributors have been suggested or implicated in the magnitude of sleep disturbances in pregnancy, including hormonal, physiologic, physical (Baratte-Beebe and Lee 1999), and behavioral changes (Buster and Carson 2003; Challis and Lye 2003). As pregnancy progress, other symptoms contribute to sleep disturbances, some of them being fetal movements, leg cramps, shortness of breath, and an inability to get comfortable (Baratte-Beebe et al. 1999). This review will extend beyond the description of how sleep patterns change during pregnancy to suggest that there is an important relationship between pregnancy-associated sleep disturbances and cytokine and hormone changes that may have relevance to maternal and fetal health. Discussion will include how disturbances experienced during pregnancy are associated with cytokine alterations and hormonal changes, and how these relationships could add to a woman's risk for developing pregnancy complications or experiencing poor pregnancy outcomes.

12.2 Pregnancy and Alterations in Sleep

12.2.1 Generalized Sleep Patterns Across Pregnancy

Alterations in sleep patterns during normal pregnancy are typically described separately for each of the three trimesters. Initially in the first trimester women complain of increased fatigue and report taking more naps; however, both polysomnographically recorded sleep and self-reports of sleep indicate that they also experience a decrease in total sleep time, take longer to fall asleep once they turn off the lights and have poorer sleep efficiency (defined as the total amount of time a person slept divided by the total amount of time spent in bed) during the nocturnal sleep period (Karacan and Williams 1970; Lee, Zaffke, and McEnany 2000; Mindell and Jacobson 2000; Manber, Colrain, and Lee 2001; Santiago, Nolledo, Kinzler, and

Santiago 2001). During the second trimester, sleep patterns improve slightly compared to those from the first trimester. The increase in wake after sleep onset (WASO) and number of awakenings observed in the first trimester, for instance, wanes in the second trimester, but never reaches prepregnancy sleep patterns (Lee et al. 2000). It is during the third trimester that the majority of sleep disturbances occur, with an increase in spontaneous nighttime awakenings compounding physiological sleep disturbances (Mauri 1990). The degree to which these changes are reported depends on the method of data collection, the sample of women evaluated and in which trimester the data stem from. Several types of data collection tools are available and each provides a variant description of the sleep of a pregnant woman. Two of the most commonly used measures are polysomnography (PSG) and subjective reports. The summary of research that follows illustrates the diversity in research methodology and questions that can be asked.

12.2.2 Sleep Measured by In-lab Polysomnography

There is agreement within the professional sleep community (American Sleep Disorders Association 1997) of a pregnancy-associated sleep disorder, however much of the data are inconsistent and contradictory. The discrepancies stem primarily from changes in technology and in the understanding of sleep over the past several decades. Early studies were able to suggest only crude estimations of sleep changes due to limitations of the EEG recording equipment used (Branchey and Petre-Quadens 1968; Karacan, Heine, and Agnew 1968), while recent studies are able to delineate more accurately between stages of sleep and analyze data more in depth via spectral analysis (Brunner, Munch, Biedermann, Huch, Huch, and Borbely 1994). Despite the current technological sophistication and an expansion of neuroscience, data from the last 35 years are still being corroborated today. In 1968, Branchey and Petre-Quadens conducted one of the earliest studies objectively using PSG to assess sleep architecture in pregnant women. Approximately 57 nocturnal sleep recordings were conducted on 17 pregnant women. The overall results revealed an increase in paradoxical sleep (analogous to REM sleep) at 25 weeks of pregnancy with a "clear-cut" *decrease* in the final 3 to 4 weeks of pregnancy (Branchey et al. 1968). Another study by Karacan et al. (1968) investigated sleep patterns during late pregnancy and showed sleep to be markedly altered during the last trimester of pregnancy even with a small, homogenous population (all middle-class Caucasians). Corroborating their initial study, Karacan et al. (1968) further demonstrated that sleep onset latency (SOL) was considerably longer and the number of awakenings was higher in women during the last month of pregnancy compared to controls, while a complete reduction in Stage 4 sleep was seen in 57% of the subjects (Karacan et al. 1968; Karacan, Williams, Hursch, McCaulley, and Heine 1969). These sleep disturbances were greater than previously reported. Additionally, the authors state that sleep disturbances could "profoundly affect the woman's physiological disposition, perhaps to the point of inducing disease" (p. 933), suggesting over 30 years ago the association between pregnancy-related sleep disturbances and immune dysregulation.

Recent PSG studies have extended and clarified the initial findings of the 1960s. Hertz, Fast, Feinsilver, Albertario, Schulman, and Fein (1992) evaluated 12 women in their third trimester of pregnancy and 10 age-matched nonpregnant controls. One night of PSG revealed that pregnant women had increased WASO and awakenings, increased %Stage 1 sleep and decreased %REM sleep compared to controls. Considering sleep across pregnancy, Driver and Shapiro conducted PSG studies in five women during each trimester and found significantly longer time spent awake following an arousal as the pregnancy progressed (Driver and Shapiro 1992). They also reported that slow wave sleep (SWS) was increased, with increased Stage 4 sleep from the first trimester to second as well as first to third trimester. Similar to the Branchey and Petre-Quadens' results, a significant decrease in rapid eye movement (REM) sleep was observed during the same time frames, unfortunately, Driver and Shapiro failed to report which week of pregnancy the women were studied. Similar to the study conducted by Driver and Shapiro, Brunner et al. (1994) investigated PSG sleep during the course of pregnancy in nine women for two consecutive nights during each trimester. Unlike the Driver and Shapiro (1992) study, the time frame of gestation in which data were collected was narrower and defined; for example, the third trimester recordings occurring during weeks 32 to 35. And unlike Branchey and Petre-Quadens (1968), no data were collected during the final few weeks of pregnancy. Despite these methodological differences, results were comparable with increased WASO mostly in the third trimester, and decreased REM sleep over the course of the pregnancy. More recently a PSG study of four pregnant women carried out by Schorr, Chawla, Devidas, Sullivan, Naef, and Morrison (1998) reported that pregnant women experienced less SWS than the nonpregnant women, but unlike the previously mentioned studies, no differences in REM sleep were noted among or between the groups. Unlike the Driver et al. study, there were no differences seen across pregnancy, only when compared to controls. These studies confirm that physiological sleep is disturbed during pregnancy and that the degree and type of disruption depends on the trimester and/or week of pregnancy.

12.2.3 Sleep Measured by Ambulatory PSG

Another method used to obtain physiological sleep data that is gaining popularity is ambulatory PSG. It allows a person to sleep in their home environment while still providing comparable data as if the subject were tested in the lab (Reichert, Bloch, Cundiff, and Votteri 2003). Recent studies assessing sleep during pregnancy have utilized this method as it allows for more women to be evaluated and in their own home environment, without the loss of important data. One of the first ambulatory PSG studies of pregnant women was conducted by Coble, Reynolds, Kupfer, Houck, Day, and Giles (1994). They used in-home PSG that transmitted over phone lines for two nights at 12, 24, and 36 weeks respectively on 34 women: 14 with a history of affective disorder but not current and 20 without a history of affective disorder. They found modest differences in sleep between early pregnancy and late pregnancy; however, there was a decrease in REM latency during the 36th week of pregnancy (Coble et al. 1994). Using ambulatory PSG to assess the sleep differences between women who had preeclampsia at 33 weeks gestation and women with healthy pregnancies at 34 weeks gestation, Edwards, Blyton, Kesby, Wilcox, and Sullivan (2000) showed greater sleep disruption in women with preeclampsia. This study goes beyond comparing pregnant to nonpregnant women and provides data that confirm that sleep is disrupted in pregnancy and in pregnancy complications. Whether sleep disturbances precede the complication or vice versa is still under investigation.

Lastly, Lee et al. (2000) studied 33 women who served as their own controls. They were studied prepregnancy and throughout pregnancy with ambulatory PSG (Lee et al. 2000). Data showed that total sleep time and awakenings increased, while sleep efficiency declined during pregnancy compared to prepregnancy. No changes in REM sleep over the course of pregnancy were observed. Studies utilizing ambulatory PSG show comparable results to traditional laboratory assessed PSG. Due to the convenience of studying pregnant women in their home environment or in the hospital, ambulatory PSG is a logical choice to acquire physiological sleep data throughout pregnancy.

12.2.4 Sleep Measured by Subjective Reports

A notable concern with using PSG is the cost and potential burden to the subject and the investigator (Reichert et al. 2003). Furthermore, physiological sleep measures do not capture the perception of the sleep by the individual, which in many instances is as crucial as the objective data (R. Bootzin, pers. commun.). Therefore, subjective reports can be useful in detecting sleep disturbances, particularly during pregnancy. Unfortunately, studies of sleep during pregnancy use various subjective sleep assessments, which result in even more outcomes. For instance, data from a subjective survey of 100 women 38 weeks pregnant reported generalized sleep disruption in 68% of the sample (Schweiger 1972). The way the author obtained these data was by retrospective questions about their sleep during pregnancy; nothing quantitative was obtained. In a similar study (Suzuki, Dennerstein, Greenwood, Armstrong, and Satohisa 1994), 192 pregnant Japanese women in various stages of pregnancy provided subjective reports of their sleep patterns. Results indicate that 88% of the sample reported disturbed sleep from their normal experience, with most of the disturbance occurring in the third trimester. These two studies are examples of basic, nonquantitative evaluations of sleep disturbance during pregnancy. Which provided dichotomous "yes or no" outcomes; they do not provide detailed information regarding the sleep patterns of pregnant women at any given point in time.

Questionnaires continue to be the most common tool used to acquire data in sleep research (Sateia, Doghramji, Hauri, and Morin 2000). They are structured to address global measurements of sleep and demonstrate high global test–retest correlations (Sateia et al. 2000), but there are numerous questionnaires in use and they vary tremendously in their level of detail and the depth of information they acquire. The caveat that not all results are analogous is necessary to keep in mind when interpreting the take home message from these studies. One study by Mindell and Jacobson (2000) cross-sectionally questioned 127 women at one of four points during pregnancy about their sleep habits and sleep disturbances using a questionnaire devised by the authors and the Epworth Sleepiness Scale. Consistent with other studies, nighttime awakenings were the most common sleep disturbance, followed by difficulty falling asleep, longer TST and more naps by the end of pregnancy. Hedman, Pohjasvaara, Tolonen, Suhonen-Malm, and Myllyla (2002) used the Basic Nordic Sleep Questionnaire to acquire sleep information from 325 women at five different time points: prior to becoming pregnancy, once during each trimester, and again during the post partum period. Their participants report an increase in TST during the first trimester, with a decrease in the second and third trimesters. There was an effect of age on TST in that older mothers had less TST at the end of pregnancy than younger mothers. Like the data derived from other studies, late pregnancy was associated with increased nighttime awakenings and restless sleep (Hedman et al. 2002). As mentioned, the questionnaires used in research studies can vary. A study by Izci, Martin, Dundas, Liston, Calder, and Douglas (2005) examined whether snoring and sleepiness were linked in pregnancy. They used a scale to determine "refreshment" upon awakening as well as the ESS. They did confirm that sleepiness is increased in the third trimester, however, sleepiness in pregnancy is not primarily a result of snoring or breathing problems (Izci et al. 2005). The take home message from this overview of using the questionnaire method to acquire sleep data during pregnancy is that the results ultimately depend on how the question(s) is asked.

A method that is used quite often and provides somewhat comparable data to PSG is the sleep diary (Baker, Simpson, and Dawson 1997). Sleep diaries are self-reports of sleep behavior and are common in sleep research due to the convenience of administration, the minimal cost associated with the measure and ease of maintenance (Lockley, Skene, and Arendt 1999). Several studies have utilized the sleep diary in describing the sleep of pregnant women. A study examining the sleep schedules of pregnant women was conducted by Fujino, Shirata, Imanaka, Nishio, Ogita, and Park (1995). The inquiries were of bedtimes, waketimes and napping in a cohort of 1968 pregnant Japanese women and 247 age-matched nonpregnant women. They report that total sleep time did not change throughout pregnancy, but bedtimes were significantly earlier by the end of pregnancy. This is consistent with PSG studies, which indicate that pregnant women spend more time in bed, yet acquire no more total sleep. Furthermore, more pregnant women took naps than nonpregnant women (Fujino et al. 1995). Shinkoda, Matsumoto, and Park (1999) had four women complete sleep logs for a 20-week period, several weeks prior to delivery and 12 weeks postpartum. Data from the sleep diaries suggested poorer sleep efficiency and longer time spent awake after sleep onset as the third trimester progressed, but nothing was statistically significant (Shinkoda et al. 1999). The fact that the sample size was extremely small implies that power may have been extremely limited. Wolfson, Crowley, Anwer, and Bassett (2003) asked 38 women who were 27 to 40 weeks pregnant to complete a sleep diary for 7 days and a depression questionnaire (Center for Epidemiologic Studies-Depression) to see if there were differences in sleep patterns between women with higher and lower depression scores. Women with higher CES-D scores had more TST, later rise times and longer nap times than those with lower CES-D scores (third trimester). Most recently, Okun and Coussons-Read (in press) compared the sleep of 35 pregnant women seen once at the end of each trimester and a sample of 41 nonpregnant controls from 2-weeks of sleep diaries. The sleep was significantly different between the pregnant and nonpregnant groups, particularly in the third trimester. Most consistent with previous studies was the greater number of awakenings and longer WASO observed across pregnancy. In regards to how sleep changed across pregnancy, WASO was higher in the first trimester than the second trimester and greatest overall in the third trimester which is in conjunction with other data; adding to the greater sleep disturbance was that they spent more time in bed and had more nighttime awakenings by the third trimester. The studies that utilize sleep diaries make an effort to acquire data comparable to that obtained from PSG. Although no information regarding sleep stages or spectral analysis can be determined, the data do provide insight into the perception of how an individual believes she is sleeping on a daily basis. This information may be better suited to understanding how comorbid conditions or interventions affect sleep.

12.3 Understanding Sleep in Pregnancy via Animal Models

Animal studies are widely used to further the understanding of how various diseases, illnesses, or situations may affect the human immune system. They can provide an opportunity to control variables that are typically not controllable in human studies. The gathering of information on the sleep patterns of animals during pregnancy has been occurring since the early 1970s due to the acknowledgement that sleep was significantly disturbed in humans (Branchey and Branchey 1970). Clearer information about sleep and wake patterns was acquired with the ability to continuously monitor sleep in controlled environments throughout pregnancy as well as into postpartum. Data from one of the first studies revealed that as early as 15 h before delivery, sleep patterns in rats were modified, reflecting increased wake time and a marked decrease in REM sleep (Branchey et al. 1970). Although no hormonal data were collected from these animals, the authors contend that the observed sleep changes are most likely due to dramatic hormonal changes that occur near or at term. Nishina, Honda, Okai, Kozuma, Inoue, and Taketani (1996) studied the sleep patterns in pregnant female rats. They found increases in the number of NREM episodes, but no increases in total time of NREM sleep. Additionally, they showed a reduction in the number of REM episodes as pregnancy advanced and a prominent decrease in REM sleep during the last 4 days before delivery (Nishina et al. 1996). A protocol by Kimura et al. hoped to further elucidate the sleep-related mechanisms occurring during pregnancy (Kimura, Zhang, and Inoue 1996). Two groups of female rats (pregnant and nonpregnant) were implanted with EEG recording equipment and measured for approximately 3 weeks. Results indicate that significant increases in sleep, primarily NREM sleep, occurred in the pregnant rats. There were also similar fluctuations throughout the pregnancy similar to that in humans, more NREM sleep during the early and late periods than during the midperiod of pregnancy (Kimura et al. 1996). REM sleep was also shown to increase over the course of gestation compared to the nonpregnant rats, but this was attributed primarily to an increase in the number of REM episodes rather than time for each episode. Importantly, the authors heed caution in comparing animal studies to human studies, even though sleep has been shown to be disturbed in pregnancy in both animal and human models.

12.4 Factors That May Contribute to Disrupted Sleep During Pregnancy

12.4.1 Psychosocial Factors

Various psychosocial factors may mediate how sleep is disrupted during pregnancy. These factors include the age of the mother, her initial and subsequent BMI, ethnicity, and health behaviors such as nutritional practices, exercise practices, smoking or consumption of alcohol. Unfortunately, few studies examining sleep in general have reported on these factors, and only one study assessing sleep during pregnancy has commented on any of these variables. Studies considering gender and/or age differences have provided some of the information. Increasing age, for instance, was found to be associated with less time asleep, poorer sleep efficiency, and more minutes awake in the last 2 h of sleep (Carrier, Land, Buysse, Kupfer, and Monk 2001). Similarly, women between the ages of 30 and 40 had less SWS percentage compared to women between 20 and 30 years of age (Ehlers and Kupfer 1997). The only study assessing how age influenced sleep during pregnancy supported the previous study: older pregnant women had less total sleep than younger pregnant women (Hedman et al. 2002). Extrapolation and comparison from most of the studies examining gender and/or age is difficult because a majority of the individuals in these studies are beyond childbearing years and subsequently may have age-associated sleep changes, such as hormonal changes.

There have been several studies that have considered sleep in women of childbearing years and variables that may affect sleep patterns. In a study of employed women, Lee (1992) found influences of age on how women self-reported their sleep disturbances. Younger women reported having poorer sleep quality, while older women stated less sleep quantity, increased difficulty initiating sleep, more midsleep awakenings and increased sleepiness during the day. A study evaluating a sleep restriction protocol in women and how age modulates its effects showed that younger women (~23 years) had more SWS than older women (~60 years), and had increased sleepiness as depicted by the Stanford Sleepiness Scale and the Maintenance of Wakefulness Test (MWT); furthermore, the percentage of SWS was well preserved in the older sample (Stenuit and Kerkhofs 2005). Recently, Tworoger, Davis, Vitiello, Lentz, and McTiernan (2005) considered various factors associated with sleep reported objectively and subjectively in a sample of young, normally cycling women (~30 years). They found that increased BMI was associated with poorer sleep efficiency and more wake time, although the sample was lean (~24 ± 3.8). No correlations were observed between alcohol consumption or exercise and physiologically recorded or subjectively reported sleep.

Another factor that has been considered to be sleep disrupting with and without a relationship to age/gender is smoking. Data stemming from the past 25 years have linked smoking with shorter sleep duration (Palmer, Harrison, and Hiorns 1980), and chronic insomnia, including difficulty initiating and maintaining sleep (Phillips and Danner 1995). It has been suggested that nocturnal nicotine withdrawals are one of the subsequent causes for nocturnal awakenings (Colrain, Trinder, and Swan 2004). These frequent nocturnal awakenings may contribute to the findings that smokers have higher complaints of excessive daytime sleepiness, more reports of minor accidents, higher reported depression and greater impairment in daytime functioning than nonsmokers (Phillips et al. 1995). There are no reports considering the relationship between smoking and sleep disturbance during pregnancy, even though various studies have assessed smoking during pregnancy (Burguet et al. 2004; Kaneita et al. 2005; Narahara and Johnston 1993; Simhan, Caritis, Hillier, and Krohn 2005).

The importance of understanding the synergistic relationship among the psychosocial variables and sleep has been a more recent trend. Stepnowsky, Moore, and Dimsdale (2003) studied the effects of ethnicity on sleep and found that both in the laboratory and at home, Blacks had less SWS than Caucasians. Considering additional factors, Redline, Kirchner, Quan, Gottlieb, Kapur, and Newman (2004) evaluated sex, age, obesity and ethnicity on sleep architecture in a cohort from the Sleep Heart Health study. Although not fully applicable to women of childbearing age as the age range was from 37 to 92, they were able to show that parameters of sleep architecture differed in individuals who were older than 61 years of age compared to those <54 years old. Specifically, sleep was comprised of an increased amount of lighter sleep stages, 1 and 2, in the older subjects. Sleep was also shown to vary with ethnicity. American Indians had more Stage 1 sleep than Caucasians or Blacks, more Stage 2 and less SWS than Caucasians, Blacks, Hispanics, or Asian Americans. Similarly, Blacks had greater percentage Stage 2 sleep than Caucasians or Hispanics. The authors also considered how BMI related to sleep architecture. They showed that as BMI increased lighter stages of sleep, i.e., Stages 1 and 2, dominated the architecture and SWS diminished. Finally, due to the large sample size, the authors were able to evaluate how smoking status affected sleep architecture. They reported that people who smoked or were ex-smokers had more Stage 1 and 2 sleep compared to those who never smoked; and SWS was highest in those who never smoked, while lowest in those who currently smoked. Percent REM sleep was highest in current smokers compared to ex-smokers. Hong, Mills, Loredo, Adler, and Dimsdale (2005) evaluated similar demographic variables and visually scored sleep parameters in a large sample of men and women. Results indicate that being older was associated with increased percentage of Stage 1 and less SWS. The range of ages was 25 to 50. Furthermore, ethnicity, in particular being African-American, was associated with reduced percentage of SWS. No relationship was observed between sleep and BMI or smoking; this is likely a result of a fairly fit sample (average BMI 25.4 ± 4.1) and few smokers (7/63). In a large cohort sample of Japanese men and women, Sekine, Chandola, Martikainen, Marmot, and Kagamimori (2006) suggested that SES did not affect self-reported sleep quality, but being unmarried was associated with poor subjective sleep quality. Several psychosocial and behavioral factors have been considered in relation to sleep patterns, but clearly no definitive answers have been derived.

12.4.2 Immunological Factors—Cytokines

The role that the immune system plays throughout pregnancy is constantly being explored. It is beyond the scope of this paper to discuss all the immunological aspects of pregnancy (see Silver, Peltier, and Branch 2004). Immunological factors, and cytokines in particular, are extremely important in all phases of gestation including the success or failure of implantation, placental development, and timing of labor (Wood 1997; Chegini and Williams 2000; Dudley 1996; Hill 2000). During the narrow implantation window, for instance, the amount of cytokine expression is correlated with success or failure of this process (Dudley 1996). IL-6, for instance, has been shown to be weakly expressed during the proliferative phase, but highly expressed during this implantation window suggesting a role for IL-6 in preparing tissue for implantation (Chegini et al. 2000). Following implantation, placental development is reliant upon a plethora of factors all working in concert. Increased levels of cytokines/growth factors, such as IL-1, IL-6, and TNF-α, are observed during placental development and are thought to derive from the large numbers of macrophages that populate placental stroma. Although unclear if they play a role in placental and fetal growth, they are considered neutral at worst (Wood 1997).

The knowledge regarding the orchestrated balance of cytokine synthesis and function during pregnancy is evolving on a regular basis. The construal of a working system that explains success or failure of pregnancy is exciting. Although very general, the Th1 (proinflammatory) versus Th2 (anti-inflammatory) working model has been undeniably helpful in understanding immune functioning, particularly in the murine model (Dudley 1996). Although readily applicable to the mouse, it is not as clear-cut in humans. IL-6, for instance, is listed among the proinflammatory cytokines because it is a mediator of host response to tissue damage; but it has properties that make it associated with a Th2 (anti-inflammatory) state because it increases IL-4 and IL-10 (Hunt 2000). Despite the prominence of proinflammatory cytokines such as TNF-α, IL-1β, and IL-6 at the beginning and the end of pregnancy, it is currently accepted that Th2-type dominance must occur for a successful pregnancy and that this regulation is overseen by the endocrine system (Hill 2000; Hunt 2000).

12.4.3 Endocrinological Factors—Hormones

Once conception occurs, numerous endocrine changes take place on an almost-daily basis. The bidirectional communication between the mother and the fetal–placental unit relies on a host of hormonal factors. The hormones with associations to sleep will be described. Progesterone is the most described and understood hormone during pregnancy and is thought necessary for the maintenance of the pregnancy. Progesterone is a steroid hormone that is produced primarily by the placenta and is synthesized by maternal cholesterol (Liu 2004). It has a variety of roles, including uterine and mammary gland development and the onset of labor (Astle, Slater, and Thornton 2003). It is also the primary contributor to whether uterine quiescence is present throughout gestation (Buster et al. 2003). Progesterone is known to have immunosuppressive properties and is thought to be an important component of successful pregnancy when high levels are present at the maternal–fetal interface; very high levels of progesterone can induce production of Th2-type cytokines such as IL-4 and IL-5 (Piccinni, Scaletti, Maggi, and Romagnani 2000). Maternal serum levels increase as pregnancy progresses and secretion sites of progesterone increase so that by the end of the third trimester most of the progesterone is secreted from the fetal-placenta unit (Buster et al. 2003; Liu 2004; Porterfield 2001; Taylor, Lebovic, and Martin-Cadieux 2001).

Estrogens, a group of steroid hormones, have various roles throughout a person's life, but during pregnancy their primary function is to regulate progesterone and maintain maturation. The pregnancy-related estrogen, estriol, is a hormone originating almost exclusively from the placenta thus its production in nonpregnant women is minimal (Buster et al. 2003). It plays a significant role in augmenting uterine blood flow and in the timing of labor stimulation (Buster et al. 2003). It increases as the pregnancy moves toward term (Porterfield 2001) and is a fundamental marker of the progression to spontaneous labor (Goodwin 1999). A steep increase in estriol output that is accompanied by a reduction in progesterone is observed at approximately 3 weeks before parturition. For women who had abnormal delivery times, i.e., past their due dates, salivary estriol levels and the ratio of estriol:progesterone were lower than for women who delivered spontaneously (Goodwin 1999). In comparing preterm and term labors, the surge of estriol has been documented to occur about 4 weeks sooner in those who experience preterm labor (McGregor et al. 1995). Furthermore, low maternal concentrations have been associated with fetal and placental development problems including fetal abnormalities and congenital derangements (Buster et al. 2003; Taylor et al. 2001).

There are a few other hormonal factors that change during pregnancy and are involved in the sleep process. During pregnancy, corticotropin-releasing hormone (CRH), which is normally synthesized by the hypothalamus, is synthesized by the placenta, fetal membranes and decidua and is physiologically identical to that of the mother (Florio, Cobellis, Woodman, Severi, Linton, and Petraglia 2002; Wadhwa, Sandman, Chicz-DeMet, and Porto 1997). CRH is a crucial element in parturition and fetal growth and varying levels have been shown to be involved with preterm labor (Inder et al. 2001; Siler-Khodr, Forthman, Khodr, Matyszczyk, Khodr, and Khodr 2003; Wadhwa et al. 1997). Prolactin is a hormone produced by the anterior pituitary gland with its primary role being initiation and sustenance of lactation. The secretion of prolactin is increased by stress and is dependent upon a woman's estrogen levels. Finally, cortisol, a well-known steroid hormone, is particularly involved in the pregnancy process. The levels of maternal cortisol are associated with the rise in estrogen production. However, the rate of secretion of cortisol by maternal adrenals is not increased in pregnancy; rather, only the rate of clearance is decreased (Goodwin 1999). Cortisol is critical for maturation of the lungs, liver, and other tissues of the developing fetus (Goodwin 1999).

12.5 Where Does Sleep Fit In? The Cytokine and Hormone Changes Associated with Sleep Deprivation/Sleep Disorders

Evaluation of the effects of sleep deprivation upon immune function has generally focused on people with disorders of excessive daytime sleepiness, healthy adult men or depressed populations. A consistent finding is that proinflammatory cytokines are elevated in these groups (Irwin, Clark, Kennedy, Christian, and Ziegler 2003; Okun, Giese, Lin, Einen, Mignot, and Coussons-Read 2004; Redwine, Dang, Hamano, and Irwin 2003; Vgontzas, Papanicolaou, Bixler, Kales, Tyson, and Chrousos 1997; Vgontzas, Zoumakis, Bixler, Lin, Follett, Kales, and Chrousos 2004). Sleep apneics and narcoleptics have higher TNF-α levels compared to controls (Vgontzas et al. 1997), while abstinent alcoholics and people partially deprived of sleep show elevations of both TNF-α and IL-6 (Irwin, Rinetti, Redwine, Motivala, Dang, and Ehlers 2004; Uthgenannt, Schoolmann, Pietrowsky, Fehm, and Born 1995; Vgontzas et al. 2004). Experimentally induced sleep deprivation has been found to alter the diurnal pattern of cellular and humoral immune functions (Dinges, Douglas, Hamarman, Zaugg, and Kapoor 1995; Heiser et al. 2000; Moldofsky, Lue, Davidson, and Gorczynski 1989) and possibly decrease overall immune function (Redwine, Hauger, Gillin, and Irwin 2000) in normal adults. More recently support for the hypothesis that sleep improves immune function comes from studies assessing the vaccination response after partial sleep deprivation (Lange, Perras, Fehm, and Born

2003; Spiegel, Sheridan, and Van Cauter 2002). It appears that sleep restriction may impair an individual to effectively respond to a vaccination.

It is suggested that immune alterations may be associated with a biological pressure for sleep (Dinges et al. 1994), and that sleep loss could produce an overall shift in immune function (Dinges et al. 1995; Irwin, Thompson, Miller, Gillin, and Ziegler 1999) that favors proinflammatory cytokine production, such as IL-1 and TNF-α (Krueger and Majde 1995). Cytokine production is different between sleep and sleep deprivation, and there exists a circadian rhythm to the production not only of cytokines, but also of the various immune cells that produce the cytokines (Born, Lange, Hansen, Molle, and Fehm 1997). This understanding guided Born and colleagues (1997) to assess the role of nocturnal sleep on normal immune regulation in a design to assess acute sleep loss rather than excessive sleep loss. Men, serving as their own controls, slept two consecutive regular sleep–wake cycles or remained awake for 24 h followed by recovery sleep. The researchers found no alteration in the absolute production of IL-1β and TNF-α between the two experimental conditions; however, the expected decrease of IL-1β and TNF-α during sleep was blocked when subjects were kept awake. Hence, there was an increase in the nocturnal production of both cytokines during the sleep deprivation period (Born et al. 1997). Other studies evaluating sleep restriction, found a delayed nocturnal release of sleep-associated cytokines, IL-1, IL-6, and TNF-α, with subsequent recuperation of normal levels on recovery nights (Moldofsky et al. 1989; Redwine et al. 2000; Vgontzas et al. 1997). This suggests that cytokines have a sleep dependent relationship. Whether experimental sleep deprivation/restriction for an acute period (Irwin et al. 2004; Vgontzas et al. 2004) or in chronic sleep disorders (Okun et al. 2004; Vgontzas et al. 1997; Vgontzas et al. 2002), proinflammatory cytokines are elevated.

Estrogen and progesterone are altered during pregnancy; they are also potential contributors to the differences in sleep patterns observed between the pregnant and nonpregnant state. Data for the influence of hormones on sleep patterns has been documented since the early 1960s. Heuser, Kales, and Jacobson found that progesterone did not decrease REM time in a pilot study of five young adult subjects (Heuser, Kales, and Jacobson 1968), and Hartmann (1966) suggested that hormonal changes in the menstrual cycle might produce an elevated need for "Dreaming Sleep" known today as REM sleep. The data that Branchey and Petre-Quadens (1968) analyzed led them to conclude that the changes in paradoxical sleep observed in their subjects could be attributed to the sexual hormones secreted during pregnancy (p. 457). Recent work intimates that estrogen enhances the total time spent in REM and reduces the latency period prior to REM sleep (Manber et al. 2001). However, the primary source for data on estrogen and its effects on sleep come from pre and postmenopausal studies. Women who experience menopause have greater sleep disruption and mood changes (Baker et al. 1997), but no studies measured estrogen levels. Women who received hormone replacement therapy in a double-blind, placebo-controlled study showed improvement in sleep parameters, but they were nonsignificant compared to the placebo group (Saletu-Zyhlarz et al. 2003). Information regarding progesterone, on the other hand, is clearer. Progesterone is secreted in high amounts by the placenta (Lee et al. 2000) and increases non-REM (NREM) sleep, shortens sleep latency, and reduces wakefulness after sleep onset (Manber et al. 2001; Santiago et al. 2001). Moldofsky et al. (1995) reported that at times of elevated progesterone in normal controls, there is a delay in onset to SWS and a decrease in Stage 4 sleep, which is associated with a reduced duration in the decline of NK activity (Moldofsky, Lue, Shahal, Jiang, and Gorczynski 1995).

12.6 Pregnancy Complications

An increasing compilation of data supports the correlation that pregnancy complications and poor outcomes are associated with an excess of certain immune-markers. Many of the immune markers identified are proinflammatory cytokines, including IL-1β, IL-6, and TNF-α. Although there currently are no data to support the effect of pregnancy-associated sleep disturbances on pregnancy complications or poor pregnancy outcomes via increased proinflammatory cytokine production, the previously discussed data suggest a possible mediating role of sleep in the occurrence of these situations.

In an attempt to reduce their frequency, various pregnancy complications or poor outcomes and their potential causes (particularly increased proinflammatory cytokine levels) have been evaluated by researchers. Increased levels of proinflammatory cytokines, such as IL-2, IL-6, TNFα, and IFN-γ, have been observed in spontaneous abortions (Clark et al. 1996), recurrent miscarriages (MacLean, Wilson, Jenkins, Miller, and Walker 2002; Raghupathy 2001), and preeclampsia (Afshari et al. 2005; Dekker and Sibai 1999), while anti-inflammatory cytokines, including IL-3, -4, and -10, have been noted to assist in the promotion, establishment, and completion of successful fetal growth (Clark et al. 1996; Dekker et al. 1999; Gaunt and Ramin 2001; Gennaro and Fehder 1996; Hanson 2000; Piccinni et al. 2000; Raghupathy 2001; Sacks, Studena, Sargent, and Redman 1998; Saito, Sakai, Sasaki, Tanebe, Tsuda, and Michimata 1999). It has been suggested that a predominance of proinflammatory cytokines at various points throughout pregnancy is incompatible with successful pregnancy (Raghupathy 2001).

About 15 to 20% of all pregnancies result in miscarriage, and the more pregnancy losses a woman experiences, the greater the risk of losing each subsequent conception (Coulam 2000). As noted previously, a healthy pregnancy is associated with alterations in the immune system, which may resemble a suppression of cell-mediated immunity. Although there is incongruity in which mechanisms are involved in the success or failure of pregnancy, information continues to be ascertained regarding the role of cytokines in this process. Studies increasingly are focusing on how immunological problems can result in recurrent spontaneous abortion or miscarriage. From an alloimmune perspective, there are two possibilities that could result in RSA: either the mother's immune system does not recognize the pregnancy or the mother develops an abnormal immunologic response to the pregnancy (Coulam 2000). A few hypotheses have been suggested and tested in order to provide some evidence as to why miscarriages occur. One idea is that elevated levels of proinflammatory cytokines may impair the trophoblast from implanting, resulting in miscarriage (Hill, Polgar, and Anderson 1995; Hill et al. 1995). However, other data suggest TNF-α may be a protector of the feto-placental unit when it is exposed to teratogenic stress (Torchinsky et al. 2003). These contradictory hypotheses suggest a more complex interaction and influence of cytokines on pregnancy: *when* and *how* the shift occurs toward an anti-inflammatory cytokine profile may be most important. Since there appears to be a role for inflammation in implantation, pregnancy is not a Th2 phenomenon. Several "windows" for certain cytokines to be up-regulated or down-regulated, accompanied by precise timing and "tuning," seem essential (Chaouat et al. 2002). No study examining the causes of miscarriage has considered sleep disturbances as a possible contributor to the improperly timed increases in proinflammatory cytokines.

The occurrence of preeclampsia in women during the latter half of gestation has been identified as a pressing medical concern. The disease process includes hypertension, altered hematology, placental insufficiency, and edema (Edwards et al. 2000; Roberts 2004). It is the leading cause of preterm birth and intrauterine growth retardation and occurs in approximately 4 to 5% of all pregnancies (Dekker et al. 1999; Roberts 2004). Like RSA, preeclampsia is believed to be a result of an inappropriate activation of the maternal inflammatory response, including activation of granulocytes and increased release of TNF-α and IL-6, although the precise etiology is unclear (Dekker et al. 1999; Sacks et al. 1998). The importance of detecting preeclampsia early and halting its destructive path has been elucidated and is underscored by the numerous studies trying to understand the immunological variations associated with this disease.

Several studies have considered changes in "sleep" and preeclampsia. Two studies have considered differences in sleep quality and sleep architecture in preeclampsia. Evaluating sleep quality from subjective questionnaires and body movements in bed, Ekholm, Polo, Rauhala, and Ekblad (1992) suggested that sleep may be impaired in women with preeclampsia due to increased body movements during sleep. Examining sleep architecture, Edwards et al. (2000) showed significant increases in SWS in a group of preeclamptic women compared to healthy pregnant women. The authors propose that a "release of cytokines" may provide an explanation for the architecture differences; however, additional work is suggested (Edwards et al. 2000). Most studies, nonetheless, have assessed sleep-disordered breathing during pregnancy (Blyton, Sullivan, and Edwards 2004; Izci et al. 2005; Yinon et al. 2006) due to the strong association of increased hypertension and hypoxia that occurs in preeclampsia (Blyton et al. 2004). These relationships are subsequently associated with poor fetal outcomes. Presently, there are no studies examining the sleep–cytokine relationship in the occurrence of preeclampsia.

Another outcome that has been considered to be potentially immune-regulated is that of preterm birth. Experts have begun to view preterm birth as the final result of many possible causes, including intrauterine infections, hormonal disturbances, fetal injury, uterine ischemia, and uterine overdistention (Dudley 1999). Several immune parameters are thought to be involved in preterm birth. Evidence exists that inflammatory cytokines, such as IL-1β, TNF-α, IL-6, IL-8, and IL-4, are involved in infection-associated preterm labor (Dudley 1999). It is mostly in animal models where a clearer demonstration has been made that inflammatory cytokines can mediate early pregnancy loss; however, it appears that cytokines are only part of a cascade of events and it is an abnormally regulated maternal immune response, not an infection, that predisposes toward early pregnancy loss (Dudley 1999). Only one study assessed sleep disturbances at 22 to 26 weeks of gestation; the relationship between preterm labor and their findings were not significant (Stinson and Lee 2003). Similar to the other pregnancy complications discussed, no investigator has considered the sleep–immune relationship in the occurrence of preterm labor.

12.7 Summary

Sleep disturbances have been identified and grossly described during pregnancy. The consensus is that during a snapshot of time during each trimester, pregnant women complain of fatigue, poor sleep quality and nocturnal sleep disturbances. Sleep loss, restriction and deprivation have been correlated with immune alterations in various populations; however, little data exists relating pregnancy-associated sleep disturbances with immune or endocrine alterations that may be related to pregnancy complications. Hence, the implications for maternal and fetal health resulting from severely disturbed sleep during pregnancy have yet to be evaluated. Studies examining stress and sleep indicate that various kinds of stress from psychosocial stressors (Paulsen and Shaver 1991) to acute lab stressors (Hall et al. 2004) can negatively affect sleep. Similarly, research on stress and pregnancy conclude that there are negative pregnancy consequences from high stress (Coussons-Read, Okun, Schmitt, and Giese 2005; Wadhwa, Culhane, Rauh, and Barve 2001). In order to effectively determine the causes of poor pregnancy outcomes, all the various contributors to the neuroimmune and neuroendocrine changes that occur during pregnancy, including poor sleep quality, duration, continuity and architecture need to be evaluated.

References

Afshari, J.T., Ghomian, N., Shameli, A., Shakeri, M.T., Fahmidehkar, M.A., Mahajer, E. et al. (2005). Determination of interleukin-6 and tumor necrosis factor-alpha concentrations in Iranian–Khorasanian patients with preeclampsia. *BMC Pregnancy Childbirth* 5, 14.

American Sleep Disorders Association (1997) *The International Classification of Sleep Disorders: A Diagnostic and Coding Manual*. (Rev. Edn.). Author, Rochester, MN.

Astle, S., Slater, D.M., and Thornton, S. (2003). The involvement of progesterone in the onset of human labour. *Eur J Obstet Gynecol Reprod Biol* 108, 177–181.

Baker, A., Simpson, S., and Dawson, D. (1997). Sleep disruption and mood changes associated with menopause. *J Psychosom Res* 43, 359–369.

Baratte-Beebe, K.R., and Lee, K. (1999). Sources of midsleep awakenings in childbearing women. *Clin Nurs* Res 8, 386–397.

Blyton, D.M., Sullivan, C.E., and Edwards, N. (2004). Reduced nocturnal cardiac output associated with preeclampsia is minimized with the use of nocturnal nasal CPAP. *Sleep* 27, 79–84.

Born, J., Lange, T., Hansen, K., Molle, M., and Fehm, H.L. (1997). Effects of sleep and circadian rhythm on human circulating immune cells. *J Immunol* 158, 4454–4464.

Branchey, M., and Branchey, L. (1970). Sleep and wakefulness in female rats during pregnancy. *Physiol Behav* 5, 365–368.

Branchey, M., and Petre-Quadens, O. (1968). A comparative study of sleep parameters during pregnancy. *Acta Neurol Psychiatr Belg* 68, 453–459.

Brunner, D.P., Munch, M., Biedermann, K., Huch, R., Huch, A., and Borbely, A.A. (1994). Changes in sleep and sleep electroencephalogram during pregnancy. *Sleep* 17, 576–582.

Burguet, A., Kaminski, M., Abraham-Lerat, L., Schaal, J.P., Cambonie, G., Fresson, J. et al. (2004). The complex relationship between smoking in pregnancy and very preterm delivery. Results of the Epipage study. *BJOG* 111, 258–265.

Buster, J.E., and Carson, S.A. (2003) Endocrinology and diagnosis of pregnancy. In: S.G. Gabbe, J.R. Niebyl, and J.L. Simpson (Eds.), *Obstetrics: Normal and Problem Pregnancies* (4th Edn.). Churchill Livingstone, New York, pp. 3–34.

Carrier, J., Land, S., Buysse, D.J., Kupfer, D.J., and Monk, T.H. (2001). The effects of age and gender on sleep EEG power spectral density in the middle years of life (ages 20–60 years old). *Psychophysiology* 38, 232–242.

Challis, J.R.G., and Lye, S.J. (2003) Physiology and endocrinology of term and preterm labor. In: S.G. Gabbe, J.R. Niebyl, and J.L. Simpson (Eds.), *Obstetrics: Normal and Problem Pregnancies* (4th Edn.). Churchill Livingstone, New York, pp. 93–106.

Chaouat, G., Zourbas, S., Ostojic, S., Lappree-Delage, G., Dubanchet, S., Ledee, N. et al. (2002). A brief review of recent data on some cytokine expressions at the materno-foetal interface which might challenge the classical Th1/Th2 dichotomy. *J Reprod Immunol* 53, 241–256.

Chegini, N., and Williams, R.S. (2000) Cytokines and growth factor networks in human endometrium from menstruation to embryo implantation. In: J.A. Hill (Ed.), *Cytokines in Human Reproduction*. John Wiley and Sons, New York, pp. 93–132.

Clark, D.A., Arck, P.C., Jalali, R., Merali, F.S., Manuel, J., Chaouat, G. et al. (1996). Psycho-neuro-cytokine/endocrine pathways in immunoregulation during pregnancy. *Am J Reprod Immunol*, 35, 330–337.

Coble, P.A., Reynolds, C.F., III, Kupfer, D.J., Houck, P.R., Day, N.L., and Giles, D.E. (1994). Childbearing in women with and without a history of affective disorder. II. Electroencephalographic sleep. *Comp Psychiatry* 35, 215–224.

Colrain, I.M., Trinder, J., and Swan, G.E. (2004). The impact of smoking cessation on objective and subjective markers of sleep: Review, synthesis, and recommendations. *Nicotine Tob Res* 6, 913–925.

Coulam, C.B. (2000). Understanding the immunobiology of pregnancy and applying it to treatment of recurrent pregnancy loss. *Early Pregnancy* 4, 19–29.

Coussons-Read, M.E., Okun, M.L., Schmitt, M.P., and Giese, S. (2005). Prenatal stress alters cytokine levels in a manner that may endanger human pregnancy. *Psychosom Med* 67, 625–631.

Dekker, G.A., and Sibai, B.M. (1999). The immunology of preeclampsia. *Semin Perinatol* 23, 24–33.

Dinges, D.F., Douglas, S.D., Hamarman, S., Zaugg, L., and Kapoor, S. (1995). Sleep deprivation and human immune function. *Adv Neuroimmunol* 5, 97–110.

Dinges, D.F., Douglas, S.D., Zaugg, L., Campbell, D.E., McMann, J.M., Whitehouse, W.G. et al. (1994). Leukocytosis and natural killer cell function parallel neurobehavioral fatigue induced by 64 hours of sleep deprivation. *J Clin Invest* 93, 1930–1939.

Driver, H.S., and Shapiro, C.M. (1992). A longitudinal study of sleep stages in young women during pregnancy and postpartum. *Sleep* 15, 449–453.

Dudley, D.J. (1996) The immunobiology of nidation and implantation: Novel regulation of innate and adaptive immunity. In: R.A. Bronson, N.J. Alexander, D. Anderson, D.W. Branch, and W.H. Kutteh (Eds.), *Reproductive Immunology*. Blackwell Science, Cambridge, pp. 359–382.

Dudley, D.J. (1999). Immunoendocrinology of preterm labor: The link between corticotropin-releasing hormone and inflammation. *Am J Obstet Gynecol* 180, S251–S256.

Edwards, N., Blyton, C.M., Kesby, G.J., Wilcox, I., and Sullivan, C.E. (2000). Pre-eclampsia is associated with marked alterations in sleep architecture. *Sleep* 23, 619–625.

Ehlers, C.L., and Kupfer, D.J. (1997). Slow-wave sleep: Do young adult men and women age differently? *J Sleep Res* 6, 211–215.

Ekholm, E.M., Polo, O., Rauhala, E.R., and Ekblad, U.U. (1992). Sleep quality in preeclampsia. *Am J Obstet Gynecol* 167, 1262–1266.

Florio, P., Cobellis, L., Woodman, J., Severi, F.M., Linton, E.A., and Petraglia, F. (2002). Levels of maternal plasma corticotropin-releasing factor and urocortin during labor. *J Soc Gynecol Invest* 9, 233–237.

Fujino, Y., Shirata, K., Imanaka, M., Nishio, J., Ogita, S., and Park, Y.K. (1995). Sleeping habits of pregnant women: A questionnaire study. *Appl Human Sci* 14, 305–307.

Gaunt, G., and Ramin, K. (2001). Immunological tolerance of the human fetus. *Am J Perinatol* 18, 299–312.

Gennaro, S., and Fehder, W.P. (1996). Stress, immune function, and relationship to pregnancy outcome. *Nurs Clin North Am* 31, 293–303.

Goodwin, T.M. (1999). A role for estriol in human labor, term and preterm. *Am J Obstet Gynecol* 180, S208–S213.

Hall, M., Vasko, R., Buysse, D., Ombao, H., Chen, Q., Cashmere, J.D. et al. (2004). Acute stress affects heart rate variability during sleep. *Psychosom Med* 66, 56–62.

Hanson, L.A. (2000). The mother-offspring dyad and the immune system. *Acta Paediatr* 89, 252–258.

Hartmann, E. (1966). Dreaming sleep (the D-state) and the menstrual cycle. *J Nerv Ment Dis* 143, 406–416.

Hedman, C., Pohjasvaara, T., Tolonen, U., Suhonen-Malm, A.S., and Myllyla, V.V. (2002). Effects of pregnancy on mothers' sleep. *Sleep Med* 3, 37–42.

Heiser, P., Dickhaus, B., Schreiber, W., Clement, H.W., Hasse, C., Hennig, J. et al. (2000). White blood cells and cortisol after sleep deprivation and recovery sleep in humans. *Eur Arch Psychiatry Clin Neurosci* 250, 16–23.

Hertz, G., Fast, A., Feinsilver, S.H., Albertario, C.L., Schulman, H., and Fein, A.M. (1992). Sleep in normal late pregnancy. *Sleep* 15, 246–251.

Heuser, G., Kales, A., and Jacobson, A. (1968). Human sleep patterns after progesterone administration. *Psychophysiology* 4, 378.

Hill, J.A. (2000). Cytokines in early pregnancy success and failure. In: J.A. Hill (Ed.), *Cytokines in Human Reproduction.* John Wiley and Sons, New York, pp. 161–169.

Hill, J.A., Polgar, K., and Anderson, D.J. (1995). T-Helper 1-type immunity to trophoblast in women with recurrent spontaneous abortion. *JAMA* 273, 1933–1936.

Hong, S., Mills, P.J., Loredo, J.S., Adler, K.A., and Dimsdale, J.E. (2005). The association between interleukin-6, sleep, and demographic characteristics. *Brain Behav Immun* 19, 165–172.

Hunt, J.S. (2000) Cytokine networks in the human placenta. In: J.A.Hill (Ed.), *Cytokines in Human Reproduction.* John Wiley and Sons, New York, pp. 203–220.

Inder, W.J., Prickett, T.C., Ellis, M.J., Hull, L., Reid, R., Benny, P.S. et al. (2001). The utility of plasma CRH as a predictor of preterm delivery. *J Clin Endocrinol Metab* 86, 5706–5710.

Irwin, M., Clark, C., Kennedy, B., Christian, G.J., and Ziegler, M. (2003). Nocturnal catecholamines and immune function in insomniacs, depressed patients, and control subjects. *Brain Behav Immun* 17, 365–372.

Irwin, M., Rinetti, G., Redwine, L., Motivala, S., Dang, J., and Ehlers, C. (2004). Nocturnal proinflammatory cytokine-associated sleep disturbances in abstinent African American alcoholics. *Brain Behav Immun* 18, 349–360.

Irwin, M., Thompson, J., Miller, C., Gillin, J.C., and Ziegler, M. (1999). Effects of sleep and sleep deprivation on catecholamine and interleukin- 2 levels in humans: Clinical implications. *J Clin Endocrinol Metab* 84, 1979–1985.

Izci, B., Martin, S.E., Dundas, K.C., Liston, W.A., Calder, A.A., and Douglas, N.J. (2005). Sleep complaints: Snoring and daytime sleepiness in pregnant and pre-eclamptic women. *Sleep Med* 6, 163–169.

Kaneita, Y., Ohida, T., Takemura, S., Sone, T., Suzuki, K., Miyake, T. et al. (2005). Relation of smoking and drinking to sleep disturbance among Japanese pregnant women. *Prev Med* 41, 877–882.

Karacan, I., Heine, M.W., and Agnew, H. (1968). Characteristics of sleep patterns during late pregnancy and the postpartum periods. *Am J Obstet Gynecol* 101, 579–586.

Karacan, I., and Williams, R.L. (1970). The relationship of sleep disturbances to psychopathology. *Int Psychiatry Clin* 7, 93–111.

Karacan, I., Williams, R.L., Hursch, C.J., McCaulley, M., and Heine, M.W. (1969). Some implications of the sleep patterns of pregnancy for postpartum emotional disturbances. *Br J Psychiatry* 115, 929–935.

Kimura, M., Zhang, S.Q., and Inoue, S. (1996). Pregnancy-associated sleep changes in the rat. *Am J Physiol* 271, R1063–R1069.

Krueger, J.M., and Majde, J.A. (1995). Cytokines and sleep. *Int Arch Allergy Immunol* 106, 97–100.

Lange, T., Perras, B., Fehm, H.L., and Born, J. (2003). Sleep enhances the human antibody response to hepatitis A vaccination. *Psychosom Med* 65, 831–835.

Lee, K.A. (1992). Self-reported sleep disturbances in employed women. *Sleep* 15, 493–498.

Lee, K.A. (2006) Sleep during pregnancy and postpartum. In: T. Lee-Chiong (Ed.), *Encyclopedia of Sleep Medicine.* John Wiley and Sons, New York, pp. 629–635.

Lee, K.A., Zaffke, M.E., and McEnany, G. (2000). Parity and sleep patterns during and after pregnancy. *Obstet Gynecol* 95, 14–18.

Liu, J.H. (2004) Endocrinology of pregnancy. In: R.K. Creasey, R. Resnik, and J.D. Iams (Eds.), *Maternal-Fetal Medicine: Principles and Practice.* 5th Edn. W.B. Saunders, Philadelphia, pp. 121–134.

Lockley, S.W., Skene, D.J., and Arendt, J. (1999). Comparison between subjective and actigraphic measurement of sleep and sleep rhythms. *J Sleep Res* 8, 175–183.

MacLean, M.A., Wilson, R., Jenkins, C., Miller, H., and Walker, J.J. (2002). Interleukin-2 receptor concentrations in pregnant women with a history of recurrent miscarriage. *Hum Reprod* 17, 219–220.

Manber, R., Colrain, I.M., and Lee, K. (2001) Sleep disorders. In: S.Kornstein and A. Clayton (Eds.), *Women's Mental Health: A Comprehensive Textbook.* Guilford Press, New York.

Mauri, M. (1990). Sleep and the reproductive cycle: A review. *Health Care Women Int* 11, 409–421.

McGregor, J.A., Jackson, G.M., Lachelin, G.C., Goodwin, T.M., Artal, R., Hastings, C. et al. (1995). Salivary estriol as risk assessment for preterm labor: A prospective trial. *Am J Obstet Gynecol* 173, 1337–1342.

Mindell, J.A., and Jacobson, B.J. (2000). Sleep disturbances during pregnancy. *J Obstet Gynecol Neonatal Nurs* 29, 590–597.

Moldofsky, H., Lue, F.A., Davidson, J.R., and Gorczynski, R. (1989). Effects of sleep deprivation on human immune functions. *FASEB J* 3, 1972–1977.

Moldofsky, H., Lue, F.A., Shahal, B., Jiang, C.G., and Gorczynski, R.M. (1995). Diurnal sleep/wake-related immune functions during the menstrual cycle of healthy young women. *J Sleep Res* 4, 150–159.

Narahara, H., and Johnston, J.M. (1993). Smoking and preterm labor: Effect of a cigarette smoke extract on the secretion of platelet-activating factor-acetylhydrolase by human decidual macrophages. *Am J Obstet Gynecol* 169, 1321–1326.

Nishina, H., Honda, K., Okai, T., Kozuma, S., Inoue, S., and Taketani, Y. (1996). Characteristic changes in sleep patterns during pregnancy in rats. *Neurosci Lett* 203, 5–8.

Okun, M.L., Giese, S., Lin, L., Einen, M., Mignot, E., and Coussons-Read, M.E. (2004). Exploring the cytokine and endocrine involvement in narcolepsy. *Brain Behav Immun* 18, 326–332.

Palmer, C.D., Harrison, G.A., and Hiorns, R.W. (1980). Association between smoking and drinking and sleep duration. *Ann Hum Biol* 7, 103–107.

Paulsen, V.M., and Shaver, J.L. (1991). Stress, support, psychological states and sleep. *Soc Sci Med* 32, 1237–1243.

Phillips, B.A., and Danner, F.J. (1995). Cigarette smoking and sleep disturbance. *Arch Intern Med* 155, 734–737.

Piccinni, M.P., Scaletti, C., Maggi, E., and Romagnani, S. (2000). Role of hormone-controlled Th1- and Th2-type cytokines in successful pregnancy. *J Neuroimmunol* 109, 30–33.

Porterfield, S.P. (2001) Endocrinology of pregnancy. In: W.Schmitt (Ed.), *Endocrine Physiology* (2nd Edn.). Mosby, St. Louis, pp. 201–225.

Raghupathy, R. (2001). Pregnancy: Success and failure within the Th1/Th2/Th3 paradigm. *Semin Immunol* 13, 219–227.

Redline, S., Kirchner, H.L., Quan, S.F., Gottlieb, D.J., Kapur, V., and Newman, A. (2004). The effects of age, sex, ethnicity, and sleep-disordered breathing on sleep architecture. *Arch Intern Med* 164, 406–418.

Redwine, L., Dang, J., Hamano, N., and Irwin, M. (2003). Disordered sleep, nocturnal cytokines, and immunity in alcoholics. *Psychosom Med* 65, 75–85.

Redwine, L., Hauger, R.L., Gillin, J.C., and Irwin, M. (2000). Effects of sleep and sleep deprivation on interleukin-6, growth hormone, cortisol, and melatonin levels in humans. *J Clin Endocrinol Metab* 85, 3597–3603.

Reichert, J.A., Bloch, D.A., Cundiff, E., and Votteri, B.A. (2003). Comparison of the NovaSom QSG, a new sleep apnea home-diagnostic system, and polysomnography. *Sleep Med* 4, 213–218.

Roberts, J.M. (2004) Pregnancy-Related hypertension. In: R.K.Creasey and R. Resnik (Eds.), *Maternal-Fetal Medicine* (5th Edn.). Saunders, Philadelphia, pp. 859–899.

Sacks, G.P., Studena, K., Sargent, K., and Redman, C.W. (1998). Normal pregnancy and preeclampsia both produce inflammatory changes in peripheral blood leukocytes akin to those of sepsis. Am J Obstet Gynecol 179, 80–86.

Saito, S., Sakai, M., Sasaki, Y., Tanebe, K., Tsuda, H., and Michimata, T. (1999). Quantitative analysis of peripheral blood Th0, Th1, Th2 and the Th1:Th2 cell ratio during normal human pregnancy and preeclampsia. *Clin Exp Immunol* 117, 550–555.

Saletu-Zyhlarz, G., Anderer, P., Gruber, G., Mandl, M., Gruber, D., Metka, M. et al. (2003). Insomnia related to postmenopausal syndrome and hormone replacement therapy: Sleep laboratory studies on baseline differences between patients and controls and double-blind, placebo-controlled investigations on the effects of a novel estrogen-progestogen combination (Climodien, Lafamme) versus estrogen alone. *J Sleep Res* 12, 239–254.

Santiago, J.R., Nolledo, M.S., Kinzler, W., and Santiago, T.V. (2001). Sleep and sleep disorders in pregnancy. *Ann Intern Med* 134, 396–408.

Sateia, M.J., Doghramji, K., Hauri, P.J., and Morin, C.M. (2000). Evaluation of chronic insomnia. An American Academy of Sleep Medicine review. *Sleep* 23, 243–308.

Schorr, S.J., Chawla, A., Devidas, M., Sullivan, C.A., Naef, R.W., and Morrison, J.C. (1998). Sleep patterns in pregnancy: A longitudinal study of polysomnography recordings during pregnancy. *J Perinatol* 18, 427–430.

Schweiger, M.S. (1972). Sleep disturbance in pregnancy. A subjective survey. *Am J Obstet Gynecol* 114, 879–882.

Sekine, M., Chandola, T., Martikainen, P., Marmot, M., and Kagamimori, S. (2006). Work and family characteristics as determinants of socioeconomic and sex inequalities in sleep: The Japanese Civil Servants Study. *Sleep* 29, 206–216.

Shinkoda, H., Matsumoto, K., and Park, Y.M. (1999). Changes in sleep-wake cycle during the period from late pregnancy to puerperium identified through the wrist actigraph and sleep logs. *Psychiatry Clin Neurosci* 53, 133–135.

Siler-Khodr, T.M., Forthman, G., Khodr, C., Matyszczyk, S., Khodr, Z., and Khodr, G. (2003). Maternal serum corticotropin-releasing hormone at midgestation in Hispanic and white women. *Obstet Gynecol* 101, 557–564.

Silver, R.M., Peltier, M.R., and Branch, D.W. (2004) The immunology of pregnancy. In: R.K.Creasey and R. Resnik (Eds.), *Maternal-Fetal Medicine: Principles and Practices* (5th Edn.). Saunders, Philadelphia, pp. 89–109.

Simhan, H.N., Caritis, S.N., Hillier, S.L., and Krohn, M.A. (2005). Cervical anti-inflammatory cytokine concentrations among first-trimester pregnant smokers. *Am J Obstet Gynecol* 193, 1999–2003.

Spiegel, K., Sheridan, J.F., and Van Cauter, E. (2002). Effect of sleep deprivation on response to immunization. *JAMA* 288, 1471–1472.

Stenuit, P., and Kerkhofs, M. (2005). Age modulates the effects of sleep restriction in women. *Sleep* 28, 1283–1288.

Stepnowsky, C.J., Jr., Moore, P.J., and Dimsdale, J.E. (2003). Effect of ethnicity on sleep: Complexities for epidemiologic research. *Sleep* 26, 329–332.

Stinson, J.C., and Lee, K.A. (2003). Premature labor and birth: Influence of rank and perception of fatigue in active duty military women. *Mil Med* 168, 385–390.

Suzuki, S., Dennerstein, L., Greenwood, K.M., Armstrong, S.M., and Satohisa, E. (1994). Sleeping patterns during pregnancy in Japanese women. *J Psychosom Obstet Gynaecol* 15, 19–26.

Taylor, R.N., Lebovic, D.I., and Martin-Cadieux, M.C. (2001) Endocriniology of pregnancy. In: F.S.Greenspan and D.G. Gardner (Eds.), *Basic and Clinical Endocrinology* (6th Edn.). Lange Medical Books/McGraw-Hill, New York, pp. 575–604.

Torchinsky, A., Shepshelovich, J., Orenstein, H., Zaslavsky, Z., Savion, S., Carp, H. et al. (2003). TNF-alpha protects embryos exposed to developmental toxicants. *Am J Reprod Immunol* 49, 159–168.

Tworoger, S.S., Davis, S., Vitiello, M.V., Lentz, M.J., and McTiernan, A. (2005). Factors associated with objective (actigraphic) and subjective sleep quality in young adult women. *J Psychosom Res* 59, 11–19.

Uthgenannt, D., Schoolmann, D., Pietrowsky, R., Fehm, H.L., and Born, J. (1995). Effects of sleep on the production of cytokines in humans. *Psychosom Med* 57, 97–104.

Vgontzas, A.N., Papanicolaou, D.A., Bixler, E.O., Kales, A., Tyson, K., and Chrousos, G.P. (1997). Elevation of plasma cytokines in disorders of excessive daytime sleepiness: Role of sleep disturbance and obesity. *J Clin Endocrinol Metab* 82, 1313–1316.

Vgontzas, A.N., Zoumakis, E., Bixler, E.O., Lin, H.M., Follett, H., Kales, A. et al. (2004). Adverse effects of modest sleep restriction on sleepiness, performance, and inflammatory cytokines. *J Clin Endocrinol Metab* 89, 2119–2126.

Vgontzas, A.N., Zoumakis, M., Papanicolaou, D.A., Bixler, E.O., Prolo, P., Lin, H.M. et al. (2002). Chronic insomnia is associated with a shift of interleukin-6 and tumor necrosis factor secretion from nighttime to daytime. *Metabolism* 51, 887–892.

Wadhwa, P.D., Culhane, J.F., Rauh, V., and Barve, S.S. (2001). Stress and preterm birth: Neuroendocrine, immune/inflammatory, and vascular mechanisms. *Matern Child Health J* 5, 119–125.

Wadhwa, P.D., Sandman, C.A., Chicz-DeMet, A., and Porto, M. (1997). Placental CRH modulates maternal pituitary adrenal function in human pregnancy. *Ann N Y Acad Sci* 814, 276–281.

Wolfson, A.R., Crowley, S.J., Anwer, U., and Bassett, J.L. (2003). Changes in sleep patterns and depressive symptoms in first-time mothers: Last trimester to 1-year postpartum. *Behav Sleep Med* 1, 54–67.

Wood, G.W. (1997) Cytokines in Reproduction: Molecular Mechanism of Fetal Allograft Survival. Chapman and Hill, New York.

Yinon, D., Lowenstein, L., Suraya, S., Beloosesky, R., Zmora, O., Malhotra, A. et al. (2006). Pre-eclampsia is associated with sleep-disordered breathing and endothelial dysfunction. *Eur Respir J* 27, 328–333.

13 Changes in Sleep and Behavior Following Experimental Immune Stimulation Using Bacterial Endotoxin in Humans

Andreas Schuld, Monika Haack, Janet Mullington, and Thomas Pollmächer

13.1 Introduction

During acute or chronic infection and inflammation complex changes in a variety of physiological systems occur. In addition to changes in temperature or the induction of fever also behavioral changes like tiredness, flu-like symptoms and impaired concentration are common. During the last 20 to 30 years there is rapidly growing interest and also knowledge about the underlying mechanisms: One of the first important findings was the observation, that the hypothalamo-pituitary-adrenal (HPA) system is stimulated following the peripheral administration of interleukin-1 (IL-1) (Besedovsky, del Rey, Sorkin, and Dinarello 1986). After this basic finding in psychoneuroimmunological research further studies focused on the inflammatory cytokines such as IL-1 as major components in the interaction of peripheral immune function and central nervous system (CNS): These peptides belong to the huge number of cytokines which are involved in this brain–immune interaction (see Hopkins and Rothwell 1995; Rothwell and Hopkins 1995). In human research mainly cytokines of the cytokine families of interleukins, interferons (IFN), or the tumor necrosis factor (TNF) family have been investigated.

The induction of fever and neuroendocrine activation have been most intensively studied using animal models to study CNS-mediated host response parameters (Kluger 1991; Turnbull and Rivier 1999). The central role of inflammatory cytokines are well accepted, IL-1 and IL-6 as well as TNF-α are the most important molecules. Based on these findings it has been suggested that these cytokines also may be responsible for changes in complex brain functions during experimental models of infection and inflammation. In rodents a typical pattern of behavioral responses including reduced motor activity, decreased food intake, reduced exploratory behavior, social and sexual interaction, less intake of food and water, altered sleep, and impaired learning occurs. This syndrome has been called "sickness behavior" (Dantzer 2001; Yirmiya et al. 2000). These behavioral responses were classically induced by the injection of bacterial endotoxin. In more specific studies it has been shown that inflammatory cytokines released peripherally, such as IL-1β, IL-6, and TNF-α, also are able to cause the "sickness behavior." It was initially unclear how these big peptides circulating in the blood could influence CNS-structures behind the blood–brain barrier, but it was later demonstrated that a variety of major pathways are involved: In the periventricular organs passive diffusion of cytokines has been demonstrated. Additionally, active transport mechanisms and specific transport peptides to the brain do exist for most of the cytokines. Moreover, an activation of neural afferents like the vagus nerve by cytokines in the periphery is an important mechanism. Systemic inflammation has been shown to induce cytokine synthesis and release mainly from microglial cells within the brain (Rothwell and Hopkins 1995). It can be assumed that these pathways are also functional in humans. Hence, cytokine signals from the periphery are very likely to modulate complex human CNS function.

13.2 The Role of the Neuroendocrine-Immune-CNS Interplay in Clinical Medicine

Acute infection and inflammation are the most evident situations that demonstrate the effects of cytokines on the brain, although very little detailed information is available on the circulating amounts of TNF-α and IL-6 during infection and inflammation besides the acute phase of sepsis. Moreover, there is a range of other clinical conditions where peripheral cytokine signals might modulate complex human brain functions. Numerous studies showed that the therapeutic administration of cytokines for the treatment of hepatitis, multiple sclerosis, or rheumathoid arthritis can induce depressive symptomatology, which widely overlaps with the syndrome of "sickness behavior" observed in animal models of infection and inflammation. However, in these clinical situations and during acute febrile infections the amounts of circulating inflammatory cytokines are huge, usually two orders of magnitude or more above baseline levels. In contrast, circulating levels are only slightly or moderately increased in the most frequent clinical situations where cytokines might play a role in inducing symptoms of depression, such as chronic infection or inflammation, cancer, cardiovascular disease, and autoimmune disorders (Yirmiya et al. 2000). Unfortunately, the studies on cytokine levels in patients with autoimmune diseases, chronic inflammation and infection are rare and conflicting, in particular with respect to longitudinal investigations (e.g., Mangge, Gallistl, and Schauenstein 1999). Independently from these disorders involving definite immunopathology, slightly increased levels of cytokines have been hypothesized to be involved in the pathophysiology of major depressive disorders in otherwise healthy people (Smith 1991). In recent years, it has become evident that inflammatory processes play an important role in cardiovascular disease and heart failure. Slightly increased TNF-α and possibly also IL-6 levels are often found in patients with heart disease (Ferrari 1999). A number of psychotropic drugs induce slight increases in the circulating levels of TNF-α, IL-6, or both, which are in the same range as those seen in patients with heart disease (Haack, Hinze-Selch, Fenzel, Kraus, Kühn, and Pollmächer 1999). Unfortunately, only few studies addressed the physiological relevance of circulating inflammatory cytokines or of slight to moderate increases in their levels with respect to central nervous function. The present review focuses on changes in sleep, memory, mood, and food intake following the experimental administration of bacterial endotoxin to healthy volunteers.

13.3 Changes in Sleep–Wake Behavior Following Endotoxin

It is known from many animal studies that a prominent increase in inflammatory cytokine levels increases non-REM sleep (Krueger, Obál, and Fang 1999). In human beings most experiments have not injected cytokines themselves but administered bacterial endotoxin to healthy people. Mostly, a highly purified preparation derived from *salmonella abortus equi* was used, an endotoxin stripped of protein, prepared specifically for use in humans. In order to facilitate the comparison of results between the different studies using different dosages and/or different times of the day, we have focused on the consecutive increases in inflammatory cytokines and the neuroendocrine activation in parallel. Very low amounts of bacterial endotoxin (0.2 ng/kg) administered to healthy volunteers at 23:00 h induced neither neuroendocrine activation nor an increase in body temperature, but an increase in TNF-α and IL-6 levels to about 200% of baseline. The administration of intermediate dosages of 0.4 ng/kg body weight in the evening or of 0.8 ng/kg in the morning were followed by more robust increases in cytokine plasma levels and by a slight, but significant activation of the HPA-system in parallel to increases of body temperature of about 0.5 to 1.0°. Finally, injections of 0.8 ng/kg bodyweight in the evening led to even higher fever and increased the levels of cytokines and hormones even stronger (Mullington et al. 2000).

Following different stimulation conditions, sleep is changed with respect to both rapid eye movement (REM) sleep and non-REM sleep: Non-REM sleep was influenced by the acute phase response in a bimodal manner: when slight increases in inflammatory cytokines occur without concomitant neuroendocrine activation, non-REM sleep is slightly increased. When immune parameters are more strongly increased and fever is induced, sleep continuity and non-REM sleep amount are reduced. REM-sleep changes are less complex, the amount of REM-sleep suppression is directly related to the induction of temperature increased (Schuld, Haack, Hinze-Selch, Mullington, and Pollmächer 2001). Thus, one can conclude that proinflammatory cytokines, which also increase non-REM sleep in animals, also tend to increase non-REM sleep in humans. When the immunological changes are more prominent, other mechanisms like, for example, the activation in HPA-system may lead to a disturbance in sleep continuity and reduce non-REM amount. This view is also supported by the fact that in humans experimental enhancement of the circulating amounts of cytokine antagonists (IL-1 receptor antagonist and soluble TNF receptors) transiently suppressed non-REM sleep (Schuld et al. 1999). Additionally, there is some evidence from psychopharmacological research that all psychotropic drugs which tend to slightly increase circulating TNF-α levels are sedating or increase non-REM sleep amount (Pollmächer, Haack, Schuld, Kraus, and Hinze-Selch 2000).

13.4 Changes in Memory, Mood, and Food Intake Following Endotoxin

As summarized above, most studies on CNS effects of endotoxin examined sleep. In addition, cognitive performance, food intake and psychological parameters have been investigated using similar experimental protocols. In one study, 0.8 ng/kg endotoxin or placebo were injected at 09:00 h. First testing was performed at a time point when only cytokine plasma levels were elevated, whereas the second measurement took place when also HPA system was stimulated. At the last measurement, most humoral parameters were again normalized. None of the subjects recognized any symptoms subjectively, nevertheless they reported increased anxiety at the first time point and increased depressive mood after 4 h, fitting to the hypothesis that immune parameters influenced the brain directly in the morning and directly and indirectly via neuroendocrine stimulation at noon time. Moreover, the experiment caused memory deficits without changing executive function or attention. Finally, food intake was reduced immediately after injection, but was increased, when HPA axis also was activated. All these changes were positively correlated to changes in TNF-plasma levels (Reichenberg et al. 2002). Thus, both the unspecific immune response system and the neuroendocrine system seem to exert separate effects on CNS function during immune response.

References

Besedovsky, H., del Rey, A., Sorkin, E., and Dinarello, C.A. (1986). Immunoregulatory feedback between interleukin-1 and glucocorticoid hormones. *Science* 233, 652–654.
Dantzer, R. (2001). Cytokine-induced sickness behavior: Where do we stand? *Brain Behav Immun* 15, 7–24.
Ferrari, R. (1999). The role of TNF in cardiovascular disease. *Pharmacol Res* 40, 97–105.
Haack, M., Hinze-Selch, D., Fenzel, T., Kraus, T., Kühn, M., Schuld, A., and Pollmächer, T. (1999). Plasma levels of cytokines and soluble cytokine receptors in psychiatric patients upon hospital admission: Effects of confounding factors and diagnosis. *J Psychiatr Res* 33, 407–418.
Hopkins, S.J., and Rothwell, N.J. (1995). Cytokines and the nervous system I: Expression and recognition. *Trends Neurosci* 18, 83–88.
Kluger, M.J. (1991). Fever: Role of pyrogens and cryogens. *Physiol Rev* 71, 93–127.
Krueger, J.M., Obál, F., Jr., and Fang, J. (1999). Humoral regulation of physiological sleep: Cytokines and GHRH. *J Sleep Res* 8(Suppl 1), 53–59.
Mangge, H., Gallistl, S., and Schauenstein, K. (1999). Long-term follow-up of cytokines and soluble cytokine receptors in peripheral blood of patients with juvenile rheumatoid arthritis. *J Interferon Cytokine Res* 19, 1005–10.
Mullington, J., Korth, C., Herman, D.M., Orth, A., Galanos, C., Holsboer, F., and Pollmächer, T. (2000). Dose-dependent effects of endotoxin on human sleep. *Am J Physiol Cell Physiol* 278, R947–R955.
Pollmächer, T., Haack, M., Schuld, A., Kraus, T., and Hinze-Selch, D. (2000). Effects of antipsychotic drugs on cytokine networks. *J Psychiatr Res* 34, 369–382.
Reichenberg, A., Yirmiya, R., Schuld, A., Kraus, T., Haack, M., Morag, A., and Pollmächer, T. (2002). Cytokine-associated emotional and cognitive disturbances in humans. *Arch Gen Psychiatry* 58, 445–452.
Rothwell, N.J., and Hopkins, S.J. (1995). Cytokines and the nervous system II: Actions and mechanisms of action. *Trends Neurosci* 18, 130–136.
Schuld, A., Haack, M., Hinze-Selch, D., Mullington, J., Pollmacher, T. (2005) Experimentelle Untersuchungen der Interaktion zwischen Schlaf und Immunsystem beim Menschen. *Psychother Psychosom Med Psychol* 55(1): 29–35.

Schuld, A., Haack, M., Hinze-Selch, D., Mullington J., and Pollmächer, T. (2001). Exoperimental studies on the interaction between sleep and the immune system in humans. *Psychother Psych Med* 55, 29–35.

Schuld, A., Mullington, J., Hermann, D., Hinze-Selch, D., Fenzel, T., Holsboer, F., and Pollmächer, T. (1999). Effects of granulocyte colony-stimulating factor on night sleep in humans. *Am J Physiol* 276, R1149–R1155.

Smith, R.S. (1991). The macrophage theory of depression. *Med Hypotheses* 35, 298–306.

Turnbull, A.V., and Rivier, C.L. (1999). Regulation of the hypothalamic-pituitary-adrenal axis by cytokines: Actions and mechanisms of action. *Physiol Rev* 79, 1–71.

Yirmiya, R., Pollak, Y., Morag, M., Reichenberg, A., Barak, O., Avitsur, R., Shavit, Y., Ovadia, H., Weidenfeld, J., Morag, A., and Pollmächer, T. (2000). Illness, cytokines and depression. *Ann N Y Acad Sci* 917, 478–487.

14 Inflammation and Sleep

Rita A. Trammell, Krishna Jhaveri, and Linda A. Toth

14.1 Inflammation

Inflammation is the physiological process by which vascularized tissues respond to an infectious agent, antigenic challenge, or physical injury. During the inflammatory process, soluble mediators and cellular components work together to contain and eliminate the agents causing physical distress. Although inflammation is crucial to maintaining the health and integrity of an individual organism, both inadequate and excessive immune responses are detrimental to the health and life of the host. When the inflammatory response is impaired, the host may be excessively susceptible to the detrimental effects of microbial or neoplastic disease. When the inflammatory response is poorly controlled, the host may experience significant or excessive tissue damage.

In addition to its local effects, inflammation also triggers a complex series of behavioral, physiologic, and biochemical systemic changes collectively known as the acute phase response (reviewed in Gabay and Kushner (1999)). The acute phase response is characterized by fever, disordered sleep patterns, alterations in metabolism, and changes in the plasma concentrations of a variety of substances (e.g., iron, C-reactive protein, fibrinogen, serum amyloid A, and albumin). Proinflammatory cytokines, including interleukin-6 (IL-6), IL-1β, tumor necrosis factor-α (TNF-α), IFN-γ, transforming growth factor-β (TGF-β) and possibly IL-8, are key regulators of the acute phase response, with IL-6 being the chief stimulator of most acute phase proteins (Gabay and Kushner 1999).

Inflammation can be characterized as acute or chronic. Acute inflammation is a complex, tightly regulated, and generally self-limiting process that is critical for defense against pathogenic microorganisms. However, in some situations, acute inflammation does not resolve or self-limit, but instead progresses into a chronic condition. An appropriate equilibrium between proinflammatory and anti-inflammatory mediators is critical for eradication of the antigenic stimulus without progression to a chronic inflammatory state. If foreign antigen is not eliminated (as in chronic infections such as tuberculosis and chronic hepatitis) or the immune response is inappropriately directed against a self-antigen (as in autoimmune disease), chronic inflammation develops. In some situations, the mechanisms responsible for the development of a chronic inflammatory state are not as clearly defined (e.g., aging, type 2 diabetes, obesity, cardiovascular disease).

14.2 Cytokines and Sleep

Other chapters in this book review of the role of cytokines in the regulation of sleep, and therefore this topic will be discussed here only briefly. A substantial body of literature provides strong evidence that cytokines, particularly the proinflammatory cytokines IL-1 and TNF-α, are powerful modulators of sleep–wake behavior. Numerous studies have demonstrated that substances or manipulations that induce IL-1 or TNF-α increase sleep, whereas substances that inhibit the synthesis or actions of these cytokines reduce sleep (Krueger and Toth 1994; Opp, Kapás, and Toth 1992). In rats, IL-1β and TNF-α mRNA (Bredow, Guba-Thakurta, Taishi, Obal, and Krueger 1997), and protein concentrations (Floyd and Krueger 1997) vary diurnally in brain, with peaks occurring during the sleep phase. In humans, plasma concentrations of TNF peak during sleep (Gudewill, Pollmächer, Vedder, Schreiber, Fassbender, and Holsboer 1992). Administration of exogenous TNF-α or IL-1β increases both time spent in slow wave sleep (SWS) and the amplitude of EEG slow waves (a measure of the depth of sleep) (De Sarro, Gareri, Sinopoli, David, and Rotiroti 1997; Fang, Wang, and Krueger 1997). Accumulating evidence suggests that the cytokine IL-6 also modulates sleep. Serum IL-6 concentrations are elevated during conditions associated with excessive daytime sleepiness (e.g., narcolepsy, obstructive sleep apnea) (Vgontzas, Papanicolaou, Bixler, Kales, Tyson, and Chrousos 1997), and vary in phase with sleep–wake behavior in humans and rats (Guan, Vgontzas, Omori, Peng, Bixler, and Fang 2005; Vgontzas, Bixler, Lin, Prolo, Trakada, and Chrousos 2005). In healthy human volunteers, prolonged wakefulness increased plasma IL-6 concentrations (Shearer et al. 2001), and administration of IL-6 decreased time in SWS during the first half of the night and increased the amount of SWS during the second half (Spath-Schwalbe et al. 1998).

   To date, the majority of studies on cytokines and sleep examined the impact of individual cytokines. However, individual cytokines are only one facet of a complex signaling network. Cells are seldom, if ever, exposed to a single cytokine, but rather experience a changing milieu of proinflammatory and anti-inflammatory cytokines, soluble cytokine receptors, receptor antagonists, hormones, and other inflammatory modulators. Furthermore, the regulation of complex traits such as sleep is itself complex. Therefore, the effects of single cytokines may diverge from those of an array of cytokines. The overall inflammatory milieu is likely to be crucial in determining the sleep response.

14.3 Acute Inflammation and Sleep

Research that began in the 1970s indicates that the host response to microbial infections alters the expression of immune-modulatory substances that also regulate sleep. Various bacterial and viral components both trigger immune responses and elicit alterations in sleep during infection. The quantitative and temporal changes that develop in sleep throughout the course of an infectious disease have been characterized in animals infected with various bacteria, viruses, and parasites. The precise characteristics of infection-related changes in sleep vary with the specific microorganism, the route of infection, the genetic background of the host, and differences in the disease process. For example, rabbits develop different alterations in sleep when infected with *Pasteurella multocida* via the intravenous route, which causes septicemia, or the intranasal route, which causes pneumonia (Toth and Krueger 1990b). In general, rabbits with bacterial and fungal infections develop an initial increase and a subsequent decrease in the amount of time spent in SWS, whereas rapid eye movement sleep (REMS) is consistently reduced (Toth and Krueger 1989; Toth and Krueger 1990a). Infected rabbits also typically develop fevers, but the fevers generally persist beyond the period of enhanced sleep. In mice, the effects of infection with the fungal organism *Candida albicans* on sleep and temperature vary depending on the genetic background of the mice (Toth and Hughes, *Compar Med*, 2006).

The administration of killed bacteria and isolated bacterial components can also alter sleep patterns, and treatment with bacteriocidal antibiotics microbially induced changes in sleep (Toth and Krueger 1988). Administration of bacterial endotoxin alters sleep in both humans and animals (Lancel, Cronlein, Muller-Preuss, and Holsboer 1995; Mathias, Schiffelholz, Linthorst, Pollmacher, and Lancel 2000; Mullington et al. 2000; Schiffelholz and Lancel 2001). In rats, endotoxin administration promotes non-REMS (NREMS) during the dark (active) phase and decreases REMS during both the light and dark phases of the diurnal cycle (Lancel et al., 1995; Mathias et al., 2000; Schiffelholz and Lancel 2001). In humans, endotoxin suppresses REMS and prolongs REMS latency (Korth, Mullington, Schreiber, and Pollmacher 1996; Mullington et al. 2000), but its effects on NREMS depend on the dose and the diurnal time of administration (Mullington et al. 2000). Increases in the amount and intensity of nocturnal NREMS consistently develop only after the administration of subpyrogenic doses of endotoxin given shortly before the normal evening onset of sleep (Korth et al., 1996; Pollmacher, Schuld, Kraus, Haack, Hinze-Selch, and Mullington 2000). Pyrogenic doses disrupt sleep and reduce NREMS (Mullington et al. 2000).

Infection with influenza virus alters sleep in people and animals. Healthy human volunteers experimentally infected with rhinovirus, influenza virus, or both show less sleep during the incubation period, but more sleep during the symptomatic period; sleep quality and the number of awakenings are not affected (Smith 1992). Influenza-infected mice develop alterations in sleep that vary qualitatively and quantitatively depending on the genetic background of the mouse (Toth, Rehg, and Webster 1995; Toth and Verhulst 2003; Toth and Williams 1999c). Several observations suggest that viral replication is necessary to cause prolonged changes in sleep and activity. For example, inoculation of mice with the avian paramyxovirus Newcastle disease virus (NDV), which undergoes only abortive replication in mice, elicits only transient sleep enhancement (Toth 1996). Similarly, inoculation of mice with killed influenza virus does not promote sleep (Toth et al., 1995; Toth and Verhulst 2003). Double-stranded RNA (dsRNA), which is produced in infected host cells during viral replication, may mediate at least some of the somnogenic effects of viral infections. Intracerebroventricular administration of dsRNA induces sleep in rabbits, and synthetic dsRNA (polyinosinic:polycytidilic acid, or poly IC) increases sleep in both rabbits and mice (Kimura-Takeuchi, Majde, Toth, and Krueger 1992a; Kimura-Takeuchi, Majde, Toth, and Krueger 1992b; Toth 1996).

14.4 Chronic Inflammation and Sleep

Disordered sleep and fatigue are common features of diseases associated with chronic inflammation. For example, patients with autoimmune diseases (rheumatoid arthritis, multiple sclerosis, type 1 diabetes, Crohn's disease, and autoimmune thyroid disease) often report disturbed sleep and excessive fatigue (Bourguignon, Labyak, and Taibi 2003; Lashley 2003). Neuroimmune influences on sleep during autoimmune disease are the focus of other chapters of this book and will not be discussed here.

14.4.1 Chronic Infections

Unlike acute or abortive infections or the administration of viral or bacterial products, chronic infections or latent viral infections generate prolonged and continuous immune stimulation of the host. The resultant continually primed secondary immune response may be a major stimulus for fatigue and excessive sleepiness during chronic disease. Growing numbers of individuals suffer from chronic viral infections (e.g., human immunodeficiency virus, Epstein–Barr virus (EBV), hepatitis C) that can be associated with abnormal sleep, excessive sleepiness, and fatigue (Cunningham et al. 1998; Darko, Mitler, and Henriksen 1995; Davis et al. 1994; Norman, Chediak, Kiel, and Cohn 1990; Tobi and Strauss 1988).

A chronic viral infection that has been studied extensively with respect to sleep is human immunodeficiency virus (HIV). Initial polysomnographic studies of asymptomatic men infected with HIV revealed a significant increase in the percentage of time spent in SWS during the second half of the night; frequent nighttime awakenings and abnormal REMS architecture were also common (Darko, McCutchan, Kripke, Gillin, and Golshan 1992; Darko et al. 1995; Darko, Mitler, and Miller 1998). However, recent analyses suggest that pain and psychologic and psychosocial comorbidity are major determinants of disturbed sleep in HIV infection (Reid and Dwyer 2005; Vance and Burrage 2005; Vosvick et al. 2004). Nonetheless, regardless of the proximal cause, sleep complaints are a common problem in HIV-infected populations (Davis 2004).

Fatigue and disturbed sleep are also common symptoms in patients who are chronically infected with hepatitis C virus (Forton, Taylor-Robinson, and Thomas 2006; Lang et al. 2006; Spiegel, Younossi, Hays, Revicki, Robbins, and Kanwal 2005b). An estimated 170 million people worldwide are infected with the hepatitis C virus (World Health Organization 1999), and 50 to 75% of infected patients develop chronic disease (Bonkovsky and Mehta 2001; Wong and Terrault 2005). In a recent study, 65% of patients with chronic hepatitis C (CHC) infections reported sleep problems as the most severe symptom (Lang et al. 2006).

An association of cirrhosis to the development of impaired glucose tolerance has been recognized for several years (Petrides and DeFronzo 1989; Petrides, Schulze-Berge, Vogt, Matthews, and Strohmeyer 1993). However, patients with CHC have increased risk of developing insulin resistance and type 2 diabetes even without cirrhosis (Knobler, Schihmanter, Zifroni, Fenakel, and Schattner 2000). Another risk factor for the development of insulin resistance and type 2 diabetes is obesity, which further increases the risk in CHC patients (Charlton, Pockros, and Harrison 2006; Knobler and Schattner 2005; Mehta, Brancati, Sulkowski, Strathdee, Szklo, and

Thomas 2000), and which is itself also associated with mild chronic inflammation (Xu et al. 2003). Several studies suggest that the chronic inflammatory state associated with both CHC and obesity may be a significant contributing factor in the development of insulin resistance. For example, TNF-α is increased in obesity, type 2 diabetes and CHC, and may link these conditions (reviewed in Knobler and Schattner (2005)). Thus, obesity, CHC and type 2 diabetes all generate changes in the amounts and balance of cytokines that influence sleep.

14.4.2 Chronic Fatigue Syndrome

Chronic fatigue syndrome (CFS) is a condition of unknown etiology that is diagnosed based on patient report of an array of clinical symptoms, generally in the absence of an organic or inciting etiology. Key diagnostic features are severe fatigue of more than 6 months duration and a subset of other symptoms that can include impairments of memory or the ability to concentrate, tender lymph nodes, myalgia, arthralgia, headache, cognitive disturbances, low-grade fever, disturbed sleep, sore throat, and postexertional fatigue (Afari and Buchwald 2003; Fukuda, Straus, Hickie, Sharpe, Dobbins, and Komaroff 1994). Most patients with CFS have significant functional impairment (Buchwald, Pearlman, Umali, Schmaling, and Katon 1996) and at least one sleep disorder (Buchwald, Pascualy, Bombardier, and Kith 1994; Morriss, Sharpe, Sharpley, Cowen, Hawton, and Morris 1993).

Because many of the symptoms of CFS are also symptoms of acute viral infections, numerous studies have explored the possibility that viral infections precipitate CFS. Some data support a role for viral infections in causing CFS (Davis et al. 1994; Martin 1996; Sairenji, Yamanishi, Tachibana, Bertoni, and Kurata 1995; White et al. 1998; White, Thomas, Sullivan, and Buchwald 2004), although others do not (Buchwald et al. 1996; Swanink, van der Meer, Vercoulen, Bleijenberg, Fennis, and Galama 1995). One virus that has been studied extensively in this regard is EBV. The initial lytic infection produced by EBV resolves within days and is followed by the establishment of a latent state that is maintained under the control of the host immune system.

Infection with or reactivation of EBV in humans is often associated with fatigue and excessive sleepiness (Ablashi 1994; Ebell 2004; Guilleminault and Mondini 1986; Josephs et al. 1991; Petersen, Thomas, Hamilton, and White 2006; Tobi and Strauss1988; White et al. 1998). Such symptoms could be related to immune stimulation or dysfunction, neural–endocrine homeostatic imbalance, or both, as produced secondary to the acute and chronic viral infection (Cleare 2003; Papanicolaou et al. 2004). Studies demonstrating that antibodies to EBV enzymes may be elevated in CFS patients support the hypothesis that EBV is involved in the pathogenesis of CFS (Glaser and Kiecolt-Glaser 1998; Jones, Williams, and Schooley 1988; Natelson et al. 1994). Synthesis of these viral enzymes may reflect stress-related reactivation of latent virus (Glaser et al. 2005). Although host responses to reactivation may abort complete viral replication, the viral enzymes themselves and the host responses they elicit may cause symptoms similar to those of both acute viral challenge and CFS (Glaser et al. 2005).

14.4.3 Obesity and Type 2 Diabetes

During the last 20 years, obesity has reached epidemic proportions in the United States and worldwide. Recent data from the National Center for Health Statistics indicate that 30% of U.S. adults 20 years of age and older (over 60 million people) are obese (Ogden, Carroll, Curtin, McDowell, Tabak, and Flegal 2006). Visceral obesity and three other pathologic conditions (dyslipidemia, hypertension, and insulin resistance) comprise the so-called "metabolic syndrome," also known as "Syndrome X." Syndrome X is a major risk factor for type 2 diabetes (T2D; also called non-insulin-dependent diabetes mellitus, or NIDDM) (Haffner, Ruilope, Dahlof, Abadie, Kupfer, and Zannad 2006).

A common feature of obesity, insulin resistance, and T2D is chronic, low-grade inflammation (Dandona, Aljada, and Bandyopadhyay 2004; Dandona, Aljada, Chaudhuri, Mohanty, and Garg 2005; Weisberg et al. 2006; Weisberg, McCann, Desai, Rosenbaum, Leibel, and Ferrante 2003; Wellen and Hotamisligil 2005). Markers of chronic subclinical inflammation (e.g., C-reactive protein and IL-6) are closely linked to insulin resistance and obesity (Finegood 2003; Temelkova-Kurktschiev, Henkel, Koehler, Karrei, and Hanefeld 2002). In addition, proinflammatory cytokines such as TNF-α, monocyte chemotactic protein-1 (MCP-1), IL-6, IL-8, and macrophage inflammatory peptide (MIP)-1α are often increased in patients with obesity, insulin resistance and T2D (Gerhardt, Romero, Cancello, Camoin, and Strosberg 2001; Mohamed-Ali et al. 1997; Sartipy and Loskutoff 2003; Takahashi et al. 2003; Uysal, Wiesbrock, Marino, and Hotamisligil 1997; Weisberg et al. 2003, 2006; Xu et al. 2003; Yudkin, Kumari, Humphries, and Mohamed-Ali 2000). In obesity, MCP-1, a member of the chemokine family, is thought to play a key role in the recruitment of monocytes/macrophages to adipose tissue, thereby contributing to the proinflammatory environment (Weisberg et al. 2003). Furthermore, adipose tissue itself secretes immune-modulatory proteins (i.e., leptin, adiponectin, and resistin) that are collectively known as adipokines. Leptin promotes macrophage production of IL-1β, IL-6, IL-12, and TNF-α and decreases the anti-inflammatory cytokine IL-10 both in vitro and in vivo (Loffreda et al. 1998). In contrast, adiponectin has anti-inflammatory properties and its plasma concentrations are positively and strongly associated with insulin sensitivity and negatively associated with TNF-α levels (Abbasi et al. 2004; Ajuwon and Spurlock 2005). Adipokines have been implicated in the pathogenesis of obesity, metabolic syndrome and T2D (reviewed in Fantuzzi (2005), Fernandez-Real (2006), Vettor, Milan, Rossato, and Federspil (2005)).

The pathogenesis of T2D is complex and multifactorial. However, recent studies support a role for chronic inflammation in the development of metabolic syndrome and T2D. Furthermore, many of the proinflammatory substances associated with chronic inflammation also modulate sleep (Cannon 2000; Elenkov, Iezzoni, Daly, Harris, and Chrousos 2005; Kapsimalis, Richardson, Opp, and Kryger 2005). A number of studies have revealed interactions between obesity and sleep in both people and mice. For example, C57BL/6J mice rendered obese via access to a high fat diet show increased total sleep, more bouts of sleep, and shorter periods of waking (Jenkins, Omori, Guan, Vgontzas, Bixler, and Fang 2005; O'Donnell et al. 1999). Obese leptin-deficient mice also show abnormal sleep architecture (i.e., more frequent arousals, more frequent but shorter bouts of sleep, greater total daily sleep time, and a flattened diurnal rhythm of sleep–wake) (Laposky, Shelton, Bass, Dugovic, Perrino, and Turek 2006).

In people, metabolic syndrome and its associated panoply of problems (obesity, dyslipidemia, insulin resistance) are also often associated with fragmented sleep. Analogously, reduced sleep is associated with metabolic perturbations, glucose intolerance, and insulin resistance. A landmark study demonstrated relative glucose intolerance in healthy men after a single night of sleep loss (Spiegel, Leproult, and Van Cauter 1999a). Additional laboratory-based and epidemiologic studies have since supported a relationship of impaired or reduced sleep to insulin resistance and T2D (Al-Delaimy, Manson, Willett, Stampfer, and Hu 2002; Ayas et al. 2003; Elmasry, Janson, Lindberg, Gislason, Tageldin, and Boman 2000; Elmasry et al. 2001; Ip, Lam, Ng, Lam, Tsang, and Lam 2002; Punjabi, Sorkin, Katzel, Goldberg, Schwartz, and Smith 2002; Resnick et al. 2003; Spiegel, Knutson, Leproult, Tasali, and Van Cauter 2005). In fact, the frequent association of sleep apnea, which causes reduced or fragmented sleep, with the other facets of "Syndrome X" has led to the suggested alternative title of "Syndrome Z" (Wilson, McNamara, and Collins 1998).

These relationships have fostered speculations that reduced, impaired or disrupted sleep, which are common in today's society, may contribute to the development of both obesity and diabetes. Recent work has established that sleep loss and sleep disorders alter normal endocrine diurnal rhythms and metabolic function (Leproult, Copinschi, Buxton, and Van Cauter 1997; Qin, Li, Wang, Wang, Xu, and Kaneko 2003; Spiegel, Tasali, Penev, and Van Cauter 2004; Van Cauter, Blackman, Roland, Spire, Refetoff, and Polonsky 1991), perhaps contributing to obesity. This clustering of pathophysiologic conditions suggests positive (reinforcing) interactive metabolic feedback mechanisms that link sleep loss, obesity, and glucose intolerance via common neuroendocrine and immune mediators (Bjorntorp and Rosmond 2000; Boethel 2002).

14.4.4 Cancer

Insomnia and daytime sleepiness (hypersomnia) are prevalent and often chronic in cancer, with sleep–wake disturbances affecting between 30 and 50% of patients (Savard, Simard, Hervouet, Ivers, Lacombe, and Fradet 2005). Disturbed sleep and fatigue in cancer patients may be attributable to both the disease and its treatment (Morrow, Andrews, Hickok, Roscoe, and Matteson 2002). For example, women with breast cancer report fatigue and disturbed sleep before beginning chemotherapy (Ancoli-Israel et al. 2006), and list insomnia and other sleep problems among the most distressing symptoms experienced during chemotherapy (Berglund, Bolund, Fornander, Rutqvist, and Sjoden 1991).

Human and mouse epithelial tumors are often infiltrated by inflammatory cells that generate a complex array of cytokines and chemokines, thereby establishing a chronic inflammatory state in the host (Kulbe, Hagemann, Szlosarek, Balkwill, and Wilson 2005; Wilson and Balkwill 2002). Cytokines associated with chronic inflammation are also potential modulators of sleep in cancer patients. The sleep regulatory cytokines IL-1, TNF-α, and IL-6 are increased in some cancer patients (Balkwill and Coussens 2004; Ben Baruch 2006; Cleeland et al. 2003), and TNF-α and IL-6 have been associated with fatigue and reduced quality of life in patients with myelodysplastic syndrome and acute leukemia (Lee et al. 2004).

Fatigue and sleep disturbances associated with cancer treatments (e.g., chemotherapy, immunotherapy, radiotherapy) are also potentially linked to immune (cytokine) imbalance in cancer patients. For example, patients undergoing IFN therapy report excessive sleepiness (Mattson et al. 1983; Smedley, Katrak, Sikora, and Wheeler 1983), although their sleep may be disrupted rather than enhanced (Spath-Schwalbe, Lange, Perras, Horst, Fehm, and Born 2000). Treatment of cancer patients with the chemotherapeutic agents paclitaxel, tamoxifen, and cisplatin increases serum levels of proinflammatory cytokines known to modulate sleep (Lee et al. 2004; Pusztai et al. 2004). Collectively, these data indicate that sleep-modulatory cytokines are increased in cancer patients, and support the hypothesis that cytokines contribute to their sleep disruption and fatigue.

14.4.5 Age-Related Inflammation

A mild chronic inflammatory state is often present in elderly persons even in the absence of overt disease (Dinarello 2006; Gerli et al. 2000; Giuliani et al. 2001). Recent studies suggest that this proinflammatory state may contribute both to age-related sleep impairments and to the pathogenesis of some age-related chronic diseases (e.g., atherosclerosis, cancer, type 2 diabetes) (Agarwal and Gotman 2001; Ershler and Keller 2000; Kritchevsky, Cesari, and Pahor 2005; Maggio, Guralnik, Longo, and Ferrucci 2006; Prinz 2004).

14.5 Genetic Variation in Sleep and Inflammation

14.5.1 Genetic Variation in the Inflammatory Response

Inflammation drives the development or severity of a large number of diseases (e.g., atherosclerosis, asthma, ulcerative colitis, postmenopausal osteoporosis, sleep disorders). Variation in genes that encode proteins regulating inflammation may underlie a large proportion of interindividual variation in susceptibility to numerous diseases. Furthermore, variation in the production or function of cytokines can cause variation in inflammatory responses that in turn influences the progression of inflammatory disease.

Examples of genetic variation in inflammation are abundant. For example, a single nucleotide polymorphism in the human caspase-12 proenzyme confers hyporesponsiveness to LPS-stimulated cytokine production that manifests clinically as increased susceptibility to severe sepsis and mortality (Saleh et al. 2004). Similarly, a polymorphism in TNF-α promoter region is associated with increased TNF-α production in response to LPS and increased risk of sepsis-related morbidity and mortality (Bessler, Osovsky, and Sirota 2004; Hartel, von Puttkamer, Gallner, Strunk, and Schultz 2004; Lorenz, Mira, Frees, and Schwartz 2002; Strunk and Burgner 2006), an IL-6 polymorphism is associated with higher risk of bacterial sepsis (Heesen, Kunz, Bachmann-Mennenga, Merk, and Bloemeke 2003), and certain IL-1β polymorphisms correlate with susceptibility to *Helicobacter pylori* infection and gastric cancer, Alzheimer's disease, and possibly rheumatoid arthritis (Cantagrel et al. 1999; El Omar, Chow, Mccoll, Fraumeni, and Rabkin 2000; Hamajima et al. 2001; Mcgeer and Mcgeer 2001). Receptors for inflammatory mediators also show genetic variation. For example, toll-like receptor 4 (TLR4), which mediates the effects of LPS, has two missense single nucleotide polymorphisms are associated with differences in susceptibility to sepsis (Agnese et al. 2002; Barber, Aragaki, Rivera-Chavez, Purdue, Hunt, and Horton 2004; Child et al. 2003).

The impact of genetic variation on the inflammatory response is perhaps best illustrated in mice. C3H/HeJ mice, for example, are hyporesponsive to *Escherichia coli* LPS because they bear a mutation in the *TLR4* gene that is not present in related strains (e.g., C3H/HeOuJ and C3H/HeN) (Qureshi et al. 1996; Vallance, Deng, Jacobson, and Finlay 2003). The cytokine responses (TNF-α, IL-1, IL-6) of C3H/HeJ mice treated with LPS are similar those of in humans with *TLR4* mutations (Kuhns, Priel, and Gallin 1997; Poltorak et al. 1998). As a second example, C57BL/6J and A/J mice injected with LPS demonstrate different numbers of infiltrating polymorphonuclear leucocytes (PMN) in liver and lungs, serum levels of IL-1β and IL-6 are higher in C57BL/6J mice than in A/J mice, A/J mice have higher levels of TNF-α, and mortality is higher in the C57BL/6J strain (De Maio, Mooney, Matesic, Paidas, and Reeves 1998). In contrast to LPS, A/J mice are far more susceptible to infection with *C. albicans* than are C57BL/6J mice (Marquis, Montplaisir, Pelletier, Auger, and Lapp 1988; Tuite, Elias, Picard, Mullick, and Gros 2005). Genetic variation in the sensitivity and immune responses of different mouse strains to microbial challenge is a likely factor in the changes in vigilance associated with infectious and inflammatory disease.

In considering the genetics of inflammation and its relationship to disease, three questions warrant attention: (a) will identified polymorphisms significantly impact the inflammatory process, (b) will the impact on inflammation be clinically relevant, and (c) does genetic information suggest therapeutic measures that may reduce clinical morbidity and mortality (Kornman 2006). Functional genetic polymorphisms in the cytokine response both under normal conditions and after inflammation may influence both the disease process and the sleep.

14.5.2 Genetic Variation in Sleep Under Normal Conditions and during Inflammatory Disease

Several sleep and circadian disorders, including narcolepsy, advanced sleep phase syndrome (ASPS), and fatal familial insomnia (FFI), have known genetic determinants. For example, canine narcolepsy is transmitted as a single autosomal recessive trait with full penetrance (Mignot et al. 1991). The critical genetic determinant for canine narcolepsy is a mutation in the hypocretin receptor gene *HCRTR2* (Lin et al. 1999). This finding spurred studies that revealed low levels of hypocretin-1 in the cerebral spinal fluid of human patients with narcolepsy (Baumann, Khatami, Werth, and Bassetti 2006; Nishino, Ripley, Overeem, Lammers, and Mignot 2000; Peyron et al. 2000; Thannickal et al. 2000). Individuals with ASPS have persistent early diurnal onset of sleep and early awakenings; this syndrome shows a strong genetic link to an autosomal dominant allele of the human period-2 gene (Jones and Ptacek 1999; Toh et al. 2001). FFI is a rare disorder characterized by progressively worsening insomnia, motor disturbances, dysautonomia, and eventual death (Lugaresi et al. 1986). A point mutation in the prion protein gene on Chr 20 has been implicated as the causative factor leading to FFI (Medori et al. 1992; Plazzi et al. 2002).

Polymorphisms in sleep-modulatory immune mediators may also alter sleep and susceptibility to sleep disorders. For example, a TNF-α (–308A) allele is significantly associated with obstructive sleep apnea (OSA) (Riha et al. 2005). Siblings with OSA were significantly more likely to carry the TNF-α (–308A) allele than siblings from a control group. OSA is also associated with allelic variation in the angiotensin converting enzyme (ACE) (Zhang et al. 2000), apolipoprotein E genotype e4 (Gottlieb et al. 2004; Larkin, Patel, Redline, Mignot, Elston, and Hallmayer 2006; Saarelainen, Lehtimaki, Nikkila, Solakivi, Nieminen, and Jaakkola 2000), and serotonin receptor type 2A and 2C (Sakai et al.2005).

Associations between *HLA* haplotypes and sleep disorders are also frequent in the literature. For example, numerous studies have revealed significant links between *HLA* haplotypes and narcolepsy in various patient populations (Juji, Satake, Honda, and Doi 1984; Lin, Hungs, and Mignot 2001). REM-sleep behavior disorder has been associated with the *HLA-DQ1* (Schenck, Garcia-Rill, Segall, Noreen, and Mahowald 1996), and Kleine–Levin syndrome with the *HLA-DQB1*0201* (Dauvilliers et al. 2002). The HLA associations found in narcolepsy, REMS disorder behavior, and Kleine–Levin syndrome suggest a likely link between sleep and immune mechanisms, mediated by cytokine, endocrine, and other factors that influence both immune function and sleep regulation (Dauvilliers, Maret, and Tafti 2005; Marshall and Born 2002).

Differences in normal physiological sleep are well known in humans, with variations occurring in the total amount of time spent in SWS and REMS, the diurnal timing of sleep, the daily amount of sleep needed for daily recuperation, and other measures. Twin studies have revealed both genetic and environmental components of sleep phenotypes. Geyer (1937) and Gedda (1951) produced the first reports of a higher concordance in the sleep habits in monozygotic versus dizygotic twins. Others have confirmed and extended these reports (Heath, Kendler, Eaves, and Martin 1990; Linkowski 1999; Partinen, Kaprio, Koskenvuo, Putkonen, and Langinvainio 1983; Webb and Campbell 1983). Moreover, a higher correlation of sleep duration occurs in monozygotic twins, whether living either together or apart, than in dizygotic twins (Gedda and Brenci 1983).

Different strains of inbred mice also exhibit consistent variations in their sleep patterns, and several independent laboratories have demonstrated genetic influences on various facets of sleep. For example, EEG delta power during NREMS has been linked to a region on chromosome 13 (*Dps1* at ~ 15cM, *Dps2, Dps3*) (Franken, Chollet, and Tafti 2001b). Similarly, the amount of time spent in REMS has been linked to loci on Chr 2, 4, 16, 17, and 19 (Tafti, Franken, Kitahama, Malafosse, Jouvet, and Valatx 1997; Toth and Williams 1999b). The *Tcp1* region of mouse Chr 17 influences the regulation of high affinity choline uptake, which is a critical step in acetylcholine (ACh) synthesis. Because cholinergic mechanisms are central to regulation of REMS, these findings suggest that differences in the rate of ACh synthesis may contribute to strain differences in REMS.

As with sleep patterns, certain features of the EEG are also under genetic regulation in mice. EEG theta and delta power vary significantly across mouse strains (Franken, Chollet, and Tafti 2001a; Franken, Malafosse, and Tafti 1998). EEG frequency during REMS is inherited as an autosomal recessive trait and is modulated by the gene *Acad* (acylcoenzyme-A-dehydrogenase) (Tafti et al. 2003). Analysis of delta power during recovery after sleep deprivation of recombinant inbred (RI) mice revealed a quantitative trait locus (QTL) on Chr 13 that accounted for 49% of the genetic variance (Franken et al. 2001a). This locus (*Dps1*, or delta power in SWS QTL1), incorporates a number of genes associated with brain energy metabolism (e.g., neurotrophic tyrosine kinase-2 receptor, growth hormone releasing hormone (GHRH), glycogen phosphorylase, adenosine deaminase). DBA/2J mice show lower delta power during SWS and predominant theta power in the EEG (Franken et al. 2001a). Furthermore, their sleep is fragmented, and sleep pressure accumulates at a slower rate in this strain as compared to other inbred strains (Franken et al. 2001a). Analysis of these traits revealed association to a polymorphism in the retinoic acid receptor beta (*Rarb*) gene (Maret, Franken, Dauvilliers, Ghyselinck, Chambon, and Tafti 2005). Some EEG variants (e.g., 16–19/s beta waves) follow a Mendelian autosomal dominant mode of inheritance in humans (Vogel 1965).

14.5.3 The Intersection of Genetic Variation in Sleep and Inflammation

A growing body of literature indicates that variation in the inflammatory response may influence sleep, and perhaps that variation in sleep may impact recuperation and prognosis. Numerous studies have shown that marked but varied alterations in sleep develop during infectious diseases and inflammatory processes. For example, in mice, influenza infection leads to increased somnolence in some strains, but impaired sleep in others (Toth and Verhulst 2003). Inbred strains of mice also vary in their sensitivity and responses to numerous infectious challenges. Strain-related differences in the cytokine milieu and other factors produced over the course of an infection could contribute to quantitative and qualitative variation in sleep during infectious diseases. For example, the sleep responses to some viral challenges, but not others, are influenced by the gene *If-1*, which regulates production of the cytokine IFN-γ in response to viral challenge (De Maeyer and De Maeyer-Guignard 1970). Congenic B6.C-H28c mice, which have the *If-1* allele for low IFN-α production on a C57BL/6J genetic background, show a C57BL/6J-like sleep phenotype after influenza infection, but a BALB/cByJ-like phenotype after infection with Newcastle disease virus. These data suggest that specific genes may influence sleep changes in response to some infectious challenges, but not others. In response to influenza infection, 7 of 13 RI strains showed a BALB/cByJ-like response (reduced SWS) during the light phase, whereas six showed a C57BL/6J-like response (normal SWS). In contrast, during the dark phase, nine RI strains showed a C57BL/6J-like response (enhanced sleep), whereas four had a BALB/cByJ-like responses (normal SWS). These data suggest that different genetic factors influence influenza-induced changes in SWS during the light and dark phases of the diurnal cycle (Toth and Williams 1999c). Linkage analysis of the light phase trait revealed a QTL (sleep response to influenza, light phase, or *Srilp* that is delineated by markers *D6Mit74* and *D6Mit188* on chromosome 6 (Toth and Williams 1999a). The 95% confidence interval that defines *Srilp* incorporates several likely candidate genes, including *Ghrhr* (growth hormone releasing hormone-receptor), *Crhr2* (corticotrophin releasing hormone receptor 2), and *Cd8a* (cytotoxic T lymphocytes epitope). Several of these have now been eliminated as candidates (Ding and Toth

2006; Toth and Hughes 2004). However, C57BL/6J-*lit/lit* mice bear a spontaneous single nucleotide point mutation in *Ghrhr* that generates a nonfunctional receptor (Godfrey, Rahal, Beamer, Copeland, Jenkins, and Mayo 1993; Lin, Lin, Gukovsky, Lusis, Sawchenko, and Rosenfeld 1993). As compared to normal C57BL/6J mice, *lit/lit* mice show less sleep both under normal conditions and during influenza infection (Alt, Obal, Traynor, Gardi, Majde, and Krueger 2003), suggesting that *Ghrhr* may contribute to the sleep phenotype. Furthermore, influenza infection altered hypothalamic mRNA for IL-1β and TNF-α in normal C57BL/6J mice but not in *lit/lit* mice (Alt et al. 2003), supporting a role for *Ghrhr* in the cytokine response, which could in turn be related to the sleep phenotype.

As with influenza, various strains of inbred mice demonstrate a variety of sleep responses and varying severity of disease after inoculation with *C. albicans* (Ashman, Fulurija, and Papadimitriou 1997; Hector, Domer, and Carrow 1982; Marquis, Montplaisir, Pelletier, Mousseau, and Auger 1986; Tuite et al. 2005) (Toth and Hughes, *Compar Med*, in press). The marked interstrain differences in patterns of both sleep and clinical disease suggest that genetic factors influence both of these pathophysiologic responses to challenge. Changes in sleep and core temperature were highly correlated in *Candida*-infected mice, as were renal *Candida* titers and blood urea nitrogen, relative neutrophilia, and serum IL-6 concentrations, suggesting that common factors may elicit these diverse responses. The immune system, like other physiologic systems, generally employs functionally overlapping mechanisms to protect homeostatic regulation. These mechanisms could, by analogy, also influence sleep and other pathophysiologic responses that develop during microbial infections.

The involvement of macrophages in the generation of peripheral signals that induce sleep during influenza infection is an intriguing possibility. Macrophages both produce and metabolize somnogenic substances after microbial challenge (Fincher, Johannsen, Kapás, Takahashi, and Krueger 1996; Johannsen, Wecke, Obál, and Krueger 1991). Furthermore, manipulations designed to impair macrophage function attenuate or prevent influenza-related sleep enhancement in C57BL/6J mice (Toth and Hughes 2004). Murine macrophages are highly susceptible to infection with influenza virus and respond to the virus by secreting an array of inflammatory substances, including TNF-α and MIP-1α (Gong et al. 1991; Hofmann et al. 1997). Targeted mutation of the gene *Ccl3* (previously named both *Scya3* and *Mip1α*) converts the C57BL/6J sleep phenotype (dark phase sleep enhancement) into a BALB/cByJ-like phenotype (normal amounts of SWS during the dark phase) (Toth and Hughes 2004). This observation suggests that *Ccl3* may underlie, at least in part, the phenotypic differences in sleep that distinguish these two mouse strains during influenza infection. Supporting a modulatory role for *Ccl3*, two *Ccl3* polymorphisms have been identified, and C57BL/6J and BALB/cAn mouse strains possess different allelic variants (Blackburn, Griffith, and Morahan 1995; Wilson et al. 1990). Furthermore, influenza infection is associated with increased monocyte and epithelial cell expression of *Mip1α* (Bussfeld, Kaufmann, Meyer, Gemsa, and Sprenger 1998). *Mip1α*-deficient mice show major abnormalities in monocyte recruitment and cytokine expression (Lu et al. 1998) and reduced pneumonitis but higher viral titers after challenge with influenza (Cook et al. 1995). However, to our knowledge, the somnogenic properties of CCL3/MIP-1α have not been evaluated to date.

14.6 Summary and Conclusions

Inability to sleep, excessive sleepiness, and fatigue are common problems that many, if not all, people experience at some time in their lives. Excessive sleepiness and fatigue, particularly when persistent, reduce the quality of life of affected individuals and also cause significant economic loss in terms of diminished productivity and employment capability. Considerable data support strong interactions between the immune response and vigilance states. Furthermore, infectious and inflammatory diseases and/or their therapies are often associated with nonrestorative sleep, excessive daytime sleepiness, and fatigue, and many individuals also suffer from chronic fatigue of unknown etiology. Identifying the mechanisms responsible for poor sleep, chronic fatigue, and excessive sleepiness and developing effective interventions for these disabling symptoms could improve the economic welfare and quality of life of many individuals.

Acknowledgments. This work was supported in part by NIH grants R01-NS40220, R01-HL70522, and K26-RR17543, and by the Southern Illinois University School of Medicine.

References

Abbasi, F., Chu, J.W., Lamendola, C., McLaughlin, T., Hayden, J., Reaven, G.M., and Reaven, P.D. (2004) Discrimination between obesity and insulin resistance in the relationship with adiponectin. *Diabetes* 53, 585–590.
Ablashi, D.V. (1994) Summary: Viral studies of chronic fatigue syndrome. *Clin Infect Dis* 18(Suppl 1), S130–S133.
Afari, N., and Buchwald, D. (2003) Chronic fatigue syndrome: A review. *Am J Psychiatry* 160, 221–236.
Agarwal, R., and Gotman, J. (2001) Computer-assisted sleep staging. *IEEE Trans Biomed Eng* 48, 1412–1423.
Agnese, D.M., Calvano, J.E., Hahm, S.J., Coyle, S.M., Corbett, S.A., Calvano, S.E., and Lowry, S.F. (2002) Human toll-like receptor 4 mutations but not CD14 polymorphisms are associated with an increased risk of gram-negative infections. *J Infect Dis* 186, 1522–1525.
Ajuwon, K.M., and Spurlock, M.E. (2005) Adiponectin inhibits LPS-induced NF-kappaB activation and IL-6 production and increases PPARgamma2 expression in adipocytes. *Am J Physiol Regul Integr Comp Physiol* 288, R1220–R1225.
Al-Delaimy, W.K., Manson, J.E., Willett, W.C., Stampfer, M.J., and Hu, F.B. (2002) Snoring as a risk factor for type II diabetes mellitus: A prospective study. *Am J Epidemiol* 155, 387–393.
Alt, J.A., Obal, F., Jr., Traynor, T.R., Gardi, J., Majde, J.A., and Krueger, J.M. (2003) Alterations in EEG activity and sleep after influenza viral infection in GHRH receptor-deficient mice. *J Appl Physiol* 95, 460–468.

Ancoli-Israel, S., Liu, L., Marler, M.R., Parker, B.A., Jones, V., Sadler, G.R., Dimsdale, J., Cohen-Zion, M., and Fiorentino, L. (2006) Fatigue, sleep, and circadian rhythms prior to chemotherapy for breast cancer. *Support Care Cancer* 14, 201–209.

Ashman, R.B., Fulurija, A., and Papadimitriou, J.M. (1997) Evidence that two independent host genes influence the severity of tissue damage and susceptibility to acute pyelonephritis in murine systemic candidiasis. *Microb Pathogen* 22, 187–192.

Ayas, N.T., White, D.P., Al-Delaimy, W.K., Manson, J.E., Stampfer, M.J., Speizer, F.E., Patel, S., and Hu, F.B. (2003) A prospective study of self-reported sleep duration and incident diabetes in women. *Diabetes Care* 26, 380–384.

Balkwill, F., and Coussens, L.M. (2004) Cancer: An inflammatory link. *Nature* 431, 405–406.

Barber, R.C., Aragaki, C.C., Rivera-Chavez, F.A., Purdue, G.F., Hunt, J.L., and Horton, J.W. (2004) TLR4 and TNF-alpha polymorphisms are associated with an increased risk for severe sepsis following burn injury. *J Med Genet* 41, 808–813.

Baumann, C.R., Khatami, R., Werth, E., and Bassetti, C.L. (2006) Hypocretin (orexin) deficiency predicts severe objective excessive daytime sleepiness in narcolepsy with cataplexy. *J Neurol Neurosurg Psychiatry* 77, 402–404.

Ben Baruch, A. (2006) Inflammation-associated immune suppression in cancer: The roles played by cytokines, chemokines and additional mediators. *Semin Cancer Biol* 16, 38–52.

Berglund, G., Bolund, C., Fornander, T., Rutqvist, L.E., and Sjoden, P.O. (1991) Late effects of adjuvant chemotherapy and postoperative radiotherapy on quality of life among breast cancer patients. *Eur J Cancer* 27, 1075–1081.

Bessler, H., Osovsky, M., and Sirota, L. (2004) Association between IL-1ra gene polymorphism and premature delivery. *Biol Neonate* 85, 179–183.

Bjorntorp, P., and Rosmond, R. (2000) The metabolic syndrome—A neuroendocrine disorder? *Br J Nutr* 83(Suppl 1), S49–S57.

Blackburn, C.C., Griffith, J., and Morahan, G. (1995) A high resolution map of the chromosomal region surrounding the nude gene. *Genomics* 26, 308–317.

Boethel, C.D. (2002) Sleep and the endocrine system: New associations to old diseases. *Curr Opin Pulmon Med* 8, 502–505.

Bonkovsky, H.L., and Mehta, S. (2001) Hepatitis C: A review and update. *J Am Acad Dermatol* 44, 159–182.

Bourguignon, C., Labyak, S.E., and Taibi, D. (2003) Investigating sleep disturbances in adults with rheumatoid arthritis. *Holist Nurs Pract* 17, 241–249.

Bredow, S., Guba-Thakurta, N., Taishi, P., Obal, F., and Krueger, J.M. (1997) Diurnal variations of tumor necrosis factor alpha mRNA and alpha-tubulin mRNA in rat brain. *Neuroimmunomodulation* 4, 84–90.

Buchwald, D., Pascualy, R., Bombardier, C., and Kith, P. (1994) Sleep disorders in patients with chronic fatigue. *Clin Infect Dis* 18(Suppl 1), S68–S72.

Buchwald, D., Pearlman, T., Umali, J., Schmaling, K., and Katon, W. (1996) Functional status in patients with chronic fatigue syndrome, other fatiguing illnesses, and healthy individuals. *Am J Med* 171, 364–370.

Bussfeld, D., Kaufmann, A., Meyer, R.G., Gemsa, D., and Sprenger, H. (1998) Differential mononuclear leukocyte attracting chemokine production after stimulation with active and inactivated influenza A virus. *Cell Immunol* 186, 1–7.

Cannon, J.G. (2000) Inflammatory cytokines in nonpathological states. *News Physiol Sci* 15, 298–303.

Cantagrel, A., Navaux, F., Loubet-Lescoulie, P., Nourhashemi, F., Enault, G., Abbal, M., Constantin, A., Laroche, M., and Mazieres, B. (1999) Interleukin-1 beta, interleukin-1 receptor antagonist, interleukin-4, and interleukin-10 gene polymorphisms—Relationship to occurrence and severity of rheumatoid arthritis. *Arthritis Rheum* 42, 1093–1100.

Charlton, M.R., Pockros, P.J., and Harrison, S.A. (2006) Impact of obesity on treatment of chronic hepatitis C. *Hepatology* 43, 1177–1186.

Child, N.J., Yang, I.A., Pulletz, M.C., Courcy-Golder, K., Andrews, A.L., Pappachan, V.J., and Holloway, J.W. (2003) Polymorphisms in Toll-like receptor 4 and the systemic inflammatory response syndrome. *Biochem Soc Trans* 31, 652–653.

Cleare, A.J. (2003) The neuroendocrinology of chronic fatigue syndrome. *Endocr Rev* 24, 236–252.

Cleeland, C.S., Bennett, G.J., Dantzer, R., Dougherty, P.M., Dunn, A.J., Meyers, C.A., Miller, A.H., Payne, R., Reuben, J.M., Wang, X.S., and Lee, B.N. (2003) Are the symptoms of cancer and cancer treatment due to a shared biologic mechanism? A cytokine-immunologic model of cancer symptoms. *Cancer* 97, 2919–2925.

Cook, D.N., Beck, M.A., Coffman, T.M., Kirby, S.L., Sheridan, J.F., Pragnell, I.B., and Smithies, O. (1995) Requirement of MIP-1 alpha for an inflammatory response to viral infection. *Science* 269, 1583–1585.

Cunningham, W.E., Shapiro, M.F., Hays, R.D., Dixon, W.J., Visscher, B.R., George, W.L., Ettl, M.K., and Beck, C.K. (1998) Constitutional symptoms and health-related quality of life in patients with symptomatic HIV disease. *Am J Med* 104, 129–134.

Dandona, P., Aljada, A., and Bandyopadhyay, A. (2004) Inflammation: The link between insulin resistance, obesity and diabetes. *Trends Immunol* 25, 4–7.

Dandona, P., Aljada, A., Chaudhuri, A., Mohanty, P., and Garg, R. (2005) Metabolic syndrome: A comprehensive perspective based on interactions between obesity, diabetes, and inflammation. *Circulation* 111, 1448–1454.

Darko, D.F., McCutchan, J.A., Kripke, D.F., Gillin, J.C., and Golshan, S. (1992) Fatigue, sleep disturbance, disability, and indices of progression of HIV infection. *Am J Psychiatry* 149, 514–520.

Darko, D.F., Mitler, M.M., and Henriksen, S.J. (1995) Lentiviral infection, immune response peptides and sleep. *Adv Neuroimmunol* 5, 57–77.

Darko, D.F., Mitler, M.M., and Miller, J.C. (1998) Growth hormone, fatigue, poor sleep, and disability in HIV infection. *Neuroendocrinology* 67, 317–324.

Dauvilliers, Y., Maret, S., and Tafti, M. (2005) Genetics of normal and pathological sleep in humans. *Sleep Med Rev* 9, 91–100.

Dauvilliers, Y., Mayer, G., Lecendreux, M., Neidhart, E., Peraita-Adrados, R., Sonka, K., Billiard, M., and Tafti, M. (2002) Kleine-Levin syndrome: An autoimmune hypothesis based on clinical and genetic analyses. *Neurology* 59, 1739–1745.

Davis, G.L., Balart, L.A., Schiff, E.R., Lindsay, K., Bodenheimer, H.C., Perrillo, R.P., Carey, W., Jacobsen, I.M., Payne, J., Dienstag, J.L., VanThiel, D.H., Tamburro, C., Martino, F.P., Sangvhi, B., and Albrecht, J.K. (1994) Assessing health-related quality of life in chronic hepatitis C using the sickness impact profile. *Clin Ther* 16, 334–343.

Davis, S. (2004) Clinical sequelae affecting quality of life in the HIV-infected patient. *J Assoc Nurses AIDS Care* 15, 28S–33S.

De Maeyer, E., and De Maeyer-Guignard, J. (1970) A gene with quantitative effect on circulating interferon induction—Further studies. *Ann N Y Acad Sci* 173, 228–238.

De Maio, A., Mooney, M.D., Matesic, L.E., Paidas, C.N., and Reeves, R.H. (1998) Genetic component in the inflammatory response induced by bacterial lipopolysaccharide. *Shock* 10, 319–323.

De Sarro, G., Gareri, P., Sinopoli, A., David, E., and Rotiroti, D. (1997) Comparative, behavioural and electrocortical effects of tumor necrosis factor-a and interleukin-1 microinjected into the locus coeruleus of rat. *Life Sci* 60, 555–564.

Dinarello, C.A. (2006) Interleukin 1 and interleukin 18 as mediators of inflammation and the aging process. *Am J Clin Nutr* 83, 447S–455S.

Ding, M., and Toth, L.A. (2006) mRNA expression in mouse hypothalamus and basal forebrain during influenza infection: a novel model for sleep regulation. *Physiol Genom* 24 (3): 225–234.

Ebell, M.H. (2004) Epstein-Barr virus infectious mononucleosis. *Am Fam Physician* 70, 1279–1287.

El Omar, E.M., Chow, W.H.M., Mccoll, K.E., Fraumeni, J.F., and Rabkin, C.S. (2000) Interleukin-1 beta-enhancing genotypes are associated with increased risk of gastric cancer and its precursors. *Gastroenterology* 118, A181.

Elenkov, I.J., Iezzoni, D.G., Daly, A., Harris, A.G., and Chrousos, G.P. (2005) Cytokine dysregulation, inflammation and well-being. *Neuroimmunomodulation* 12, 255–269.

Elmasry, A., Janson, C., Lindberg, E., Gislason, T., Tageldin, M.A., and Boman, G. (2000) The role of habitual snoring and obesity in the development of diabetes: A 10-year follow-up study in a male population. *J Intern Med* 248, 11–20.

Elmasry, A., Lindberg, E., Berne, C., Janson, C., Gislason, T., Awad Tageldin, M., and Boman, G. (2001) Sleep-disordered breathing and glucose metabolism in hypertensive men: A population-based study. *J Intern Med* 249, 153–161.

Ershler, W.B., and Keller, E.T. (2000) Age-associated increased interleukin-6 gene expression, late-life diseases, and frailty. *Annu Rev Med* 51, 245–270.

Fang, J., Wang, Y., and Krueger, J.M. (1997) Mice lacking the TNF 55 kDa receptor fail to sleep more after TNFalpha treatment. *J Neurosci* 17, 5949–5955.

Fantuzzi, G. (2005) Adipose tissue, adipokines, and inflammation. *J Allergy Clin Immunol* 115, 911–919.

Fernandez-Real, J.M. (2006) Genetic predispositions to low-grade inflammation and type 2 diabetes. *Diabetes Technol Ther* 8, 55–66.

Fincher, E.F., Johannsen, L., Kapás, L., Takahashi, S., and Krueger, J.M. (1996) Microglia digest *Staphylococcus aureus* into low molecular weight biologically active compounds. *Am J Physiol* 271, R149–R156.

Finegood, D.T. (2003) Obesity, inflammation and type II diabetes. *Int J Obes Relat Metab Disord* 27 Suppl 3, S4–S5.

Floyd, R.A., and Krueger, J.M. (1997) Diurnal variation of TNF alpha in the rat brain. *NeuroReport* 3, 915–918.

Forton, D.M., Taylor-Robinson, S.D., and Thomas, H.C. (2006) Central nervous system changes in hepatitis C virus infection. *Eur J Gastroenterol Hepatol* 18, 333–338.

Franken, P., Chollet, D., and Tafti, M. (2001a) The homeostatic regulation of sleep need is under genetic control. *J Neurosci* 21, 2610–2621.

Franken, P., Chollet, D., and Tafti, M. (2001b) The homeostatic regulation of sleep need is under genetic control. *J Neurosci* 21, 2610–2621.

Franken, P., Malafosse, A., and Tafti, M. (1998) Genetic variation in EEG activity during sleep in inbred mice. *Am J Physiol* 275, R1127–R1137.

Fukuda, K., Straus, S.E., Hickie, I., Sharpe, M.C., Dobbins, J.G., and Komaroff, A.L. (1994) The chronic fatigue syndrome: A comprehensive approach to its definition and study (International Chronic Fatigue Study Group). *Ann Intern Med* 121, 953–999.

Gabay, C., and Kushner, I. (1999) Acute-phase proteins and other systemic responses to inflammation. *N Engl J Med* 340, 448–454.

Gedda, L. (1951) *Studio dei Gemelli*. Orizzonte Medico, Roma.

Gedda L., and Brenci G. (1983) Twins living apart test: Progress report. *Acta Genet Med Gemellol (Roma)* 32, 17–22.

Gerhardt, C.C., Romero, I.A., Cancello, R., Camoin, L., and Strosberg, A.D. (2001) Chemokines control fat accumulation and leptin secretion by cultured human adipocytes. *Mol Cell Endocrinol* 175, 81–92.

Gerli, R., Monti, D., Bistoni, O., Mazzone, A.M., Peri, G., Cossarizza, A., Di Gioacchino, M., Cesarotti, M.E., Doni, A., Mantovani, A., Franceschi, C., and Paganelli, R. (2000) Chemokines, sTNF-Rs and sCD30 serum levels in healthy aged people and centenarians. *Mech Ageing Dev* 121, 37–46.

Geyer, H. (1937) Ueber den Schlaf von Zwillingen. *Z Indukt Abstamm Verebungsl* 78, 524–527.

Giuliani, N., Sansoni, P., Girasole, G., Vescovini, R., Passeri, G., Passeri, M., and Pedrazzoni, M. (2001) Serum interleukin-6, soluble interleukin-6 receptor and soluble gp130 exhibit different patterns of age- and menopause-related changes. *Exp Gerontol* 36, 547–557.

Glaser, R., and Kiecolt-Glaser, J.K. (1998) Stress-associated immune modulation: Relevance to viral infections and chronic fatiigue syndrome. *Am J Med* 105, 35S–42S.

Glaser, R., Padgett, D.A., Litsky, M.L., Baiocchi, R.A., Yang, E.V., Chen, M., Yeh, P.E., Klimas, N.G., Marshall, G.D., Whiteside, T., Herberman, R., Kiecolt-Glaser, J., and Williams, M.V. (2005) Stress-associated changes in the steady-state expression of latent Epstein-Barr virus: Implications for chronic fatigue syndrome and cancer. *Brain Behav Immun* 19, 91–103.

Godfrey, P., Rahal, J.O., Beamer, W.G., Copeland, N.G., Jenkins, N.A., and Mayo, K.E. (1993) GHRH receptor of little mice contains a missense mutation in the extracellular domain that disrupts receptor function. *Nat Genet* 4, 227–232.

Gong, J.H., Sprenger, H., Hinder, F., Bender, A., Schmidt, A., Horch, S., Nain, M., and Gemsa, D. (1991) Influenza A virus infection of macrophages. Enhanced tumor necrosis factor-alpha (TNF-alpha) gene expression and lipopolysaccharide-triggered TNF-alpha release. *J Immunol* 147, 3507–3513.

Gottlieb, D.J., DeStefano, A.L., Foley, D.J., Mignot, E., Redline, S., Givelber, R.J., and Young, T. (2004) APOE epsilon4 is associated with obstructive sleep apnea/hypopnea: The Sleep Heart Health Study. *Neurology* 63, 664–668.

Guan, Z., Vgontzas, A.N., Omori, T., Peng, X., Bixler, E.O., and Fang, J. (2005) Interleukin-6 levels fluctuate with the light-dark cycle in the brain and peripheral tissues in rats. *Brain Behav Immun* 19, 526–529.

Gudewill, S., Pollmächer, T., Vedder, H., Schreiber, W., Fassbender, K., and Holsboer, F. (1992) Nocturnal plasma levels of cytokines in healthy men. *Eur Arch Psychiatr Clin Neurosci* 242, 53–56.

Guilleminault, C., and Mondini, S. (1986) Mononucleosis and chronic daytime sleepiness: A longterm follow-up study. *Arch Intern Med* 146, 1333–1335.

Haffner, S.M., Ruilope, L., Dahlof, B., Abadie, E., Kupfer, S., and Zannad, F. (2006) Metabolic syndrome, new onset diabetes, and new end points in cardiovascular trials. *J Cardiovasc Pharmacol* 47, 469–475.

Hamajima, N., Matsuo, K., Saito, T., Tajima, K., Okuma, K., Yamao, K., and Tominaga, S. (2001) Interleukin 1 polymorphisms, lifestyle factors, and Helicobacter pylori infection. *Jpn J Cancer Res* 92, 383–389.

Hartel, C., von Puttkamer, J., Gallner, F., Strunk, T., and Schultz, C. (2004) Dose-dependent immunomodulatory effects of acetylsalicylic acid and indomethacin in human whole blood: Potential role of cyclooxygenase-2 inhibition. *Scand J Immunol* 60, 412–420.

Heath, A.C., Kendler, K.S., Eaves, L.J., and Martin, N.G. (1990) Evidence for genetic influences on sleep disturbance and sleep pattern in twins. *Sleep* 13, 318–335.

Hector, R.F., Domer, J.E., and Carrow, E.W. (1982) Immune response to *Candida albicans* in genetically distinct mice. *Infect Immun* 38, 1020–1028.

Heesen, M., Kunz, D., Bachmann-Mennenga, B., Merk, H.F., and Bloemeke, B. (2003) Linkage disequilibrium between tumor necrosis factor (TNF)-alpha-308 G/A promoter and TNF-beta NcoI polymorphisms: Association with TNF-alpha response of granulocytes to endotoxin stimulation. *Crit Care Med* 31, 211–214.

Hofmann, P., Sprenger, H., Kaufmann, A., Bender, A., Hasse, C., Nain, M., and Gemsa, D. (1997) Susceptibility of mononuclear phagocytes to influenza A virus infection and possible role in the antiviral response. *J Leukoc Biol* 61, 408–414.

Ip, M.S.M., Lam, B., Ng, M.M.T., Lam, W.K., Tsang, K.W.T., and Lam, K.S.L. (2002) Obstructive sleep apnea is independently associated with insulin resistance. *Am J Respir Crit Care Med* 166, 670–676.

Jenkins, J.B., Omori, T., Guan, Z., Vgontzas, A.N., Bixler, E.O., and Fang, J. (2005) Sleep is increased in mice with obesity induced by high-fat food. *Physiol Behav* 87, 255–262.

Johannsen, L., Wecke, J., Obál, F., and Krueger, J.M. (1991) Macrophages produce somnogenic and pyrogenic muramyl peptides during digestion of staphylococci. *Am J Physiol* 260, R126–R133.

Jones, C.R., and Ptacek, L.J. (1999) Familial advanced sleep phase syndrome: An autosomal dominant circadian rhythm variant in humans. *Neurology* 52, A109–A110.

Jones, J.F., Williams, M., and Schooley, R.T. (1988) Antibodies to Epstein-Barr virus-specific DNase and DNA polymerase in the chronic fatigue syndrome. *Arch Intern Med* 148, 1957–1960.

Josephs, S.F., Henry, B., Balachandran, N., Strayer, D., Peterson, D., Komaroff, A.L., and Ablashi, D.V. (1991) HHV-6 reactivation in chronic fatigue syndrome. *Lancet* 337, 1346–1347.

Juji, T., Satake, M., Honda, Y., and Doi, Y. (1984) HLA antigens in Japanese patients with narcolepsy. All the patients were DR2 positive. *Tissue Antigens* 24, 316–319.

Kapsimalis, F., Richardson, G., Opp, M.R., and Kryger, M. (2005) Cytokines and normal sleep. *Curr Opin Pulm Med* 11, 481–484.

Kimura-Takeuchi, M., Majde, J.A., Toth, L.A., and Krueger, J.M. (1992a) Influenza virus-induced changes in rabbit sleep and acute phase responses. *Am J Physiol* 263, R1115–R1121.

Kimura-Takeuchi, M., Majde, J.A., Toth, L.A., and Krueger, J.M. (1992b) The role of double-stranded RNA in induction of the acute-phase response in an abortive influenza virus infection model. *J Infect Dis* 166, 1266–1275.

Knobler, H., and Schattner, A. (2005) TNF-{alpha}, chronic hepatitis C and diabetes: A novel triad. *QJM* 98, 1–6.

Knobler, H., Schihmanter, R., Zifroni, A., Fenakel, G., and Schattner, A. (2000) Increased risk of type 2 diabetes in noncirrhotic patients with chronic hepatitis C virus infection. *Mayo Clin Proc* 75, 355–359.

Kornman, K.S. (2006) Interleukin 1 genetics, inflammatory mechanisms, and nutrigenetic opportunities to modulate diseases of aging. *Am J Clin Nutr* 83, 475S–483S.

Korth, C., Mullington, J., Schreiber, W., and Pollmacher, T. (1996) Influence of endotoxin on daytime sleep in humans. *Infect Immun* 64, 1110–1115.

Kritchevsky, S.B., Cesari, M., and Pahor, M. (2005) Inflammatory markers and cardiovascular health in older adults. *Cardiovasc Res* 66, 265–275.

Krueger, J.M., and Toth, L.A. (1994) Cytokines as regulators of sleep. *Ann N Y Acad Sci* 739, 299–310.

Kuhns, D.B., Priel, D.A.L., and Gallin, J.I. (1997) Endotoxin and IL-1 hyporesponsiveness in a patient with recurrent bacterial infections. *J Immunol* 158, 3959–3964.

Kulbe, H., Hagemann, T., Szlosarek, P.W., Balkwill, F.R., and Wilson, J.L. (2005) The inflammatory cytokine tumor necrosis factor-alpha regulates chemokine receptor expression on ovarian cancer cells. *Cancer Res* 65, 10355–10362.

Lancel, M., Cronlein, J., Muller-Preuss, P., and Holsboer, F. (1995) Lipopolysaccharide increases EEG delta activity within non-REM sleep and disrupts sleep continuity in rats. *Am J Physiol* 268, R1310–R1318.

Lang, C.A., Conrad, S., Garrett, L., Battistutta, D., Cooksley, W.G., Dunne, M.P., and Macdonald, G.A. (2006) Symptom prevalence and clustering of symptoms in people living with chronic hepatitis C infection. *J Pain Symptom Manage* 31, 335–344.

Laposky, A.D., Shelton, J., Bass, J., Dugovic, C., Perrino, N., and Turek, F.W. (2006) Altered sleep regulation in leptin-deficient mice. *Am J Physiol* 290, R1–R10.

Larkin, E.K., Patel, S.R., Redline, S., Mignot, E., Elston, R.C., and Hallmayer, J. (2006) Apolipoprotein E and obstructive sleep apnea: Evaluating whether a candidate gene explains a linkage peak. *Genet Epidemiol* 30, 101–110.

Lashley, F.R. (2003) A review of sleep in selected immune and autoimmune disorders. *Holist Nurs Pract* 17, 65–80.

Lee, B.N., Dantzer, R., Langley, K.E., Bennett, G.J., Dougherty, P.M., Dunn, A.J., Meyers, C.A., Miller, A.H., Payne, R., Reuben, J.M., Wang, X.S., and Cleeland, C.S. (2004) A cytokine-based neuroimmunologic mechanism of cancer-related symptoms. *Neuroimmunomodulation* 11, 279–292.

Leproult, R., Copinschi, G., Buxton, O., and Van Cauter, E. (1997) Sleep loss results in an elevation of cortisol levels in the next evening. *Sleep* 20, 865–870.

Lin, L., Faraco, J., Li, R., Kadotani, H., Rogers, W., Lin, X.Y., Qiu, X.H., de Jong, P.J., Nishino, S., and Mignot, E. (1999) The sleep disorder canine narcolepsy is caused by a mutation in the hypocretin (orexin) receptor 2 gene. *Cell* 98, 365–376.

Lin, L., Hungs, M., and Mignot, E. (2001) Narcolepsy and the HLA region. *J Neuroimmunol* 117, 9–20.

Lin, S.-C., Lin, C.R., Gukovsky, I., Lusis, A.J., Sawchenko, P.E., and Rosenfeld, M.G. (1993) Molecular basis of the *little* mouse phenotype and implications for cell type-specific growth. *Nature* 364, 208–213.

Linkowski, P. (1999) EEG sleep patterns in twins. *J Sleep Res* 8(Suppl 1), 11–13.

Loffreda, S., Yang, S.Q., Lin, H.Z., Karp, C.L., Brengman, M.L., Wang, D.J., Klein, A.S., Bulkley, G.B., Bao, C., Noble, P.W., Lane, M.D., and Diehl, A.M. (1998) Leptin regulates proinflammatory immune responses. *FASEB J* 12, 57–65.

Lorenz, E., Mira, J.P., Frees, K.L., and Schwartz, D.A. (2002) Relevance of mutations in the TLR4 receptor in patients with gram-negative septic shock. *Arch Intern Med* 162, 1028–1032.

Lu, B., Rutledge, B.J., Gu, L., Fiorillo, J., Lukacs, N.W., Kunkel, S.L., North, R., Gerard, C., and Rollins, B.J. (1998) Abnormalities in monocyte recruitment and cytokine expression in monocyte chemoattractant protein 1-deficient mice. *J Exp Med* 187, 601–608.

Lugaresi, E., Medori, R., Montagna, P., Baruzzi, A., Cortelli, P., Lugaresi, A., Tinuper, P., Zucconi, M., and Gambetti, P. (1986) Fatal familial insomnia and dysautonomia with selective degeneration of thalamic nuclei. *N Engl J Med* 315, 997–1003.

Maggio, M., Guralnik, J.M., Longo, D.L., and Ferrucci, L. (2006) Interleukin-6 in aging and chronic disease: A magnificent pathway. *J Gerontol A Biol Sci Med Sci* 61, 575–584.

Maret, S., Franken, P., Dauvilliers, Y., Ghyselinck, N.B., Chambon, P., and Tafti, M. (2005) Retinoic acid signaling affects cortical synchrony during sleep. *Science* 310, 111–113.

Marquis, G., Montplaisir, S., Pelletier, M., Auger, P. and Lapp, W.S. (1988) Genetics of resistance to infection with candida-albicans in mice. *Br J Exp Pathol* 69, 651–660.

Marquis, G., Montplaisir, S., Pelletier, M., Mousseau, S., and Auger, P. (1986) Strain-dependent differences in susceptibility of mice to experimental candidosis. *J Infect Dis* 154, 906–909.

Marshall, L., and Born, J. (2002) Brain-immune interactions in sleep. *Int Rev Neurobiol* 52, 93–131.

Martin, W.J. (1996) Severe stealth virus encephalopathy following chronic-fatigue-syndrome-like illness: Clinical and histopathological features. *Pathobiol* 64, 1–8.

Mathias, S., Schiffelholz, T., Linthorst, A.C., Pollmacher, T., and Lancel, M. (2000) Diurnal variations in lipopolysaccharide-induced sleep, sickness behavior and changes in corticosterone levels in the rat. *Neuroendocrinology* 71, 375–385.

Mattson, K., Niiranen, A., Iivanainen, M., Farkkila, M., Bergstrom, L., Holsti, L.R., Kauppinen, H.L., and Cantell, K. (1983) Neurotoxicity of interferon. *Cancer Treat Rep* 67, 958–961.

Mcgeer, P.L., and Mcgeer, E.G. (2001) Polymorphisms in inflammatory genes and the risk of Alzheimer disease. *Arch Neurol* 58, 1790–1792.

Medori, R., Tritschler, H.J., Leblanc, A., Villare, F., Manetto, V., Chen, H.Y., Xue, R., Leal, S., Montagna, P., Cortelli, P., Tinuper, P., Avoni, P., Mochi, M., Baruzzi, A., Hauw, J.J., Ott, J., Lugaresi, E., Autiliogambetti, L., and Gambetti, P. (1992) Fatal familial insomnia, a prion disease with a mutation at Codon-178 of the prion protein gene. *N Engl J Med* 326, 444–449.

Mehta, S.H., Brancati, F.L., Sulkowski, M.S., Strathdee, S.A., Szklo, M., and Thomas, D.L. (2000) Prevalence of type 2 diabetes mellitus among persons with hepatitis C virus infection in the United States. *Ann Intern Med* 133, 592–599.

Mignot, E., Wang, C., Rattazzi, C., Gaiser, C., Lovett, M., Guilleminault, C., Dement, W.C., and Grumet, F.C. (1991) Genetic-linkage of autosomal recessive canine narcolepsy with A Mu-immunoglobulin heavy-chain switch-like segment. *Proc Natl Acad Sci U S A* 88, 3475–3478.

Mohamed-Ali, V., Goodrick, S., Rawesh, A., Katz, D.R., Miles, J.M., Yudkin, J.S., Klein, S., and Coppack, S.W. (1997) Subcutaneous adipose tissue releases interleukin-6, but not tumor necrosis factor-alpha, in vivo. *J Clin Endocrinol Metab* 82, 4196–4200.

Morriss, R., Sharpe, M., Sharpley, A.L., Cowen, P.J., Hawton, K., and Morris, J. (1993) Abnormalities of sleep in patients with the chronic fatigue syndrome. *Br Med J* 306, 1161–1164.

Morrow, G.R., Andrews, P.L., Hickok, J.T., Roscoe, J.A., and Matteson, S. (2002) Fatigue associated with cancer and its treatment. *Support Care Cancer* 10, 389–398.

Mullington, J., Korth, C., Hermann, D.M., Orth, A., Galanos, C., Holsboer, F., and Pollmacher, T. (2000) Dose-dependent effects of endotoxin on human sleep. *Am J Physiol Regul Integr Comp Physiol* 278, R947–R955.

Natelson, B.H., Ye, D., Moul, D.E., Jenkins, F.J., Oren, D.A., Tapp, W.N., and Cheng, Y.-C. (1994) High titers of anti-Epstein-Barr virus DNA polymerase are found in patients with severe fatiguing illness. *J Med Virol* 42, 42–46.

Nishino, S., Ripley, B., Overeem, S., Lammers, G.J., and Mignot, E. (2000) Hypocretin (orexin) deficiency in human narcolepsy. *Lancet* 355, 39–40.

Norman, S.E., Chediak, H.D., Kiel, M., and Cohn, M.A. (1990) Sleep disturbances in HIV-infected homosexual men. *AIDS* 4, 775–781.

O'Donnell, C.P., Schaub, C.D., Haines, A.S., Berkowitz, D.E., Tankersley, C.G., Schwartz, A.R., and Smith, P.L. (1999) Leptin prevents respiratory depression in obesity. *Am J Respir Crit Care Med* 159, 1477–1484.

Ogden, C.L., Carroll, M.D., Curtin, L.R., McDowell, M.A., Tabak, C.J., and Flegal, K.M. (2006) Prevalence of overweight and obesity in the United States, 1999–2004. *JAMA* 295, 1549–1555.

Opp, M.R., Kapás, L., and Toth, L.A. (1992) Cytokine involvement in the regulation of sleep. *Proc Soc Exp Biol Med* 201, 16–27.

Papanicolaou, D.A., Amsterdam, J.D., Levine, S., McCann, S.M., Moore, R.C., Newbrand C.H., Allen, G., Nisenbaum, R., Pfaff, D.W., Tsokos, G.C., Vgontzas, A.N., and Kales, A. (2004) Neuroendocrin aspects of chronic fatigue syndrome. *Neuroimmunomodulation* 11, 66–74.

Partinen, M., Kaprio, J., Koskenvuo, M., Putkonen, P., and Langinvainio, H. (1983) Genetic and environmental determination of human sleep. *Sleep* 6, 179–185.

Petersen, I., Thomas, J.M., Hamilton, W.T., and White, P.D. (2006) Risk and predictors of fatigue after infectious mononucleosis in a large primary-care cohort. *QJM* 99, 49–55.

Petrides, A.S., and DeFronzo, R.A. (1989) Glucose and insulin metabolism in cirrhosis. *J Hepatol* 8, 107–114.

Petrides, A.S., Schulze-Berge, D., Vogt, C., Matthews, D.E., and Strohmeyer, G. (1993) Glucose resistance contributes to diabetes mellitus in cirrhosis. *Hepatology* 18, 284-291.

Peyron, C., Faraco, J., Rogers, W., Ripley, B., Overeem, S., Charnay, Y., Nevsimalova, S., Aldrich, M., Reynolds, D., Albin, R., Li, R., Hungs, M., Pedrazzoli, M., Padigaru, M., Kucherlapati, M., Fan, J., Maki, R., Lammers, G.J., Bouras, C., Kucherlapati, R., Nishino, S., and Mignot, E. (2000) A mutation in a case of early onset narcolepsy and a generalized absence of hypocretin peptides in human narcoleptic brains. *Nat Med* 6, 991–997.

Plazzi, G., Montagna, P., Beelke, M., Nobili, L., De Carli, F., Cortelli, P., Vandi, S., Avoni, P., Tinuper, P., Gambetti, P., Lugaresi, E., and Ferrillo, F. (2002) Does the prion protein gene

129 codon polymorphism influence sleep? Evidence from a fatal familial insomnia kindred. *Clin Neurophysiol* 113, 1948–1953.

Pollmacher, T., Schuld, A., Kraus, T., Haack, M., Hinze-Selch, D., and Mullington, J. (2000) Experimental immunomodulation, sleep, and sleepiness in humans. *Ann N Y Acad Sci* 917, 488–499.

Poltorak, A., He, X.L., Smirnova, I., Liu, M.Y., Van Huffel, C., Du, X., Birdwell, D., Alejos, E., Silva, M., Galanos, C., Freudenberg, M., Ricciardi-Castagnoli, P., Layton, B., and Beutler, B. (1998) Defective LPS signaling in C3H/HeJ and C57BL/10ScCr mice: Mutations in Tlr4 gene. *Science* 282, 2085–2088.

Prinz, P.N. (2004) Age impairments in sleep, metabolic and immune functions. *Exp Gerontol* 39, 1739–1743.

Punjabi, N.M., Sorkin, J.D., Katzel, L.I., Goldberg, A.P., Schwartz, A.R., and Smith, P.L. (2002) Sleep-disordered breathing and insulin resistance in middle-aged and overweight men. *Am J Respir Crit Care Med* 165, 677–682.

Pusztai, L., Mendoza, T.R., Reuben, J.M., Martinez, M.M., Willey, J.S., Lara, J., Syed, A., Fritsche, H.A., Bruera, E., Booser, D., Valero, V., Arun, B., Ibrahim, N., Rivera, E., Royce, M., Cleeland, C.S., and Hortobagyi, G.N. (2004) Changes in plasma levels of inflammatory cytokines in response to paclitaxel chemotherapy. *Cytokine* 25, 94–102.

Qin, L.Q., Li, J., Wang, Y., Wang, J., Xu, J.Y., and Kaneko, T. (2003) The effects of nocturnal life on endocrine circadian patterns in healthy adults. *Life Sci* 73, 247–2475.

Qureshi, S.T., Lariviere, L., Sebastiani, G., Clermont, S., Skamene, E., Gros, P., and Malo, D. (1996) A high-resolution map in the chromosomal region surrounding the Lps locus. *Genomics* 31, 283–294.

Reid, S., and Dwyer, J. (2005) Insomnia in HIV infection: A systematic review of prevalence, correlates, and management. *Psychosom Med* 67, 260–269.

Resnick, H.E., Redline, S., Shahar, E., Gilpin, A., Newman, A., Walter, R., Ewy G.A., Howard, B.V., and Punjabi, N.M. (2003) Diabetes and sleep disturbances. *Diabetes Care* 26, 702–709.

Riha, R.L., Brander, P., Vennelle, M., McArdle, N., Kerr, S.M., Anderson, N.H., and Douglas, N.J. (2005) Tumour necrosis factor-alpha (-308) gene polymorphism in obstructive sleep apnoea-hypopnoea syndrome. *Eur Respir J* 26, 673–678.

Saarelainen, S., Lehtimaki, T., Nikkila, M., Solakivi, T., Nieminen, M.M., and Jaakkola, O. (2000) Association between apolipoprotein E alleles and autoantibodies against oxidised low-density lipoprotein. *Clin Chem Lab Med* 38, 477–478.

Sairenji, T., Yamanishi, K., Tachibana, Y., Bertoni, G., and Kurata, T. (1995) Antibody responses to Epstein-Barr virus, human herpesvirus 6 and human herpesvirus 7 in patients with chronic fatigue syndrome. *Intervirol* 38, 269–273.

Sakai, K., Takada, T., Nakayama, H., Kubota, Y., Nakamata, M., Satoh, M., Suzuki, E., Akazawa, K., and Gejyo, F. (2005) Serotonin-2A and 2C receptor gene polymorphisms in Japanese patients with obstructive sleep apnea. *Intern Med* 44, 928–933.

Saleh, M., Vaillancourt, J.P., Graham, R.K., Huyck, M., Srinivasula, S.M., Alnemri, E.S., Steinberg, M.H., Nolan, V., Baldwin, C.T., Hotchkiss, R.S., Buchman, T.G., Zehnbauer, B.A., Hayden, M.R., Farrer, L.A., Roy, S., and Nicholson, D.W. (2004) Differential modulation of endotoxin responsiveness by human caspase-12 polymorphisms. *Nature* 429, 75–79.

Sartipy, P., and Loskutoff, D.J. (2003) Monocyte chemoattractant protein 1 in obesity and insulin resistance. *Proc Natl Acad Sci U S A* 100, 7265–7270.

Savard, J., Simard, S., Hervouet, S., Ivers, H., Lacombe, L., and Fradet, Y. (2005) Insomnia in men treated with radical prostatectomy for prostate cancer. *Psychooncology* 14, 147–156.

Schenck, C.H., Garcia-Rill, E., Segall, M., Noreen, H., and Mahowald, M.W. (1996) HLA class II genes associated with REM sleep behavior disorder. *Ann Neurol* 39, 261–263.

Schiffelholz, T., and Lancel, M. (2001) Sleep changes induced by lipopolysaccharide in the rat are influenced by age. *Am J Physiol Regul Integr Comp Physiol* 280, R398–R403.

Shearer, W.T., Reuben, J.M., Mullington, J.M., Price, N.J., Lee, B.N., Smith, E.O., Szuba, M.P., Van Dongen, H.P.A., and Dinges, D.F. (2001) Soluble TNF-a receptor 1 and IL-6 plasma levels in humans subjected to the sleep deprivation model of space flight. *J Allergy Clin Immunol* 107, 165–170.

Smedley, H., Katrak, M., Sikora, K., and Wheeler, T. (1983) Neurological effects of human recombinant interferon. *Br Med J* 286, 262–265.

Smith, A. (1992) Sleep, colds, and performance. In: R.J. Broughton and R.D. Ogilvie (Eds.), *Sleep, Arousal and Performance*. Birkhauser, Boston, pp. 233–242.

Spath-Schwalbe, E., Hansen, K., Schmidt, F., Schrezenmeier, H., Marshall, L., Burger, K., Fehm, H.L., and Born, J. (1998) Acute effects of recombinant human interleukin-6 on endocrine and central nervous sleep functions in healthy men. *J Clin Endocrinol Metab* 83, 1573–1579.

Spath-Schwalbe, E., Lange, T., Perras, B., Horst, B., Fehm, H.L., and Born, J. (2000) Interferon-a acutely impairs sleep in healthy humans. *Cytokine* 12, 518–521.

Spiegel, K., Knutson, K., Leproult, R., Tasali, E., and Van Cauter, E. (2005a) Sleep loss: A novel risk factor for insulin resistance and Type 2 diabetes. *J Appl Physiol* 99, 2008–2019.

Spiegel, K., Leproult, R., and Van Cauter, E. (1999) Impact of sleep debt on metabolic and endocrine function. *Lancet* 354, 1435–1439.

Spiegel, K., Tasali, E., Penev, P., and Van Cauter, E. (2004) Sleep curtailment in healthy young men is associated with decreased leptin levels, elevated ghrelin levels, and increased hunger and appetite. *Ann Intern Med* 141, 846–850.

Spiegel, B.M., Younossi, Z.M., Hays, R.D., Revicki, D., Robbins, S., and Kanwal, F. (2005b) Impact of hepatitis C on health related quality of life: A systematic review and quantitative assessment. *Hepatology* 41, 790–800.

Strunk, T., and Burgner, D. (2006) Genetic susceptibility to neonatal infection. *Curr Opin Infect Dis* 19, 259–263.

Swanink, C.M.A., van der Meer, J.W.M., Vercoulen, J.H.M.M., Bleijenberg, G., Fennis, J.F.M., and Galama, J.M.D. (1995) Epstein-Barr virus (EBV) and the chronic fatigue syndrome: Normal virus load in blood and normal immunologic reacitivity in the EBV regression assay. *Clin Infect Dis* 20, 1390–1392.

Tafti, M., Franken, P., Kitahama, K., Malafosse, A., Jouvet, M., and Valatx, J.L. (1997) Localization of candidate genomic regions influencing paradoxical sleep in mice. *Neuroreport* 8, 3755–3758.

Tafti, M., Petit, B., Chollet, D., Neidhart, E., de Bilbao, F., Kiss, J.Z., Wood, P.A., and Franken, P. (2003) Deficiency in short-chain fatty acid beta-oxidation affects theta oscillations during sleep. *Nat Genet* 34, 320–325.

Takahashi, K., Mizuarai, S., Araki, H., Mashiko, S., Ishihara, A., Kanatani, A., Itadani, H., and Kotani, H. (2003) Adiposity elevates plasma MCP-1 levels leading to the increased CD11b-positive monocytes in mice. *J Biol Chem* 278, 46654–46660.

Temelkova-Kurktschiev, T., Henkel, E., Koehler, C., Karrei, K., and Hanefeld, M. (2002) Subclinical inflammation in newly detected Type II diabetes and impaired glucose tolerance. *Diabetologia* 45, 151.

Thannickal, T.C., Moore, R.Y., Nienhuis, R., Ramanathan, L., Gulyani, S., Aldrich, M., Cornford, M., and Siegel, J.M. (2000) Reduced number of hypocretin neurons in human narcolepsy. *Neuron* 27, 469–474.

Tobi, M., and Strauss, S.E. (1988) Chronic mononucleosis—A legitimate diagnosis. *Postgrad Med* 83, 69–78.

Toh, K.L., Jones, C.R., He, Y., Eide, E.J., Hinz, W.A., Virshup, D.M., Ptacek, L.J., and Fu, Y.H. (2001) An hPer2 phosphorylation site mutation in familiar advanced sleep phase syndrome. *Science* 291, 1040–1043.

Toth, L.A. (1996) Strain differences in the somnogenic effects of interferon inducers in mice. *J Interferon Cytokine Res* 16, 1065–1072.

Toth, L.A., and Hughes, L.F. (2004) Macrophage participation in influenza-induced sleep enhancement in C57BL/6J mice. *Brain Behav Immun* 18, 375–389.

Toth, L.A., Hughes, L.F. (2006) Sleep and temperature responses of inbred mice with Candida albicans-induced pyelonephritis. *Comp Med* 56(4): 252–261.

Toth, L.A., and Krueger, J.M. (1988) Alteration of sleep in rabbits by Staphylococcus aureus infection. *Infect Immun* 56, 1785–1791.

Toth, L.A., and Krueger, J.M. (1989) Effects of microbial challenge on sleep in rabbits. *FASEB J* 3, 2062–2066.

Toth, L.A., and Krueger, J.M. (1990a) Infectious disease, cytokines and sleep. In: M. Mancia and G. Marini (Eds.), *The Diencephalon and Sleep*. Raven Press, New York, pp. 331–341.

Toth, L.A., and Krueger, J.M. (1990b) Somnogenic, pyrogenic and hematologic effects of experimental pasteurellosis in rabbits. *Am J Physiol* 258, R536–R542.

Toth, L.A., Rehg, J.E., and Webster, R.G. (1995) Strain differences in sleep and other pathophysiological sequelae of influenza virus infection in naive and immunized mice. *J Neuroimmunol* 58, 89–99.

Toth, L.A., and Verhulst, S.J. (2003) Strain differences in sleep patterns of healthy and influenza-infected inbred mice. *Behav Genet* 33, 325–336.

Toth, L.A., and Williams, R.W. (1999a) A quantitative genetic analysis of locomotor activity in CXB recombinant inbred mice. *Behav Genet* 29, 319–328.

Toth, L.A., and Williams, R.W. (1999b) A quantitative genetic analysis of slow-wave sleep and rapid-eye movement sleep in CXB recombinant inbred mice. *Behav Genet* 29, 329–337.

Toth, L.A., and Williams, R.W. (1999c) A quantitative genetic analysis of slow-wave sleep in influenza-infected CXB recombinant inbred mice. *Behav Genet* 29, 339–348.

Tuite, A., Elias, M., Picard, S., Mullick, A., and Gros, P. (2005) Genetic control of suceptibility to Candida albicans in susceptible A/J and resistant C57BL/6J mice. *Genes Immun* 6, 672–682.

Uysal, K.T., Wiesbrock, S.M., Marino, M.W., and Hotamisligil, G.S. (1997) Protection from obesity-induced insulin resistance in mice lacking TNF-alpha function. *Nature* 389, 610–614.

Vallance, B.A., Deng, W., Jacobson, K., and Finlay, B.B. (2003) Host susceptibility to the attaching and effacing bacterial pathogen Citrobacter rodentium. *Infect Immun* 71, 3443–3453.

Van Cauter, E., Blackman, J.D., Roland, D., Spire, J.P., Refetoff, S., and Polonsky, K.S. (1991) Modulation of glucose regulation and insulin secretion by circadian rhythmicity and sleep. *J Clin Invest* 88, 834–942.

Vance, D.E., and Burrage, J.W., Jr. (2005) Sleep disturbances and psychomotor decline in HIV. *Percept Mot Skills* 100, 1004–1010.

Vettor, R., Milan, G., Rossato, M., and Federspil, G. (2005) Review article: Adipocytokines and insulin resistance. *Aliment Pharmacol Ther* 22(Suppl 2), 3–10.

Vgontzas, A.N., Bixler, E.O., Lin, H.M., Prolo, P., Trakada, G., and Chrousos, G.P. (2005) IL-6 and its circadian secretion in humans. *Neuroimmunomodulation* 12, 131–140.

Vgontzas, A.N., Papanicolaou, D.A., Bixler, E.O., Kales, A., Tyson, K., and Chrousos, G.P. (1997) Elevation of plasma cytokines in disorders of excessive daytime sleepiness: Role of sleep disturbance and obesity. *J Clin Endocrinol Metab* 82, 1313–1316.

Vogel, F. (1965) ["14 and 6/sec positive spikes" in the sleep EEG of young mono-and dizygotic twins]. *Humangenetik* 1, 390–391.

Vosvick, M., Gore-Felton, C., Ashton, E., Koopman, C., Fluery, T., Israelski, D., and Spiegel, D. (2004) Sleep disturbances among HIV-positive adults: The role of pain, stress, and social support. *J Psychosom Res* 57, 459–463.

Webb, W.B., and Campbell, S.S. (1983) Relationships in sleep characteristics of identical and fraternal twins. *Arch Gen Psychiatry* 40, 1093–1095.

Weisberg, S.P., Hunter, D., Huber, R., Lemieux, J., Slaymaker, S., Vaddi, K., Charo, I., Leibel, R.L., and Ferrante, A.W., Jr. (2006) CCR2 modulates inflammatory and metabolic effects of high-fat feeding. *J Clin Invest* 116, 115–124.

Weisberg, S.P., McCann, D., Desai, M., Rosenbaum, M., Leibel, R.L., and Ferrante, A.W., Jr. (2003) Obesity is associated with macrophage accumulation in adipose tissue. *J Clin Invest* 112, 1796–1808.

Wellen, K.E., and Hotamisligil, G.S. (2005) Inflammation, stress, and diabetes. *J Clin Invest* 115, 1111–1119.

White, P.D., Thomas, J.M., Amess, J., Crawford, D.H., Grover, S.A., Kangro, H.O., and Clare, A.W. (1998) Incidence, risk and prognosis of acute and chronic fatigue syndromes and psychiatric disorders after glandular fever. *Br J Psychiatr* 173, 475–481.

White, P.D., Thomas, J.M., Sullivan, P.F., and Buchwald, D. (2004) The nosology of sub-acute and chronic fatigue syndromes that follow infectious mononucleosis. *Psychol Med* 34, 499–507.

Wilson, J., and Balkwill, F. (2002) The role of cytokines in the epithelial cancer microenvironment. *Semin Cancer Biol* 12, 113–120.

Wilson, S.D., Billings, P.R., D'Eustachio, P., Fournier, R.E., Geissler, E., Lalley, P.A., Burd, P.R., Housman, D.E., Taylor, B.A., and Dorf, M.E. (1990) Clustering of cytokine genes on mouse chromosome 11. *J Exp Med* 171, 1301–1314.

Wilson, I., McNamara, S.G., and Collins, F.L. (1998) "Syndrome Z": The interaction of sleep apnoea, vascular risk factors and heart disease. *Thorax* 53 (Suppl 3), S25–S28.

Wong, W., and Terrault, N. (2005) Update on chronic hepatitis C. Clin. *Gastroenterol Hepatol* 3, 507–520.

World Health Organization (1999) Global surveillance and control of hepatitis C. Report of a WHO Consultation organized in collaboration with the Viral Hepatitis Prevention Board, Antwerp, Belgium. *J Viral Hepat* 6, 35–47.

Xu, H., Barnes, G.T., Yang, Q., Tan, G., Yang, D., Chou, C.J., Sole, J., Nichols, A., Ross, J.S., Tartaglia, L.A., and Chen, H. (2003) Chronic inflammation in fat plays a crucial role in the development of obesity-related insulin resistance. *J Clin Invest* 112, 1821–1830.

Yudkin, J.S., Kumari, M., Humphries, S.E., and Mohamed-Ali, V. (2000) Inflammation, obesity, stress and coronary heart disease: Is interleukin-6 the link? *Atherosclerosis* 148, 209–214.

Zhang, J., Zhao, B., Gesongluobu, Sun, Y.H., Wu, Y., Pei, W.D., Ye, J., Hui, R.T., and Liu, L.S. (2000) Angiotensin-converting enzyme gene insertion/deletion (I/D) polymorphism in hypertensive patients with different degrees of obstructive sleep apnea. *Hypertension Res* 23, 407–411.

15 Neuroimmune Activation in Sleep Apnea

Paul J. Mills, Michael G. Ziegler, and Joel E. Dimsdale

15.1 Introduction

Obstructive sleep apnea (OSA) is a highly prevalent sleep disorder, characterized by repeated disruptions of breathing during sleep. The sleep fragmentation and the accompanying hypoxemia lead to many negative consequences including cardiovascular diseases, cognitive impairment, daytime sleepiness, fatigue, and depressive symptoms (Parish and Somers 2004; Reimer and Flemons 2003). Originally viewed as an interesting but rare malady, OSA is now recognized as a common disorder that is associated with major morbidity and mortality (Newman et al. 2001).

Noradrenergic activation, as indicated by augmented sympathetic neural activity, elevated circulating norepinephrine (NE) levels, and elevated urinary NE excretion, is a hallmark of OSA (Carlson, Hedner, Elam, Ejnell, Sellgren, and Wallin 1993; Dimsdale, Coy, Ziegler, Ancoli-Israel, and Clausen 1995; Fletcher 2003; Eisensehr et al. 1998).

There are numerous physiologically relevant consequences of this noradrenergic activation, including adrenergic receptor desensitization (Grote, Hraiczi, and Hedner 2000; Mills, Dimsdale, Coy, Ancoli-Israel, Clausen, and Nelesen 1995; Mills, Dimsdale, Ancoli-Israel, Clausen, and Loredo 1998; Nelesen, Dimsdale, Mills, Clausen, and Ancoli-Israel 1996), high blood pressure (Norman et al. 2006; Robinson, Stradling, and Davies 2004), and elevations in proinflammatory cytokine levels (Vgontzas et al. 2000; Yokoe et al. 2003).

This chapter will review studies that our group and others have conducted documenting neuroimmune effects of OSA, as well as studies documenting the effects of successful treatment of OSA on these physiological systems.

15.2 Adrenergic Agonists and Receptors

Numerous studies repeatedly show that circulating NE levels and urinary NE excretion are elevated in OSA (Carlson et al. 1993; Dimsdale et al. 1995; Eisensehr et al. 1998). Typical values are shown in Fig. 15.1. In our studies over the years, we have defined OSA as the presence of a respiratory disturbance index (RDI) of >20 or, in more recent studies, an apnea hypopnia index (AHI) >15. RDI is the frequency of abnormal respiratory events per hour of sleep. AHI is defined as the total number

of apneas (brief pauses in breathing) and hypopneas (reductions in airflow) per hour
of sleep.

A reasonable question to pose is "What is the source of sympathetic nervous
system activation in OSA?" Is it a response to the respiratory disturbance itself, the

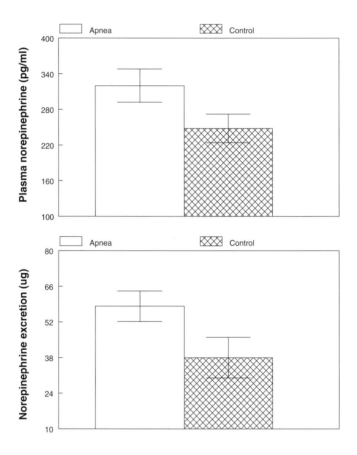

FIGURE 15.1. Plasma NE levels ($p < 0.01$) (top) and 24-h NE excretion ($p < 0.01$) (bottom) are
elevated in OSA (Dimsdale, Coy, Ancoli-Israel, Mills, Clausen, and Ziegler 1997; Ziegler,
Nelesen, Mills, Ancoli-Israel, Kennedy, and Dimsdale 1997).

resulting hypoxia, or the repeated disruption of sleep? We examined this question and found that 24-h NE excretion correlated with RDI ($r = 0.39$, $p < 0.01$) and with mean oxygen saturation ($r = -0.36$, $p < 0.05$) (Dimsdale et al. 1997). In a multiple regression model, these variables, together with amount of time spent in slow-wave sleep, accounted for a statistically significant (but modest percentage) of the variance in NE excretion ($R^2 = 0.19$, $p < 0.05$). Other studies examining this issue find evidence that the repeated hypoxia and sleep disturbance cause sympathetic activation and a subsequent increase in blood pressure. In a study of healthy subjects who performed voluntary end-expiratory (hypoxic) apneas, muscle sympathetic nerve activity (MSNA via peroneal microneurography) and blood pressure increased significantly and remained elevated (Leuenberger, Brubaker, Quraishi, Hogeman, Imadojemu, and Gray 2005). In subjects who had performed repetitive non-end-expiratory apneas, MSNA and blood pressure did not change. Thus, in this study, it was the repetitive end-expiratory hypoxic apneas that resulted in sustained sympathetic activation. Another study found that the degree of arousal in OSA was more significantly correlated with increased blood pressure than oxygen saturation (Yoon and Jeong 2001). Respiratory disturbance duration, on the other hand, was not related to the increase in blood pressure. The authors concluded that it is the degree of arousal, not hypoxia, that causes sympathetic activation in OSA. Thus, studies of diverse designs indicate that respiratory disturbance and hypoxia contribute to sympathetic activation in OSA.

NE exerts its physiological effects via β- and α-adrenergic receptors. We have used both in vitro and in vivo techniques to assess the effects of elevated circulating NE levels in OSA on adrenergic receptors. We have shown that as a result of elevated NE levels, individuals with OSA exhibit decreased lymphocyte β_2-adrenergic receptor sensitivity (Mills et al. 1995) (Fig. 15.2). Lymphocytes are often used as in vitro models of the β_2-adrenergic receptor because lymphocyte β_2-adrenergic receptors correlate well with β_2-adrenergic receptors on tissue such as the heart and lung (Brodde, Michel, Gordon, Sandoval, Gilbert, and Bristow 1989). In this in vitro technique, lymphocytes are isolated from whole blood and stimulated with a maximal dose of isoproterenol. β_2-Adrenergic receptor sensitivity is inferred from the amount of isoproterenol-stimulated cyclic AMP generated by the cell over nonstimulated levels.

An in vivo technique for assessing the β_1- and β_2-adrenergic receptors involves intravenously infusing incremental doses of isoproterenol (0.10, 0.25, 0.50, 1.0, 2.0, and 4.0 μg), measuring the resultant heart rate, and inferring the dose necessary to increase heart rate by 25 beats/min; this dose is termed the "Chronotropic 25 Dose" (CD_{25}) (Mills et al. 1998). We infused isoproterenol in a group of apneics and nonapneic controls while the subject was supine and while breathing room air. OSA patients had a higher CD_{25} than nonapneics, meaning it took more isoproterenol to increase their heart rate, equating to reduced cardiac β-adrenergic receptor sensitivity (Fig. 15.2).

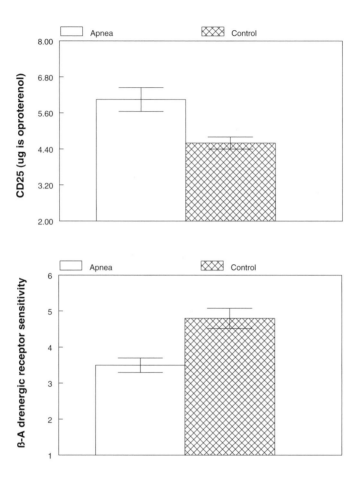

FIGURE 15.2. Lymphocyte β₂-adrenergic receptors are desensitized in OSA ($p \leq 0.01$) (top). CD_{25} is increased in OSA ($p \leq 0.01$) (bottom) indicating reduced cardiac β-adrenergic receptor sensitivity (Mills et al. 1995; Mills et al. 1998).

We concluded from these observations that adrenergic activation and repeated hypoxic events in apnea lead to β-adrenergic receptor desensitization in OSA.

In addition to the effects of catecholamines, there is a considerable literature on the effects of hypoxia on the β-adrenergic receptor. Studies on the effects of ischemia on cardiac β_1- and β_2-adrenergic receptors, for example, show that hypoxia down-regulates ventricular β-adrenergic receptors approximately 50% (Marsh and Sweeney 1989; Bernstein, Doshi, Huang, Strandness, and Jasper 1992). Based on this literature and the hypoxia typical of OSA, we extended our β_2-adrenergic receptor studies to examine the effects of hypoxia on β-adrenergic receptor sensitivity in OSA. We determined the CD_{25} while subjects breathed either room air (21% O_2, 79% N_2) or a hypoxic gas mixture (15% O_2, 85% N_2) for 10 min and then during the isoproterenol infusion protocol, which took approximately 25 min (Mills et al. 1998). Under normoxic conditions, apnea patients showed a significantly higher CD_{25} (lower β-adrenergic receptor sensitivity) as compared to controls. In response to hypoxia, apnea patients showed no change in CD_{25} while controls showed a significant increase in CD_{25} (β-adrenergic receptor desensitization) from 4.6 μg to a value comparable to the apneics (5.8 μg) ($p < 0.01$) (Fig. 15.2).

15.3 Norepinephrine Release and Clearance

What is the source of elevated NE levels in OSA? Venous and arterial NE levels reflect not only release, but metabolic degradation reuptake, diffusion, and regional and local circulation. Catecholamines are cleared by reuptake into nerves (referred to as uptake 1) and by uptake into nonneuronal tissue (referred to as uptake 2). NE is cleared by uptake 1 more than other catecholamines.

Tracer amounts of radiolabeled NE can be used to measure the release rate and the clearance rate of NE, which provide insight into the factors that contribute to circulating levels. We measured NE kinetics to determine whether the elevated NE levels in OSA result from an enhanced NE release rate and/or decreased NE clearance rate.

The basis of the NE kinetics technique is that if the rate at which NE is cleared from the plasma is known and the plasma level of NE is known, then the rate at which NE appears in the plasma can be calculated. This is known as the NE spillover or release rate. A relatively simple method for evaluating NE clearance rate and half-life is to infuse radiolabeled NE ($[^3H]$-NE) until constant blood levels are attained and then measure the disappearance rate of the $[^3H]$-NE in sequential blood samples (Ziegler, Kennedy, Morrissey, and O'Connor 1990). The release and clearance of NE can be calculated by the following formulae:

$$\text{NE clearance (l/min)} = \frac{^3\text{H-NE infused / min}}{^3\text{H-NE per liter plasma}}$$

$$\text{NE release rate(ng/min)} = \text{clearance} \times \text{plasma NE(ng/l)}$$

We measured NE clearance and release rates in supine OSA subjects using ^{3}H-NE of >98% radiochemical purity (New England Nuclear, Boston, MA). ^{3}H-NE was infused into an antecubital vein at 1.5 µCi/min for 10 min. The infusion rate was then decreased to 0.78 µCi/min. Plateau levels are obtained in less than 1 h with this technique. Blood samples were drawn from the antecubital vein of the contralateral arm to measure plasma NE and ^{3}H-NE levels during the infusion and for 16 min after the infusion was terminated. Plasma NE levels were measured by the radioenzymatic method of Kennedy and Ziegler (1990).

Since patients with OSA are often hypertensive, we measured NE clearance and release rate in a group of 65 apneics and hypertensives. Subjects were studied while breathing room air and a hypoxic gas mixture (Ziegler et al. 1997).

Consistent with prior studies, OSA patients had higher plasma NE across all conditions compared to non-OSA subjects ($p \leq 0.01$). We found that NE clearance increased significantly from 3.2 to 3.9 l/min when subjects breathed the hypoxic gas mixture ($p < 0.001$). Hypoxia also increased the NE-release rate from 892 to 1042 ng/min ($p < 0.001$) and increased the NE-release rate more among OSA patients than nonapneics ($p < 0.001$).

Normotensive apneics had the largest increase in NE release during hypoxia ($p < 0.01$). Apneics tended to have lower rates of NE clearance than nonapneics ($p < 0.08$) (Fig. 15.3).

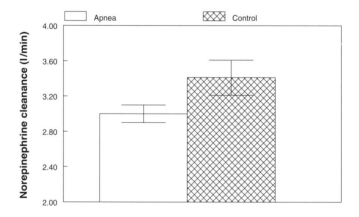

FIGURE 15.3. NE clearance in OSA and nonapnea. Patients with OSA tended toward lower rates of NE clearance than nonapneic controls (Ziegler et al. 1997).

The results of these NE kinetics studies suggested that individuals with OSA are subject to transient increases in sympathetic activity, that hypertensive apneics maintain increased sympathetic nervous release of NE in the daytime, and that individuals with OSA tend to have reduced NE clearance.

15.4 Cytokines in OSA

Akin to neurotransmitters of the nervous system (such as NE), cytokines are key mediators of vast immune and neuroimmune interactions. There are many features of OSA that argue for the importance of studying cytokines, including neuroimmune interactions, mood changes (such as fatigue and depression), and behaviors that directly affect the course of the disorder (such as caffeine consumption, smoking, diet, and difficulties with adherence to treatment). Before reviewing studies of OSA and cytokines, we first briefly discuss cytokines in normal sleep in order to put into context effects seen as a result of the disrupted sleep of OSA.

15.4.1 Cytokines and Normal Sleep

The contributions of cytokines to sleep are far from straightforward. They are both sleep inducing (e.g., IL-1β, TNF-α) and sleep inhibiting (e.g., IL-10 and IL-4), depending on the cytokine, the dose, and the circadian phase (Opp 2002).

Sleep inhibiting cytokines may exert their effects through antagonizing somnogenic cytokines. IL-10 and IL-4, for example, may inhibit sleep by inhibiting the production of IL-1β and TNF-α (Kelley et al. 2003; Krueger, Obal, Fang, Kubota and Taishi 2001). Administration of TNF-α, IL-1β, or IL-18 increases the amount of nonrapid eye movement (NREM) sleep time and decreases the duration of REM sleep. TNF-α or IL-1β also increase the amplitude of slow wave EEG, while administration of IL-10 and IL-4 on the other hand inhibit NREM. While circulating IL-2 levels increase during sleep, there is little evidence that there is a direct sleep promoting effect of IL-2. Although the precise mechanisms of the somnogenic or antisomnogenic effects of cytokines have yet to be fully elucidated, growth hormone releasing hormone, corticotrophin releasing hormone, prostaglandins, and molecular intermediates (e.g., activation of the DNA transcription binding protein NF-κB) have been implicated.

Like TNF-α, IL-6 is a somnogenic proinflammatory cytokine associated with disturbed sleep and with fatigue. IL-6 negatively correlates with the amount of sleep as well as the depth of sleep. Better sleep is associated with decreased daytime secretion of IL-6, while nocturnal sleep disturbances are associated with increased daytime levels of IL-6 and TNF-α. In older adults, elevated IL-6 levels are associated with poor sleep and sleep disturbances, particularly when accompanied by elevations in cortisol levels (Vgontzas et al. 2003). It has been postulated that whether somnogenic and fatigue-inducing proinflammatory cytokines lead to sleepiness and deep sleep or to fatigue and poor sleep depends upon whether there is simultaneous hypothalamic pituitary adrenal (HPA) axis activation (Vgontzas et al. 2003; Opp 1995).

15.4.2 Cytokines and OSA

Inflammation is common in OSA, including elevated levels of proinflammatory cytokines. We compared plasma IL-6 levels obtained in the early morning from 53 patients with OSA with levels obtained from 97 nonapneic individuals of similar age and weight. Apneics had significantly higher IL-6 levels ($p < 0.001$) (Fig. 15.4).

Other studies show that in addition to IL-6, TNF-α levels are elevated in OSA independent of obesity (Vgontzas et al. 2000; Vgontzas, Papanicolaou, Bixler, Kales, Tyson, and Chrousos 1997) and the circadian rhythm of TNF-α is disrupted (Entzian, Linnemann, Schlaak, and Zabel 1996).

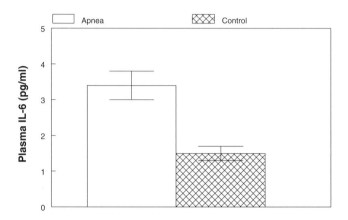

FIGURE 15.4. Plasma IL-6 levels in OSA and nonapnea. Patients with OSA have higher circulating IL-6 levels ($p < 0.001$).

In addition to proinflammatory cytokines, OSA leads to elevations of other mediators of inflammation, including intercellular adhesion molecule-1 (ICAM-1) and C-reactive protein (CRP) (Yokoe et al. 2003). CRP, a member of the pentraxin family of proteins, is an acute phase reactant marker of inflammation synthesized primarily in hepatocytes in response to IL-6. It is intimately involved in atherogenesis, and stimulates the release of proinflammatory cytokines, thereby inducing the expression of adhesion molecules such as ICAM-1. The elevated levels of TNF-α and IL-6 in OSA have been shown to be correlated with elevated CRP levels. Elevated levels of the soluble form of ICAM-1 (sICAM-1) and IL-6 are independently associated with increased morbidity and mortality independent of other established risk factors (Ridker, Hennekens, Roitman-Johnson, Stampfer, and Allen 1998). Thus, the disruption of sleep seen in OSA activates several different inflammatory pathways that may lead to increased susceptibility to cardiovascular diseases.

15.4.3 What Features of OSA Might Lead to Elevations of Inflammatory Cytokines?

As reviewed earlier in this chapter, studies repeatedly demonstrate that OSA is associated with sympathetic activation during both day and nighttime. Given the well-established link between sympathetic activation and elevated proinflammatory cytokines and systemic inflammation, perhaps the elevated proinflammatory cytokines levels observed in OSA are a byproduct of the increased sympathetic activity. As will be discussed later in the chapter, continuous positive airway pressure (CPAP) is a common treatment for OSA and reverses many of the clinical complaints and physiological features of OSA, including elevated NE, CRP, and IL-6 (Yokoe 2003).

Elevated cytokine levels in OSA might also be the result of hypoxia. OSA is associated with repetitive transient episodes of partial or complete obstruction of the upper airway during sleep and with transient drops in oxyhemoglobin saturation. High altitude studies show that exposure to chronic hypoxia leads to the release of inflammatory cytokines; including a several fold elevation of IL-6 levels (Hartmann et al. 2000).

The inflammatory reactions occurring in hypoxia, perhaps related to a recurring cycle of hypoxia/reoxygenation stress reminiscent of ischemia/reperfusion injury, might initiate a vicious cycle whereby the response is further amplified.

OSA is commonly associated with obesity, and abdominal fat is a major reservoir of cytokines. OSA is commonly seen in middle-aged obese men. Vgontzas and colleagues (1997) showed that OSA patients have a greater amount of visceral fat compared to obese non-OSA controls and higher levels of the adipose tissue-derived hormone leptin. The authors concluded that there is a strong independent association among OSA, visceral obesity, and elevated IL-6, TNF-α levels. Other findings from these investigators support the view that sleep apnea in obese patients may be a manifestation of the Metabolic Syndrome, in that cytokines and insulin resistance are mediators of excessive daytime sleepiness and sleep apnea in humans. They propose a model of a "bi-directional, feed forward, pernicious association between sleep apnea, sleepiness, inflammation, and insulin resistance, all promoting atherosclerosis and cardiovascular disease" (Vgontzas, Bixler, and Chrousos 2005).

Finally, in addition to sympathetic activation, studies suggest that the other branch of the autonomic nervous system, the parasympathetic nervous system, is related to inflammatory processes and elevated proinflammatory cytokines, but in the opposite fashion; i.e., parasympathetic activation supports anti-inflammatory pathways. Studies show that direct in vivo stimulation of the vagus nerve significantly attenuates the release of IL-6, TNF-α, and IL-1β (Borovikova et al. 2000). Our prior studies show that patients with sleep apnea have attenuated vagal activity (Nelesen, Yu, Ziegler, Mills, Clausen, and Dimsdale 2001). Thus, apneics may be susceptible to increased cytokine levels from both sides of the autonomic nervous system—increased sympathetic activity as well as decreased parasympathetic activity.

15.4.4 Cytokines in the Context of Mood and Quality of Life in OSA

We conclude this section of the chapter with a discussion of possible consequences of elevated cytokine levels and inflammation on the poor mood and low quality of life often observed in OSA.

OSA patients typically consult their doctor not because of "an elevated oxygen desaturation index" but because they are very sleepy, depressed or worried about their snoring. These complaints are subjective and thus aspects of personality may greatly influence patients' referral for treatment as well as compliance with treatment. Here we summarize some of our findings on the complexities of subjective report and mood in patients with this sleep disorder.

There is a growing literature linking cytokines and depressive symptoms. Perhaps some of the psychological symptoms experienced by OSA patients may be related to the elevated levels of proinflammatory cytokines. Apneics are noted for their increased incidence of depressive complaints, which may reflect elevations in cytokines.

Irwin (2001) has shown that the degree of sleep disturbance in depression is related to the degree of immune dysfunction. We have found that avoidant coping style, age, body mass index (BMI), and hypertension status all contribute to the depressive self-reports in OSA (Bardwell, Ancoli-Israel, and Dimsdale 2001).

OSA patients complain of fatigue. We find that OSA patients with high levels of depressive symptoms on the Center for Epidemiologic Studies Depression Scale (CESD > 16) report twice as much fatigue (Profile of Mood States) as OSA patients with fewer depressive symptoms.

We wondered if depressive symptoms in patients with OSA would account for some of the fatigue beyond that explained by OSA severity (Bardwell, Moore, Ancoli-Israel, and Dimsdale 2003). We found that higher levels of depressive symptoms in OSA are dramatically and independently associated with greater levels of fatigue: whereas the respiratory disturbance and the oxygen desaturation together accounted for 4.2% of the variance in fatigue scores in OSA patients, depressive symptoms accounted for 10 times the variance (i.e., an additional 42.3%) in fatigue scores. Thus, assessment and treatment of mood symptoms, not just treatment of the disordered breathing itself, might reduce the fatigue experienced by patients with OSA.

Another hallmark sign of OSA is debilitating sleepiness. Opp (2002) reviewed the literature on cytokines and sleep promotion, citing elevated IL-1, TNF and possibly IL-6 as mechanisms of excessive daytime sleepiness in OSA. Observations that the circadian rhythm of TNF-α is disrupted in OSA, with a loss of the nocturnal physiologic peaks and the presence of an additional daytime peak (Vgontzas et al. 1997; Entzian et al. 1996) support the notion that disrupted cytokines in OSA contribute to daytime sleepiness but not necessarily to nighttime sleep disruptions.

We have found that plasma sTNF-R1 levels are significantly correlated with sleepiness and impaired neuropsychological functioning in OSA. For instance, decreased performance on the Brief Visuospatial Memory Test and Trailmaking A Test is associated with increased levels of sTNF-R1 in OSA.

15.5 Nasal Continuous Positive Airway Pressure

A common and effective treatment for OSA is nasal continuous positive airway pressure (CPAP). When air is pumped at positive pressure through a facemask, the patient's airway is kept open, reducing the number of apneas (respiratory suspensions) and hypopneas (reductions in airflow and oxygenation during sleep) With CPAP, the majority of patients are able to get their first good night's sleep in years.

We have conducted several therapeutic treatment studies using CPAP to see if the positive effects on sleep are also accompanied by a reversal of the many negative physiological effects of OSA, including elevated NE levels and adrenergic receptor desensitization.

Patients were assessed prior to and following either 1 or 2 weeks of treatment. CPAP treatment begins with a manual overnight CPAP titration of increasing steps of 1 to 2 cm H_2O until unequivocal obstructive apneas or hypopneas are controlled.

We typically compare CPAP treatment to what we call placebo-CPAP (CPAP at an ineffective positive pressure to control apneic events). We have also compared CPAP to oxygen supplementation.

Nocturnal supplemental oxygen has been suggested by some as an alternative therapy in the nonsomnolent or the CPAP noncompliant OSA patient (Phillips, Schmitt, Berry, Lamb, Amin, and Cook 1990; Landsberg, Friedman, and Ascher-Landsberg 2001).

Equipment for the three treatment arms was similar and consisted of a CPAP generator (Aria LX CPAP System, Respironics Inc., Murrysville, PA), CPAP mask (Profile Light, Respironics Inc., Murrysville, PA) and tubing, heated humidifier (Fisher and Pykel HC199, Aukland, New Zealand) and oxygen concentrator (Alliance, Healthdyne Technologies Model 505, Marietta, GA). The concentrator can be switched to produce room air. The supplemental gas (room air or oxygen) is introduced into the CPAP system at the level of the humidifier.

In order to maintain the blind to treatment, subjects randomized to therapeutic CPAP receive active CPAP plus an oxygen concentrator that provides room air. Subjects randomized to placebo-CPAP receive subtherapeutic CPAP (<1 cm H_2O at the mask) plus an oxygen concentrator providing room air. The placebo-CPAP treatment consists of a CPAP mask with 10 one-fourth inch drill holes for adequate room air exchange with pressure set at a constant 3 cm H_2O. A pressure reducer is placed in the tubing between the CPAP unit and the modified mask.

With this system, the pressure at the mask is 0.5 cm H_2O at end-expiration and 0 cm H_2O during inspiration and the patients feel a gentle breeze at the nose. Those assigned to nocturnal oxygen receive placebo-CPAP plus an oxygen concentrator delivering oxygen at 3 l/min (FiO_2 of 32 to 34% at the mask). Subjects are first tested on our Clinical Research Center, then sent home with their equipment until returning to the research center for their repeat testing.

15.5.1. CPAP Effects on NE levels and Adrenergic Receptors

We have found that CPAP treatment is effective in normalizing sympathetic nervous system activity in OSA, including the elevated levels of NE and NE excretion, and restoration of desensitized β-adrenergic receptors (Ziegler, Mills, Loredo, Ancoli-Israel, and Dimsdale 2001).

We conducted a randomized, placebo-controlled trial of CPAP on sympathetic nervous activity in 38 patients with OSA (Ziegler et al. 2001). All of the patients had OSA. In this study, patients were randomized blindly to either CPAP or placebo CPAP treatment for 10 days. As shown in Fig. 15.5, CPAP, but not placebo CPAP, lowered daytime plasma NE levels by 23% (Ziegler et al. 2001).

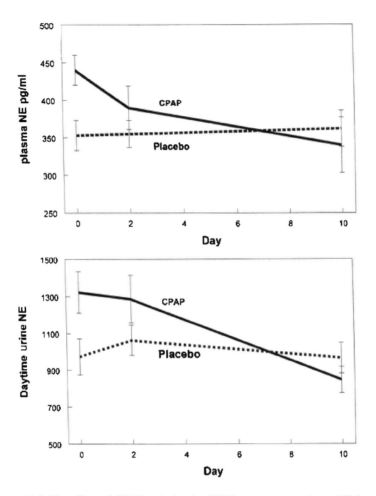

FIGURE 15.5. The effect of CPAP and placebo CPAP treatment on plasma NE levels and daytime urinary NE levels. CPAP decreased NE levels ($p < 0.04$), differing from the effect of placebo treatment Urine NE excretion fell in the CPAP group ($p < 0.001$), differing from the effect of placebo (Ziegler et al. 2001).

CPAP also led to a lowering of daytime urine NE excretion levels by 36% (Fig. 15.5). The effect of CPAP treatment on nighttime urine NE levels did not differ from placebo treatment (data not shown).

CPAP treatment also helped normalize the sensitivity of lymphocyte β_2-adrenergic receptors. Recall earlier we showed that β_2-adrenergic receptors were desensitized in OSA. We found that CPAP increased β-receptor sensitivity from 4.8 to 5.2 while β-receptor sensitivity was slightly decreased in the placebo CPAP group (from 5.1 to 4.9) ($p \leq 0.01$).

15.5.2 CPAP Effects on NE Kinetics

Given the findings that CPAP reduces elevated NE levels and NE excretion, we wondered whether CPAP has these effects by altering the release and/or the clearance rate of NE (Mills, Kennedy, Loredo, Dimsdale, and Ziegler 2006). Considering prior observations that patients with OSA have a tendency toward diminished NE clearance and that CPAP reduces NE levels, we suspected that CPAP treatment would lead to an increase in NE clearance or perhaps a decrease in NE release rate. We randomized OSA patients to a 2-week therapeutic trial of CPAP, placebo CPAP or oxygen supplementation.

NE excretion and blood pressure were also assessed in these subjects. We evaluated the adequacy of urine collection by measures of volume and creatinine excretion. Urinary NE excretion was expressed in micrograms excreted per hour during wake (16 h, 06:00 to 22:00 h) and sleep (8 h, 22:00 to 06:00 h).

We concluded from these studies that CPAP is an effective treatment to restore more normal levels of sympathetic activity in OSA.

Prior to CPAP treatment, we found that the AHI was related to NE release rate ($r = 0.385$; $p < 0.01$) and day ($r = 0.381$; $p < 0.01$) and night ($r = 0.313$; $p < 0.05$) NE excretion rates. $SaO_2 < 90\%$ (percent time less than 90% SaO_2) correlated with NE release rate ($r = 0.463$; $p < 0.01$), plasma NE levels ($r = 0.319$, $p < 0.05$) and day ($r = 0.583$ $p < 0.01$) and night ($r = 0.667$; $p < 0.01$) NE excretion rates.

AHI was significantly reduced by CPAP ($F = 28.9$, $p < 0.001$) but not by oxygen or placebo CPAP treatment. $SaO_2 < 90\%$ was significantly reduced by CPAP and oxygen ($ps < 0.05$) but not by placebo CPAP. Systolic ($p \leq 0.013$) and diastolic ($p \leq 0.026$) blood pressures were decreased in response to CPAP but not the oxygen or placebo CPAP.

Regarding the effects of CPAP on NE kinetics, 2 weeks of CPAP led to a significant increase in NE clearance ($p \leq 0.01$) (Fig. 15.6). NE clearance, volume of distribution, and half-life were unchanged following either oxygen or placebo CPAP.

As previously reported plasma NE levels were reduced following CPAP ($p \leq 0.018$) but unchanged following either oxygen or placebo CPAP. Daytime NE excretion was reduced following 2 weeks of CPAP treatment ($p < 0.001$) and oxygen treatment ($p < 0.01$) but unchanged following placebo CPAP.

Nighttime NE excretion was reduced following CPAP treatment only ($p < 0.05$). NE release rate was unchanged with any treatment (data not shown) (Mills et al. 2006).

In addition to examining the direct effects of treatment on NE kinetics, in an effort to better understand the role of NE in blood pressure responses to treatment, we conducted a series of multiple regression analyses examining possible predictors of posttreatment levels of systolic and diastolic blood pressure. Dependent variables were entered into the regression in blocks as follows—block 1: age, BMI, gender, and diagnosis of hypertension; block 2: pretreatment AHI, $SaO_2 < 90\%$, and the respective pretreatment blood pressure; block 3: pretreatment NE clearance, NE release rate, supine plasma NE, and daytime and nighttime NE excretion; block 4: posttreatment AHI, $SaO_2 < 90\%$; block 5: posttreatment NE clearance, NE release rate, supine plasma NE, and daytime and nighttime NE excretion.

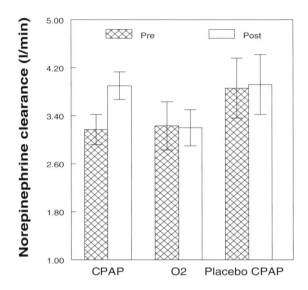

FIGURE 15.6. Plasma NE clearance in patients with sleep apnea following 14 days of CPAP, oxygen (O_2) or placebo CPAP. CPAP led to a significant increase in NE clearance ($p \leq 0.01$) whereas there was no change following O_2 supplementation or placebo CPAP (Mills et al. 2006).

As shown in Table 15.1, posttreatment systolic blood pressure was predicted by pretreatment systolic blood pressure, posttreatment NE clearance, and posttreatment NE release rate, yielding a regression model of $r^2 = 0.687$, $F = 16.1$, $p < 0.001$, with all other predictor variables dropping out as not significant. Thus, a lower posttreatment systolic blood pressure was associated with a lower pretreatment systolic blood pressure, a higher posttreatment NE clearance, and a lower posttreatment NE release rate.

TABLE 15.1. Predictors of systolic blood pressure following CPAP.

Pretreatment systolic blood pressure ($\beta = 0.450$, $p = 0.005$)

Posttreatment NE clearance ($\beta = -0.836$, $p = 0.001$)

Posttreatment NE release rate ($\beta = 0.717$, $p = 0.005$)

Full regression model: $r^2 = 0.687$, $F = 16.1$, $p < 0.001$

Posttreatment diastolic blood pressure was best predicted by pretreatment diastolic blood pressure (full regression model: $r^2 = 0.284$, $F = 10.5$, $p < 0.01$).

To summarize this treatment study, we observed that in patients treated with CPAP, in addition to reductions in circulating NE and NE excretion, daytime NE clearance was increased. We did not observe a change in daytime NE release rate, although prior studies have shown that OSA patients have increased sympathetic nerve activity in the daytime and that CPAP leads to a reduction in sympathetic neural activity during sleep. Our measures help specify where CPAP induces changes in daytime sympathetic nerve activity. The major site for NE release is in muscle vascular beds and we did not find a change in NE release with CPAP. On the other hand, there was a clear reduction in urine NE with CPAP. NE in the urine comes from the blood and from NE released by sympathetic nerves in the kidney. Our findings suggest that CPAP causes a reduction in renal sympathetic neuronal activity. CPAP also led to an increase in the volume of distribution for NE. This could be due to more active diffusion of NE out of the bloodstream, enhanced reversible transport of NE by uptake-1 and uptake-2 mechanisms, or enhanced binding of NE by plasma proteins and cellular structures. Our study did not indicate which of these potential mechanisms predominates, but we have previously found a suggestion of enhanced binding to receptors following CPAP (Ziegler et al. 2001).

15.5.3 CPAP Effects on Inflammation

In addition to reducing sympathetic activation, studies show that CPAP is effective in reducing inflammatory markers. The elevations of circulating levels of IL-6 and TNF-α and the production of TNF-α by monocytes in OSA are significantly reduced by CPAP treatment (Minoguchi et al. 2004; Kobayashi et al. 2006). In these studies, all patients were assigned to active CPAP, with no nonactive treatment arm (e.g., placebo CPAP) for comparison. CPAP has also been shown to reduce circulating levels of soluble CD40 ligand (sCD40L) (Yokohama 2006). The reduction of sCD40L is significant because it suggests a potential mechanism of CPAP's reduction of inflammation. CD40L is expressed on many cell types, including endothelial cells, monocytes, and macrophages. CD40L signaling triggers the expression of many proinflammatory mediators including cytokines IL-6 and TNF-α and adhesion molecules ICAM-1 and VCAM-1 (Yokohama 2006).

15.6 Summary

OSA is characterized by sleep disruption and hypoxia, leading to noradrenergic activation, elevated blood pressure, β-adrenergic receptor desensitization, and increased levels of proinflammatory cytokines. Successful treatment of OSA with CPAP reverses many of these adverse effects, including elevated inflammatory markers, potentially by increasing the clearance of NE.

Acknowledgments. This work was supported by grants HL57265, HL36005, HL44915 and HL40102 from the National Institutes of Health and the UCSD General Clinical Research Center (MO1RR00827).

References

Bardwell, W.A., Ancoli-Israel S., and Dimsdale J. (2001) Types of coping strategies are associated with increased depressive symptoms in patients with obstructive sleep apnea. *Sleep* 24, 905–909.

Bardwell, W.A., Moore, P., Ancoli-Israel, S., and Dimsdale J. (2003) Fatigue in obstructive sleep apnea: Driven by depressive symptoms instead of apnea severity? *Am J Psychiatry* 160, 350–355.

Bernstein, D., Doshi, R., Huang, S., Strandness, E., and Jasper, J. R. (1992) Transcriptional regulation of left ventricular beta-adrenergic receptors during chronic hypoxia. *Circ Res* 71, 1465–1471.

Borovikova, L.V., Ivanova, S., Ivanova, M., Yang, H., Botchkina, G., Watkins, L.R., Wang, H., Abumrad, N., Eaton, J.W., and Tracey, K.J. (2000) Vagus nerve stimulation attenuates the systemic inflammatory response to endotoxin. *Nature* 405, 458–462.

Brodde, O.E., Michel, M., Gordon, E.P., Sandoval, A., Gilbert, E.M., and Bristow, M.R. (1989) Beta-adrenoceptor regulation in the human heart: Can it be monitored in circulating lymphocytes? *Eur Heart J* 10, SB2–SB10.

Carlson, J.T., Hedner, J., Elam, M., Ejnell, H., Sellgren, J., and Wallin, B.G. (1993) Augmented resting sympathetic activity in awake patients with obstructive sleep apnea. *Chest* 103, 1763–1768.

Dimsdale, J., Coy, T., Ziegler, M.G., Ancoli-Israel, S., and Clausen, J. (1995) The effect of sleep apnea on plasma and urinary catecholamines. *Sleep* 18, 377–381.

Dimsdale, J.E., Coy, T., Ancoli-Israel, S., Mills, P.J., Clausen, J., and Ziegler, M. (1997) Sympathetic nervous system alterations in sleep apnea. The relative importance of respiratory disturbance, hypoxia, and sleep quality. *Chest* 111, 639–642

Eisensehr, I., Ehrenberg, B., Noachtar, S., Korbett, K., Byrne, A., McAuley, A., and Palabrica, T. (1998) Platelet activation, epinephrine, and blood pressure in obstructive sleep apnea syndrome. *Neurology* 51, 188–195.

Entzian, P., Linnemann, K., Schlaak, M., and Zabel, P. (1996) Obstructive sleep apnea syndrome and circadian rhythms of hormones and cytokines. *Am J Respir Crit Care Med* 153, 1080–1086.

Fletcher, E.C. (2003) Sympathetic over activity in the etiology of hypertension of obstructive sleep apnea. *Sleep* 26, 15–19.

Hartmann, G., Tschop, M., Fischer, R., Bidlingmaier, C., Riepl, R., Tschop, K., Hautmann, H., Endres, S., and Toepfer, M. (2000) High altitude increases circulating interleukin-6, interleukin-1 receptor antagonist and C-reactive protein. *Cytokine* 12, 246–252.

Irwin, M. (2001) Neuroimmunology of disordered sleep in depression and alcoholism. *Neuropsychopharmacology* 5, S45–S49.

Kelley, K.W., Bluthe, R.M., Dantzer, R., Zhou, J.H., Shen, W.H., Johnson, R.W., and Broussard, S.R. (2003) Cytokine-induced sickness behavior. *Brain Behav Immun* S1, S112–S118.

Kennedy, B., and Ziegler, M.G. (1990) A more sensitive and specific radioenzymatic assay for catecholamines. *Life Sci* 47, 2143–2153.

Kobayashi, K., Nishimura, Y., Shimada, T., Yoshimura, S., Funada, Y., Satouchi, M., and Yokohama, M. (2006) Effect of continuous positive airway pressure on soluble CD40 ligand in patients with obstructive sleep apnea syndrome. *Chest* 129, 632–637.

Krueger, J.M., Obal, F.J., Fang, J., Kubota, T., and Taishi, P. (2001) The role of cytokines in physiological sleep regulation. *Ann N Y Acad Sci* 933, 211–221.

Landsberg, R., Friedman, M., and Ascher-Landsberg, J. (2001) Treatment of hypoxemia in obstructive sleep apnea. *Am J Rhinol* 15, 311–313.

Leuenberger, U.A., Brubaker, D., Quraishi, S., Hogeman, C., Imadojemu, V.A., and Gray, K.S. (2005) Effects of intermittent hypoxia on sympathetic activity and blood pressure in humans. *Auton Neurosci* 121, 87–93.

Marsh, J.D., and Sweeney, K.A. (1989) Beta-adrenergic receptor regulation during hypoxia in intact cultured heart cells. *Am J Physiol* 256, H275–H281.

Mills, P.J., Dimsdale, J., Coy, T., Ancoli-Israel, S., and Clausen, J. (1995) β_2-Adrenergic receptor characteristics in sleep apnea patients. *Sleep* 18, 39–42.

Mills, P.J., Dimsdale, J.E., Ancoli-Israel, S., Clausen, J., and Loredo, J. (1998) The effects of hypoxia and sleep apnea on isoproterenol sensitivity. *Sleep* 21, 731–735.

Mills, P.J., Kennedy, B.P., Loredo, J., Dimsdale, J.E., and Ziegler, M.G. (2006) Effects of nasal continuous positive airway pressure and oxygen on norepinephrine kinetics and cardiovascular responses in obstructive sleep apnea. *J Appl Physiol* 100, 343–348.

Minoguchi, K., Tazaki, T., Yokoe, T., Minoguchi, H., Watanabe, Y., Yamamoto, M., and Adachi, M. (2004) Elevated production of tumor necrosis factor-alpha by monocytes in patients with obstructive sleep apnea syndrome. *Chest 2004* 126, 1473–1479.

Nelesen, R.A., Yu, H., Ziegler, M.G., Mills, P.J., Clausen, J.L., and Dimsdale, J.E (2001). Continuous positive airway pressure normalizes cardiac autonomic and hemodynamic responses to a laboratory stressor in apneic patients. *Chest* 119, 1092–1101.

Nelesen, R.A., Dimsdale, J., Mills, P.J., Clausen, J., Ziegler, M.G., and Ancoli-Israel, S. (1996) Altered cardiac contractility in sleep apnea. *Sleep* 19, 139–144.

Newman, A.B., Nieto, F.J., Guidry, U., Lind, B.K., Redline, S., Pickering, T.G., and Quan, S.F. (2001) Sleep Heart Health Study Research Group. Relation of sleep-disordered breathing to cardiovascular disease risk factors: The Sleep Heart Health Study. *Am J Epidemiol* 154, 50–59.

Norman, D., Loredo, J., Nelesen, R., Ancoli-Israel, S., Mills, P.J., Ziegler, M.G., and Dimsdale, J.E. (2006). Effects of continuous positive airway pressure versus supplemental oxygen on 24-hour ambulatory blood pressure. *Hypertension* 47, 840–845.

Opp, M.R. (1995) Corticotropin-releasing hormone involvement in stressor-induced alterations in sleep and in the regulation of waking. *Adv Neuroimmunol* 5, 127–143.

Opp, M.R. (2002) Cytokines and sleep promotion: A potential mechanism for disorders of excessive daytime sleepiness. In: A. Pack (Ed.), *Pathogenesis, Diagnosis and Treatment of Sleep Apnea*. Marcel Dekker, New York, pp. 327–351.

Parish, J.M., and Somers, V.K. (2004) Obstructive sleep apnea and cardiovascular disease. *Mayo Clin Proc* 79, 1036–1046.

Phillips, B.A., Schmitt, F., Berry, D.T., Lamb, D.G., Amin, M., and Cook, Y.R. (1990) Treatment of obstructive sleep apnea. A preliminary report comparing nasal CPAP to nasal oxygen in patients with mild OSA. *Chest* 98, 325–330.

Reimer, M.A., and Flemons, W.W. (2003) Quality of life in sleep disorders. *Sleep Med Rev* 7, 335–349.

Ridker, P.M., Hennekens, C.H., Roitman-Johnson, B., Stampfer, M.J., and Allen, J. (1998) Plasma concentration of soluble intercellular adhesion molecule 1 and risks of future myocardial infarction in apparently healthy men. *Lancet* 351, 88–92.

Robinson, G.V., Stradling, J.R., and Davies, R. (2004) Obstructive sleep apnoea/hypopnoea syndrome and hypertension. *Thorax* 59, 1089–1094.

Vgontzas, A.N., Papanicolaou, D.A., Bixler, E.O., Kales, A., Tyson, K., and Chrousos, G.P. (1997) Elevation of plasma cytokines in disorders of excessive daytime sleepiness: Role of sleep disturbance and obesity. *J Clin Endocrinol Metab* 82, 1313–1316.

Vgontzas, A.N., Papanicolaou, D.A., Bixler, E.O., Hopper, K., Lotsikas, A., Lin, H.M., Kales, A., and Chrousos, G.P. (2000) Sleep apnea and daytime sleepiness and fatigue: Relation to visceral obesity, insulin resistance, and hypercytokinemia. *J Clin Endocrinol Metab* 85, 1151–1158.

Vgontzas, A.N., Zoumakis, M., Bixler, E.O., Lin, H.M., Prolo, P., Vela-Bueno, A., Kales, A., and Chrousos, G.P. (2003) Impaired nighttime sleep in healthy old versus young adults is associated with elevated plasma interleukin-6 and cortisol levels: Physiologic and therapeutic implications. *J Clin Endocrinol Metab* 88, 2087–2095.

Vgontzas, A.N., Bixler, E., and Chrousos, G.P. (2005) Sleep apnea is a manifestation of the metabolic syndrome. *Sleep Med Rev* 9, 211–224.

Yokoe, T., Minoguchi, K., Matsuo, H., Oda, N., Minoguchi, H., Yoshino, G., Hirano, T., and Adachi, M. (2003) Elevated levels of C-reactive protein and interleukin-6 in patients with obstructive sleep apnea syndrome are decreased by nasal continuous positive airway pressure. *Circulation* 107, 1129–1134.

Yoon, I.Y., and Jeong, D.U. (2001) Degree of arousal is most correlated with blood pressure reactivity during sleep in obstructive sleep apnea. *J Korean Med Sci* 16, 707–711.

Ziegler, M., Kennedy, B, Morrissey, E., and O'Connor, D. (1990) Norepinephrine clearance, chromogranin A and dopamine beta hydroxylase in renal failure. *Kidney Int* 37, 1357–1362.

Ziegler, M.G., Nelesen, R., Mills, P.J., Ancoli-Israel, S., Kennedy, B., and Dimsdale, J. (1997) Sleep apnea, norepinephrine-release rate, and daytime hypertension. *Sleep* 20, 224–231.

Ziegler, M.G., Mills, P.J., Loredo, J., Ancoli-Israel, S., and Dimsdale, J. (2001) Effect of continuous positive airway pressure and placebo treatment on sympathetic nervous activity in patients with obstructive sleep apnea. *Chest* 120, 887–893.

16 Role and Circadian Rhythms of Proinflammatory Cytokines, Cortisol, and Melatonin in Children with Obstructive Sleep Apnea Syndrome

Luana Nosetti and Luigi Nespoli

16.1 Obstructive Sleep Apnea Syndrome in Children: Clinical Features

Obstructive sleep apnea syndrome (OSAS) is characterized by episodes of partial or complete upper airway obstruction that occur during sleep, usually associated with a reduction in oxyhemoglobin saturation and/or hypercarbia.

OSAS in children differs significantly from that in adults. Excessive daytime sleepiness (EDS) and snoring with apnea are essential diagnostic elements for OSAS in adults. EDS appears to be uncommon in children with equally severe OSAS. Some children are obese like adults, but most are not. Some have large tonsils and adenoids, while some with enormous adenoids or tonsils have only mild OSAS or are completely asymptomatic. In adults enlarged tonsils and adenoids are uncommon. OSAS in adults occurs predominantly in males and postmenopausal females. In children there isn't a significant difference between males and females.

OSAS is more common in adults than in children, and the prevalence of OSAS in adults increases with age (Carroll and Loughlin 1995). Habitual snoring not associated with obstructive apnea, hypoxia, or hypoventilation is common in childhood, and occurs in 13% of preschool and school-aged children (Castronovo et al. 2003). OSAS is present in approximately 2% of 4- to 5-year-old children. OSAS occurs in children of all age. The peak incidence occurs between 3 and 6 years of age, mirroring the peak age of adenotonsillar hypertrophy.

Symptoms reported in children with OSAS to be present on awakening in the morning include dry mouth, grogginess, disorientation, confusion, headaches, and mouth breathing. During the day, children with OSAS may present hyperactivity, decreased intellectual performance, and learning problems. During sleep some children with OSAS snore loudly and habitually. Sometimes they have grunting, snorting, gasping, or other form of noisy breathing.

Frequently they have respiratory retractions and episodes of increased respiratory effort associated with lack of airflow. Cyanosis or pallor may occur during sleep. Children with OSAS frequently sleep in positions to promote airways patency, such as prone, seated, or with hyperextension of the neck. The complications of OSAS result from chronic nocturnal hypoxia, acidosis, and sleep fragmentation. Pulmonary hypertension is a major cause of morbidity in patients with OSAS, and if untreated will progress to cor pulmonare (Brouillette, Fernbach, and Hunt 1982). Failure to thrive is a frequent complication of OSAS in children. Causes for poor growth include anorexia or dysphagia secondary to adenotonsillar hypertrophy, increased work of breathing, hypoxia, or abnormal nocturnal growth hormone secretion (Marcus, Koerner, Pysik, and Loughlin 1994).

16.2 Pathophysiology of OSAS

The etiology of OSAS is uncertain and multifactorial. For OSAS to occur there must probably be a combination of three factors: altered airway structure, diminished neuromuscular control, and miscellaneous factors, such as genetic and hormonal influences. Thus, one child with a narrow airway due to adenotonsillar hypertrophy with a high ventilatory drive may not develop OSAS. An intense local and systemic inflammation tends to presents in these patients. In the upper airway, this process may promote oropharyngeal inspiratory muscle dysfunction and amplify both upper airway narrowing and collapsibility thereby worsening the frequency and duration of apneas during sleep. Adenotonsillar hypertrophy is the commonest condition associated with childhood OSAS. The lymphoid tissue in the upper airway increases in volume from birth to 12 years of age with a peak between 3 and 6 years of age. This coincides with the peak incidence of childhood OSAS. Craniofacial anomalies due to narrowing of the upper airway can be associated to OSAS. It is most likely to occur when the patient has nasal obstruction, micrognathia, macroglossia, midfacial hypoplasia, associated obesity, or hypotonia. OSAS has been reported to occur in Treacher Collins syndrome, Down syndrome, Achondroplasia, Crouzon syndrome, Apert syndrome, Arnold–Chiari malformation, and Prader–Willi syndrome.

Sound of snoring originates in the collapsible part of the airway, where there is no rigid support, which implicates a primary role of the nasopharyngeal inlet, pharynx, and tongue. The key force that promotes the closure of the upper airway is the negative pressure applied during inspiration, which is determined by the inspiratory effort and the physiological dimensions of the upper airway. The primary force holding the airway is the activity of the dilator muscles that give tone and tension to the pharyngeal wall. Another cause of OSAS is nasal airway obstruction that contributes to nearly 40% of total airway resistance in healthy children. Causes of nasal airway obstruction are many, including enlarged adenoids as well as local inflammation.

Most studies suggest that the patients with OSAS have smaller pharyngeal airway, the knowledge of the detailed microanatomy of pharyngeal tissue might yield insights into the pathogenesis of OSAS. Sekosan, Zakkar, Wenig, and Olopade (1996) demonstrated the presence of inflammation, characterized by plasma cell infiltration and interstitial edema. The presence of inflammation in this area may conceivably increase the thickness of the mucosa, resulting in upper airway narrowing. Paulsen et al. (2002) revealed a significant diffuse infiltration of leukocytes, mainly T cells, inside the lamina propria of patients with OSAS. The immunohistochemical staining with antibodies against epithelial cytokeratins showed difference in the expression pattern of cytokeratin 13. In the nasal lavage fluid of OSAS patients polymorphonuclear leukocytes and concentrations of bradykinin an vasoactive intestinal peptide are increased (Rubinstein 1995).

Goldbart, Krishna, Li, Serpero, and Gozal (2006) studied inflammatory mediators LTB4 (leukotriene B4) and cysteinyl leukotrienes (cys-LTs; leukotriene C4 [LTC4]/leukotriene D4 [LTD4]/leukotriene E4 [LTE4]) in exhaled breath condensate and observed values more elevated in children with OSAS. In contrast, PGE2 concentrations were similar in normal children and in OSAS patients. O'Brien, Serpero, Tauman, Gozal, and O'Brien (2006) observed that children with sleep-disordered breathing (SDB) have plasma elevations of P-selectin, a marker of platelet activation, lending support to premise that inflammatory processes are elicited by SDB in children and may contribute to accelerated risk for cardiovascular morbidity. These inflammatory changes are postulated to occur, in part due to snoring that evokes vibration frequencies associated with soft-tissue damage (Cohn, Helsa, and Kiel 1986). In addition to local inflammation, evidence of systemic inflammation is present in patients with OSAS.

To understand OSAS it is also important to analyze the physiological stimuli that initiate the arousal and the increased inspiratory effort associated with the end of apneic episode. First, there is an initial decrease in the drive to breathe caused by decreased sensitivity of the peripheral and central chemoreceptors. As a result, the airway collapses, leading to the apneic event that limits gas exchange at the lungs. The subsequent hypoxic–hypercapnic state increases the drive to breathe. The breathing effort, however, is impeded by the obstructed airway, which causes further impairment of gas exchange. The severe hypoxia and hypercapnia produce a breathing effort adequate to terminate the apneic event. This effort often elicits arousal, which disrupts the progression of sleep, and results in disturbed sleep architecture (Cutler, Hamdan, Hamdan, Ramaswamy, and Smith 2002).

The changes in heart rate and blood pressure associated with OSAS are thought to be primarily a result of alterations in the autonomic nervous system during each apneic episode. During each apneic event, there is a progressive increase in sympathetic nerve activity throughout the episode, reaching a peak at termination of the apnea, after which there is a marked decrease during recovery. The increase in sympathetic nerve activity during apnea is mainly the result of acute hypoxia and hypercapnia. The increase in sympathetic nerve activity is accompanied by several changes in the cardiovascular system. An increase in activity at the end of each apneical event leads to vasoconstriction, increasing peripheral vascular resistance in the systemic and pulmonary vasculature, finally resulting in a progressive rise in arterial pressure.

16.3 OSAS Neuroimmunology

The links between immune system and sleep were first identified in the 1970s, when a sleep-inducing factor was isolated and chemically characterized from human urine: Factor S, a muramyl peptide derived from bacterial peptoglycan. Subsequently muramyl dipeptide and Factor S-related peptidoglycans were all shown to induce the key immunoregulatory cytokine. Then interleukin (IL)-1β was shown to be a potent somnogen, as well as a potent pyrogen. In fact, IL-1β is one of the most neurologically active molecules known. Subsequent studies revealed that bacterial LPS, LPS components, and viral synthetic dsRNA, as well as killed and living bacteria can induce IL-1, tumor necrosis factor (TNF)-α, IL-6, and IL-10 (Majde and

Krueger 2005). The presence of systemic inflammation, characterized by an elevation of certain potent proinflammatory cytokines, such as IL-1, IL-6, IL-10, and TNF-α may predispose patients with OSAS to develop cardiovascular complications. Alberti, Sarchielli, Gallinella, Floridi, Mazzotta, and Gallai (2003) reported a prevailing activation of the Th1-type cytokine pattern in OSAS patients, which is not associated with the severity and duration of OSAS. A factor critical for the development of an effective immune response is the cytokine balance between T helper (Th1) and T helper 2 (Th2) cells, determining the selection of the effector mechanisms of type 1 or type 2 immunity. Th1 cells releasing mainly interferon-γ (INF-γ), aside from other cytokines including IL-2 and TNF-α become activated in response to intracellular viral and bacterial challenges and support various cellular (type 1) responses, including macrophage activation and antigen presentation. In contrast, the cytokines characteristic of Th2 immunity, IL-4 as well as IL-5, IL-10, and IL-13 tend to drive humoral (type 2) defense via stimulating mast cells, eosinophils and B cells against extracellular pathogens. Predominance of type 2 cytokines, has been associated with a reduced response to vaccination (Ginaldi et al. 1999) on the other hand (Petrovsky and Harrison 1997) suggest that nocturnal sleep could favor a shift towards Th1 mediate immune defense. They found a circadian peak of the ratio of IFN-γ/IL-10 production in whole blood samples during nocturnal sleep. This peak was completely abolished following administration of cortisone acetate al 9:00 p.m. in the preceding evening, suggesting that the suppression of endogenous cortisol release during early sleep plays a mediating role for this Th1 shift. However, SWS not only suppresses the release of glucocorticoids but also promotes the release of growth hormone (GH) and prolactin, which appears to support Th1-cell-mediated immunity (Chikanza and Dimitrov 1999).

16.4 Melatonin Role and Circadian Rhythms in Children with OSAS

Melatonin is produced by pineal gland. Light inhibits its secretion. Hence, a rhythmical secretion pattern is seen in all species, including humans. In humans, the highest melatonin levels are found in 2- to 5-year-old children; from that age, secretion decrease progressively (Waldhauser, Weiszenbacher, and Tatzer 1988). It has long been known that melatonin has sleep-promoting properties and regulates the sleep–wake cycle. Melatonin has a strong circadian rhythm with high values during the nighttime and low values in the afternoon. Sleep disorders impair the quality of life for children with neurodevelopmental disabilities and their families. Melatonin, a natural regulator of sleep, may be an effective therapy in this area (Wassmer and Whitehouse 2006).

Sleep a disordered breathing may change the circadian rhythms of melatonin, which may have diagnostic implications. One major feature in patients with OSAS is EDS chiefly resulting from disturbed sleep at night. It has been shown that treatment with continuous positive airway pressure (CPAP) may neutralize OSAS completely and abolish the sleep disturbance. Wikner, Svanborg, Wetterberg, and Röjdmark (1997) reported similar nocturnal melatonin levels in patients with OSAS and healthy controls. Sleep apnea recordings were normalized during CPAP treatment and daytime sleepiness disappeared in all patients, neither melatonin secretion nor urinary excretion changed significantly as a result of the CPAP treatment.

Ulfberg, Micic, and Strøm (1998), in another study, showed that in comparison with normal controls, patients suffering from OSAS had significantly higher serum-melatonin levels in the afternoon. However, determination of afternoon serum-melatonin, as a diagnostic test for OSAS in patients with sleep-disordered breathing, showed a low sensitivity but high specificity. Determination of afternoon serum-melatonin alone or together with a scoring of daytime sleepiness does not identify OSAS patients in a heterogeneous population of patients complaining of heavy snoring and EDS (Ulfberg et al. 1998). OSAS in children differs significantly from that in adults. EDS and snoring with apnea are essential diagnostic elements for OSAS in adults, EDS appears to be uncommon in children with equally severe OSAS. Children with OSAS, during the day, can present hyperactivity, decreased intellectual performance and learning problems. We studied circadian rhythms of melatonin in children with severe OSAS. Serum melatonin concentrations were determined every second hour between 8.00 p.m. and 8.00 a.m. during a 12 channels polisomnography. Melatonin secretion among OSAS pediatric patients did not differ from that found in healthy controls. We did not found significant correlation between melatonin circadian secretion, age, or BMI. Only pediatric patients with many body movements during the night presented a significant increment of melatonin secretion (Nosetti et al. 2006) (Fig. 16.1).

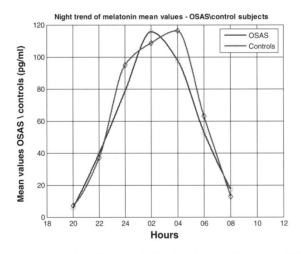

FIGURE 16.1. Comparison of the night trend of melatonin (pg/ml) between OSAS patients and control subjects.

16.5 Cortisol Role and Circadian Rhythms in OSAS

Cortisol provides an important link between the immune system, sleep, and psychological stress. High levels of cortisol suppress the immune system, so excessively tired people are more susceptible to illness.

Sleep disruption and sustained psychological stress increase cortisol concentration in the blood. Indeed, one night of lost sleep can raise cortisol concentrations by almost 50% by the following evening. High levels of cortisol suppress the immune system, so excessively tired people are more susceptible to illness (Foster and Wulff 2005). Indeed, one night of lost sleep can raise cortisol concentrations by almost 50% by the following evenings (Leprout, Copinschi, Buxton, and Van Cauter 1997). We studied circadian rhythms of cortisol in children with severe OSAS. Cortisol concentrations were determined every second hour between 8.00 p.m. and 8.00 a.m. during a 12 channels polisomnography. Cortisol among OSAS pediatric patients presented a significant growth in the interval included between 0.00 and 4.00 a.m. compared to healthy controls. Sleep disruption and hypoxia in patients with OSAS increase cortisol concentration in the blood. We found a significant correlation between cortisol circadian secretion and BMI (Nosetti et al. 2006) (Fig. 16.2).

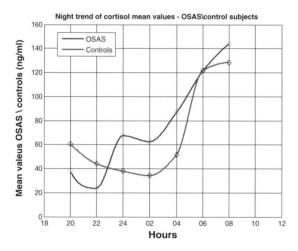

FIGURE 16.2. Comparison of the night trend of cortisol (ng/ml) between OSAS patients and control subjects.

Obese pediatric patient with OSAS have an increased risk of insulin resistance, hypertension, and increase in inflammatory markers. These metabolic abnormalities improve with treatment of OSAS, indicating that OSAS plays a direct role in their development (de la Eva, Baur, Donague, and Waters 2002).

16.6 Proinflammatory Cytokines Role and Circadian Rhythms in OSAS

Cytokines are produced during infection and concurrent with changes in sleep pattern, they are good candidates for having a direct role in sleep regulation. In addition, several cytokines are produced in central nervous system. With respect to sleep the most studied cytokines are: TNF, IL-1β.

16.6.1 TNF, IL-1β

In animals, increases in the intracerebral or plasma levels of either TNF on IL-1β result in an increase in SWS duration, whereas decreases in the intracerebral levels of TNF or IL-1β inhibit SWS. The injection of cytokine into the right or left lobe affects changes in SWS and fever, demonstrating that these responses are independent of each other. Antagonizing either of these cytokines (by pretreatment with antibodies) leads to decreased sleep, indicating that they have a role in the physiological regulation of sleep. Interestingly, each of these cytokines influences the effect the other has on sleep. Pretreatment with a fragment of the receptor for IL-1β (type 1 IL-1R, IL-1R1) attenuates TNF-induced SWS enhancement, and TNF antagonists inhibit IL-1β-induced increases in SWS duration. TNF and IL-1β both stimulate the transcriptional activity of nuclear factor-κB (NF-κB) and enhance sleep. Factors that inhibit NF-κB activation, such as IL-4, IL-10 and inhibitor of NF-κB (IκB), inhibit sleep. NF-κB itself promotes the production on TNF and IL-1β, forming a positive-feedback loop, possibly to promote the homeostatic drive for sleep. This is supported by the finding that sleep deprivation causes increased levels of NF-κB in the central nervous system (Bryant, Trinder, and Curtis 2004).

16.6.2 Other Cytokines

Although there is good evidence to support a role for TNF and IL-1β in the impact of infections on sleep, the role of other cytokines is more ambiguous. This might reflect either the lack of a direct role or problems with experimental design. IL-2, IL-5, IL-18, INF-α, and INF-γ usually increase the duration of sleep, whereas IL-4, IL-10, and IL-13 usually inhibit spontaneous sleep. The effect of IL-6 is less conclusive. In rats, it has been reported to be either somnogenic or to have no effect on sleep. By contrast, in humans, it decreases SWS in the first half of the night and increases SWS in the second half, with REM sleep decreased throughout. Shearer et al. (2001) studied soluble TNF-α receptor 1 (s'TNF-α R 1) and IL-6 plasma levels in humans subjected to the sleep deprivation model of spaceflight and observed that total sleep loss produced significant increases in plasma levels of s'TNF-α R 1 and IL-6, messengers that connect the nervous, endocrine, and immune systems. Carpagnano et al. reported higher concentrations of IL-6 in exhaled breath in OSAS patients than in healthy control subjects. These findings suggest that inflammation and oxidative stress are characteristic in the airways of OSAS patients and that their levels depend on the severity of the OSAS (Carpagnano, Karitov, Resta, Foschino-Barbaro, Gramiccioni, and Barnes 2002). The effects of all these cytokines might be due to their direct action on sleep or an indirect effect through other cytokines, in particular

TNF and/or IL-1β. Therefore, one possible explanation for the variation in sleep patterns that occur after different infections is that the various microorganisms elicit different host cytokine responses. Most proinflammatory cytokines seem to be somnogenic, whereas most anti-inflammatory cytokines are not. The inhibitory actions of the anti-inflammatory cytokines on sleep might be the result of their inhibition of brain proinflammatory cytokines (Bryant et al. 2004).

16.6.3 Proinflammatory Cytokines Circadian Rhythms in OSAS

It is an important aspect of sleep circadian rhythmicity, so diurnal variation in the immune response might support a link between sleep and immune system. Circulating lymphocytes and monocytes in the blood reach maximal values during the night (Born, Lange, Hansen, Molle, and Fehm 1997). Natural killer (NK) cells have very high level in the afternoon and decrease in number and activity just after midnight (Heiser 2000). There are many studies that investigated the variations in cytokine levels that occur during the sleep–wake cycle; but, it is difficult to measure them because endogenous cytokines levels are low. We studied circadian rhythms of IL-1, IL-6, and INF in children with severe OSAS, every second hour between 8.00 p.m. and 8.00 a.m., during a 12 channels polisomnography. The circadian rhythms of IL-1, IL-6, and INF were not significantly different from normal subjects (Nosetti et al. 2006). A preliminary study concludes that pediatric sleep-disordered breathing is associated with increasing circulating levels of TNF-α that closely correlate with the degree of sleep fragmentation.

Etzian et al. observed in adults with OSAS, plasma TNF levels peak during sleep, and circadian rhythm of TNF release disrupted by OSAS: The night peak of this cytokine had almost disappeared and an additional daytime peak had developed. Circadian variations in IL-1, IL-6, and INF did not differ from those in the controls. Because TNF-α is a known modulator of sleep, and CPAP therapy did not normalize TNF rhythms; TNF-α could play a pathophysiologic role in OSAS (Etzian, Linneman, Schlaak, and Zabel 1996).

Plasma IL-1β levels also have a diurnal variation, being highest at the onset of SWS (Gudewil 1992). The levels of other cytokines (including IL-2, IL-6, IL-10 and IL-12) and the production of T cells response to mitogens also change during the sleep–wake cycle (Redwine, Dang, Hall, and Irwin 2003). The production of macrophage-related cytokines (such as TNF) increases during sleep and this happens with the rise in monocyte numbers in the blood. The T-cell-related cytokines (like IL-2) increase during sleep, independent of migratory changes in T-cell distribution. All of these observed diurnal changes could be related to the effects of sleep or associated with the circadian oscillator. Sleep and the immune system therefore share regulatory molecules. These are interested in both physiological sleep and sleep in the acute-phase response to infection. This shows that sleep and the immune system are closely interconnected. It is likely that sleep settles the immune system through the action of centrally produced cytokines that are regulated during sleep. These endogenous cytokines are known to function through the autonomic nervous system and the neuroendocrine axis, however other studies could show that other system are used (Bryant et al. 2004).

References

Alberti, A., Sarchielli, P., Gallinella, E., Floridi, A., Mazzotta, G., and Gallai, V. (2003) Plasma cytokine levels in patients with obstructive sleep apnea syndrome: A preliminary study. *J Sleep Res* 12, 305–311.

Born, J., Lange, T., Hansen, K., Molle, M., and Fehm, H.L. (1997) Effects of sleep and circadian rhythm on human circulating immune cells. *J immunol* 158, 4454–4464.

Brouillette, R.T., Fernbach, S.K., and Hunt, C.E. (1982) Obstructive sleep apnea in infant and children. *J Pediatr* 100, 31–40.

Bryant, P., Trinder, J., and Curtis, N. (2004) Sick and tired: Does sleep have a vital role in immune system? *Nat Rev Immunol* 4, 457–467.

Carpagnano, G., Karitov, S., Resta, O., Foschino-Barbaro, M.F., Gramiccioni, E., and Barnes, J. (2002) Increased 8-Isoprostano and Interleukin-6 in breath condensate of obstructive sleep apnea patients. *Chest* 122(4), 1162–1167.

Carroll, J.L., and Loughlin, G.M. (1995) Obstructive sleep apnea syndrome in infants and children: Clinical features and pathophysiology. In R. Ferber and M. Kryger (Eds.), *Principles and Practice of Sleep Medicine in the Child*, W.B. Saunders, Philadelphia, pp. 163–193.

Castronovo, V., Zucconi, M., Nosetti, L., Marazzini, C., Hensley, M., Veglia, F., Nespoli, L, and Ferini-Strambi, L. (2003) Prevalence of habitual snoring and sleep-disordered breathing in preschool-aged children in an Italian community. *J Pediatr* 142(4), 364–382.

Chikanza, I.C., Dimitrov, S., and Ann, N.Y. (1999) Prolactin and neuroimmunomodulation: In vitro and in vivo observation. *Acad Sci* 876, 119–130.

Cohn, M., Helsa, P.E., and Kiel, M (1986) Vibration frequency of snoring in obstructive sleep apnea syndrome. *Chest* 89, 529S.

Cutler, M.J., Hamdan, A.L., Hamdan, M., Ramaswamy, K., and Smith, M.L. (2002) *JABFP* 15(2), 128–141.

de la Eva, R.C., Baur, L.A., Donague, K.C., and Waters, K.A. (2002) Metabolic correlates with obstructive sleep apnea in obese subjects. *J Pediatr* 140,654–659.

Etzian, P., Linneman, K., Schlaak, M., and Zabel, P. (1996) Obstructive sleep apnea syndrome and circadian rythms of hormones and cytokines. *Am J Respir Crit Care Med* 153, 1080–1086.

Foster, R.G., and Wulff, K. (2005) The rhythm of rest and excess. *Nature* 6, 407–414.

Ginaldi, L., Martinis, M., D'Ostilio, A., Marini, L., Loreto, M.F., Corsi, M.P., and Quaglino, D. (1999) The immune system in the elderly: Specific humoral immunity. *Immunol Res* 20, 101–108.

Goldbart, A.D., Krishna, J., Li, R.C., Serpero, L.D., and Gozal, D. (2006) Inflammatory mediators in exhaled breath condensate of children with obstructive sleep apnea syndrome. *Chest* 130, 143–148.

Gudewil, S. (1992) Nocturnal plasma levels of cytokines in healthy men. *Eur Arch Psychiatry Clin Neurosci* 48, 309–318.

Heiser, P. (2000) White blood cells and cortisol after sleep deprivation and recovery sleep in humans *Eur Arch Psychiatry Clin Neurosci* 250, 16–23.

Leprout, R., Copinschi, G., Buxton, O., and Van Cauter, E. (1997) Sleep loss results in an elevation of cortisol levels the next evening. *Sleep* 10, 865–870.

Majde, J.A., and Krueger, J. (2005) Links between the innate immune system and sleep. *J Allergy Clin Immunol* 116 (6), 1188–1198.

Marcus, C.L., Koerner, C.B., Pysik, P, and Loughlin, G.M. (1994) Determinants of growth failure in children with the obstructive sleep apnea syndrome. *J Pediatr* 125, 506–511.

Nosetti, L., Ciglia, F., Luce, A., Maestroni, G.J.M., Veronelli, E., de Simone, S., Spica Russotto, V., and Nespoli, L. (2006) Circadian rhythms of melatonin and cortisol in children with obstructive sleep apnea syndrome (OSA) *Eur Respir J* 4963.

O'Brien, L., Serpero, L., Tauman, R., and Gozal, D. (2006) Plasma adhesion molecules in children with sleep-disordered breathing. *Chest* 129, 947–953.

Paulsen, F.P., Steven, P, Tsokos, M., Jungmann, K, Müller, A, Verse, T, and Pirsing, W (2002) Upper airway epithelial structural changes in obstructive sleep-disordered breathing. *Am J Respir Crit Care Med* 166, 501–509.

Petrovsky, N., and Harrison. L.C. (1997) Diurnal rhythmicity of human cytokines production: A dynamic disequilibrium in T helper cell type 1/T helper cell type 2 balance? *J Immunol* 158, 5163–5168.

Redwine, L., Dang, J., Hall, M., and Irwin, M. (2003) Disordered sleep nocturnal cytokines, and immunity in alcoholics. *Psychosom Med* 65, 75–85.

Rubinstein, I. (1995) Nasal inflammation in patients with obstructive sleep apnea. *Laryngoscope* 105, 175–177.

Sekosan, M., Zakkar, M., Wenig, B.L., and Olopade, C.O. (1996) Inflammation in the uvula mucosa of patients with obstructive sleep apnea. *Laryngoscope* 106, 1018–1020.

Shearer, W.T., Reuben, J.M., Mullington, J.M., Price, N.J., Lee, B.-N., Smith, O, Szuba, M.P., Van Dongen, P.A., and Dingers, D.F. (2001) Soluble TNF-α receptor 1 and IL-6 plasma levels in humans subjected to the sleep deprivation model of spaceflight. *J Allergy Clin Immunol* 107(1), 165–170.

Ulfberg, J., Micic, S., and Strøm, J. (1998) Afternoon serum-melatonin in sleep disordered breathing. *J Intern Med* 244, 163–168.

Waldhauser, F., Weiszenbacher, G., and Tatzer, E. (1988) Alterations in nocturnal serum melatonin levels in humans with growth and aging. *J Clin Endocrinol Metab* 66, 648–652.

Wassmer, E., and Whitehouse, W.P. (2006) Melatonin and sleep in children with neurodevelopmental disabilities and sleep disorders. *Curr Paediatr* 16, 132–138.

Wikner, J., Svanborg, E., Wetterberg, L., and Röjdmark, S. (1997) Melatonin secretion and excretion in patients with obstructive sleep apnea syndrome. *Sleep* 20(11), 1002–1007.

17 Neuroendocrine-Immune Correlates of Sleep and Traumatic Brain Injury (TBI)

Paolo Prolo and Anna N. Taylor

17.1 Introduction

Unexpected crises have set the rhythm of our days after the fall of the Berlin Wall, the dismemberment of the Soviet Union and the dissolution of the Warsaw Pact. In a world that still needs to find a clear path to be open toward the future and to think about the new and the possible, as recently stated by philosopher Bodei (2006), we have unfortunately witnessed the events of September 11, 2001 in New York, March 11, 2004 in Madrid, July 19, 2005 in London and a series of armed conflicts, from Kosovo to Sudan, from East Timor to Afghanistan, Iraq, and the Middle East. All these changes in the world political climate have brought an increase in the incidence of traumatic brain injury (TBI) and spinal cord injury (SCI) both in soldiers and in civilians from war zones and terrorist attacks.

TBI has proven unusually common in the current conflicts in Iraq and Afghanistan, in which new body armor saves soldiers from injuries that would have killed them in the past but cannot keep their brains from banging against the walls of their skulls (*San Francisco Chronicle*, July 19, 2004; USA Today, March 4–6, 2005). The weapons preferred by those attacking U.S. troops and those attacking ordinary people—suicide terrorists, car bombs, land mines, improvised explosive devices, mortars—deliver precisely the kind of concussive blast that can injure the brain in addition to other body parts or systems resulting in physical, cognitive, psychological, or psychosocial impairments and functional disability, i.e., polytrauma (Lew 2005). There is little information on the pattern of neuroendocrine and immune impairments after TBI, and their influence on sleep. Moreover, closed brain injuries and mild TBI may not be readily apparent, particularly in the presence of externally evident injuries or other life-threatening conditions requiring immediate attention. However, once a diagnosis of TBI is obvious, rehabilitation may be a lengthy process, encompassing a range of behavioral and physiological functions that direct the adaptive function of regulating bodily systems in response to past, present and future challenges.

17.2 TBI and the Neuroendocrine System

Trauma to the hypothalamus frequently occurs in severe head injuries (Crompton 1971; Rudy 1980). The hypothalamus can be injured as a result of direct or indirect damage, small hemorrhages, or ischemia (cf. Lighthall, Gochgarian, and Pinderski 1990). Sudden head movement can shear blood vessels supplying this part of the brain (Mitchell, Steffenson, and Davenport 1997). Regulation of the neuroendocrine systems involves numerous pathways and centers, including afferent neural pathways, the brainstem, the cortex, corticohypothalamic pathways, hypothalamic integrative centers, the pituitary gland, and efferent autonomic pathways. TBI may disturb any part of this complex system, depending on the severity and the location of primary and secondary injuries (Yuan and Wade 1991). Patterns of neuroendocrine abnormalities will vary according to the site of the injury and the extent of injury-transmitted hypothalamic–pituitary damage. For example, damage just above the base of the brain in a distinct region of the hypothalamus called the suprachiasmatic nucleus (SCN) will affect the primary circadian pacemaker that resides there. Three major neural pathways relay information to the SCN that may all be affected after TBI (see Fig. 17.1).

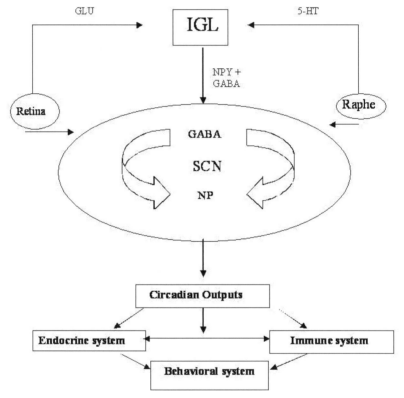

FIGURE 17.1. Three major neural pathways bring information to the suprachiasmatic nucleus (SCN), which can be compromised after TBI: (1) a direct pathway from the retinal ganglion cells; (2) a pathway from a region in the thalamus called the intergeniculate leaflet (IGL); (3) a pathway from the neurons of the raphe nuclei of the midbrain, important in the regulation of mood, arousal, sleep and other behavioral aspects. This pathway is connected to the SCN and the IGL. The SCN interacts with and contains a number of chemicals important in cellular communication, such gamma-aminobutyric acid (GABA), other neuropeptides (NP) like NPY, glutamate (GLU), and serotonin (5-HT).

Endocrine dysfunction after TBI affecting all hypothalamic–pituitary axes (i.e., corticotropin, growth hormone, gonadotropin, thyrotropin, prolactin, and vasopressin) has been described in clinical studies (Yuan and Wade 1991; Childers, Rupright, Jones, and Merveille 1998; Kelly, Gaw Gonzalo, Cohan, Berman, Swerdloff, and Wang 2000; Benvenga, Campenni, Ruggeri, and Trimarchi 2000; Lieberman, Oberoi, Gilkison, Masel, and Urban 2001; Agha, Phillips, Kelly, Tormey, and Thompson 2005). Indeed, hypopituitarism has been identified in up to half of the long-term survivors (6 to 36 months) of moderate to severe head injury (Kelly et al. 2000; Lieberman et al. 2001; Agha et al. 2005). Specifically, with respect to the HPA axis, a high incidence of ACTH and adrenal insufficiencies has been reported (Benvenga et al. 2000; Cohan et al. 2005). These abnormalities, which occur soon after TBI, are transient in some patients, while the majority shows recovery at 6 months (Agha et al. 2005). Additionally, the extent of neuroendocrine impairment has been found to correlate with the severity of the neurological insult as assessed by the Glasgow Coma Scale (GCS) (Cernak, Savic, Lazarov, Joksimovic, and Markovic 1999). For example, plasma cortisol levels increase during the early post-TBI period, but only in patients with minor to moderate injuries. In contrast, patients with severe trauma exhibit a significant decline in cortisol (Cernak et al. 1999). TBI-induced alterations in circadian rhythmicity of HPA axis function may also contribute to the high incidence of sleep disturbances in patients with TBI, including insomnia, excessive daytime somnolence and alteration of the sleep–wake schedule (Frieboes, Muller, Murch, von Cramon, Holsboer, and Steiger 1999; Mahmood, Rapport, Hanks, and Fichtenberg 2004; Rao and Rollings 2002).

Experimental studies have demonstrated that corticotrophin-releasing hormone (CRH) mRNA is up-regulated at 2 and 4 h after fluid percussion injury (FPI) (Roe, McGowan, and Rothwell 1998; Grundy, Harbuz, Jessop, Lightman, and Sharples 2001) and ACTH and cortisol are increased up to 6 h after cortical contusion injury (CCI) (McCullers, Sullivan, Scheff, and Herman 2002). However, in experimental models, beyond the acute activation of the HPA axis after the surgical intervention, the effects of TBI on neuroendocrine function at later time-points remain unknown. Furthermore, the longer-term clinical studies have defined alterations in baseline, but not stress-induced neuroendocrine function.

Our recent data indicate that CCI produces dysregulation of the stress-induced HPA response at 4 weeks postinjury in male rats. Moreover, the direction of the dysregulation differs depending upon the location of the CCI (Taylor et al. submitted).

Given that pituitary insufficiency may have serious consequences and may aggravate the physical and neuropsychiatric morbidity observed after TBI (Agha et al. 2005), frequent assessment of the subject's endocrine status is essential. Moreover, with the more elevated and prolonged risk for psychiatric illness, including depression, posttraumatic stress disorder, anxiety and sleep/wakefulness, that are persistent symptoms of TBI (Fann, Burington, Leonetti, Jaffe, Katon, and Thompson 2004; Dikmen, Bobmardier, Machamer, Fann, and Temkin 2004; Ryan and Warden 2003) and the evidence that the neuroendocrine stress system and depression share common neural pathways and hormonal mediators (Gold and Chrousos 2002), measurement of the allostatic HPA stress response should prove to be a relevant biomarker in TBI.

17.3 TBI and Immune Surveillance

In addition to central neuropathological consequences, TBI has humoral neural signs, such as elevated circulating levels of catecholamines, which contribute to the immunosuppression observed in patients with TBI by inducing the release of the immunosuppressive cytokine, IL-10. Therefore, the central insult provoked by TBI has clinically significant *sequelae* in that it puts the patients at increased risk for depressed immune surveillance, infection and sepsis (Plata-Salaman 1998; Woiciechowsky et al. 1998). In point of fact, inflammation is a critical aspect of the central and peripheral response to TBI. The process by which the central and the peripheral responses interact constitutes an important part of the pathophysiology of TBI, and is regulated in large part by the blood–brain barrier. Numerous immune mediators released within minutes of the primary injury determine and guide the neuroimmune sequence of events that follows, and its protraction in time (Morganti-Kossmann, Rancan, Stahel, and Kossmann 2002). Cytokines, such as interleukin (IL)-1β, tumor necrosis factor (TNF)-α and IL-6, released from antigen-activated immune cells, are the chief messengers of afferent signals that cross the blood--brain barrier and play a key role as regulators of the host-defense response (Besedovsky and Del Rey 2001) as well as sleep disorders (Vgontzas, Bixler, Lin, Prolo, Trakada, and Chrousos 2005).

Like the hormonal response to stress, the cytokine reaction is fundamental for survival. Through a variety of mechanisms, the brain detects activation of the peripheral immune system. The brain responds to infection by altering physiological processes and complex behavior, including sleep. These changes in physiology and behavior collectively function to support the immune system, and under normal circumstances the health of the host is restored (Prolo and Chappelli in press). Several of these cytokines, and their receptors, are present in normal healthy brain. Other cytokines, such as IL-6, regulate sleep under physiological conditions, in the absence of infection or immune challenge.

For example, IL-1 directly alters discharge patterns of neurons in hypothalamic and brainstem circuits implicated in the regulation of sleep–wake behavior (Alam et al. 2004). Other cytokines, such as IL-6, modulate sleep because they interact with neurotransmitter, peptide, and/or hormone systems to initiate a cascade of responses that subsequently alter sleep–wake behavior. Blood levels of IL-6 correlate with sleep and sleepiness, and its circadian pattern mirrors the homeostatic drive to sleep (Vgontzas et al. 2005). Because cytokines regulate/modulate sleep–wake behavior in the absence of immune challenge, and cytokine concentrations and profiles are altered during infection, it is likely that cytokines mediate infection-induced alterations in sleep.

17.4 TBI and Sleep

Close to 25% of patients with TBI develop some form of psychopathological or psychiatric disorder, including posttraumatic stress disorder, physical, cognitive-memory, and behavioral complications, such as sleep disturbances (Frieboes et al.,

1999). Insomnia is a prevalent condition after TBI, although most reports rely on self-administered questionnaires rather than laboratory data (Steele, Rajaratnam, Redman, and Ponsford 2005). Sleep disruption is a defining feature of anxiety and is associated with increased arousal at nighttime leading to disrupted sleep. However, it is difficult to ascertain the direction of the relation between anxiety and sleep complaints, because anxiety may be both a cause and a consequence of sleep disorder.

A recent study showed that higher anxiety is associated with longer time since injury and milder injury severity in TBI (Parcell, Ponsford, Rajaratnam, and Redman 2006). In this study, the TBI subjects showed increased nighttime awakenings and longer sleep onset latency, especially in patients with milder injury severity. Increased symptoms of anxiety and depression were associated with increased reporting of sleep changes.

Sleep disorders after TBI require more clinical and scientific attention as they may have important repercussions on rehabilitation. Altered circadian rhythms and sleep disorders and the role played by HPA hormones and proinflammatory cytokines in sleep deprivation and insomnia that we previously described (Vgontzas et al. 1999; Vgontzas et al. 2001; Vgontzas et al. 2005; Prolo, Iribarren, Neagos, and Chiappelli 2006) can certainly be transposed to TBI as well.

As a matter of fact, the elevation of secreted interleukin (IL)-6 and possibly its flattened circadian rhythm is a common feature of brain injury (Shafer, McNulty, and Young 2002). However, some issues arise when the same invasive methods used in our studies on insomnia (24-h continuous blood drawing, sleep lab setting, sleep deprivation) are applied to subjects recovering after TBI.

A convenient way of measuring sleep–wake cycles is by using continuous wrist activity recording. Actigraphy is a method used to study sleep–wake patterns and circadian rhythms by assessing movement, most commonly of the wrist (Littner et al. 2002). Actigraphy provides a means for clinicians and researchers to measure sleep and wake periods in a noninvasive, ambulatory way, thus not requiring that the subject be confined in a hospital room (i.e., sleep lab).

This approach is very convenient when the focus is on subjects that are outpatients, and are following a strict rehabilitation program. TBI subjects supposedly spend more time in bed, have more disrupted nighttime sleep, sleep more during the day, and have less robust circadian rhythms of activity.

Telemetric recording of spontaneous activity is, therefore, a reliable indicator of the circadian sleep–wake cycle. A graph obtained by continuous monitoring of a healthy individual is shown in Fig. 17.2.

Taken together, circadian rhythm alterations in sleep disorders include diminished amplitude, phase shifts, period changes, and erratic peaks and troughs in endocrine, metabolic, immunological, and rest-activity cycles.

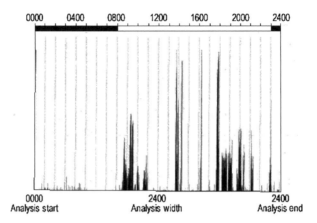

FIGURE 17.2. Typical actigraph of a healthy 42-year-old male (dark line = sleep time).

In TBI, the common understanding is that there is a deviation of the rhythm period from 24 h so that the rhythms found in these patients are free-running, meaning that head injured patients are not synchronized with their surroundings (Kropyvnytskyy, Saunders, Pols, and Zarowski 2001), similar to a situation of long-lasting jetlag. Given the high incidence of sleep disturbances in patients with TBI, including insomnia, excessive daytime somnolence and alteration of the sleep–wake schedule (Frieboes et al. 1999; Mahmood et al. 2004; Rao and Rollings 2002), we expect that TBI subjects will show less robust circadian rhythms of activity than normal subjects.

Since sleep disorders are associated with cognitive deficits even in healthy adults, these alterations may be particularly disruptive to patients who have sustained a TBI. Sleep disturbance may disproportionately affect cognitive functioning among patients with TBI. According to several studies (Ron, Algom, Harry, and Cohen 1980; Cohen, Oksenberg, Snir, Stern, and Groswasser 1992; Fichtenberg, Zafonte, Putnam, Mann, and Millard 2002), the time elapsed since injury is inversely related to the severity of cognitive deficits and is an important factor in influencing the nature of sleep complaints in patients with TBI. Although variables such as independence and locomotion improved at rates that were not correlated with sleep measures, rapid eye movement (REM) sleep frequency and cognition seemed to improve concurrently. In contrast, other researchers found that patients with severe brain injuries are less likely than those with mild brain injuries to complain of sleep problems (Beetar, Guilmette, and Sparadeo 1996). One explanation for this finding is that patients with severe head injuries may underreport sleep disturbance because of a lack of awareness or limited memory, while subjects with mild brain injuries might be more easily accessible to fill in self reported questionnaires. Thus, there is likely to be a discrepancy between subjective reports and objective measures of sleep disturbance in TBI.

17.5 Conclusions

It is clear that there is a relationship between sleep disorders and TBI, although the nature of this association is complex and not fully investigated. A brief look at the relevant literature shows that the presence of sleep disturbances and TBI varies with the characteristics of the subject studied, the nature of comorbidity with psychiatric disorders, and timing and severity after injury.

Sleep disorders complicate the course following TBI. Insomnia, excessive daytime somnolence and disturbances of the sleep–wake cycle are common disturbances that affect the course of recovery and prognosis in TBI survivors. Few studies, however, have looked at the diagnosis and management of these disturbances in TBI. Early treatment of sleep disorders must be considered an integral part of the rehabilitation process. Recognition and management of concurrent medical or surgical diseases, assessment and treatment of associated psychiatric disorders and awareness of other psychosocial stressors are fundamental steps in the management of sleep disturbances following TBI.

Among the primary mediators that must be carefully investigated are the hormones of the HPA axis, catecholamines, and cytokines, which interact with each other to create a network of reciprocal effects for maintaining bodily systems, e.g., immune, cardiovascular, metabolic, and nervous, in balance.

Acknowledgments. The authors acknowledge support by the Fondazione Cassa di Risparmio di Cuneo, Wilshire Rotary Foundation, and the Veterans Administration Medical Research Service.

References

Agha, A., Phillips, J.O., Kelly, P., Tormey, W., and Thompson, C.J. (2005) The natural history of post-traumatic hypopituitarism: Implications for assessment and treatment. *Am J Med* 118, 1416.

Alam, M.N., McGinty, D., Bushir, T., Kamar, S., Imeri, L., Opp, M.R and Scymusiak, R. (2004) Interleukin-1 beta modulates state-dependent discharge activity of preoptic area and basal forebrain neurons: Role in sleep regulation. *Eur J Neurosci* 20, 207–216.

Beetar, J.T., Guilmette, T.L., and Sparadeo, F.R. (1996) Sleep and pain complaints in symptomatic traumatic brain injury and neurological populations. *Arch Phys Med Rehabil* 77, 1298–1302.

Benvenga, S., Campenni, A., Ruggeri, R.M., and Trimarchi, F. (2000) Hypopituitarism secondary to head trauma. *J Clin Endocr Metab* 85, 1353–1361.

Bodei, R. (2006) *We, the Divided* (translated by Jeremy Parzen and Aaron Thomas). Agincourt Press, New York, NY.

Cernak, I., Savic, V.J., Lazarov, A., Joksimovic, M., and Markovic, S. (1999) Neuroendocrine responses following graded traumatic brain injury in male adults. *Brain Injury* 13, 1005–1015.

Childers, M.K., Rupright, J., Jones, P.S., and Merveille, O. (1998) Assessment of neuroendocrine dysfunction following traumatic brain injury. *Brain Injury* 12, 547–553.

Cohan, P., Wang, C., McArthur, D.L., Cook, S.W., Dusick, J.R., Armin, B., Swerdloff, R., Vespa, P., Muizelaar, J.P., Cryer, H.G., Christenson, P.D., and Kelly, D.F. (2005) Acute secondary adrenal insufficiency after traumatic brain injury: A prospective study. *Crit Care Med* 233, 2358–2366.

Cohen, M., Oksenberg, A., Snir, D., Stern, M., and Groswasser, Z. (1992). Temporally related changes of sleep complaints in traumatic brain injured patients. *J Neurol Neurosurg Psychiatry* 55, 313–315.

Crompton, M.R. (1971) Hypthalamic lesions following closed head injury. *Brain* 94, 165–172.

Dikmen, S.S., Bobmardier, C.H., Machamer, J.E., Fann, J.R., and Temkin, N.R. (2004) Natural history of depression in traumatic brain injury. *Arch Phys Med Rehabil* 85, 1457–1464.

Fann, J.R., Burington, B., Leonetti, A., Jaffe, K., Katon, W.J., and Thompson, R.S. (2004) Psychiatric illness following traumatic brain injury in an adult health maintenance organization population. *Arch Gen Psychiatry* 61, 53–61.

Fichtenberg, N., Zafonte, R., Putnam, S., Mann, N., and Millard, A. (2002) Insomnia in a post-acute brain injury sample. *Brain Injury* 16, 197–206.

Frieboes, R.-M., Muller, U., Murch, H., von Cramon, D.Y., Holsboer, F., and Steiger, A. (1999) Nocturnal hormone secretion and the sleep EEG in patients several months after traumatic brain injury. *J Neuropsychiatry Clin Neurosci* 11, 354–360.

Gold, P.W., and Chrousos, G.P. (2002) Organization of the stress system and its dysregulation in melancholic and atypical depression: High vs low CRH/NE states. *Mol Psychiatry* 7, 254–275.

Grundy, P.L., Harbuz, M.S., Jessop, D.S., Lightman, S.L., and Sharples, P.M. (2001) The hypothalamo-pituitary-adrenal axis response to experimental traumatic brain injury. *J Neurotrauma* 18, 1373–1381.

Kelly, D.F., Gaw Gonzalo, I.T., Cohan, P., Berman, N., Swerdloff, R., and Wang, C. (2000) Hypopituitarism following traumatic brain injury and aneurysmal subarachnoid hemorrhage: A preliminary report. *J Neurosurg* 93, 743–752.

Kropyvnytskyy, I., Saunders, F., Pols, M., and Zarowski, C. (2001) Circadian rhythm of temperature in head injury. *Brain Injury* 15, 511–518.

Lew, H.L. (2005) Guest Editorial: Rehabilitation needs of an increasing population of patients: Traumatic brain injury, polytrauma, and blast-related injuries. *J Rehabil Res Dev* 42, xiii–xvi.

Lieberman, S.A., Oberoi, A.L., Gilkison, C.R., Masel, B.E., and Urban, R.J. (2001) Prevalence of neuroendocrine dysfunction in patients recovering from traumatic brain injury. *J Clin Endocrinol Metab* 86, 2752–2756.

Lighthall, J.W., Gochgarian, H.G., and Pinderski, C.R. (1990) Characterization of axonal injury produced by controlled cortical impact. *J Neurotrauma* 7, 65–76.

Littner, M., Kushida, C.A., McDowell Anderson, W., Bailey, D., Berry, R.B., Davila, D.G., Hirshkowitz, M., Kapen, S., Kramer, M., Loube, D., Wise, M., and Johnson, S.F. (2002) Standards of Practice Committee of the American Academy of Sleep Medicine. Practice parameters for the role of actigraphy in the study of sleep and circadian rhythms: An update for 2002. *Sleep* 26, 337–341.

Mahmood, O., Rapport, L.J., Hanks, R.A., and Fichtenberg, N.L. (2004) Neuropsychological performance and sleep disturbance following traumatic brain injury. *J Head Trauma Rehabil* 19, 378–390.

McCullers, D.L., Sullivan, P.G., Scheff, S.W., and Herman, J.P. (2002) Mifepristone protects CA1 hippocampal neurons following traumatic brain injury in rat. *Neuroscience* 109, 219–230.

Mitchell, A., Steffenson, N., and Davenport, K. (1997) Hypopituitarism due to traumatic brain injury: A case study. *Crit Care Nurse* 17, 34–51.

Morganti-Kossmann, M.C., Rancan, M., Stahel, P.F., and Kossmann, T. (2002) Inflammatory response in acute traumatic brain injury: A double-edged sword. *Curr Opin Crit Care* 8, 101–105.

Parcell, D.L., Ponsford, J.L., Rajaratnam, S.M., and Redman, J.R. (2006) Self-reported changes to nighttime sleep after traumatic brain injury. *Arch Phys Med Rehabil* 87, 278–285.

Plata-Salaman, C.R. (1998) Brain injury and immunosuppression. *Nat Med* 4, 768–813.

Prolo, P., Iribarren, F.J., Neagos, N., and Chiappelli F. (2006) Role of pro-inflammatory cytokines in sleep disorders. In.: D.P. Cardinali and S.R. Pandi-Perumal (Eds.), *Neuroendocrine Correlates of Sleep/Wakefulness*. Springer Science, New York, pp. 537–552.

Prolo, P., and Chiappelli, F. (in press) Immune suppression. In: S. Vanstone, G. Chrousos, I. Craig, R. de Kloet, G. Feuerstein, B. McEwen, N. Rose, R. Rubin, and A. Steptoe (Eds.), *Encyclopaedia of Stress II*. Elsevier, London.

Rao, V., and Rollings, P. (2002) Sleep disturbances following traumatic brain injury. *Curr Treat Options Neurol* 4, 77–87.

Roe, S.Y., McGowan, E.M., and Rothwell, N.J. (1998) Evidence for the involvement of corticotrophin-releasing hormone in the pathogenesis of traumatic brain injury. *Eur J Neurosci* 10, 553–559.

Ron, S., Algom, D., Harry, D., and Cohen, M. (1980) Time-related changes in the distribution of sleep stages in brain injured patients. *Electroencephalogr Clin Neurophysiol* 48, 432–441.

Rudy, T.A. (1980) Pathogenesis of fever associated with cerebral trauma and intracranial hemorrhage. In: *Thermoregulatory Mechanisms and Their Therapeutic Implications. 4th International Symposium on the Pharmacology of Thermoregulation*. Oxford, pp. 75–81.

Ryan, L.M., and Warden, D.L. (2003) Post concussion syndrome. *Int Rev Psychiatry* 15, 310–316.

Shafer, L.L., McNulty, J.A., and Young, M.R. (2002) Brain activation of monocute-lineage cells: Involvement of interleukin-6. *Neuroimmunomodulation* 10, 295–304.

Steele, D.L., Rajaratnam, S.M., Redman, J.R., and Ponsford, J.L. (2005) The effect of traumatic brain injury on the timing of sleep. *Chronobiol Int* 22, 89–105.

Taylor, A.N., Rahman, S.U., Tio, D.L., Sanders, M.J., Bando, J.K., Truong, A.H., and Prolo, P. (2006) Lasting neuroendocrine-immune effects of traumatic brain injury in rats. *J Neurotrauma* 23(12): 1802–1813.

Vgontzas, A.N., Papanicolaou, A.N., Bixler, E.O., Lotsikas, A., Zachman, K., Kales, A., Wong, M.-L., Licinio, J., Prolo, P., Gold, P.W., Hermida, R.C., Mastorakos, G., and Chrousos, G.P. (1999) Circadian interleukin-6 secretion and quantity and depth of sleep. *J Clin Endocr Metab* 84, 2603–2607.

Vgontzas, A.N., Bixler, E.O., Lin, H.M., Prolo, P., Trakada, G., and Chrousos, G.P. (2005) IL-6 and its circadian secretion in humans. *Neuroimmunomodulation* 12, 131–140.

Woiciechowsky, C., Asadullah, K., Nestler, D., Eberhardt, B., Platzer, C., Schoning, B., Glockner, F., Lanksch, W.R., Volk, H.D., and Docke, W.D. (1998) Sympathetic activation triggers systemic interleukin-10 release in immunodepression induced by brain injury. *Nat Med* 4, 808–813.

Yuan, X.-Q., and Wade, C.E. (1991) Neuroendocrine abnormalities in patients with traumatic brain injury. *Front Neuroendocrinol* 12, 209–230.

18 Neuroimmune Correlates of Sleep in Depression: Role of Cytokines

J. Szelényi and E.S. Vizi

18.1 Introduction

Sleep is an integrate part of mammalian life, although the sleeping patterns of different species significantly differ in amounts and pattern. However, it is not an independent variable because, during sleep, almost every physiologic parameter changes relative to wakefulness (Lavie 2001). Insomnia is rather frequent; its prevalence is between 10 and 50% of the general population, depending on the methods used to assess sleep problems and on the population studied (Ohayon and Caulet 1996). In addition, many psychiatric disorders (e.g., depression) as well as various medical conditions including infectious, malignant and inflammatory diseases, are associated with disordered sleep (Benca and Quintas 1997; Bloom, Owens, McGuinn, Nobile, Schaeffer, and Alario 2002).

Whether physiologic or pathologic, sleep is divided into two states: non-rapid eye movement sleep (NREMS; quiet sleep) and rapid eye movement sleep (REMS; dream sleep). These two phases are different in many aspects and appear to be regulated by anatomically and biologically different factors, among them cytokines. In the study of the cross talk between cytokines and the central nervous system (CNS), sleep is a very important area involving a complex phenomenon of tonic regulation by the neural, endocrine, and immune systems. There are a number of approaches to studying the role of cytokines in sleep, among others the examination of the naturally occurring changes in cytokine levels during normal sleep. This chapter will briefly summarize and update such approaches.

18.2 Cytokines and the CNS

18.2.1 Cross Talk Between the Innate Immune Response and Sleep

Cytokines are multifunctional pleiotropic proteins that are involved not only in the immune response but also in a variety of physiological and pathological processes, including events in the CNS. On the other hand, cytokine production is under the tonic control of the CNS and the cytokine balance can also be modulated by the action of neurotransmitters in various brain regions (del Rey, Besedovsky, Sorkin, da Prada, and Arrenbrecht 1981; Vizi 1998; Elenkov, Wilder, Chrousos, and Vizi 2000; Szelenyi, Kiss, Puskas, Szelenyi, and Vizi 2000b).

Cytokines might exert their effect in the CNS both directly and indirectly (Besedovsky et al. 1991; Plata-Salaman 1991; Szelenyi 2001). Direct action means that cytokines themselves are present in the brain, in and/or around the various neuronal cells while secondary effects that are the result of cytokine action on other targets represent the indirect pathways.

The various cytokines directly affecting the CNS have two possible origins.

Cytokines that originate from the peripheral immune organs and cross the blood–brain barrier (Watkins, Maier, and Goehler 1995) even in healthy, basal conditions. There is also a convincing evidence for active, saturable, and specific transport of certain cytokines across the blood–brain barrier (Banks, Ortiz, Plotkin, and Kastin 1991).

Cytokines might be produced by the neuronal cells within the CNS. Most of the cytokines and their receptors have been demonstrated and/or postulated in various cell types of the CNS in both healthy and diseased states. Cytokines produced by a cascade of neurons and glial cells within the brain may participate in the complex autonomic, neuroendocrine, metabolic, and behavioral responses to infection, inflammation, ischemia, and other brain injuries (Breder et al. 1994; Sternberg 1997; Woiciechowsky et al. 1998).

Once in the brain, there is a CNS cytokine network that is made up of its cells (neurons and glial elements). These cells not only produce cytokines and express cytokine receptors, but also amplify cytokine signals, which in turn can have profound effects on the neurotransmitter and the CRH function, as well as on the behavior (Breder et al. 1994; Sternberg 1997; Woiciechowsky et al. 1998; Raison and Miller 2003; Dantzer 2004).

The neuroimmune interactions are bidirectional; the cytokines and other mediators released due to the activation of the innate immune system can modulate events in the CNS (Rothwell 1997; Sternberg 1997), while the cytokine production of the immunocytes is under the tonic control of the peripheral and CNSs and their balance can be modulated by neurotransmitters (Vizi 2000).

Although a number of parameters are involved in sleep regulation, we should recognize that they are acting in concert with each other forming negative and positive loops. First of all, it should be defined which molecule can be considered as a sleep regulatory agent and whether or not cytokines fulfill these criteria.

The criteria for a presumed sleep regulatory molecule were summarized by Krueger and others (Krueger, Obal, Fang, Kubota, and Taishi 2001; Mills and Dimsdale 2004) as: "(1) The molecule should induce physiological sleep; (2) The substance and its receptors should be present in the organism; (3) The concentration or turnover of the substance or its receptor should vary with the circadian rhythm; (4) Induction of the substance should induce sleep; (5) Inactivation of the substance or its receptor should reduce spontaneous sleep; (6) Inactivation of the substance should reduce sleep induced by somnogenic stimuli; and (7) Other biological actions of the substance should be separable, in part, from its sleep-promoting actions" (Krueger et al. 2001).

Cytokines as immune-mediators match with these criteria, however, their contribution is not direct for they might be sleep-inducing (enhance NREMS) and sleep-inhibiting (decrease NEMRS), depending on their dose and circadian phase (Maestroni, Conti, and Pierpaoli 1986, 1987; Maestroni et al. 1998).

Generally, the proinflammatory (Th1-type) cytokines (e.g., IL-1β, TNF-α, IL-2, IL-6) are mainly sleep-inducing while the noninflammatory (Th2-type) cytokines (e.g., IL-10, IL-4, and TGF-β) are mostly sleep-inhibiting (Table 18.1). Other mediators that promote sleep are also inflammatory (NO, PGs), and growth-promoting (BDNF, NGF) substances whose amount is dependent on the neuronal activity (Fig. 18.1).

Table 18.1. Somnogenic characteristics of pro and anti-inflammatory cytokines.

Cytokine	Type of cytokine	Effect on NREM sleep	References
IL-1-α,β	Proinflammatory	Promotes	(Moldofsky et al. 1986)
IL-2	Proinflammatory	Promotes	(Mills and Dimsdale 2004) (Krueger et al. 2001)
IL-6	Proinflammatory	Promotes	(Mills and Dimsdale 2004) (Redwine et al. 2000)
IL-8	Proinflammatory	Promotes	(Garcia-Garcia, Yoshida, and Krueger 2004)
IL-18	Proinflammatory	Promotes	(Krueger et al. 2001)
TNF-α	Proinflammatory	Promotes	(Mills and Dimsdale 2004) (Krueger et al. 2001)
IFN α.γ	Proinflammatory	Promotes	(Mills and Dimsdale 2004) (Krueger et al. 2001)
IL-4	Antiinflammatory	Inhibits	(Mills and Dimsdale 2004) (Mills and Dimsdale 2004)
IL-10	Antiinflammatory	Inhibits	(Krueger et al. 2001)
IL-13	Antiinflammatory	Inhibits	(Mills and Dimsdale 2004) (Mills and Dimsdale 2004)
TGF-β	Antiinflammatory	Inhibits	(Mills and Dimsdale 2004)

Sleep-inhibiting cytokines may exert their effects through antagonizing somnogenic cytokines. IL-10 and IL-4, for example, may inhibit sleep by inhibiting the production of IL-1β and TNF-α. It is also important to recognize that some proinflammatory cytokines (e.g., TNF-α and IL-1β) are involved in the physiologic regulation of sleep, even unrelated to inflammation (Krueger et al. 2001). It was shown that IL-1β levels in the brain correlated with sleep proneness, being highest at sleep onset and together with other proinflammatory cytokines were also able to regulate physiologic body temperature and appetite (Plata-Salaman 1991; Szelenyi 2001).

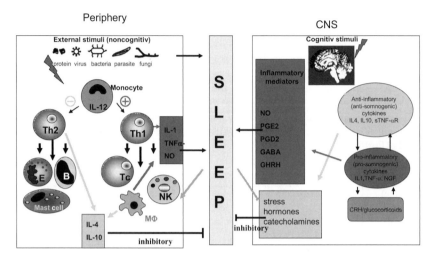

FIGURE 18.1. Bidirectional interactions between CNS and cytokines involved in sleep regulation. A variety of physical or psychological stimuli activates the production of proinflammatory cytokines and that of other inflammatory mediators both in the immune- and the CNS. Physical stimuli (e.g., infections) induces cytokine production usually in the periphery (left panel), and their effect on sleep (central panel) and on the CNS (right panel) prevail via various mechanisms. These cytokines also promote the production of each other, change the level of stress hormones and are under the regulation of the HPA axis. These interactions might be either synergistic or attenuated, and result in somnogenic (see red left and right boxes) and antisomnogenic effects (see green left and right boxes). Many of those mediators that promote sleep (e.g., IL-1, NGF, TNF-α, NO, etc.) are up-regulated by neuronal activity. The activity of the immune system and the production of the sleep-regulatory substances are also under diurnal control, therefore, type and pattern of sleep provide a feed-back control on its own activity by tuning the levels of cytokines, hormones and other mediators (yellow lines). Collectively, these actions form a biochemical cascade involved in physiological and pathological sleep regulation, as well as responses to external or internal challenges (*Abbreviations*: CNS, central nervous system; Th1, T helper -1 lymphocyte; Th2, T helper -2 lymphocyte; Tc, T-killer lymphocyte; NK, natural killer cell; M , macrophage; E, eosinophyls; B, B-lymphocyte, HPA axis, hypothalamic-pituitary-adrenal axis; IL-1, interleukin-1; IL-4, interleukin-4; IL-10, interleukin-10; TNF-α tumor necrosis factor-alpha; sTNF-αR soluble tumor necrosis factor-alpha-receptor; NGF, nerve growth factor; NO, nitric oxide; GABA, gamma-aminobutyric acid; PGE2, prostaglandin E2; PGD2, prostaglandin D2; GHRH, growth hormone-releasing hormone; CRH, corticotropin-releasing hormone).

18.3 Administering Different Types and Dosages of Cytokines and Their Effects on Sleep

Administration of TNF-α, IL-1β, or IL-18 increases the length of NREM sleep time and decreases the duration of REM sleep. Patients treated with recombinant or purified cytokines for specific forms of cancers or viral infections such as hepatitis B or C become somnolent (Papanicolaou, Wilder, Manolagas, and Chrousos 1998).

IL-1 and TNF-α have been shown to be somnogenic in every species thus far tested (e.g., mice, rats, cats, monkeys, and rabbits) and are effective whether given directly into the brain or after intraperitoneal or intravenous injections. On the other hand, the effect of cytokine administration is dependent not only on the mood and time of administration but also on the dosage. For instance, IL-1 and TNF-α appear to induce physiological sleep, however, high doses of IL-1 rather inhibit than promote sleep. Rabbits given IL-1 at dark onset sleep about 3 extra hours during the first 12 h after the injection but normal sleep patterns are maintained. Administration of TNF-α or IL-1β also increases the amplitude of slow wave EEG, while that of IL-10 and IL-4, inhibits NREM (Krueger et al. 2001).

The administration of cytokines results in pleitropic action in the CNS, e.g., IFN-γ, IL-1β, and TNF-α have not only somnogenic but also lethargic, pyrogenic, and anorectic effects (Rothwell 1997; Sternberg 1997). These effects are dose-dependent, i.e., both IL-1 and TNF-α induce increased NREMS at doses lower than those needed to induce fever. At slightly higher doses they are both somnogenic and pyrogenic, however, these actions can be separated by coadministration of an antipyretic that blocks IL-1-induced fever but not IL-1-induced NREMS (Krueger, Walter, Dinarello, Wolff, and Chedid 1984).

IL-2 that is also used as a therapeutic agent for treatment of various forms of cancer has been shown to produce partial remission in some patients with renal cell carcinoma or metastatic melanoma. It has also some neuropsychiatric side effects corresponding to symptoms associated with depression, such as fatigue, increased sleepiness and difficulty sleeping, irritability, anorexia, weight loss, and apathy (Krigel et al. 1990; Collier and Chapman 2001).

Cytokines are also used in combination in cancer therapy. For instance, IL-2 is administered in combination with IFN-α in order to increase the response rate. Although many studies agree that depression is worsened during immunotherapy, there are many differences in the prevalence of these side effects (Mulder, Ang, Chapman, Ross, Stevens, and Edgar, 2000; Musselman et al. 2001). The possible explanation for the discrepancies might be that neuropsychiatric side effects appear to be dose-related (Denicoff et al. 1987; Dusheiko 1997). In addition, the duration of therapy may also influence the occurrence of sleep disturbances.

Although the exact mechanisms of the somnogenic or antisomnogenic effects of cytokines have not yet been fully elucidated, the cascade has also been shown to involve other factors, such as growth hormone-releasing hormone, corticotrophin-releasing hormone, nitric oxide synthase, prostaglandins, and components of signaling mechanisms leading to activation of the transcription nuclear factor kappa B (NF-κB) (Mills and Dimsdale 2004).

18.4 Effect of Deprivation of Cytokines on the Duration and Pattern of Sleep

A number of immunological disorders including autoimmune diseases and certain infections are associated with an increased release of proinflammatory cytokines. Deprivation of proinflammatory cytokines may be achieved by including an administration of monoclonal antibodies against the cytokine, by administration of antireceptor antibodies, soluble receptors that inhibit the activity of their cytokines and receptor antagonists. A good example for the latter is the study where the effectiveness of administration of the naturally occurring IL-1 receptor antagonist that binds to the IL-1 receptor but fails to transduce a signal was demonstrated (Dinarello 1998). Another approach is the usage of physiological down-regulators of proinflammatory cytokine production, i.e., the usage of anti-inflammatory cytokines (e.g., IL-10) in the treatment of inflammatory and autoimmune diseases (Braat, Peppelenbosch, and Hommes 2003; Asadullah, Sabat, Friedrich, Volk, and Sterry 2004; Hofstetter, Flondor, Hoegl, Muhl, and Zwissler 2005). Cytokine production in the brain exhibit diurnal rhythms with peaks during major sleep periods. The plasma proinflammatory cytokines in humans exhibit a rhythm with peaks during the night. Accordingly, deprivation of proinflammatory cytokines results in disturbances in the amount and depth of sleep (Redwine, Hauger, Gillin, and Irwin 2000).

18.5 Studying the Effects of Sleep Deprivation/Disruption on Cytokine Production

Recent studies have focused on the role of sleep in the homeostatic regulation of the sympathetic nervous and immune systems, and used experimental sleep deprivation to probe these relationships (Dinges, Douglas, Hamarman, Zaugg, and Kapoor 1995; Irwin, Thompson, Miller, Gillin, and Ziegler 1999; Rogers, Szuba, Staab, Evans, and Dinges 2001). During sleep, sympathetic tone and circulating levels of sympathetic catecholamines decline, whereas sleep deprivation leads to nocturnal increases in norepinephrine (NE) and epinephrine (Dodt, Breckling, Derad, Fehm, and Born 1997; Irwin et al. 1999). Acute sleep loss is also associated with alterations in the expression of cytokines and declination in cellular immune responses (Redwine et al. 2000). For instance, decreased activity of natural killer (NK) cell, an immune cell that is considered to be an important marker of immunological defense against tumors responses, was reported due to acute sleep loss (Shakhar and Ben-Eliyahu 1998).

Patients with primary insomnia or with depression show abnormalities in sleep continuity (Bonnet and Arand 1995; Thase, Simons, and Reynolds 1996), but the relationships between disordered sleep and changes in nocturnal sympathetic catecholamines are not well defined in either of these groups. Insomnia is thought to be associated with chronic sympathetic hyperactivity as evidenced by elevated heart rate, body temperature, sympathetic tone, and whole body metabolic rate. Subjects with primary insomnia show nocturnal elevations of circulating levels of norepinephrine and a reduction in NK activity. It was shown that difficulties with sleep maintenance were associated with increases in nocturnal norepinephrine in the insomniacs and partial night sleep deprivation increased nocturnal levels of norepinephrine and decreased NK cell responses (Irwin, McClintick, Costlow, Fortner, White, and Gillin 1996). The above data implicate sleep continuity in the regulation of the immune and sympathetic nervous systems. Impairments of sleep efficiency correlated with nocturnal elevations of norepinephrine in the insomniacs but not in the depressives or controls. These results indicate that insomnia is associated with nocturnal sympathetic arousal and declines of natural immunity.

18.6 Regulation of Cytokine Balance by Monoamines in Healthy and in Depressed States

The major source of cytokine production is the peripheral immune system, although cytokines are also produced in the CNS. It is now well established that immune activation triggers the sympathetic nervous system (SNS) to release its neurotransmitters NE, epinephrine, and dopamine (Besedovsky, del Rey, Sorkin, and Dinarello 1986; Akiyoshi, Shimizu, and Saito 1990; Shimizu, Hori, and Nakane 1994).

The sympathetic nervous system innervates immune organs and, when activated, releases its signaling molecules in the vicinity of immune cells. Accordingly, cytokine balance can be modulated by sympathetic neurotransmitters both in the CNS and in the periphery. Immune cells express various neurotransmitter receptors that are sensitive to monoamines, and the production of cytokines (and other immune/inflammatory mediators (chemokines and free radicals) is modulated by activation of these receptors (Elenkov et al. 2000).

Once the neurotransmitters have reached the target cells, they occupy their appropriate receptors and the initiated signal transduction modulates the cytokine production in the cell. There is ample evidence that in this modulatory effect, cAMP plays one of the key roles (Bourne, Lichtenstein, Melmon, Henney, Weinstein, and Shearer 1974; Kambayashi, Jacob, Zhou, Mazurek, Fong, and Strassmann 1995; Vizi 1998). Thus, occupation of neurotransmitter receptors that stimulate/or inhibit adenylate-cyclase influences the cytokine profile of the system. NE and adrenergic drugs may influence the immune response directly, through the adrenergic receptors expressed on macrophages and also on other immunologically competent cells, as well as indirectly via alteration of the endogenous NE level by influencing the activity of release-regulating presynaptic $\alpha2$-adrenoceptors ($\alpha2$-AR) located on sympathetic nerve terminals (Elenkov and Vizi 1991; Vizi 1998; Szelenyi et al. 2000b; Szelenyi, Kiss, and Vizi 2000a). Activation of the presynaptic $\alpha2$-adrenoceptors results in a negative feed back effect on NE release, leading to decreased extracellular NE concentration (Vizi 1979; Kiss et al. 1995). The majority of the direct effect of NE is prevailed via the $\beta2$-adrenoceptors expressed by various immune cells (Hasko, Szabo, Nemeth, Salzman, and Vizi 1998a; Szelenyi et al. 2000a).

The effect of catecholamines and adrenergic drugs in inflammation and sepsis is one of the most thoroughly studied examples of neuroimmunomodulation. Adrenoceptors are known to be coupled to various G-proteins, either stimulating (Gs) or inhibiting (Gi) adenylate-cyclase, and there is evidence that among the three major adrenoceptor groups ($\alpha1$-, $\alpha2$-, and the β-adrenoceptor subtypes, i.e., $\beta1$-, $\beta2$-, and $\beta3$-adrenoceptors), at least $\alpha2$-, $\beta1$-, and $\beta2$-adrenoceptors play major roles in the regulation of cytokine balance.

Several groups demonstrated that the activation of $\beta2$-adrenoceptor on macrophages resulted in suppression of the LPS-induced proinflammatory cytokine production (Hasko, Szabo, Nemeth, Salzman, and Vizi 1998b; Izeboud, Mocking, Monshouwer, van Miert, and Witkamp 1999; Elenkov et al. 2000; Szelenyi, Kiss, Puskas, Selmeczy, Szelenyi, and Vizi 2000c). Stimulation of $\beta2$-adrenoceptor is the classical example of activation of adenylyl cyclase via stimulatory G protein (Gs) resulting in the subsequent increase in intracellular cAMP. Since cAMP is generally proved to suppress the inflammatory immune response, i.e., the sympathetic activation, it inhibits the innate immune response (van der Poll, Jansen, Endert, Sauerwein, and van Deventer 1994; Elenkov et al. 2000; Chong, Shin, and Suh 2003). Recently, however, evidence was given that the modulatory effect of β2-adrenoceptor on the proinflammatory/antiinflammatory cytokine balance was not necessarily immunosuppressive but just the opposite depending on the applied stimulus. However, the exact molecular mechanisms have not yet been fully understood (Szelenyi, Selmeczy, Brozik, Medgyesi, and Magocsi 2006).

As both α2- and β2-adrenoceptors are expressed on the surface of various immune cells (Abrass, O'Connor, Scarpace, and Abrass 1985; Spengler, Allen, Remick, Strieter, and Kunkel 1990), they can exert a direct regulatory effect on cytokine production via modulation of the cAMP level in the cell. This means that ligand binding of α2-adrenoceptors decreases while activation of β-adrenoceptors increases the cAMP level in the cell, resulting in an increasing effect of proinflammatory cytokine production for α2-, and in an opposite effect in the case of β-adrenoceptors.

Macrophagic α2-adrenoceptors have only a minor role, although, there is evidence that it also may have a substantial modulatory role in cytokine production both in vitro and in vivo. Occupation of α2-adrenoceptors on macrophages results in the suppression of intracellular cAMP levels because these receptors are associated with a Gi-type protein (Bylund et al. 1994). Since there is a negative correlation between cAMP levels and inflammatory cytokine production in macrophages (van der Pouw Kraan, Boeije, Smeenk, Wijdenes, and Aarden 1995; Nemeth, Hasko, Szabo, and Vizi 1997; Hasko et al. 1998b) the decreased levels of cAMP may explain the increased TNF-α production. Generally, those catecholamine receptors that upon activation increase the accumulation of cAMP were shown to reduce the synthesis of TNF-α (Katakami, Nakao, Koizumi, Katakami, Ogawa, and Fujita 1988; Elenkov, Hasko, Kovacs, and Vizi 1995), IL-2 (Novogrodsky, Patya, Rubin, and Stenzel 1983), IFNγ (Ivashkiv, Ayres, and Glimcher 1994), and IL-12 (van der Pouw Kraan et al. 1995), while the increase in cAMP levels stimulates IL-4 (Lacour, Arrighi, Muller, Carlberg, Saurat, and Hauser 1994), IL-5 (Siegel, Zhang, Ray, and Ray 1995), IL-6 (Surprenant, Rassendren, Kawashima, North, and Buell 1996), and IL-10 (Platzer, Meisel, Vogt, Platzer, and Volk 1995). Thus, activation of neurotransmitter receptors that stimulate adenylate-cyclase leads to a shift toward T helper 2 (Th2)-type responses being anti-inflammatory and protective, whereas downregulation of intracellular cAMP stimulates a T helper 1 (Th1)-type response, resulting in inflammatory and cell-destructive effects.

Consistent with pathophysiology of depression, it was demonstrated that proinflammatory cytokine-induced behavioral changes were associated with alterations in the metabolism of monoamine neurotransmitters; serotonin, norepinephrine, and dopamine. Changes in catecholamine metabolism in brain regions being essential to the regulation of emotion including the limbic system (amygdala, hippocampus, and nucleus accumbens), might influence sickness behavior (Dunn, Wang, and Ando 1999; Gao, Jiang, Wilson, Zhang, Hong, and Liu 2002). In addition to the effects on neurotransmitter metabolism, inflammatory cytokines exert profound stimulatory effects on the HPA axis hormones as well as on

CRH (mRNA and protein), both in the hypothalamus and in the amygdala, a brain region that has an important role in fear and anxiety (Besedovsky and del Rey 1996; Capuron and Miller 2004). These effects are, in large part, mediated by the cross talk of cytokines and their receptors within the HPA axis tissues that facilitate the integration of cytokine signals (Silverman, Pearce, Biron, and Miller 2005). The cytokine signal transduction pathways that include mitogen-activated protein kinases (MAPKs) (Kaminska 2005; Szelenyi et al. 2006) and NF-κB, are also able to disrupt glucocorticoid receptor signaling (Wang, Wu, and Miller 2004), and thus might contribute to altered glucocorticoid-mediated feedback regulation of both CRH and of further proinflammatory cytokine release. In addition, activation of p38 MAPKs might contribute to disturbances in neurotransmitter function through effects on the serotonin transporter (Zhu, Carneiro, Dostmann, Hewlett, and Blakely 2005).

The first monoamine hypothesis of depression was based on the observation that depression was associated with a decrease in noradrenergic neurotransmission (Schildkraut 1965) accompanied with supersensitivity of the inhibitory, presynaptic α2-ARs leading to postsynaptic β-adrenoreceptor up-regulation. This was first of all based on the observed antidepressant effect of the monoamine transporter inhibitors that increase the biophase level of the monoamine in question. In the meantime, however, a number of studies have found significantly higher plasma, urine, and cerebrospinal fluid (CSF) NE levels in patients with depression than in controls (Potter et al. 1985; Maes, Minner, Suy, Vandervorst, and Raus 1991). Consequently, this catecholamine-depletion hypothesis has become strongly disputed. Considering the complexity of the phenomenon, a "dysregulation hypothesis" was proposed, in other words, an impaired negative feedback on the presynaptic neuron might cause increased/decreased NE release. Evidence was also described for a subsensitivity instead of supersensitivity of the presynaptic α2-ARs (Kafka and Paul 1986; Maes, Van Gastel, Delmeire, and Meltzer 1999a). This subsensitivity of the central α2-ARs may be the cause of an impaired negative feedback on the presynaptic catecholaminergic neuron, which in turn may induce the loss of inhibition to external stimuli of noradrenergic output in response to any activation. In contrast, in animal depression models (see later) the enhancing effect of α2-AR antagonist on the endogenous release of NE could be demonstrated, suggesting that the negative regulatory α2-ARs were not desensitized (Vizi, Zsilla, Caron, and Kiss 2004; Gilsbach et al. 2006). Thus, dysregulation of the NE system may also contribute to the pathophysiology of depression (Maes et al. 1999b), while a number of other theories are also trying to understand the background of this heterogeneous disorder.

Effects of cytokines on the noradrenergic system have also been described. IL-1 was shown to stimulate hypothalamic and preoptic noradrenergic neurotransmission (Dunn 1988), similar to the effects observed after administration of various forms of IL-1 (Dunn et al. 1999). There are inconsistent data for other proinflammatory cytokines on their influence on noradrenergic neurotransmission. IL-2 showed similar effects as IL-1, while other studies reported that TNF-α inhibited NE release from the median eminence (Elenkov, Kovacs, Duda, Stark, and Vizi 1992).

The immunomodulatory effects of dopamine (DA) are not as definite and well studied as those of NE. It should be emphasized that, as for NE, the nature of the immunomodulatory effect of DA is determined by the type/subtype of the receptor occupied by it. For example, it has been shown that the suppressive action of DA on

IL-12 production in cultured macrophages prevails partly through β2-adrenoceptors (Hasko, Szabo, Nemeth, and Deitch 2002), whereas a role for dopamine D2 receptors has been suggested for the observed in vivo suppression of interferon-γ (IFN-γ) and TNF-α (Sternberg, Wedner, Leung, and Parker 1987; Ritchie, Ashby, Knight, and Judd 1996).

Disturbances in the serotonergic neurotransmission also play a causal role in the pathophysiology of depression. Several neurochemical changes in the 5-HT system are found in depressed people. Tryptophan (Trp), the precursor of 5-HT, has to compete with other competing amino acids and certain changes in 5-HT metabolism have been shown. The effect of 5HT is also dependent on the type/subtype of the receptors involved in its action. 5-HTT is located on the presynaptic membrane whereas 5-HT is located in cell bodies in the raphe nuclei (RN) where it regulates 5-HT levels in the synaptic cleft by modulating the reuptake of 5-HT into the presynaptic cell (Staley, Malison, and Innis 1998). Other central changes of the 5-HT system in depression include changes in 5-HT2 and 5-HT1A brain receptors (Lesch and Mossner 1998).

Concerning the serotonergic system, cytokines have substantial effects on its function in the brain and in the periphery. Peripheral and central administration of IL-1β, IFN-γ, and TNF-α significantly increase extracellular serotonin (5-hydroxy-triptamine, 5-HT) concentrations in several brain areas of rats (Clement et al. 1997).

Cytokines, such as IL-1, IFN-γ, and TNF-α were shown to stimulate 5-HT neurotransmission and to reduce the production of 5-HT (Heyes et al. 1992). These proinflammatory cytokines have also been shown to up-regulate the 5-HTT, causing a depletion in extracellular 5-HT (Morikawa, Sakai, Obara, and Saito 1998; Mossner, Heils, Stober, Okladnova, Daniel, and Lesch, 1998). No changes were found after IL-6 injection. Among the anti-inflammatory cytokines, IL-4 was shown to induce a reduction in 5-HT uptake (Mossner, Daniel, Schmitt, Albert, and Lesch 2001).

Finally, we should emphasize that the monoamine theory of depression is only one, oversimplified version of the known possibilities, although its importance in the therapy of depression is not questionable and the regulatory role of the extracellular monoamine level both on sleep and on inflammatory immune response is of great importance.

Nevertheless, it is beyond the scope of this chapter, and depression is known to be a pleiotropic entity with a very complex background that should be studied in its whole diversity.

18.7 Sleep Disturbances in Diseased States Associated with Altered Cytokine Profile

18.7.1 Infection

Although it is well recognized that infection is commonly associated with somnolence and fatigue, the mechanism of such behaviors has only been investigated within the last 10 years. For instance, it is now believed that muramyl peptide, a common microbial product, and bacterial endotoxin (LPS) stimulate the release of proinflammatory cytokines that interact with specific neurohormones and neurotransmitters in the brain to produce somnogenic activity (Krueger and Majde 1994).

Some human studies suggest that plasma levels of IL-1 peak at the onset of slow-wave sleep in healthy human volunteers and levels of IL-1 in cerebrospinal fluid increase during sleep (Moldofsky, Lue, Eisen, Keystone, and Gorczynski 1986; Lue, Bail, Jephthah-Ochola, Carayanniotis, Gorczynski, and Moldofsky 1988). Specific cytokines therefore seem to play a role in sleep regulation, particularly during an infection.

Although most research in this area have focused on IL-1, there is a growing evidence that TNF-α and interferons may also have somnogenic activities while IL-2 and IL-6 probably have not (Kronfol and Remick 2000).

Sleep regulation is a complex phenomenon and involves interactions between neuropeptides, biogenic amines, and other neurotransmitters. The exact role played by specific cytokines remains to be determined.

18.7.2 Depression

Sleep disturbances are an integral part of depressive disorder. Insomnia is a particularly frequent complaint, and it is reported by more than 90% of depressed patients (Thase 1999). Depressive disorders have multiple etiologies and might be well characterized as conditions of immune activation, especially hyperactivity of innate immune inflammatory responses. Increasing amounts of data suggest that inflammatory responses play an important role in the pathophysiology of depression. Depressed patients have been found to have higher levels of proinflammatory cytokines, acute phase proteins, chemokines and cellular adhesion molecules (Maes et al. 1991, 1995; Wichers and Maes 2002).

In addition, therapeutic administration of the cytokine interferon-α, leads to depression in up to 50% of patients. Moreover, proinflammatory cytokines have been proven to interact with many of the pathophysiological domains that characterize depression, including neurotransmitter metabolism, neuroendocrine function, synaptic plasticity, and behavior. Stress, which can precipitate depression, is also able to promote inflammatory responses through effects on sympathetic and parasympathetic nervous system pathways (Pavlov and Tracey 2005).

Major depression and the stress response share a number of similar phenomena, mediators and circuitries. Psychological stress is a common risk factor for the development of major depression in every culture examined, and most initial episodes of major depression are preceded by an identifiable stressor (Kendler, Thornton, and Gardner 2000). Consistent with the notion that stress might provide a link between depression and inflammation, more and more data indicate that psychological stress activates proinflammatory cytokines and their signaling pathways both in the periphery and in the CNS.

Major depression, however, is a complex disorder that can be divided into two major subtypes: melancholic and atypical. The features of melancholic depression include insomnia (most often early morning awakening), loss of appetite, weight loss, accompanied with negative behavioral changes, while the second major subtype is major depression with mostly opposite, atypical features characterized partly by hypersomnia, hyperphagia, lethargy, and fatigue.

It was supposed that melancholic depression represents activation of the principal effectors of the stress resulting in higher NE and CRH levels. Centrally, NE acts as a major alarm-producing neurotransmitter in the brain that inhibits sleeping. It was shown, that depressed patients had significantly higher NE as well as plasma and cerebrospinal fluid (CSF) cortisol levels, while despite their hypercortisolism, depressed patients had normal levels of plasma ACTH and that of CRH in their CSF (Wong et al. 2000; Leonard 2001b). Symptoms of depression can often be observed after treatment of patients with high doses of IFN-α or IL-2 (e.g., in chronic hepatitis or in malignant melanoma) (Capuron and Miller 2004). However, great differences exist in the prevalence of the development of depressive symptoms across studies. These symptoms include abnormal sleep patterns, irritability, anxiety, cognitive impairments, lethargy, and anorexia (Kronfol and Remick 2000; Corcos, Guilbaud, Hjalmarsson, Chambry, and Jeammet 2002; Wichers and Maes 2002). In major depression, increased serum levels of somnogenic cytokines, such as IL-1, IL-6 were demonstrated. Increases in the plasma concentration and in vitro production of IL-1, IL-6, soluble IL-2 receptors, soluble IL-6 receptors, and acute phase proteins were reported in patients with major depression (Maes, Bosmans, Meltzer, Scharpe, and Suy 1993; Maes et al. 1995). It was concluded from these data that the increase in proinflammatory cytokines in patients with major depression seemed to correlate with the severity of the illness and measures of HPA hyperactivity.

Unfortunately, these observations have not been replicated consistently. Most investigators could confirm the increase in the plasma levels of acute phase proteins (Leonard 2001a), while others found a reduction in the secretion of IL-1β, IL-2, and in IL-3-like activity in depressed patients as compared to control subjects (Weizman, Laor, Podliszewski, Notti, Djaldetti, and Bessler 1994). This finding and other reports of mild leukocytosis, neutrophilia with elevated C-reactive protein and complement components (Kronfol and House 1989; Leonard 2001a) suggests a mild inflammatory response in depression, possibly initiated by cytokines since these elevated proinflammatory mediator levels induce hypercortisolemia and a hypernoradrenergic state (Wong et al. 2000; Gold and Chrousos 2002).

Growing evidence suggests that overactivation of innate immune responses following stress and during depression might come at the expense of decreased cellular and humoral acquired immune responses (Raison and Miller 2003). Activation of the stress system might promote cytokine production through several mechanisms (Fig. 18.2). Despite of suppression certain immune process, activation of the sympathetic nervous system (SNS) has been linked in several studies to proinflammatory activation in the periphery, which might, in turn, influence inflammatory processes in the CNS. It was demonstrated that stress-induced activation of NF-κB in peripheral blood mononuclear cells appeared to be dependent on norepinephrine and can be abrogated by α1-adrenoceptor blockade (Bierhaus et al. 2003). Chronic β-adrenoceptor blockade reduces plasma levels of IL-6 in concert with symptomatic improvement in patients with congestive heart failure (Murray, Prabhu, and Chandrasekar 2000; Mayer, Holmer, Hengstenberg, Lieb, Pfeifer, and Schunkert 2005). Decrease in vagal activity in response to stress might also promote inflammation, given the evidence that efferent vagal activity inhibits NF-κB activation (and the release of TNF-α from macrophages) via cholinergic signaling through the α-7 subunit of the nicotinic acetylcholine receptor (Pavlov and Tracey

2005). Finally, chronic stress promotes the development of glucocorticoid resistance, which is associated with increased cytokine production and might also release the sympathetic nervous system from inhibitory control, further promoting inflammatory activation (Leonard 2001b). The exact role of cytokines in the pathophysiology of mood disorder however, remains to be clarified (Raison, Capuron, and Miller 2006).

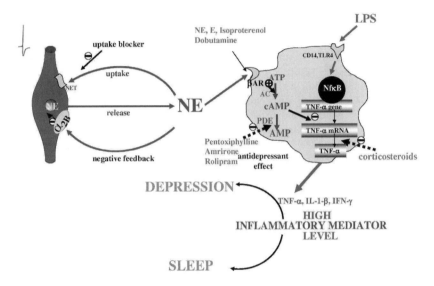

FIGURE 18.2. The inflammation–sleep-depression loop. An inadequate response to cognitive or noncognitive stimuli shifts the balance both between the release and uptake of monoamines and between the production of proinflammatory and anti-inflammatory cytokines. Norepinephrine (NE) release is negatively regulated by the presynaptic α2B-adrenoceptor, while for its uptake the norepinephrine transporter (NET) is responsible. Altered release, uptake, or metabolism of relevant neurotransmitters such as NE, are known to be involved in the development of depression. Uptake of NE is inhibited by monoamine-transporter blocking antidepressants. Activation of the innate immune system is linked to activation of nuclear-factor-kappa-B (NF-κB) through Toll-like receptors (TLR) during immune challenge (e.g., with LPS). This leads to an inflammatory response including the release of the proinflammatory cytokines TNF-α, IL-1, and IL-6. These cytokines are able to access the brain via various mechanisms and their signals participate in pathways known to be involved in sleep regulation and in the development of depression. Cytokine production might also be increased within the brain by stress. NE provides a dampening signal on LPS-induced inflammatory response via α2-adrenoceptors on macrophages by increasing intracellular cAMP levels. Drugs, inhibiting cAMP degradation (PDE inhibitors, e.g., rolipram) exhibit antidepressant effects. Corticosteroids alleviate the inflammatory response by regulating gene transcription via interfering with transcription factor binding to DNA response elements, such as NF-κB. High levels of inflammatory mediators and dysregulation of monoamine neurotransmitters play important roles both in physiological and pathological sleep regulation and in behavioral changes. Feedback mechanisms occur at several levels and participate in the common regulatory cascades forming the inflammation–sleep-depression loop.

18.8 Cytokine Balance and Antidepressants

18.8.1 Effect of Multitarget Drugs (e.g., Monoamine Transporter Blockers)

More recently, a number of studies focused on the pro and anti-inflammatory cytokine balance in major depressive disorders. A dysregulation within the cytokine balance could induce depressive symptoms due to lower levels of anti-inflammatory cytokines and higher levels of proinflammatory cytokines. Anti-inflammatory cytokines or cytokine receptors are known to evoke an antiinflammatory state both on their own (IL-10, TGFβ) receptors and also by the blockade of the binding of proinflammatory stimuli to their cell surface receptors (IL-1ra, soluble TNF receptors II). Growing evidence supports the idea of an anti-inflammatory action of antidepressants, which could explain their efficiency on depressive symptoms. In some clinical studies, it has been reported that antidepressants may attenuate the effects of proinflammatory cytokines by increasing the production of anti-inflammatory cytokines such as IL-10 and IL-1ra. Most antidepressant treatments could significantly reduce the IFNγ/IL-10 ratio and have negative immunoregulatory effects (Maes et al. 1999b).

Currently, the most important and widely used antidepressants are the inhibitors of monoamine-transporters, among them the tricyclic antidepressants (TCAs) and their improved counterparts. The correlation between antidepressant treatment and immunological changes demonstrated that treatment with TCAs had complex time-dependent immunoregulatory effects. Short-term administration of TCAs was shown to stimulate cell-mediated immunity and to increase IL-1 bioactivity, while prolonged treatment significantly increased the secretion of IL-10 (Maes et al. 1999b). In contrast, other groups concluded that the effect of chronic treatment with antidepressants was probably not mediated by the effect of TCAs on peripheral cytokine production, because even a 5-week-long treatment with TCAs did not alter the LPS-induced expression of TNF-α, while IL-1β release is unknown (Yirmiya et al. 2001).

In our earlier studies, it was shown that cytokine production was under tonic, sympathetic control (Vizi 1998; Elenkov et al. 2000). In an animal depression model achieved by reserpine treatment, a significantly higher TNF-α response was observed (Szelenyi et al. 2000a). Recently it was demonstrated that both an acute treatment of mice with the antidepressant desipramine (an inhibitor of norepinephrine transporter (NET)), and genetic modifications of mice that lead to NET deficiency (NET-KO) resulted in a significant decrease in the LPS-evoked TNF-α response (Szelenyi and Selmeczy 2002; Selmeczy, Szelenyi, and Vizi 2003).

The extent and/or the duration of monoamine neurotransmitter action are one of the key points in the development of therapies for depression. The extracellular monoamine concentration is basically determined by the balance of its release and uptake. The release of NE, e.g., is controlled by the negative feedback mechanism of presynaptic α2-ARs (Vizi 1979; Kiss et al. 1995), as it was discussed above. For the rapid removal of norepinephrine released from sympathetic neurons, its reuptake via NET is responsible. Antidepressants have been reported to inhibit the activity of monoamine transporters that result in an increased biophase level of monoamines. A similar effect could be achieved by genetic removal of the transporter gene as it was realized in NET-KO (Xu et al. 2000), DAT-KO (Moron, Brockington, Wise, Rocha, and Hope 2002), and 5HTT-KO mice (Alexandre et al. 2006). These animals behave like mice chronically treated with antidepressants, exhibiting high extracellular monoamine levels. Long-term inhibition of the NET by antidepressants has been reported to change the density and function of pre and postsynaptic $\alpha2$-ARs, which may contribute to the antidepressant effects of NET inhibitors such as desipramine. In the NET-KO animals, it was demonstrated that the density of $\alpha2$-AR is up-regulated in the brainstem, hippocampus, and striatum (Gilsbach et al. 2006). In these mice the $\alpha2$-AR autoreceptors are not desenzitized and the inhibitory tone on NE release is stronger as the consequence of elevated extracellular NE concentration (Vizi et al. 2004).

Since the extracellular levels of monoamines is highly dependent on the activity of their transporters, it was assumed that, in acute treatment of mice, not only with desipramine but also other monoamine transporter inhibitors exhibiting antidepressant characteristics (dopamine transporter and serotonin transporter inhibitors), had modulatory effects on the inflammatory immune response. Whereas an approximately 3-week long chronic treatment is necessary to start the therapeutic antidepressant effect of these drugs, it was studied whether the immunomodulatory effect was also present after chronic treatments.

It was revealed recently (Szelenyi, Selmeczy, and Vizi 2004) that the immunomodulatory effect was more significant on the LPS-induced cytokine production in the acute than that of chronic treatments; showing that the high extracellular monoamine level, i.e., the presence of the monoamine transporter inhibitor was necessary during LPS induction. All the three monoamine transporter inhibitors were effective to different extents in our experiments (Fig. 18.3), which supported the hypothesis that the immunomodulatory effect of these drugs was correlated with the actual increased in extracellular monoamine levels evoked by them. However, this was not in direct correlation with their mood effects. Concerning the crucial role that cAMP plays in cytokine production, it may be supposed that TCAs produce their immunomodulatory effects through this signaling mechanism. The antidepressant effects of phosphodiesterase inhibitors (like rolipram) (Fig. 18.2) (Zhu, Mix, and Winblad 2001) and the overexpression of the cAMP response element binding protein (CREB) in the hippocampus after chronic antidepressant administration (Chen, Shirayama, Shin, Neve, and Duman 2001), might support this assumption. An increasing amount of evidence is available, however, that chronic treatment with TCAs has pleiotropic effects in addition to blocking the monoamine uptake systems, which contribute to exert their antidepressive action in the CNS.

Clinical observations indicate that TCAs may attenuate the adverse effects of glucocorticoids, and consequently the release of proinflammatory cytokines is decreased (Chen et al. 2001). As well as the interactive relationship between NE release regulating $\alpha2$-adrenoceptor sensitivity and the TNF-α levels in the CNS (Nickola, Ignatowski, Reynolds, and Spengler 2001) TCAs may exert their effect also through interaction with other monoamine uptake systems, for example, the dopaminergic and serotonergic systems, that were shown to be promiscuous (Moron et al. 2002; Selmeczy et al. 2003; Vizi et al. 2004).

chronic antidepressant administration

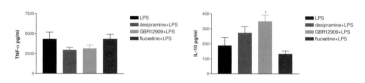

acute antidepressant administration

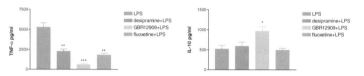

FIGURE 18.3. Comparison of the effects of acute and chronic antidepressant treatment on the cytokine balance. In acute experiments, Balb/c mice were injected intraperitoneally (i.p.) with the drugs (10 mg/kg), 60 min before the administration of lipopolysaccharide (LPS) (10 mg/kg i.p.). In the chronic protocol the same procedure was daily repeated for 3 weeks but LPS-induction was performed only on the last day. Control animals received the same diluent as vehicle and the same volume of saline. Blood sample was taken 90 min after the LPS administration, and the plasma TNF-α or IL-10 content was determined by ELISA technique.

The suppressive effect of increased serotonin release on the LPS-induced proinflammatory cytokine secretion cannot be attributed to the serotonin-releasing properties of selective serotonin reuptake inhibitors (SSRIs) (Connor, Dennedy, Harkin, and Kelly 2001). Sleep continuity is improved by the rise in the synaptic levels of serotonin. Among tricyclic antidepressants trimipramine and amitriptyline are the best to improve sleep disorders. Stimulation of serotonin-2 (5-HT2) receptors is thought to underlie insomnia and changes in sleep architecture are seen with selective serotonin reuptake inhibitors (SSRIs) or serotonin-norepinephrine reuptake inhibitors (SNRIs) (Thase 1999; Aszalos 2006).

The plasma concentration of proinflammatory cytokines might also be in correlation with the treatment responsiveness. Increased plasma concentrations of IL-6 and acute-phase proteins have been found in patients with a history of nonresponse to antidepressants when compared with treatment-responsive patients (Maes, Bosmans, De Jongh, Kenis, Vandoolaeghe, and Neels 1997). Similarly, patients with evidence of increased inflammatory activity before treatment were reported to be less responsive to antidepressants or sleep deprivation (a potent short-term mood elevator) (Lanquillon, Krieg, Bening-Abu-Shach, and Vedder 2000; Benedetti, Lucca, Brambilla, Colombo, and Smeraldi 2002). However, if the efficiency of antidepressants is primarily realized by their anti-inflammatory effects, anti-inflammatory drugs should also be effective in the treatment of major depressive disorders. Our results suggest that other signaling routes may also participate in the immunomodulatory role of antidepressant drugs, in other words, they have multiple targets that might explain the lack of correlation between their effectiveness and specificity in blocking a certain monoamine transporter, as well as, the limited antidepressant efficacy of anti-inflammatory drugs (Szelenyi et al. 2004).

18.9 Conclusions

Cytokines are capable of acting as neuromodulators within the CNS. As such, they affect important brain activities such as sleep, appetite, and neuroendocrine regulation. An inadequate response to either internal or external challenge may induce inflammation, depression and sleep abnormalities by increasing the production of proinflammatory cytokines that are normally present in very low concentrations in the brain. An increased amount of inflammatory mediators affects the central and peripheral monoaminergic neurotransmitter systems, the activity of the HPA axis, and the innate immune response. Proinflammatory cytokines alter the cross talk and mutual interactions among the participants of these systems. As a consequence, a disturbance in one of these systems will also affect the other systems resulting in a long loop among them.

The inhibitory effect of TCAs both on the various monoamine uptake systems and on the production of proinflammatory cytokines also supports the idea that depression, sleep abnormalities, and some other disorders affecting the CNS, are closely linked to immunological alterations. This is not surprising as cytokines are involved not only in the immune response but also in a variety of other physiological and pathological processes, including the events in the peripheral nervous system and the CNS, that is, they are both immunoregulators and neuromodulators. There are currently no conclusive data to prove whether the inhibition of proinflammatory cytokines or the increase of biophase level of monoamines is sufficient for antidepressive effects, because of the problems of studying each in isolation. In addition, the heterogeneity of depressive and sleep disorders suggests that their study and therapy need a more complex approach. Therefore, we would like to call the attention to the importance of multiple targets in drug development rather than specificity. The observation however, that inflammation plays a crucial role both in psychiatric and sleep disturbances calls the attention to their common features, that should have therapeutic consequences and might also offer new targets in the development of antidepressants and sleep ameliorating drugs.

Acknowledgments. This work was supported by grants from the Hungarian Research Fund (OTKA) No. T-46896 (J. Sz.); No. TS-049868 (E.S.V.), and that of from the Scientific Committee of Ministry of Health No. 123/2003 (E.S.V.).

References

Abrass, C.K., O'Connor, S.W., Scarpace, P.J., and Abrass, I.B. (1985) Characterization of the beta-adrenergic receptor of the rat peritoneal macrophage. *J Immunol* 135, 1338–1341.

Akiyoshi, M., Shimizu, Y., and Saito, M. (1990) Interleukin-1 increases norepinephrine turnover in the spleen and lung in rats. *Biochem Biophys Res Commun* 173, 1266–1270.

Alexandre, C., Popa, D., Fabre, V., Bouali, S., Venault, P., Lesch, K.P., Hamon, M., and Adrien, J. (2006) Early life blockade of 5–hydroxytryptamine 1A receptors normalizes sleep and depression-like behavior in adult knock-out mice lacking the serotonin transporter. *J Neurosci* 26, 5554–5564.

Asadullah, K., Sabat, R., Friedrich, M., Volk, H.D., and Sterry, W. (2004) Interleukin-10: An important immunoregulatory cytokine with major impact on psoriasis. *Curr Drug Targets Inflamm Allergy* 3, 185–192.

Aszalos, Z. (2006) Effects of antidepressants on sleep. *Orv Hetil* 147, 773–783.

Banks, W.A., Ortiz, L., Plotkin, S.R., and Kastin, A.J. (1991) Human interleukin (IL) 1 alpha, murine IL-1 alpha and murine IL-1 beta are transported from blood to brain in the mouse by a shared saturable mechanism. *J Pharmacol Exp Ther* 259, 988–996.

Benca, R.M., and Quintas, J. (1997) Sleep and host defenses: A review. *Sleep* 20, 1027–1037.

Benedetti, F., Lucca, A., Brambilla, F., Colombo, C., and Smeraldi, E. (2002) Interleukine-6 serum levels correlate with response to antidepressant sleep deprivation and sleep phase advance. *Prog Neuropsychopharmacol Biol Psychiatry* 26, 1167–1170.

Besedovsky, H., del Rey, A., Sorkin, E., and Dinarello, C.A. (1986) Immunoregulatory feedback between interleukin-1 and glucocorticoid hormones. *Science* 233, 652–654.

Besedovsky, H.O., and del Rey, A. (1996) Immune-neuro-endocrine interactions: Facts and hypotheses. *Endocr Rev* 17, 64–102.

Besedovsky, H.O., del Rey, A., Klusman, I., Furukawa, H., Monge Arditi, G., and Kabiersch, A. (1991) Cytokines as modulators of the hypothalamus-pituitary-adrenal axis. *J Steroid Biochem Mol Biol* 40, 613–618.

Bierhaus, A., Wolf, J., Andrassy, M., Rohleder, N., Humpert, P.M., Petrov, D., Ferstl, R., von Eynatten, M., Wendt, T., Rudofsky, G., Joswig, M., Morcos, M., Schwaninger, M., McEwen, B., Kirschbaum, C., and Nawroth, P.P. (2003) A mechanism converting psychosocial stress into mononuclear cell activation. *Proc Natl Acad Sci U S A* 100, 1920–1925.

Bloom, B.J., Owens, J.A., McGuinn, M., Nobile, C., Schaeffer, L., and Alario, A.J. (2002) Sleep and its relationship to pain, dysfunction, and disease activity in juvenile rheumatoid arthritis. *J Rheumatol* 29, 169–173.

Bonnet, M.H., and Arand, D.L. (1995) 24-Hour metabolic rate in insomniacs and matched normal sleepers. *Sleep* 18, 581–588.

Bourne, H.R., Lichtenstein, L.M., Melmon, K.L., Henney, C.S., Weinstein, Y., and Shearer, G.M. (1974) Modulation of inflammation and immunity by cyclic AMP. *Science* 184, 19–28.

Braat, H., Peppelenbosch, M.P., and Hommes, D.W. (2003) Interleukin-10-based therapy for inflammatory bowel disease. *Expert Opin Biol Ther* 3, 725–731.

Breder, C.D., Hazuka, C., Ghayur, T., Klug, C., Huginin, M., Yasuda, K., Teng, M., and Saper, C.B. (1994) Regional induction of tumor necrosis factor alpha expression in the mouse brain after systemic lipopolysaccharide administration. *Proc Natl Acad Sci U S A* 91, 11393–11397.

Bylund, D.B., Eikenberg, D.C., Hieble, J.P., Langer, S.Z., Lefkowitz, R.J., Minneman, K.P., Molinoff, P.B., Ruffolo, R.R., Jr., and Trendelenburg, U. (1994) International Union of Pharmacology nomenclature of adrenoceptors. *Pharmacol Rev* 46, 121–136.

Capuron, L., and Miller, A.H. (2004) Cytokines and psychopathology: Lessons from interferon-alpha. *Biol Psychiatry* 56, 819–824.

Chen, A.C., Shirayama, Y., Shin, K.H., Neve, R.L., and Duman, R.S. (2001) Expression of the cAMP response element binding protein (CREB) in hippocampus produces an antidepressant effect. *Biol Psychiatry* 49, 753–762.

Chong, Y.H., Shin, Y.J., and Suh, Y.H. (2003) Cyclic AMP inhibition of tumor necrosis factor alpha production induced by amyloidogenic C-terminal peptide of Alzheimer's amyloid precursor protein in macrophages: Involvement of multiple intracellular pathways and cyclic AMP response element binding protein. *Mol Pharmacol* 63, 690–698.

Clement, H.W., Buschmann, J., Rex, S., Grote, C., Opper, C., Gemsa, D., and Wesemann, W. (1997) Effects of interferon-gamma, interleukin-1 beta, and tumor necrosis factor-alpha on the serotonin metabolism in the nucleus raphe dorsalis of the rat. *J Neural Transm* 104, 981–991.

Collier, J., and Chapman, R. (2001) Combination therapy with interferon-alpha and ribavirin for hepatitis C: Practical treatment issues. *BioDrugs* 15, 225–238.

Connor, T.J., Dennedy, M.C., Harkin, A., and Kelly, J.P. (2001) Methylenedioxymethamphetamine-induced suppression of interleukin-1beta and tumour necrosis factor-alpha is not mediated by serotonin. *Eur J Pharmacol* 418, 147–152.

Corcos, M., Guilbaud, O., Hjalmarsson, L., Chambry, J., and Jeammet, P. (2002) Cytokines and depression: An analogic approach. *Biomed Pharmacother* 56, 105–110.

Dantzer, R. (2004) Cytokine-induced sickness behaviour: A neuroimmune response to activation of innate immunity. *Eur J Pharmacol* 500, 399–411.

del Rey, A., Besedovsky, H.O., Sorkin, E., da Prada, M., and Arrenbrecht, S. (1981) Immunoregulation mediated by the sympathetic nervous system, II. *Cell Immunol* 63, 329–334.

Denicoff, K.D., Rubinow, D.R., Papa, M.Z., Simpson, C., Seipp, C.A., Lotze, M.T., Chang, A.E., Rosenstein, D., and Rosenberg, S.A. (1987) The neuropsychiatric effects of treatment with interleukin-2 and lymphokine-activated killer cells. *Ann Intern Med* 107, 293–300.

Dinarello, C.A. (1998) Interleukin-1, interleukin-1 receptors and interleukin-1 receptor antagonist. *Int Rev Immunol* 16, 457–499.

Dinges, D.F., Douglas, S.D., Hamarman, S., Zaugg, L., and Kapoor, S. (1995) Sleep deprivation and human immune function. *Adv Neuroimmunol* 5, 97–110.

Dodt, C., Breckling, U., Derad, I., Fehm, H.L., and Born, J. (1997) Plasma epinephrine and norepinephrine concentrations of healthy humans associated with nighttime sleep and morning arousal. *Hypertension* 30, 71–76.

Dunn, A.J. (1988) Systemic interleukin-1 administration stimulates hypothalamic norepinephrine metabolism parallelling the increased plasma corticosterone. *Life Sci* 43, 429–435.

Dunn, A.J., Wang, J., and Ando, T. (1999) Effects of cytokines on cerebral neurotransmission. Comparison with the effects of stress. *Adv Exp Med Biol* 461, 117–127.

Dusheiko, G. (1997) Side effects of alpha interferon in chronic hepatitis C. *Hepatology* 26, 112S–121S.

Elenkov, I.J., and Vizi, E.S. (1991) Presynaptic modulation of release of noradrenaline from the sympathetic nerve terminals in the rat spleen. *Neuropharmacology* 30, 1319–1324.

Elenkov, I.J., Hasko, G., Kovacs, K.J., and Vizi, E.S. (1995) Modulation of lipopolysaccharide-induced tumor necrosis factor-alpha production by selective alpha- and beta-adrenergic drugs in mice. *J Neuroimmunol* 61, 123–131.

Elenkov, I.J., Wilder, R.L., Chrousos, G.P., and Vizi, E.S. (2000) The sympathetic nerve–An integrative interface between two supersystems: The brain and the immune system. *Pharmacol Rev* 52, 595–638.

Elenkov, I.J., Kovacs, K., Duda, E., Stark, E., and Vizi, E.S. (1992) Presynaptic inhibitory effect of TNF-alpha on the release of noradrenaline in isolated median eminence. *J Neuroimmunol* 41, 117–120.

Gao, H.M., Jiang, J., Wilson, B., Zhang, W., Hong, J.S., and Liu, B. (2002) Microglial activation-mediated delayed and progressive degeneration of rat nigral dopaminergic neurons: Relevance to Parkinson's disease. *J Neurochem* 81, 1285–1297.

Garcia-Garcia, F., Yoshida, H., and Krueger, J.M. (2004) Interleukin-8 promotes non-rapid eye movement sleep in rabbits and rats. *J Sleep Res* 13, 55–61.

Gilsbach, R., Faron-Gorecka, A., Rogoz, Z., Bruss, M., Caron, M.G., Dziedzicka-Wasylewska, M., and Bonisch, H. (2006) Norepinephrine transporter knockout-induced up-regulation of brain alpha2A/C-adrenergic receptors. *J Neurochem* 96, 1111–1120.

Gold, P.W., and Chrousos, G.P. (2002) Organization of the stress system and its dysregulation in melancholic and atypical depression: High vs low CRH/NE states. *Mol Psychiatry* 7, 254–275.

Hasko, G., Szabo, C., Nemeth, Z.H., and Deitch, E.A. (2002) Dopamine suppresses IL-12 p40 production by lipopolysaccharide-stimulated macrophages via a beta-adrenoceptor-mediated mechanism. *J Neuroimmunol* 122, 34–39.

Hasko, G., Szabo, C., Nemeth, Z.H., Salzman, A.L., and Vizi, E.S. (1998a) Stimulation of beta-adrenoceptors inhibits endotoxin-induced IL-12 production in normal and IL-10 deficient mice. *J Neuroimmunol* 88, 57–61.

Hasko, G., Szabo, C., Nemeth, Z.H., Salzman, A.L., and Vizi, E.S. (1998b) Suppression of IL-12 production by phosphodiesterase inhibition in murine endotoxemia is IL-10 independent. *Eur J Immunol* 28, 468–472.

Heyes, M.P., Saito, K., Crowley, J.S., Davis, L.E., Demitrack, M.A., Der, M., Dilling, L.A., Elia, J., Kruesi, M.J., Lackner, A., and et al. (1992) Quinolinic acid and kynurenine pathway metabolism in inflammatory and non-inflammatory neurological disease. *Brain* 115 (Pt 5), 1249–1273.

Hofstetter, C., Flondor, M., Hoegl, S., Muhl, H., and Zwissler, B. (2005) Interleukin-10 aerosol reduces proinflammatory mediators in bronchoalveolar fluid of endotoxemic rat. *Crit Care Med* 33, 2317–2322.

Irwin, M., Thompson, J., Miller, C., Gillin, J.C., and Ziegler, M. (1999) Effects of sleep and sleep deprivation on catecholamine and interleukin-2 levels in humans: Clinical implications. *J Clin Endocrinol Metab* 84, 1979–1985.

Irwin, M., McClintick, J., Costlow, C., Fortner, M., White, J., and Gillin, J.C. (1996) Partial night sleep deprivation reduces natural killer and cellular immune responses in humans. *FASEB J* 10, 643–653.

Ivashkiv, L.B., Ayres, A., and Glimcher, L.H. (1994) Inhibition of IFN-gamma induction of class II MHC genes by cAMP and prostaglandins. *Immunopharmacology* 27, 67–77.

Izeboud, C.A., Mocking, J.A., Monshouwer, M., van Miert, A.S., and Witkamp, R.F. (1999) Participation of beta-adrenergic receptors on macrophages in modulation of LPS-induced cytokine release. *J Recept Signal Transduct Res* 19, 191–202.

Kafka, M.S., and Paul, S.M. (1986) Platelet alpha 2-adrenergic receptors in depression. *Arch Gen Psychiatry* 43, 91–95.

Kambayashi, T., Jacob, C.O., Zhou, D., Mazurek, N., Fong, M., and Strassmann, G. (1995) Cyclic nucleotide phosphodiesterase type IV participates in the regulation of IL-10 and in the subsequent inhibition of TNF-alpha and IL-6 release by endotoxin-stimulated macrophages. *J Immunol* 155, 4909–4916.

Kaminska, B. (2005) MAPK signalling pathways as molecular targets for anti-inflammatory therapy–From molecular mechanisms to therapeutic benefits. *Biochim Biophys Acta* 1754, 253–262.

Katakami, Y., Nakao, Y., Koizumi, T., Katakami, N., Ogawa, R., and Fujita, T. (1988) Regulation of tumour necrosis factor production by mouse peritoneal macrophages: The role of cellular cyclic AMP. *Immunology* 64, 719–724.

Kendler, K.S., Thornton, L.M., and Gardner, C.O. (2000) Stressful life events and previous episodes in the etiology of major depression in women: An evaluation of the "kindling" hypothesis. *Am J Psychiatry* 157, 1243–1251.

Kiss, J.P., Zsilla, G., Mike, A., Zelles, T., Toth, E., Lajtha, A., and Vizi, E.S. (1995) Subtype-specificity of the presynaptic alpha 2-adrenoceptors modulating hippocampal norepinephrine release in rat. *Brain Res* 674, 238–244.

Krigel, R.L., Padavic-Shaller, K.A., Rudolph, A.R., Konrad, M., Bradley, E.C., and Comis, R.L. (1990) Renal cell carcinoma: Treatment with recombinant interleukin-2 plus beta-interferon. *J Clin Oncol* 8, 460–467.

Kronfol, Z., and House, J.D. (1989) Lymphocyte mitogenesis, immunoglobulin and complement levels in depressed patients and normal controls. *Acta Psychiatr Scand* 80, 142–147.

Kronfol, Z., and Remick, D.G. (2000) Cytokines and the brain: Implications for clinical psychiatry. *Am J Psychiatry* 157, 683–694.

Krueger, J.M., and Majde, J.A. (1994) Microbial products and cytokines in sleep and fever regulation. *Crit Rev Immunol* 14, 355–379.

Krueger, J.M., Walter, J., Dinarello, C.A., Wolff, S.M., and Chedid, L. (1984) Sleep-promoting effects of endogenous pyrogen (interleukin-1). *Am J Physiol* 246, R994–R999.

Krueger, J.M., Obal, F.J., Fang, J., Kubota, T., and Taishi, P. (2001) The role of cytokines in physiological sleep regulation. *Ann N Y Acad Sci* 933, 211–221.

Lacour, M., Arrighi, J.F., Muller, K.M., Carlberg, C., Saurat, J.H., and Hauser, C. (1994) cAMP up-regulates IL-4 and IL-5 production from activated CD4+ T cells while decreasing IL-2 release and NF-AT induction. *Int Immunol* 6, 1333–1343.

Lanquillon, S., Krieg, J.C., Bening-Abu-Shach, U., and Vedder, H. (2000) Cytokine production and treatment response in major depressive disorder. *Neuropsychopharmacology* 22, 370–379.

Lavie, P. (2001) Sleep-wake as a biological rhythm. *Annu Rev Psychol* 52, 277–303.

Leonard, B.E. (2001a) Changes in the immune system in depression and dementia: Causal or co-incidental effects? *Int J Dev Neurosci* 19, 305–312.

Leonard, B.E. (2001b) The immune system, depression and the action of antidepressants. *Prog Neuropsychopharmacol Biol Psychiatry* 25, 767–780.

Lesch, K.P., and Mossner, R. (1998) Genetically driven variation in serotonin uptake: Is there a link to affective spectrum, neurodevelopmental, and neurodegenerative disorders? *Biol Psychiatry* 44, 179–192.

Lue, F.A., Bail, M., Jephthah-Ochola, J., Carayanniotis, K., Gorczynski, R., and Moldofsky, H. (1988) Sleep and cerebrospinal fluid interleukin-1-like activity in the cat. *Int J Neurosci* 42, 179–183.

Maes, M., Van Gastel, A., Delmeire, L., and Meltzer, H.Y. (1999a) Decreased platelet alpha-2 adrenoceptor density in major depression: Effects of tricyclic antidepressants and fluoxetine. *Biol Psychiatry* 45, 278–284.

Maes, M., Minner, B., Suy, E., Vandervorst, C., and Raus, J. (1991) Coexisting dysregulations of both the sympathoadrenal system and hypothalamic-pituitary-adrenal-axis in melancholia. *J Neural Transm Gen Sect* 85, 195–210.

Maes, M., Bosmans, E., Meltzer, H.Y., Scharpe, S., and Suy, E. (1993) Interleukin-1 beta: A putative mediator of HPA axis hyperactivity in major depression? *Am J Psychiatry* 150, 1189–1193.

Maes, M., Bosmans, E., De Jongh, R., Kenis, G., Vandoolaeghe, E., and Neels, H. (1997) Increased serum IL-6 and IL-1 receptor antagonist concentrations in major depression and treatment resistant depression. *Cytokine* 9, 853–858.

Maes, M., Meltzer, H.Y., Bosmans, E., Bergmans, R., Vandoolaeghe, E., Ranjan, R., and Desnyder, R. (1995) Increased plasma concentrations of interleukin-6, soluble interleukin-6, soluble interleukin-2 and transferrin receptor in major depression. *J Affect Disord* 34, 301–309.

Maes, M., Song, C., Lin, A.H., Bonaccorso, S., Kenis, G., De Jongh, R., Bosmans, E., and Scharpe, S. (1999b) Negative immunoregulatory effects of antidepressants: Inhibition of interferon-gamma and stimulation of interleukin-10 secretion. *Neuropsychopharmacology* 20, 370–379.

Maestroni, G.J., Conti, A., and Pierpaoli, W. (1986) Role of the pineal gland in immunity. Circadian synthesis and release of melatonin modulates the antibody response and antagonizes the immunosuppressive effect of corticosterone. *J Neuroimmunol* 13, 19–30.

Maestroni, G.J., Conti, A., and Pierpaoli, W. (1987) Role of the pineal gland in immunity: II. Melatonin enhances the antibody response via an opiatergic mechanism. *Clin Exp Immunol* 68, 384–391.

Maestroni, G.J., Cosentino, M., Marino, F., Togni, M., Conti, A., Lecchini, S., and Frigo, G. (1998) Neural and endogenous catecholamines in the bone marrow. Circadian association of norepinephrine with hematopoiesis? *Exp Hematol* 26, 1172–1177.

Mayer, B., Holmer, S.R., Hengstenberg, C., Lieb, W., Pfeifer, M., and Schunkert, H. (2005) Functional improvement in heart failure patients treated with beta-blockers is associated with a decline of cytokine levels. *Int J Cardiol* 103, 182–186.

Mills, P.J., and Dimsdale, J.E. (2004) Sleep apnea: A model for studying cytokines, sleep, and sleep disruption. *Brain Behav Immun* 18, 298–303.

Moldofsky, H., Lue, F.A., Eisen, J., Keystone, E., and Gorczynski, R.M. (1986) The relationship of interleukin-1 and immune functions to sleep in humans. *Psychosom Med* 48, 309–318.

Morikawa, O., Sakai, N., Obara, H., and Saito, N. (1998) Effects of interferon-alpha, interferon-gamma and cAMP on the transcriptional regulation of the serotonin transporter. *Eur J Pharmacol* 349, 317–324.

Moron, J.A., Brockington, A., Wise, R.A., Rocha, B.A., and Hope, B.T. (2002) Dopamine uptake through the norepinephrine transporter in brain regions with low levels of the dopamine transporter: Evidence from knock-out mouse lines. *J Neurosci* 22, 389–395.

Mossner, R., Daniel, S., Schmitt, A., Albert, D., and Lesch, K.P. (2001) Modulation of serotonin transporter function by interleukin-4. *Life Sci* 68, 873–880.

Mossner, R., Heils, A., Stober, G., Okladnova, O., Daniel, S., and Lesch, K.P. (1998) Enhancement of serotonin transporter function by tumor necrosis factor alpha but not by interleukin-6. *Neurochem Int* 33, 251–254.

Mulder, R.T., Ang, M., Chapman, B., Ross, A., Stevens, I.F., and Edgar, C. (2000) Interferon treatment is not associated with a worsening of psychiatric symptoms in patients with hepatitis C. *J Gastroenterol Hepatol* 15, 300–303.

Murray, D.R., Prabhu, S.D., and Chandrasekar, B. (2000) Chronic beta-adrenergic stimulation induces myocardial proinflammatory cytokine expression. *Circulation* 101, 2338–2341.

Musselman, D.L., Lawson, D.H., Gumnick, J.F., Manatunga, A.K., Penna, S., Goodkin, R.S., Greiner, K., Nemeroff, C.B., and Miller, A.H. (2001) Paroxetine for the prevention of depression induced by high-dose interferon alfa. *N Engl J Med* 344, 961–966.

Nemeth, Z.H., Hasko, G., Szabo, C., and Vizi, E.S. (1997) Amrinone and theophylline differentially regulate cytokine and nitric oxide production in endotoxemic mice. *Shock* 7, 371–375.

Nickola T.J., Ignatowski, T.A., Reynolds, J.L., and Spengler, R.N. (2001) Antidepressant drug-induced alterations in neuron-localized tumor necrosis factor-alpha mRNA and alpha(2)-adrenergic receptor sensitivity. *J Pharmacol Exp Ther* 297, 680–687.

Novogrodsky, A., Patya, M., Rubin, A.L., and Stenzel, K.H. (1983) Agents that increase cellular cAMP inhibit production of interleukin-2, but not its activity. *Biochem Biophys Res Commun* 114, 93–98.

Ohayon, M.M., and Caulet, M. (1996) Psychotropic medication and insomnia complaints in two epidemiological studies. *Can J Psychiatry* 41, 457–464.

Papanicolaou, D.A., Wilder, R.L., Manolagas, S.C., and Chrousos, G.P. (1998) The pathophysiologic roles of interleukin-6 in human disease. *Ann Intern Med* 128, 127–137.

Pavlov, V.A., and Tracey, K.J. (2005) The cholinergic anti-inflammatory pathway. *Brain Behav Immun* 19, 493–499.

Plata-Salaman, C.R. (1991) Immunoregulators in the nervous system. *Neurosci Biobehav Rev* 15, 185–215.

Platzer, C., Meisel, C., Vogt, K., Platzer, M., and Volk, H.D. (1995) Up-regulation of monocytic IL-10 by tumor necrosis factor-alpha and cAMP elevating drugs. *Int Immunol* 7, 517–523.

Potter, W.Z., Scheinin, M., Golden, R.N., Rudorfer, M.V., Cowdry, R.W., Calil, H.M., Ross, R.J., and Linnoila, M. (1985) Selective antidepressants and cerebrospinal fluid. Lack of specificity on norepinephrine and serotonin metabolites. *Arch Gen Psychiatry* 42, 1171–1177.

Raison, C.L., and Miller, A.H. (2003) When not enough is too much: The role of insufficient glucocorticoid signaling in the pathophysiology of stress-related disorders. *Am J Psychiatry* 160, 1554–1565.

Raison, C.L., Capuron, L., and Miller, A.H. (2006) Cytokines sing the blues: Inflammation and the pathogenesis of depression. *Trends Immunol* 27, 24–31.

Redwine, L., Hauger, R.L., Gillin, J.C., and Irwin, M. (2000) Effects of sleep and sleep deprivation on interleukin-6, growth hormone, cortisol, and melatonin levels in humans. *J Clin Endocrinol Metab* 85, 3597–3603.

Ritchie, P.K., Ashby, M., Knight, H.H., and Judd, A.M. (1996) Dopamine increases interleukin 6 release and inhibits tumor necrosis factor release from rat adrenal zona glomerulosa cells in vitro. *Eur J Endocrinol* 134, 610–616.

Rogers, N.L., Szuba, M.P., Staab, J.P., Evans, D.L., and Dinges, D.F. (2001) Neuroimmunologic aspects of sleep and sleep loss. *Semin Clin Neuropsychiatry* 6, 295–307.

Rothwell, N.J. (1997) Sixteenth Gaddum Memorial Lecture December 1996. Neuroimmune interactions: The role of cytokines. *Br J Pharmacol* 121, 841–847.

Schildkraut, J.J. (1965) The catecholamine hypothesis of affective disorders: A review of supporting evidence. *Am J Psychiatry* 122, 509–522.

Selmeczy, Z., Szelenyi, J., and Vizi, E.S. (2003) Intact noradrenaline transporter is needed for the sympathetic fine-tuning of cytokine balance. *Eur J Pharmacol* 469, 175–181.

Shakhar, G., and Ben-Eliyahu, S. (1998) In vivo beta-adrenergic stimulation suppresses natural killer activity and compromises resistance to tumor metastasis in rats. *J Immunol* 160, 3251–3258.

Shimizu, N., Hori, T., and Nakane, H. (1994) An interleukin-1 beta-induced noradrenaline release in the spleen is mediated by brain corticotropin-releasing factor: An in vivo microdialysis study in conscious rats. *Brain Behav Immun* 8, 14–23.

Siegel, M.D., Zhang, D.H., Ray, P., and Ray, A. (1995) Activation of the interleukin-5 promoter by cAMP in murine EL-4 cells requires the GATA-3 and CLE0 elements. *J Biol Chem* 270, 24548–24555.

Silverman, M.N., Pearce, B.D., Biron, C.A., and Miller, A.H. (2005) Immune modulation of the hypothalamic-pituitary-adrenal (HPA) axis during viral infection. *Viral Immunol* 18, 41–78.

Spengler, R.N., Allen, R.M., Remick, D.G., Strieter, R.M., and Kunkel, S.L. (1990) Stimulation of alpha-adrenergic receptor augments the production of macrophage-derived tumor necrosis factor. *J Immunol* 145, 1430–1434.

Staley, J.K., Malison, R.T., and Innis, R.B. (1998) Imaging of the serotonergic system: Interactions of neuroanatomical and functional abnormalities of depression. *Biol Psychiatry* 44, 534–549.

Sternberg, E.M. (1997) Neural-immune interactions in health and disease. *J Clin Invest* 100, 2641–2647.

Sternberg, E.M., Wedner, H.J., Leung, M.K., and Parker, C.W. (1987) Effect of serotonin (5-HT) and other monoamines on murine macrophages: Modulation of interferon-gamma induced phagocytosis. *J Immunol* 138, 4360–4365.

Surprenant, A., Rassendren, F., Kawashima, E., North, R.A., and Buell, G. (1996) The cytolytic P2Z receptor for extracellular ATP identified as a P2X receptor (P2X7). *Science* 272, 735–738.

Szelenyi, J. (2001) Cytokines and the central nervous system. *Brain Res Bull* 54, 329–338.

Szelenyi, J., and Selmeczy, Z. (2002) Immunomodulatory effect of antidepressants. *Curr Opin Pharmacol* 2, 428–432.

Szelenyi, J., Kiss, J.P., and Vizi, E.S. (2000a) Differential involvement of sympathetic nervous system and immune system in the modulation of TNF-alpha production by alpha2- and beta-adrenoceptors in mice. *J Neuroimmunol* 103, 34–40.

Szelenyi, J., Selmeczy, Z., and Vizi, E.S. (2004) Effect of monoamine transporter inhibitors on the LPS-induced cytokine production. *Fundam Clin Pharmacol* 18, 95.

Szelenyi, J., Kiss, J.P., Puskas, E., Szelenyi, M., and Vizi, E.S. (2000b) Contribution of differently localized alpha 2- and beta-adrenoceptors in the modulation of TNF-alpha and IL-10 production in endotoxemic mice. *Ann N Y Acad Sci* 917, 145–153.

Szelenyi, J., Selmeczy, Z., Brozik, A., Medgyesi, D., and Magocsi, M. (2006) Dual beta-adrenergic modulation in the immune system: Stimulus-dependent effect of isoproterenol on MAPK activation and inflammatory mediator production in macrophages. *Neurochem Int* 49, 94–103.

Szelenyi, J., Kiss, J.P., Puskas, E., Selmeczy, Z., Szelenyi, M., and Vizi, E.S. (2000c) Opposite role of alpha2- and beta-adrenoceptors in the modulation of interleukin-10 production in endotoxaemic mice. *Neuroreport* 11, 3565–3568.

Thase, M.E. (1999) Antidepressant treatment of the depressed patient with insomnia. *J Clin Psychiatry* 60(Suppl 17), 28–31; discussion 46–28.

Thase, M.E., Simons, A.D., and Reynolds, C.F., 3rd. (1996) Abnormal electroencephalographic sleep profiles in major depression: Association with response to cognitive behavior therapy. *Arch Gen Psychiatry* 53, 99–108.

van der Poll, T., Jansen, J., Endert, E., Sauerwein, H.P., and van Deventer, S.J. (1994) Noradrenaline inhibits lipopolysaccharide-induced tumor necrosis factor and interleukin 6 production in human whole blood. *Infect Immun* 62, 2046–2050.

van der Pouw Kraan, T.C., Boeije, L.C., Smeenk, R.J., Wijdenes, J., and Aarden, L.A. (1995) Prostaglandin-E2 is a potent inhibitor of human interleukin 12 production. *J Exp Med* 181, 775–779.

Vizi, E.S. (1979) Presynaptic modulation of neurochemical transmission. *Prog Neurobiol* 12, 181–290.

Vizi, E.S. (1998) Receptor-mediated local fine-tuning by noradrenergic innervation of neuroendocrine and immune systems. *Ann N Y Acad Sci* 851, 388–396.

Vizi, E.S. (2000) Role of high-affinity receptors and membrane transporters in nonsynaptic communication and drug action in the central nervous system. *Pharmacol Rev* 52, 63–89.

Vizi, E.S., Zsilla, G., Caron, M.G., and Kiss, J.P. (2004) Uptake and release of norepinephrine by serotonergic terminals in norepinephrine transporter knockout mice: Implications for the action of selective serotonin reuptake inhibitors. *J Neurosci* 24, 7888–7894.

Wang, X., Wu, H., and Miller, A.H. (2004) Interleukin 1alpha (IL-1alpha) induced activation of p38 mitogen-activated protein kinase inhibits glucocorticoid receptor function. *Mol Psychiatry* 9, 65–75.

Watkins, L.R., Maier, S.F., and Goehler, L.E. (1995) Cytokine-to-brain communication: A review and analysis of alternative mechanisms. *Life Sci* 57, 1011–1026.

Weizman, R., Laor, N., Podliszewski, E., Notti, I., Djaldetti, M., and Bessler, H. (1994) Cytokine production in major depressed patients before and after clomipramine treatment. *Biol Psychiatry* 35, 42–47.

Wichers, M., and Maes, M. (2002) The psychoneuroimmuno-pathophysiology of cytokine-induced depression in humans. *Int J Neuropsychopharmacol* 5, 375–388.

Woiciechowsky, C., Asadullah, K., Nestler, D., Eberhardt, B., Platzer, C., Schoning, B., Glockner, F., Lanksch, W.R., Volk, H.D., and Docke, W.D. (1998) Sympathetic activation triggers systemic interleukin-10 release in immunodepression induced by brain injury. *Nat Med* 4, 808–813.

Wong, M.L., Kling, M.A., Munson, P.J., Listwak, S., Licinio, J., Prolo, P., Karp, B., McCutcheon, I.E., Geracioti, T.D., Jr., DeBellis, M.D., Rice, K.C., Goldstein, D.S.,

Veldhuis, J.D., Chrousos, G.P., Oldfield, E.H., McCann, S.M., and Gold, P.W. (2000) Pronounced and sustained central hypernoradrenergic function in major depression with melancholic features: Relation to hypercortisolism and corticotropin-releasing hormone. *Proc Natl Acad Sci U S A* 97, 325–330.

Xu, F., Gainetdinov, R.R., Wetsel, W.C., Jones, S.R., Bohn, L.M., Miller, G.W., Wang, Y.M., and Caron, M.G. (2000) Mice lacking the norepinephrine transporter are supersensitive to psychostimulants. *Nat Neurosci* 3, 465–471.

Yirmiya, R., Pollak, Y., Barak, O., Avitsur, R., Ovadia, H., Bette, M., Weihe, E., and Weidenfeld, J. (2001) Effects of antidepressant drugs on the behavioral and physiological responses to lipopolysaccharide (LPS) in rodents. *Neuropsychopharmacology* 24, 531–544.

Zhu, C.B., Carneiro, A.M., Dostmann, W.R., Hewlett, W.A., and Blakely, R.D. (2005) p38 MAPK activation elevates serotonin transport activity via a trafficking-independent, protein phosphatase 2A-dependent process. *J Biol Chem* 280, 15649–15658.

Zhu, J., Mix, E., and Winblad, B. (2001) The antidepressant and antiinflammatory effects of rolipram in the central nervous system. *CNS Drug Rev* 7, 387–398.

19 Sleep and Neuroimmune Function in Chronic Fatigue Syndrome

Fumiharu Togo, Benjamin H. Natelson, and Neil Cherniack

19.1 Introduction

Chronic fatigue syndrome (CFS) is a medically unexplained condition characterized by persistent or relapsing fatigue for at least 6 months, which substantially interferes with normal activities. It is largely a diagnosis of exclusion. The fatigue is unrelieved by rest and exacerbated by exercise. While depression may be associated with fatigue, chronic fatigue syndrome can occur in the absence of depressive disorders; approximately 60% of patients studied in our research center were negative for depression. The illness affects women twice as often as men (0.52% versus 0.29%) (Jason et al. 1999; Path, Scherbaum, and Bornstein 2000) and is often disabling.

In addition to severe fatigue, diagnosis requires the report of constant or recurring problems with four of the following symptoms: lists impaired memory or concentration, sore throat, tender cervical or axillary nodes, muscle pain, joint pain unaccompanied by redness or swelling, new headaches, unrefreshing sleep, and postexertion malaise (Fukuda et al. 1994). About a quarter of our study patients report an acute flu-like illness onset, that suggests a viral infection, but a specific agent has not been consistently identified.

It is generally agreed that CFS is a heterogeneous disease and does not have a single cause. However, many patients have abnormalities in central nervous system function. It may be that some patients with CFS have a mild encephalopathy with varying adverse effects on different areas of the brain and different aspects of brain function producing with varying severity, abnormalities in information processing, sympathetic/parasympathetic nervous system balance, and/or endocrine function.

Another manifestation of a CFS encephalopathy may be impaired regulation of the neural immune system. The common complaint of unrefreshing sleep seems to be caused by poor sleep efficiency and multiple arousals that lead to chronic sleep deprivation. Disturbed sleep in some CFS patients may be the result of an immune regulatory disorder or more specifically as we have proposed by an imbalance in sleep promoting and sleep inhibiting cytokines.

In this chapter we will summarize studies on brain function and structure in CFS and critically review studies on sleep and immune function in CFS patients.

19.2 Neurological and Neuropsychological Dysfunction in CFS

CFS patients have subtle problems in neuropsychological processing particularly in motor speed (Busichio, Tiersky, DeLuca, and Natelson 2004). A premovement readiness potential is reduced in CFS (Gordon, Michalewski, Nguyen, Gupta, and Starr 1999) and in addition, brain activity during performance of a motor task is significantly greater than that for controls. We did functional imaging in a group of patients with no measurable cognitive deficit and asked them to perform a relatively simple cognitive task: their brains showed the same pattern of activation as seen in healthy subjects given a much more difficult task (Lange et al. 2005).

In another brain imaging study, CFS patients with no psychiatric comorbidity had a significantly higher rate of brain MRI abnormalities than normals (67% versus 30%; Lange, DeLuca, Maldjian, Lee, Tiersky, and Natelson 1999). For the most part, these were small areas of T-2 uptake in frontal white matter. Data exist to support our interpretation that these abnormalities may play a role in the symptom complex of CFS in that patients with MRI abnormalities reported poorer functional status than those with normal MRI studies (Cook, Lange, DeLuca, and Natelson 2001). Stratifying patients based on presence or absence of psychiatric comorbidity is important because psychiatric illnesses such as depression are known to produce similar small lesions.

Two groups have noted decreased gray matter volume in CFS patients compared to controls (De Lange, Kalkman, Bleijenberg, Hagoort, van der Meer, and Toni 2005; Okada, Tanaka, Kuratsune, Watanabe, and Sadato 2004) although they did not agree on the areas affected. This has prompted studies of cerebral perfusion, which might account for a reduction in brain volume. Early studies using SPECT (single photon emission computed tomography) done in nonstratified patient samples revealed global decreases in cerebral flow (Ichise et al. 1992; Schwartz et al. 1994), but this result was not confirmed in later studies using selected patients without psychiatric comorbidity (Costa, Tannock, and Brostoff 1995; Fischler et al. 1996; Lewis et al. 2001; MacHale et al. 2000) or in twins where one twin had CFS (Lewis et al. 2001). Our own studies of absolute brain blood flow did find wide reductions in regional flow bilaterally in CFS (Yoshiuchi, Farkas, and Natelson 2006). Importantly, one of the previous SPECT studies did find actual increases in thalamic flow (MacHale et al. 2000); these could reflect local increases in metabolism at this site. Of interest is another study reporting increased metabolism in basal ganglia, which connect directly to thalamus (Chaudhuri, Condon, Gow, Brennan, and Hadley 2003).

Two groups have used positron emission tomography (PET) to evaluate the metabolism of the brain in CFS patients free of psychiatric disorders (Siessmeier, Nix, Hardt, Schreckenberger, Egle, and Bartenstein 2003; Tirelli et al. 1998). The first found normal metabolism in half their subjects but hypometabolism in cingulate and adjacent mesial cortical areas (Siessmeier et al. 2003). The second study found reduced metabolism in brain stem, thus supporting the earlier SPECT study, which had noted reduced flow there (Tirelli et al. 1998).

TABLE 19.1. Polygraphic characteristics of healthy people and CFS patients during their second night in the sleep lab.

	Healthy ($n = 15$)	CFS ($n = 13$)
Total sleep time (min)	376 ± 42	347 ± 40*
Sleep efficiency (%)	87.3 ± 6.8	80.0 ± 7.0*
Sleep latency (min)	16 ± 16	17 ± 12
Total duration (min) of:		
Awake	40 ± 30	71 ± 29*
Stage 1	53 ± 14	61 ± 28
Stage 2	213 ± 37	194 ± 25
Stage 3	28 ± 17	33 ± 27
Stage 4	4 ± 10	3 ± 5
Stages 3/4	32 ± 25	35 ± 30
Stage REM	77 ± 18	57 ± 26

Values are mean ± SD. Sleep efficiency is the percentage of time asleep relative to the time spent in bed. REM, rapid eye movement. *Significantly ($P < 0.05$) different from healthy people.

We have performed lumbar punctures on 44 CFS patients and 13 healthy controls. Although none of the controls had elevations in cell counts or spinal fluid protein levels, 30% of the patients did. The patient group with abnormal spinal fluid had increased levels of IL-10, an anti-inflammatory cytokine.

19.3 Sleep Studies in CFS

One of the symptoms used for diagnosing CFS is unrefreshing sleep. Data strongly suggest that CFS-like symptoms develop following sleep disruption in otherwise healthy volunteers. Normal controls with disrupted sleep produced experimentally by sound pulses every 2 min but with normal total sleep time had a decrease in daytime energy levels. Moreover, their ability to do complex auditory monitoring tasks was also impaired (Martin, Engleman, Deary, and Douglas 1996). In addition, three studies in healthy volunteers (Hakkionen, Alloui, Gross, Eschallier, and Dubray 2001; Lentz, Landis, Rothermel, and Shaver 1999; Moldofsky and Scarisbrick 1976) have reported increases in musculoskeletal pain and/or decreases in pain threshold after a period of sleep disruption or deprivation, while one study did not find this result (Older et al. 1998). These data indicate that partial sleep deprivation can produce the hallmark symptoms of CFS, namely, marked daytime fatigue, cognitive problems, and musculoskeletal impairment.

Several early studies suggested that as many as one-half of individuals with CFS have mild sleep apnea syndrome (five or more episodes per hour of apnea/hypopnea), periodic leg movements or the restless leg syndrome Other studies with more stringent criteria for these disorders either did not find this result (Krupp, Jandorf, Coyle, and Mendelson 1993; Le Bon et al. 2000; Sharpley, Clements, Hawton, and Sharpe 1997). However, one recent study (Gold, Dipalo, Gold, and Broderick 2004) bearing confirmation did report a high rate of sleep disturbed breathing in patients with fibromyalgia, a pain syndrome often occurring in CFS patients.

In contrast to studies on sleep pathology, a host of studies strongly suggest that the pattern of sleep is abnormal in many CFS patients. The most consistent abnormality is significantly reduced sleep efficiency when compared to controls (Fischler, Le Bon, Hoffmann, Cluydts, Kaufman, and De Meirleir 1997; Krupp et al. 1993; Morriss, Sharpe, Sharpley, Cowen, Hawton, and Morris 1993; Sharpley et al. 1997); the reported average values range from clearly abnormal (i.e., 76.5%) (Fischler et al. 1997) to those within the normal range (i.e., 90%) (Morriss et al. 1993). From one study providing data on individual patients' sleep efficiencies, one can estimate that 75% of CFS patients have reduced sleep efficiency (Krupp et al. 1993). Sleep disturbance in these patients is obvious because they often show increases in time needed to fall asleep (Morriss et al. 1993; Sharpley et al. 1997) and multiple periods of awakenings or arousals (Fischler et al. 1997; Morriss et al. 1993; Sharpley et al. 1997). Decreases in sleep stage 4 have also been reported (Fischler et al. 1997).

We have begun a preliminary evaluation of sleep structure in 13 patients with CFS and 15 healthy controls during the second night of recorded sleep in a sleep laboratory (see Table 19.1). Total sleep time and sleep efficiency were significantly decreased in patients compared to healthy people as is shown in previous studies. Awake period after sleep onset is significantly increased in patients, however, sleep latency (time from lights-out to sleep onset) is not significantly different in patients when compared to healthy people in our study. Total duration of rapid eye movement sleep (REM) is decreased in patients. Total duration of stages 3, 4, and 3+4 sleep did not differ between patients and controls in our hands.

Exercise elevates core body temperature and increases total duration of deep sleep in the night following exercise in healthy people (Horne and Staff 1983). To our knowledge, no one has compared sleep in CFS patients before and after exercise. Exertion is a particularly interesting thing to study in CFS because CFS patients have the unique complaint that even minimal exertion produces a marked worsening of the entire symptom complex (Komaroff and Buchwald 1991). We have (Brimacombe, Helmer, and Natelson 2002a) compared symptom patterns between patients fulfilling the more demanding 1988 case definition (Holmes et al. 1988) and those fulfilling the less restrictive 1994 case definition (Fukuda et al. 1994). We found patient reports of this one symptom to be the most frequent of all symptoms in either case definition (99.4% and 93.1%, respectively; unpublished data). Indeed Jason has shown that the existence of this particular symptom can be used to differentiate CFS patients from patients with other causes for their fatigue (Jason and Taylor 2002). We have some evidence from a pilot study, which we recently performed to suggest that exertion may produce a transient disorder in chronobiological sleep regulation (Ohashi, Yamamoto, and Natelson 2002). In this study, we studied activity patterns and computed mean circadian period (MCP) on 6 days of data before and 6 days of data after a maximal exercise test. We found that MCP increased for the first few days after exercise in CFS patients, but not controls. We thought this might reflect delayed time to go to sleep for CFS patients and interpreted these results to indicate that exhaustive exercise interferes with normal entrainment to 24 h.

We have used a typical cardiac stress test to probe this symptom in other studies too. First we found that CFS patients reported more fatigue as much as four days after the exercise stress test (Sisto et al. 1996). Next, we used actigraphy to monitor activity before and after exercise and found that activity levels also were significantly lower four days after exercise (Sisto et al. 1998). In other work, we found that cognitive abilities were poorer than those of controls the day after the exercise probe (LaManca et al. 1998). We interpreted these changes in activity and cognitive function to support the patient complaint of worsening of symptoms induced by exercise or effort. In addition to these analyses, we used the actigraphic data from that study to do the chronobiological analysis noted above (Ohashi et al. 2002). The most parsimonious explanation for the transient alterations in the circadian activity rhythm noted is that the patients altered their normal sleep–wake rhythm with several days of even more impaired sleep than usual. No such effect occurs in healthy people and, in fact, some reports, although anecdotal, suggest that acute exercise can actually improve sleep (Youngstedt, O'Connor, and Dishman 1997) Finally, we have just completed an electronic diary study of the number of self-reported arousals from sleep each night for the week before and the week after a maximal exercise test. We found both a significant group effect and an interaction ($P < 0.03$ for both): CFS patients reported more nocturnal arousals than controls, but the number of arousals for controls did not change after exercise, whereas it increased in CFS patients. The sample size was too small to allow us to stratify into patient groups with normal versus disturbed sleep, but the results support our interpretation that exercise further disrupts sleep in CFS patients but not in controls.

Although data strongly suggest that CFS occurs in some patients because of poor sleep, there is no accepted way of objectively evaluating sleep quality (Davies et al. 1997). Even though it seems reasonable that sleep quality should diminish as arousals increase, frequency of cortical arousals including cortical microarousals (American Sleep Disorders Association 1992; Martin, Engleman, Kingshott, and Douglas 1997) correlates poorly with this complaint (Bennett, Langford, Stradling, and Davies 1998; Kingshott, Engleman, Deary, and Douglas 1998).

A task force of the American Sleep Disorders Association (ASDA) defined a set of scoring (American Sleep Disorders Association 1992) as a return to alpha or fast frequency electroencephalogram (EEG) activity, well differentiated from the background, lasting at least 3 s that is conventionally used in examining sleep structure. Cortical microarousals are briefer arousals lasting at least 1.5 s (Martin et al. 1997). In identifying microarousals during REM sleep, an increase in electromyogram amplitude is required.

While the major focus of sleep researchers studying arousals has been on EEG measures, one group (Pitson and Stradling 1998) suggested that non-EEG markers might be an important and even more reliable sign of arousals than cortical arousal as reflected by the EEG. For example, it is known that somatosensory and auditory stimulation during sleep can produce alterations in cardiac, respiratory, and somatic measures without overt EEG desynchronization (Carley, Applebaum, Basner, Onal, and Lopata 1997; Halasz 1993; Winkelman 1999). This is thought to reflect activation of the brainstem arousal system without affecting the cortex. Hence current thinking is that there are different levels of arousal responses generated from subcortical and cortical areas of the brain (Sforza, Jouny, and Ibanez 2000, 2002).

These subcortical arousals may be clinically relevant. Such arousals are a common feature of normal aging as well. Hence analysis of the microstructure of the sleep profile has become more important not only in developing more sensitive measures of arousal but also in revealing the sleep-arousal mechanisms (Halasz,

Terzano, Parrino, and Bodizs 2004). One study (Sforza et al. 2000) showed that bursts of K-complexes and delta waves, expressions of a subcortical arousal, represent a real arousal response inducing cardiac activation similar to that found during cortical arousals (microarousal and phases of transitory activity). We have unpublished data on sleep microstructure in young healthy men with no sleep complaints. We found that increases in delta wave power in both cortical and subcortical arousals; delta power increases might be an even better measure of arousals than alpha wave changes.

Symptoms of unrefreshing sleep are reported to be greater when the cyclic alternating pattern (CAP, periodic appearance of delta waves and K-complexes) occupies a greater percent of sleep (Terzano and Parrino 2000). And importantly when looked at the other way, patients with fibromyalgia, a syndrome of medically unexplained pain that overlaps with CFS, have increased amounts of CAP, more so in the more severely symptomatic patients (Rizzi et al. 2004).

These data led us to an evaluation of sleep EEG microstructure in 18 patients with CFS compared to 20 normal controls during the first night that we recorded polysomnography data from them in the sleep lab. As might be expected during their first exposure to a sleep lab, sleep efficiency, wake period after sleep onset, and total duration of stages 3, 4, 3+4 did not differ between patients and controls. Nonetheless, we did find more subtle differences. Figure 19.1 plots the cumulative changes in delta wave (0.75 to 4.5 Hz) power divided by total power calculated each second after sleep onsets during the night of the individual subjects. It can be seen that the CFS patients have less relative delta wave activity than normal, especially later in the night. This suggests that a simple increase in both cortical and subcortical arousals does not explain the unrefreshing sleep of CFS patients.

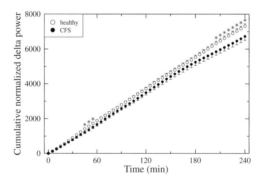

FIGURE 19.1. Mean (±SE) curves of cumulative normalized delta power in electro-encephalogram for healthy people (unfilled circle) and chronic fatigue syndrome (CFS) patients (filled circle) from sleep onset to the shortest sleep time for all participants. *Significantly ($P < 0.05$) different from CFS patients.

19.4 The Neural Immune System in CFS

Because CFS often has an acute flu-like presentation, can follow severe viral infection (Hotopf, Noah, and Wessely 1996; White, Thomas, Amess, Grover, Kangro, and Clare 1995) and often presents with fever, sore throat and tender lymph nodes, a major hypothesis as to its cause is that it represents a form of chronic smoldering infection, perhaps by one of the family of herpes viruses. Unfortunately, our own work has not found support for the hypothesis that active viral infection (Natelson 2001; Wallace, Natelson, Gause, and Hay 1999) has an important role in CFS.

The early idea that CFS represented a form of chronic Epstein–Barr infection was dropped when data were reported indicating that elevated EBV titers B reflecting prior infection are not uncommon in healthy people (Gold et al. 1990). Adherents for chronic infection by other agents including HHV-6, mycoplasma, and Astealth viruses continue to advocate their beliefs, but convincing data remain to be seen. Infection can certainly trigger the onset of CFS. Rates of ~9% of developing CFS have been reported following infectious mononucleosis (White et al. 1998), Lyme disease and severe viral infection (Hotopf et al. 1996). Thus, postviral fatigue exists, but persistence of an infectious agent has not been demonstrated.

If persistent infection is not the cause, another hypothesis is that it is infection triggers abnormal processes in the immune system. A number of papers have reported immune activation in CFS (for review, see Strober (1994)) but it is not clear whether these change are the cause or the result of changes brought about by the CFS such as inactivity, disturbed sleep and chronic stress. Some data do support some underlying immunological problem: first, some CFS patients appear to have an antibody against a specific nuclear antigen (Von Mikecz, Konstantinov, Buchwald, Gerace, and Tan 1997); second, patients have a dysregulated 2,5 RNase L antiviral defense pathway (Suhadolnik et al. 1999); and third, treatment with an immune active agent, mismatched RNA, may reduce disability (Strayer et al. 1994) (a study to replicate this outcome has recently been completed). The immune dysregulation hypothesis was further supported by reports from two prominent groups that found evidence for immunological activation in CFS (Landay, Jessop, Lennette, and Levy 1991; Straus et al. 1993), as well as studies that suggest cytokines such as IL-6 and TNF-α could produce many of the symptoms of CFS. However, as we will point out below, these studies used broad groups of study patients with chronic fatigue and compared their data to control subjects who may have been very active. Importantly, activity is known to up-regulate the immune system and thus differences in activity might explain these results.

Reported studies of sleep in CFS patient have not considered the well-established effects of cytokines on sleep (Krueger, Fang, Taishi, Chen, Kushikata, and Gardi 1998; Kubota, Kushikata, Fang, and Krueger 2000; Vitkovic, Bockaert, and Jacque 2000). Some cytokines prolong non-REM (NREM) sleep, while others seem to shorten or interrupt it. Imbalances in the effects of these two sorts of cytokines may be involved in sleep disturbances in CFS. Of the cytokines in the sleep-producing group, the most extensively studied and best established are IL-1β and TNF-α, which are proinflammatory cytokines that produce fever in addition to sleep (Mullington, Korth, Hinze-Selch, Schreiber, Galanos, and Pollmächer 1998). The febrile effects of these cytokines can be blocked without altering their soporific actions. Besides both having circadian rhythms with similar nocturnal peaks, TNF-α and IL-1β seem to act cooperatively in animals to prolong NREM sleep. IL-1 consists of a family of peptides the biological activity of which depends on the net balance of the activity of agonists such as IL-1β, and several natural counter regulatory factors particularly IL-1 receptor antagonist. Blocking TNF-α and IL-1 by a variety of means alters sleep regulation. The mechanism is unknown, but increases in either of these cytokines lead to enhanced production of the other via induction of transcription factor NF kappa B (Krueger et al. 1998). Experimental data from animals indicate that these cytokines increase in the brain during sleep deprivation (Krueger, Obál, Fang, Kubota, and Taishi 2000). Other cytokines that are reported to increase NREM sleep include IL-2, IL-6, IL-8, and IL-18 (Kapsimalis, Richardson, Opp, and Kryger 2005).

Cytokines that disrupt slow wave sleep (stages 3+4) have been less well examined, but the best-established are IL-4 and IL-10. These cytokines function by inhibiting the production of IL-1 and TNF-α, probably via inhibition of NF kappa B activation (Krueger et al. 2000). Intracerebral injections in rabbits of a cell permeable inhibitor peptide of the transcription factor, NF kappa B inhibit both spontaneous and IL-1β induced sleep (Kubota et al. 2000). Also intracerebral injections of TNF-α and IL-1 inhibitors in rabbits significantly reduced spontaneous NREM sleep, whereas pretreatment with inhibitors of these cytokines significantly attenuated sleep rebound after sleep deprivation. IL-4 has been shown in animals to reduce slow wave sleep (Kushikata, Fang, Wang, and Krueger 1998). IL-10 is spontaneously produced by both lymphocytes and monocytes, inhibits the production of TNF-α, and inhibits slow wave sleep in rabbits (Kushikata, Fang, and Krueger 1999). In contrast, IL-10 knockout mice have increased slow wave sleep (Toth and Opp 2001). Other cytokines reported to have sleep inhibiting effects are IL-13 and TGF-β.

The evidence suggestive of abnormalities in immune regulation in CFS stimulated our group to do an intensive study of immunological cell populations and cytokine gene message in a carefully delineated group of CFS patients matched with *sedentary* healthy controls. We believe that one of the reasons for inconsistent data on the immune system in CFS lies in the fact that extremely sedentary CFS patients are compared to active healthy controls, and level of fitness can affect immunological activity. Looking at the data using a number of classical statistical approaches, we could find no consistent differences between the groups (Natelson et al. 1998; Zhang et al. 1999). Moreover, we recently completed a comprehensive review of all controlled studies of immunity in CFS and were unable to find uniformly consistent results (Natelson, Haghighi, and Ponzio 2002). However, using a new neural nets methodology designed specifically to identify small differences between groups, we did find preliminary evidence that IL-4 has a role in CFS (Hanson, Gause, and Natelson 2001). The higher IL-4 levels in CFS suggest a Type 2 shift in cytokines. It may be that these subtle differences in IL-4 would be magnified if blood were samples collected in sleeping CFS patients.

We have, however, found clear cytokine abnormalities in veterans of the first Gulf War with CFS: sick veterans had higher levels of gene expression for IL-2, IL-10, INF-γ, and TNF-α than healthy veteran controls, and factor analysis showed that

Th-2-related cytokines mediated the cognitive dysfunction reported by Gulf veterans with CFS (Brimacombe, Zhang, Lange, and Natelson 2002b). We do not interpret the results of these positive studies to mean that CFS in Gulf veterans is necessarily different from that in nonveterans with CFS. Instead, we believe these differences emerged because of our finding that Gulf veterans are more immunologically homogeneous than nonveterans (Zhang et al. 1999). Such homogeneity would allow subtle but important differences to be detected.

Robust circadian rhythms are known to exist for the cytokines. In fact, IL-1, and TNF-α, cytokines which are known to enhance sleep, peak during nighttime sleep and are low during the day (Krueger et al. 1998; Moldofsky, Lue, Eisen, Keystone, and Gorczynski 1986). Thus, sampling these cytokines in daytime could, in and of itself, be one reason why more striking cytokine abnormalities have not been found in CFS.

It may be that sleep-disrupting cytokines are relatively up-regulated and sleep-producing cytokines relatively down-regulated in some patients with CFS, and that these changes in cytokines lead to sleep disturbance throughout the night. Hence there might not be abnormalities in the absolute levels of either sleep promoting or sleep inhibiting cytokines (both of which might be in the normal range) but rather in their ratio or difference. We believe this postulated cytokine imbalance is exacerbated by exertion—leading to further disruption of sleep and increased CFS symptoms.

Exercise and level of fitness are certainly known to influence cytokines. In our study of cytokine gene expression 10 min after an acute stress test in seven sedentary men, we found levels of IL-1β, IL-4, IL-6, and IL-10 did not change while levels of TNF-α decreased (Natelson et al. 1996). Our data for IL-1β, IL-4, IL-6, and IL-10 are consistent with the literature (except for IL-1β which increases in plasma in trained athletes doing high intensity exercise) (Moldoveanu, Shephard, and Shek 2001). For TNF-α, most studies were of highly trained athletes doing long duration exercise, and these reported increases in plasma levels with no change in gene expression (Moldoveanu, Shephard, and Shek 2000).

There are a few studies on CFS patients after exercise. We followed cytokine gene expression before and after a standard maximal exertion test in CFS and sedentary matched healthy controls (LaManca et al. 1999). We found no effect of exercise on IL-4 or IL-10 for either group. Exercise produced a decrease in TNF-α in both groups, but the CFS group did not differ from the controls (LaManca et al. 1999). Three studies besides our own also evaluated the effect of exercise on cytokines, including IL-1β, INF-α, TNF-α, and TGF-β (Cannon et al. 1997; Cannon, Angel, Ball, Abad, Fagioli, and Komaroff 1999; Lloyd, Gandevia, Brockman, Hales, and Wakefield 1994; Peterson et al. 1994); again none showed that CFS patients responded differently than controls. However, there are two major limitations to these studies. They only assessed cytokines shortly after exercise and never during the night, and none examined plasma cytokine levels, gene message and cytokine secretion, concurrently.

Also the complexity of the immune network and its positive and negative interactions with the HPA and with peptides involved in energy homeostasis have thus far not been taken into account in evaluating neuroimmune modulation of sleep in CFS. For example, growth hormone is sleep promoting while corticotrophin releasing hormone enhances wakefulness. Levels of both are affected by the release of cytokines such as IL-6, which in turn helps regulate their level. For example, IL-6 stimulates the HPA axis and increases levels of cortisol, but cortisol inhibits IL-6 release (Path et al. 2000). Ghrelin, an endogenous ligand of growth hormone, produced in the stomach, which stimulates both appetite and promotes sleep and in excess can give rise to obesity, has not been studied in CFS. Obesity in turn is now recognized as an important cause of fatigue and to be characteristic of some types of CFS particularly those associated with disturbed breathing during sleep (Vollmer-Conna, Aslakson, and White 2006). The role of these multiple factors on sleep is unlikely to be equal, suggesting that solving the calculus of relating sleep to neuropeptides and cytokines is likely to be quite challenging.

Another issue is methodological and has to do with the assessment of cytokines themselves. There simply is no "gold standard" assay that assures that one is getting a good representation of the biological activity of any one cytokine. The problems here are many. First cytokines are often bound to carrier proteins in the circulation that can interfere with measurement and may influence biological activity; these could consist of multifunctional liver-derived binding proteins, cytokine-specific soluble receptors or cytokine-specific soluble receptor antagonists (Cannon 2000). These issues suggest there is a problem in looking only at plasma cytokines. Another problem is that plasma cytokines usually circulate in picomolar concentrations. Thus, what one finds in the plasma may not be representative of what is happening in the immunological microenvironment. One can get a sense of this microenvironment by turning to cytokine gene expression. Some cytokines are produced mainly by cells of a single lineage or phenotype, whereas others can be produced by many. The selection of a specific cell type for study can affect yield and introduce bias. The cells directly involved in pathogenesis may be sequestered in tissues, and may be of unknown phenotype. In this situation, biologically relevant data can often be obtained following stimulation of peripheral cells in vitro with an appropriate stimulus. Such a functional assessment permits accurate measurement of cytokine production, consumption and neutralization. To deal with the lack of a "gold standard" to quantify cytokines, it may be necessary to use multiple approaches including quantifying serum levels of cytokines, measurements of RNA message, and an assessment of the frequencies of cytokine-producing cells.

References

American Sleep Disorders Association (1992) EEG arousals: Scoring rules and examples: A preliminary report from the Sleep Disorders Atlas Task Force of the American Sleep Disorders Association. *Sleep* 15, 173–184.

Bennett, L.S., Langford, B.A., Stradling, J.R., and Davies, R.J. (1998) Sleep fragmentation indices as predictors of daytime sleepiness and nCPAP response in obstructive sleep apnea. *Am J Respir Crit Care Med* 158, 778–786.

Brimacombe, M., Helmer, D., and Natelson, B.H. (2002a) Clinical differences exist between patients fulfilling the 1988 and 1994 case definitions of chronic fatigue syndrome. *J Clin Psychol Med Settings* 9, 309–314.

Brimacombe, M., Zhang, Q., Lange, G., and Natelson, B.H. (2002b) Immunological variables mediate cognitive dysfunction in Gulf veterans but not civilians with CFS. *Neuroimmunomodulation* 10, 93–100.

Busichio, K., Tiersky, L.A., DeLuca, J., and Natelson, B.H. (2004) Neuropsychological deficits in patients with chronic fatigue syndrome. *J Int Neuropsychol Soc* 10, 1–8.

Cannon, J.G. (2000) Inflammatory cytokines in nonpathological states. *News Physiol Sci* 15, 298–303.

Cannon, J.G., Angel, J.B., Abad, L.W., Vannier, E., Mileno, M.D., Fagioli, L., Wolff, S.M., and Komaroff, A.L. (1997) Interleukin-1β, interleukin-1 receptor antagonist, and soluble interleukin-1 receptor type II secretion in chronic fatigue syndrome. *J Clin Immunol* 17, 253–261.

Cannon, J.G., Angel, J.B., Ball, R.W., Abad, L.W., Fagioli, L., and Komaroff, A.L. (1999) Acute phase responses and cytokine secretion in chronic fatigue syndrome. *J Clin Immunol* 19, 414–421.

Carley, D.W., Applebaum, R., Basner, R.C., Onal, E., and Lopata, M. (1997) Respiratory and arousal responses to acoustic stimulation. *Chest* 112, 1567–1571.

Chaudhuri, A., Condon, B.R., Gow, J.W., Brennan, D., and Hadley, D.M. (2003) Proton magnetic resonance spectroscopy of basal ganglia in chronic fatigue syndrome. *Neuroreport* 14, 225–228.

Cook, D.B., Lange, G., DeLuca, J., and Natelson, B.H. (2001) Relationship of brain MRI abnormalities and physical functional status in CFS. *Int J Neurosci* 107, 1–6.

Costa, D.C., Tannock, C., and Brostoff, J. (1995) Brainstem perfusion is impaired in chronic fatigue syndrome. *Q J Med* 88, 767–773.

Davies, R.J., Bennett, L.S., and Stradling, J.R. (1997) What is an arousal and how should it be quantified? *Sleep Med Rev* 1, 87–95.

De Lange, F.P., Kalkman, J.S., Bleijenberg, G., Hagoort, P., van der Meer, J.W., and Toni, I. (2005) Gray matter volume reduction in the chronic fatigue syndrome. *Neuroimage* 26, 777–781.

Fischler, B., D'Haenen, H., Cluydts, R., Michiels, V., Demets, K., Bossuyt, A., Kaufman, L., and De Meirleir, K. (1996) Comparison of ^{99m}Tc HMPAO SPECT scan between chronic fatigue syndrome, major depression and healthy controls: An exploratory study of clinical correlates of regional cerebral blood flow. *Neuropsychobiology* 34, 175–183.

Fischler, B., Le Bon, O., Hoffmann, G., Cluydts, R., Kaufman, L., and De Meirleir, K. (1997) Sleep anomalies in the chronic fatigue syndrome—A comorbidity study. *Neuropsychobiology* 35, 115–122.

Fukuda, K., Straus, S.E., Hickie, I., Sharpe, M.C., Komaroff, A., Schluederberg, A. et al. (1994) The chronic fatigue syndrome: A comprehensive approach to its definition and study. *Ann Intern Med* 121, 953–959.

Gold, A.R., Dipalo, F., Gold, M.S., and Broderick, J. (2004) Inspiratory airflow dynamics during sleep in women with fibromyalgia. *Sleep* 27, 459–466.

Gold, D., Bowden, R., Sixbey, J., Riggs, R., Katon, W.J., Ashley, R., Obrigewitch, R., and Corey, L. (1990) Chronic fatigue: A prospective clinical and virologic study. *JAMA* 264, 48–53.

Gordon, R., Michalewski, H.J., Nguyen, T., Gupta, S., and Starr, A. (1999) Cortical motor potential alterations in chronic fatigue syndrome. *Int J Molec Med* 4, 493–499.

Hakkionen, S., Alloui, A., Gross, A., Eschallier, A., and Dubray, C. (2001) The effects of total sleep deprivation, selective sleep interruption and sleep recovery on pain tolerance in healthy subjects. *J Sleep Res* 10, 35–42.

Halasz, P. (1993) Arousals without awakening—Dynamic aspect of sleep. *Physiol Behav* 54, 795–802.

Halasz, P., Terzano, M., Parrino, L., and Bodizs, R. (2004) The nature of arousal in sleep. *J Sleep Res* 13, 1–23.

Hanson, S.J., Gause, W.C., and Natelson, B.H. (2001) Detection of immunologically significant factors for chronic fatigue syndrome using neural network classifiers. *Clin Diagn Lab Immunol* 8, 658–662.

Holmes, G.P., Kaplan, J.E., Gantz, N.M., Komaroff, A.L., Schonberger, L.B., Straus, S.E. et al. (1988) Chronic fatigue syndrome: A working case definition. *Ann Intern Med* 108, 387–389.

Horne, J.A., and Staff, L.H. (1983) Exercise and sleep: Body-heating effects. *Sleep* 6, 36–46.

Hotopf, M., Noah, N., and Wessely, S. (1996) Chronic fatigue and minor psychiatric morbidity after viral meningitis: A controlled study. *J Neurol Neurosurg Psychiatry* 60, 504–509.

Ichise, M., Salit, I.E., Abbey, S.E., Chung, D.-G., Gray, B., Kirsh, J.C., and Freedman, M. (1992) Assessment of regional cerebral perfusion by $^{99}Tc^m$-HMPAO SPECT in chronic fatigue syndrome. *Nucl Med Commun* 13, 767–772.

Jason, L.A., Richman, J.A., Rademaker, A.W., Jordan, K.M., Plioplys, A.V., Taylor, R.R., McCready, W., Huang, C.-F., and Plioplys, S. (1999) A community-based study of chronic fatigue syndrome. *Arch Intern Med* 159, 2129–2137.

Jason, L.A., and Taylor, R.R. (2002) Applying cluster analysis to define a typology of chronic fatigue syndrome in a medically-evaluated, random community sample. *Psychol Health* 1, 1–15.

Kapsimalis, F., Richardson, G., Opp, M.R., and Kryger, M. (2005) Cytokines and normal sleep. *Curr Opin Pulm Med* 11, 481–484.

Kingshott, R.N., Engleman, H.M., Deary, I.J., and Douglas, N.J. (1998) Does arousal frequency predict daytime function? *Eur Respir J* 12, 1264–1270.

Komaroff, A.L., and Buchwald, D. (1991) Symptoms and signs of chronic fatigue syndrome. *Rev Infect Dis* 13, S8–S11.

Krueger, J.M., Fang, J., Taishi, P., Chen, Z., Kushikata, T., and Gardi, J. (1998) Sleep. A physiologic role for IL-1 beta and TNF-alpha. *Ann N Y Acad Sci* 856, 148–159.

Krueger, J.M., Obál, F., Fang, J., Kubota, T., and Taishi, P. (2000) The role of cytokines in physiological sleep regulation. *Ann N Y Acad Sci* 917, 211–221.

Krupp, L.B., Jandorf, L., Coyle, P.K., and Mendelson, W.B. (1993) Sleep disturbance in chronic fatigue syndrome. *J Psychosom Res* 37, 325–331.

Kubota, T., Kushikata, T., Fang, J., and Krueger, J.M. (2000) Nuclear factor-kappaB inhibitor peptide inhibits spontaneous and interleukin-1beta-induced sleep. *Am J Physiol Regul Integr Comp Physiol* 279, R404–R413.

Kushikata, T., Fang, J., and Krueger, J.M. (1999) Interleukin-10 inhibits spontaneous sleep in rabbits. *J Interferon Cytokine Res* 19, 1025–1030.

Kushikata, T., Fang, J., Wang, Y., and Krueger, J.M. (1998) Interleukin-4 inhibits spontaneous sleep in rabbits. *Am J Physiol* 275, R1185–R1191.

LaManca, J., Sisto, S., DeLuca, J., Johnson, S.K., Lange, G., Pareja, J., Cook, S., and Natelson, B.H. (1998) The influence of exhaustive treadmill exercise on cognitive function in patients with chronic fatigue syndrome. *Am J Med* 105, 59S–65S.

LaManca, J.J., Sisto, S., Ottenweller, J.E., Cook, S., Peckerman, A., Zhang, Q., Denny, T.N., Gause, W.C., and Natelson, B.H. (1999) Immunological response in chronic fatigue syndrome following a graded exercise test to exhaustion. *J Clin Immunol* 19, 135–142.

Landay, A.L., Jessop, C., Lennette, E.T., and Levy, J.A. (1991) Chronic fatigue syndrome: Clinical condition associated with immune activation. *Lancet* 338, 707–712.

Lange, G., DeLuca, J., Maldjian, J.A., Lee, H.J., Tiersky, L.A., and Natelson, B.H. (1999) Brain MRI abnormalities exist in a subset of patients with chronic fatigue syndrome. *J Neurol Sci* 171, 3–7.

Lange, G., Steffener, J., Bly, B.M., Christodoulou, C., Liu, W.-C., DeLuca, J., and Natelson, B.H. (2005) Chronic fatigue syndrome affects verbal working memory: A BOLD fMRI study. *Neuroimage* 26, 513–524.

Le Bon, O., Fischler, B., Hoffmann, G., Murphy, J.R., De Meirleir, K., Cluydts, R., and Pelc, I. (2000) How significant are primary sleep disorders and sleepiness in the chronic fatigue syndrome? *Sleep Res Online* 3, 43–48.

Lentz, M.J., Landis, C.A., Rothermel, J., and Shaver, J.L. (1999) Effects of selective slow wave sleep disruption on musculoskeletal pain and fatigue in middle aged women. *J Rheumatol* 26, 1586–1592.

Lewis, D.H., Mayberg, H.S., Fischer, M.E., Goldberg, J., Ashton, S., Graham, M.M., and Buchwald, D. (2001) Monozygotic twins discordant for chronic fatigue syndrome: Regional cerebral blood flow SPECT. *Radiology* 219, 766–773.

Lloyd, A., Gandevia, S., Brockman, A., Hales, J., and Wakefield, D. (1994) Cytokine production and fatigue in patients with chronic fatigue syndrome and healthy control subjects in response to exercise. *Clin Infect Dis* 18(Suppl 1), S142–S146.

MacHale, S.M., Lawrie, S.M., Cavanagh, J.T.O., Glabus, M.F., Murray, G.L., Goodwin, G.M., and Ebmeier, K.P. (2000) Cerebral perfusion in chronic fatigue syndrome and depression. *Br J Psychiatry* 176, 550–556.

Martin, S.E., Engleman, H.M., Deary, I.J., and Douglas, N.J. (1996) The effect of sleep fragmentation on daytime function. *Am J Respir Crit Care Med* 153, 1328–1332.

Martin, S.E., Engleman, H.M., Kingshott, R.N., and Douglas, N.J. (1997) Microarousals in patients with sleep apnoea/hypopnoea syndrome. *J Sleep Res* 6, 276–280.

Moldofsky, H., and Scarisbrick, P. (1976) Induction of neurasthenic musculoskeletal pain syndrome by selective sleep stage deprivation. *Psychosom Med* 38, 35–44.

Moldoveanu, A.I., Shephard, R.J., and Shek, P.N. (2000) Prolonged exercise elevates plasma levels but not gene experssion of IL-1b, IL-6, and TNFa in circulating mononuclear cells. *J Appl Physiol* 80, 452–460.

Moldoveanu, A.I., Shephard, R.J., and Shek, P.N. (2001) The cytokine response to physical activity and training. *Sports Med* 31, 115–144.

Morriss, R., Sharpe, M., Sharpley, A.L., Cowen, P.J., Hawton, K., and Morris, J. (1993) Abnormalities of sleep in patients with the chronic fatigue syndrome. *BMJ* 306, 1161–1164.

Mullington, J., Korth, C., Hinze-Selch, D., Schreiber, W., Galanos, C., and Pollmächer, T. (1998) Dose-dependent effects of endotoxin on human sleep. *Am J Physiol Regul Integr Comp Physiol* 278, 947–955.

Natelson, B.H. (2001) Chronic fatigue syndrome. *JAMA* 285, 2557–2559.

Natelson, B.H., Haghighi, M.H., and Ponzio, N.M. (2002) Evidence for the presence of immune dysfunction in chronic fatigue syndrome. *Clin Diagn Lab Immunol* 9, 747–752.

Natelson, B.H., LaManca, J.J., Denny, T., Vladutiu, A.C., Oleske, J., Hill, M., Bergen, M.T., Korn, L., and Hay, J. (1998) Immunological parameters in chronic fatigue syndrome, major depression, and multiple sclerosis. *Am J Med* 105, 43S–49S.

Natelson, B.H., Zhou, X., Ottenweller, J.E., Bergen, M.T., Sisto, S.A., Drastal, S., Tapp, W.N., and Gause, W.L. (1996) Effect of acute exhausting exercise on cytokine gene expression in men. *Int J Sports Med* 17, 299–302.

Ohashi, K., Yamamoto, Y., and Natelson, B.H. (2002) Activity rhythm degrades after strenuous exercise in chronic fatigue syndrome. *Physiol Behav* 77, 39–44.

Okada, T., Tanaka, M., Kuratsune, H., Watanabe, Y., and Sadato, N. (2004) Mechanisms undelrying fatigue: A voxel-based morphometric study of chronic fatigue syndrome. *BMC Neurol* 4.

Older, S.A., Battafarano, D.F., Danning, C.L., Ward, J.A., Grady, E.P., Derman, S., and Russell, I.J. (1998) The effects of delta wave sleep interruption on pain thresholds and fibromyalgia-like symptoms in healthy subjects: Correlations with insulin-like growth factor I. *J Rheumatol* 25, 1180–1186.

Path, G., Scherbaum, W.A., and Bornstein, S.R. (2000) The role of interleukin-6 in the human adrenal gland. *Eur J Clin Invest* 30(Suppl 3), 91–95.

Peterson P.K., Sirr, S.A., Grammith, F.C., Schenck, C.H., Pheley, A.M., Hu, S., and Chao, C.C. (1994) Effects of mild exercise on cytokines and cerebral blood flow in chronic fatigue syndrome patients. *Clin Diagn Lab Immunol* 1, 222–226.

Pitson, D.J., and Stradling, J.R. (1998) Autonomic markers of arousal during sleep in patients undergoing investigation for obstructive sleep apnoea, their relationship to EEG arousals, respiratory events and subjective sleepiness. *J Sleep Res* 7, 53–59.

Rizzi, M., Sarzi-Puttini, P., Atzeni, F., Capsoni, F., Andreoli, A., Pecis, M., Colombo, S., Carrabba, M., and Sergi, M. (2004) Cyclic alternating pattern: A new marker of sleep alteration in patients with fibromyalgia? *J Rheumatol* 31, 1193–1199.

Schwartz, R.B., Komaroff, A.L., Garada, B.M., Gleit, M., Doolittle, T.H., Bates, D.W., Vasile, R.G., and Holman, B.L. (1994) SPECT imaging of the brain: Comparison of findings in patients with chronic fatigue syndrome, AIDS dementia complex, and major unipolar depression. *Am J Roentgenol* 162, 943–951.

Sforza, E., Jouny, C., and Ibanez, V. (2000) Cardiac activation during arousal in humans: Further evidence for hierarchy in the arousal response. *Clin Neurophysiol* 111, 1611–1619.

Sforza, E., Juony, C., and Ibanez, V. (2002) Time-dependent variation in cerebral and autonomic activity during periodic leg movements in sleep: Implications for arousal mechanisms. *Clin Neurophysiol* 113, 883–891.

Sharpley, A., Clements, A., Hawton, K., and Sharpe, M. (1997) Do patients with "pure" chronic fatigue syndrome (neurasthenia) have abnormal sleep? *Psychosom Med* 59, 592–596.

Siessmeier, T., Nix, W.A., Hardt, J., Schreckenberger, M., Egle, U.T., and Bartenstein, P. (2003) Observer independent analysis of cerebral glucose metabolism in patients with chronic fatigue syndrome. *J Neurol Neurosurg Psychiatry* 74, 922–928.

Sisto, S., LaManca, J., Cordero, D.L., Bergen, M.T., Drastal, S., Boda, W.L., Tapp, W.L., and Natelson, B.H. (1996) Metabolic and cardiovascular effects of a progressive exercise test in patients with chronic fatigue syndrome. *Am J Med* 100, 634–640.

Sisto, S.A., Tapp, W.N., LaManca, J.J., Ling, W., Korn, L.R., Nelson, A.J., and Natelson, B.H. (1998) Physical activity before and after exercise in women with chronic fatigue syndrome. *Q J Med* 91, 465–473.

Straus, S.E., Fritz, S., Dale, J.K., Gould, B., and Strober, W. (1993) Lymphocyte phenotype and function in the chronic fatigue syndrome. *J Clin Immunol* 13, 30–40.

Strayer, D.R., Carter, W.A., Brodsky, I., Cheney, P., Peterson, D., Salvato, P., Thompson, C., Loveless, M., Shapiro, D.E., Elsasser, W., and Gillespie, D.H. (1994) A controlled clinical trial with a specifically configured RNA drug, poly(I)·poly(C12$_U$), in chronic fatigue syndrome. *Clin Infect Dis* 18(Suppl 1), S88–S95.

Strober, W. (1994) Immunological function in chronic fatigue syndrome. In: S. Straus (Ed.), *Chronic Fatigue Syndrome*. Marcel Dekker, New York, pp. 207–237.

Suhadolnik, R.J., Peterson, D.L., Cheney, P.R., Horvath, S.E., Reichenbach, N.L., Brien, K., Lombardi, V., Welsch, S., Furr, E.G., Charubala, R., and Pfleiderer, W. (1999) Biochemical dysregulation of the 2–5A synthetase/RNase L antiviral defense pathway in chronic fatigue syndrome. *J Chr Fatigue Syndr* 5, 223–242.

Terzano, M.G., and Parrino, L. (2000) Origin and significance of the cyclic alternating pattern (CAP). (Review article). *Sleep Med Rev* 4, 101–123.

Tirelli, U., Chierichetti, F., Tavio, M., Simonelli, C., Bianchin, G., Zanco, P., and Ferlin, G. (1998) Brain positron emission tomography (PET) in chronic fatigue syndrome: Preliminary data. *Am J Med* 105, 54S–58S.

Toth, L.A., and Opp, M.R. (2001) Cytokine- and microbially induced sleep responses of interleukin-10 deficient mice. *Am J Physiol Regul Integr Comp Physiol* 280, R1806–R1814.

Vitkovic, L., Bockaert, J., and Jacque, C. (2000) "Inflammatory" cytokines: Neuromodulators in normal brain? *J Neurochem* 74, 457–471.

Vollmer-Conna, U., Aslakson, E., and White, P.D. (2006) An empirical delineation of the heterogeneity of chronic unexplained fatigue in women. *Pharmacogenomics* 7, 355–364.

Von Mikecz, A., Konstantinov, K., Buchwald, D.S., Gerace, L., and Tan, E.M. (1997) High frequency of autoantibodies to insoluble cellular antigens in patients with chronic fatigue syndrome. *Arthritis Rheum* 40, 295–305.

Wallace, II, H.L., Natelson, B.H., Gause, W.C., and Hay, J. (1999) An evaluation of human herpesviruses in chronic fatigue syndrome. *Clin Diagn Lab Immunol* 6, 216–223.

White, P.D., Thomas, J., Amess, J., Grover, S.A., Kangro, H.O., and Clare, A.W. (1995) The existence of a fatigue syndrome after glandular fever. *Psychol Med* 25, 907–924.

White, P.D., Thomas, J.M., Amess, J., Crawford, D.H., Grover, S.A., Kangro, H.O., and Clare, A.W. (1998) Incidence, risk and prognosis of acute and chronic fatigue syndromes and psychiatric disorders after glandular fever. *Br J Psychiatry* 173, 475–481.

Winkelman, J.W. (1999) The evoked heart rate response to periodic leg movements of sleep. *Sleep* 22, 575–580.

Yoshiuchi, K., Farkas, J., and Natelson, B.H. (2006) Patients with chronic fatigue syndrome have reduced absolute cortical blood flow. *Clin Physiol Funct Imaging* 26, 83–86.

Youngstedt, S.D., O'Connor, P.J., and Dishman, R.K. (1997) The effects of acute exercise on sleep: A quantitative synthesis. *Sleep* 20, 203–214.

Zhang, Q., Zhou, X., Denny, T., Ottenweller, J., Lange, G., LaManca, J.J., Lavietes, M.H., Pollet, C., Gause, W.C., and Natelson, B.H. (1999) Changes in immune parameters in Gulf War veterans but not in civilians with chronic fatigue syndrome. *Clin Diagn Lab Immunol* 6, 6–13.

20 Narcolepsy with Cataplexy: Hypocretin and Immunological Aspects

Yves Dauvilliers

20.1 Epidemiology and Clinical Features

Narcolepsy is a disabling sleep disorder characterized by severe excessive daytime sleepiness and abnormal rapid eye movement (REM) sleep manifestations including cataplexy, sleep paralysis, hypnagogic hallucinations, and sleep onset REM periods (American Academy of Sleep Medicine 2005; Dauvilliers, Billiard, and Montplaisir 2003b; Scammell 2003). Recent advances in pathophysiology demonstrated that narcolepsy is caused by the loss of hypothalamic neurons producing hypocretin (Mignot et al. 2002; Dauvilliers et al. 2003a; Peyron et al. 2000; Thannickal et al. 2000; Crocker et al. 2005; Blouin, Thannickal, Worley, Baraban, Reti, and Siegel 2005). Epidemiological data with a young and bimodal age at onset, frequent triggering factors and a tight HLA DQB1*0602 association, suggest an autoimmune hypothesis (Mignot, Tafti, Dement, and Grumet 1995; Carlander, Dauvilliers, and Billiard 2001; Chabas, Taheri, Renier, and Mignot 2003). Acting on a specific genetic background, an autoimmune process targeting hypocretin neurons, in response to yet unknown environmental factors, is the most probable hypothesis. The treatment of narcolepsy has evolved significantly over the last few years, but current therapies are only symptom-based. Hypocretin-based therapies and immune-based therapies are part of the research projects for the development of new treatments in narcolepsy (Dauvilliers and Tafti 2006).

The prevalence of narcolepsy with cataplexy was estimated between 0.05 and 0.067% in the United States (Dement, Carskadon, and Ley 1973), with same results in other countries (Hublin et al. 1994; Ohayon, Priest, Zulley, Smirne, and Paiva 2002), except in Japan with a higher prevalence (Honda 1979) and in Israel with a lower prevalence (Lavie and Peled 1987). One study reported an incidence of 1.37 in 100,000 inhabitants per year (Silber, Krahn, Olson, and Pankratz 2002). There are epidemiological evidences of an immune etiology in narcolepsy, the young and bimodal age at onset (Dauvilliers et al. 2001a), the frequent onset of symptoms after a major circumstance such as psychological stress or an abrupt change of sleep schedules (Orellana, Villemin, Tafti, Carlander, Besset, and Billiard 1994), the low concordance rate of monozygote twins (Dauvilliers et al. 2004c; Mignot 1998) and the tight association with HLA DQB1*0602 (see below) (Mignot, et al. 2001). Recently, few studies have revealed a March peak and a September trough in the birth pattern of narcolepsy patients with clear-cut and frequent cataplexy (Dauvilliers et al. 2003c; Picchioni, Mignot, and Harsh 2004). Therefore, environmental events during early development may influence narcolepsy severity or the likelihood of developing the disease that also may suggest an autoimmune etiology. In contrast to most other well-defined autoimmune disorders, there is no clear association with other autoimmune diseases, no predominance of female and any blood or CSF inflammatory markers (Dauvilliers, Billiard, and Montplaisir 2003b; Mignot et al. 1995; Dauvilliers and Tafti 2006).

20.1.1 Excessive Daytime Sleepiness

Daytime sleepiness is the most severe symptom, the most frequent cause for consultation and mostly the first symptom to appear (Dauvilliers et al. 2003b; Scammell 2003). Daytime sleepiness occurs daily, recurring typically at 2-h intervals, exacerbated during inactivity, with large variability between patients. Most of those sleep episodes are irresistible, of short duration and frequently associated with dreaming. Except in cases of children, the refreshing value of short naps is of significant diagnostic value. Severe sleepiness can also lead to unconscious microsleep episodes or lapses.

20.1.2 Cataplexy

Cataplexy is a sudden loss of voluntary muscular tone, without any alteration of consciousness, in relation with strong emotive reactions such as laughter, joking, …. All striated muscles (but not the diaphragm) can be affected leading to a progressive collapse of the subject. Cataplexy is specific to narcolepsy and is the best diagnostic marker of the disease (Dauvilliers et al. 2003b; Scammell 2003). The duration of cataplexy varies from a second to one or two minutes and its frequency varies from less than one episode per year to several episodes per day. Patients may also rarely experience "status cataplecticus" characterized by long episodes of cataplexy lasting to several hours. Several neurophysiological and pharmaceutical findings suggest that cataplexy is equivalent to the loss of muscle tone of REM sleep occurring during the waking state (Dauvilliers et al. 2003b; Scammell 2003).

20.1.3 Associated Features

Other symptoms of dissociated REM sleep experienced by narcoleptic patients include sleep-related hallucinations and sleep paralysis characterized by an inability to move the limbs or the head either at sleep onset or upon awakening. Auditory, visual, somesthetic hypnagogic (at sleep onset) or hypnopompic (upon awakening) hallucinations, or sleep paralysis are present in around 50% of narcoleptic patients (Dauvilliers et al. 2003b; Scammell 2003), therefore more frequent than in the general population (Ohayon 2000).

Nocturnal sleep is also disrupted with frequent sleep awakenings, usually worsening with advancing age (Dauvilliers et al. 2003b). Finally, patients with narcolepsy are frequently associated with a high body mass index and with a rapid weight gain at onset of the condition (Kotagal, Krahn, and Slocumb 2004).

20.1.4 Diagnostic Criteria

The diagnosis of narcolepsy with cataplexy is essentially clinical. In the recent International Classification of Sleep Disorders (American Academy of Sleep Medicine 2005), the diagnosis of narcolepsy with cataplexy is based on the presence of both excessive daytime sleepiness (occurring almost daily for at least 3 months) and clear-cut history of cataplexy. Unfortunately, cataplexy rarely occurs in the presence of the clinician, and a clear history is not always possible. It is therefore recommended to confirm the clinical diagnosis with a nocturnal polysomnography followed by a daytime multiple sleep latency test (MSLT) (American Academy of Sleep Medicine 2005), during which sleep latency should be below 8 min with at least two sleep onset REM period (SOREMP). This criterion may, however, be absent in young children, elderly and in rare patients with clear-cut cataplexy (Dauvilliers, Gosselin, Paquet, Touchon, Billiard, and Montplaisir 2004b).

The presence of the human leukocyte antigen HLA DQB1*0602 genotype, as it is neither sensitive nor specific in narcolepsy (see below), is only a supportive criterion. HLA DQB1*0602 is neither necessary, nor sufficient for the development of the disease. CSF hypocretin-1 levels lower than 110 pg/ml or one-third of mean normal control values are alternatively proposed as a definite diagnostic criteria (American Academy of Sleep Medicine 2005).

20.2 Pathophysiology

20.2.1 Hypocretin (Orexin) System

Hypocretin peptides have been identified in 1998 and called hypocretins due to their hypothalamic localization and similarities with the hormone secretin (de Lecea et al. 1998). Hypocretins 1 and 2 are produced exclusively by a well-defined group of neurons localized in the lateral hypothalamus. The neurons project to the olfactory bulb, cerebral cortex, thalamus, hypothalamus and brainstem, particularly the locus coeruleus, raphe nucleus and bulbar reticular formation (Peyron et al. 1998). Within the same time in 1998, endogenous ligands to two orphan receptors with homologous structures had been identified and called orexin A and B (Sakurai et al. 1998). Orexin peptides are the same as hypocretins. Based on its primary location within the hypothalamus, the initial function of the hypocretin was believed to be appetite regulation.

In the 1970s, a condition similar to human narcolepsy was found in different species of dogs, first in the Dachshund, then in the Poodle, and later in the Doberman pinscher and Labrador retriever. Several genetic studies have been conducted in the canine model of narcolepsy. It has been shown that in certain breeds, the Doberman pinschers and Labrador retrievers, the disease is transmitted as a recessive autosomal trait with complete penetrance, whereas in other breeds, the Poodles and the Beagles, it is polygenic and/or determined by environmental factors (Mignot et al. 1991; Ripley, Fujiki, Okura, Mignot, and Nishino 2001). Following intensive genetic studies, mutations of the gene encoding the type 2 hypocretin/orexin receptor were found to be responsible for the disease in familial canine narcolepsy (Lin et al.

1999). This G-protein-coupled receptor has a high affinity for the hypocretin neuropeptides 1 and 2.

Genome sequencing showed 226 base pairs inserted upstream of the spliced site of 4th exon in narcoleptic Doberman pinschers, and a G to A mutation in the 5• splice site sequence of the intron–exon 5 boundary in narcoleptic Labrador retrievers. Those mutations produced abnormal mRNA slicing that lead to a complete loss of function for the hypocretin receptor-2 (Lin et al. 1999). In contrast, no mutation in hypocretin genes was found in sporadic canine narcolepsies (Ripley et al. 2001). The latter species were associated with a major decrease level of hypocretin-1 in the CSF and in the brain. All those findings suggest that sporadic and familial canine narcolepsies have distinct etiologies, but result both from alteration in hypocretin transmission.

Another group reported that knockout mice for prepro-hypocretin (precursor ligand) for this receptor developed brief episodes of cataplexy, SOREMPs and a decrease REM sleep latency (Chemelli et al. 1999). In addition, hypocretin/ataxin 3 transgenic mice characterized by more than 90% hypocretin neuron degeneration present a phenotype similar to human narcolepsy with cataplectic episodes, SOREMPs and obesity despite being hypophagic (Hara et al. 2001). Hypocretin receptor-2 knockout mice were also created with a similar but less severe narcolepsy-like phenotype in comparison to preprohypocretin knockout mice (Willie et al. 2003). Interestingly, hypocretin receptor-1 knockout mice do not exhibit episodes of cataplexy, but double receptor-1 and -2 knockout mice resulted in a phenotype similar to prepro-hypocretin knockout mice (Willie et al. 2003).

After the discovery of the effects of hypocretin gene mutation and transgene in different species, the hypocretin system was also studied in human narcolepsy. Mutations screening in hypocretin system were conducted in all forms of narcolepsy, with and without cataplexy, SOREMPs, family history of narcolepsy, and HLA DQB*0602 (HLA, see below) (Peyron et al. 2000).

Unfortunately, only one atypical (HLA DQB1*0602 negative) but severe narcoleptic patient with cataplexy presented a leucine to arginine substitution in the signal sequence of the prepro-hypocretin gene (Peyron et al. 2000). Pre-prohypocretin/orexin human gene is located on chromosome 17q21 and is synthesized by neurons located exclusively in the latero-posterior hypothalamus. So far, no mutation is found in the hypocretin receptor 1 or 2 genes and several studies demonstrated that the gene loci for prepro-hypocretin, hypocretin receptor-1 and -2 do not contribute to narcolepsy susceptibility (Peyron et al. 2000; Hungs, Lin, Okun, and Mignot 2001).

Although most cases of human narcolepsy with cataplexy are not caused by hypocretin gene mutation, the hypocretin system is involved in its pathophysiology. In situ hybridization, immunohistochemistry and radioimmunological assays of peptides performed in postmortem brain tissue of few narcoleptic patients revealed undetectable levels of prepro-hypocretin RNA and a selective loss of hypocretin neurons (Fig. 20.1) (Peyron et al. 2000; Thannickal et al. 2000).

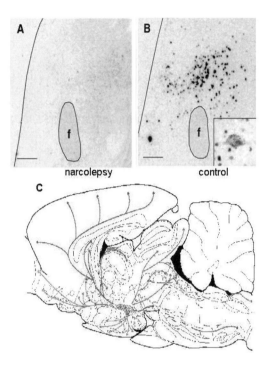

FIGURE 20.1. (A) Distribution of hypocretin-labeled somas within the lateral hypothalamus region (each dot represents a hypocretin-producing cell) in narcoleptic subjects (A) and controls (B). Hypocretin transcripts remain detectable only in controls. (C) Schematic drawing of hypocretin neuron pathways in the rat brain. Hypocretins are produced exclusively by neurons localized in the lateral hypothalamus with wide projections to the olfactory bulb, cerebral cortex, thalamus, hypothalamus, and brainstem, and particularly densely to the locus coeruleus, tuberomammillary nucleus, raphe nucleus and bulbar reticular formation. (Reprinted with permission from Peyron, C. et al., *J Neuroscience*, Vol. 18, 1998, copyright 1998 by the Society for Neuroscience.)

However, melanin-concentrating hormone (MCH) neurons, located within the same region, remain intact. Recent studies have also examined postmortem expression of prodynorphin and neuronal activity-regulated pentraxin (NARP), markers coexpressed with hypocretin within the posterior hypothalamus (Crocker et al. 2005; Blouin et al. 2005). A clear reduction (80 to 95%) in the number of neurons co-producing these markers was also found, paralleling the loss of the hypocretin signal seen in postmortem samples. In addition, Narp staining still persists in other brain areas in narcoleptics as well as in controls, therefore Narp-positive neurons were not globally affected in narcoleptic brains (Blouin et al. 2005).

All these recent findings confirm that narcolepsy results from the selective damage of hypocretin-containing neurons and not from a simple failure to produce hypocretin. However, only limited number of narcoleptic brains was available in those studies with frequent low knowledge of the clinical conditions. In addition, a long interval between symptoms onset and brain examination exist with the potential effects of narcoleptic treatments in between. But the main question still persists, what is the cause of this focal hypocretin neurodegeneration? Following those studies, several groups measured CSF hypocretin-1 levels in large series of narcoleptics, other hypersomnia conditions, healthy subjects and patients with other neurological disorders (Mignot et al. 2002; Dauvilliers et al. 2003b; Nishino and Kanbayashi. 2005). CSF hypocretin-1 levels were almost always above 200 pg/ml in controls or subjects with other sleep and neurological disorders. CSF hypocretin-1 levels lower than 110 pg/ml had a high (94%) positive predictive value for narcolepsy–cataplexy (Mignot et al. 2002; Dauvilliers et al. 2003b; Nishino and Kanbayashi. 2005). Most of sporadic narcoleptics with clear-cut cataplexy and with HLA DQB1*0602 have undetectable CSF hypocretin-1 levels (Fig. 20.2).

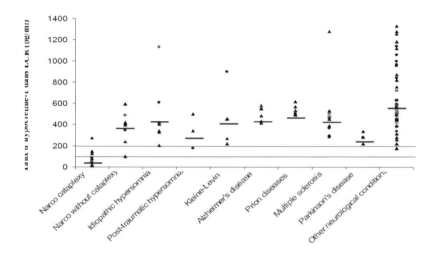

FIGURE 20.2. CSF hypocretin-1 levels among 138 patients including narcolepsy with and without cataplexy, idiopathic hypersomnia, posttraumatic hypersomnia, and other neurological conditions. Each point represents the crude concentration of CSF hypocretin-1 in a subject. The cutoffs for normal (>200 pg/ml) and low (<110 pg/ml) hypocretin-1 levels are represented. Median values are shown as a horizontal bar for each group. We may note that CSF hypocretin-1 levels are undetectable in most of narcoleptic with cataplexy subjects, while the values are clearly higher in all other conditions. Among the 26 narcoleptic patients with cataplexy, 23 were associated with undetectable CSF hypocretin-1 levels. One patient with narcolepsy with cataplexy but with HLA DQB1*0602 negative presented normal CSF hypocretin-1 and the two others within the intermediate range (between 110 and 200 pg/ml) were from familial cases. Low CSF hypocretin-1 is highly specific (99.1%) and sensitive (88.5%) for narcolepsy with cataplexy in our population. (Original data obtained from results of Dauvilliers, Y. et al., *J Neurol Neurosurg Psychiatry*, Vol. 74, 1667–1673, 2003.)

20.2.2 Genetic and Environmental Factors in Human Narcolepsy

Up to 10% of narcolepsy is familiar and the risk for first-degree relative is 20 to 40 times higher than for the general population (Mignot 1998). However, in several families (10 to 20%), first or second-degree relatives of narcoleptic patients are affected with an attenuated phenotype characterized by isolated recurrent daytime naps and/or lapses into sleep (Mignot 1998; Billiard, Pasquie-Magnetto, Heckman, et al. 1994; Dauvillierset al. 2004a). This indicates a genetic influence on the development of the disease. However, twin studies revealed only 25 to 31% of concordance, clearly indicating the major importance of environmental factors (Dauvilliers et al. 2004c; Mignot 1998; Khatami et al. 2004).

We recently found monozygotic twins who were homozygous for HLA DQB1*0602 but discordant for narcolepsy and CSF hypocretin-1 levels (Dauvilliers et al. 2004c). Genetic backgrounds are therefore not sufficient to develop hypocretin-deficiency. Narcolepsy–cataplexy is one of the diseases that most tightly associates with a specific HLA allele. A first study in 1983 reported that 100% of Japanese narcoleptic patients presented HLA DR2 (Honda, Asaka, Tanaka, and Juji 1983), an association later confirmed in Europe and North America (Mignot et al. 2001; Billiard and Seignalet 1985). The HLA complex maps on chromosome 6. It is divided into three subregions, HLA classes I, II, and III. It plays a key role in the recognition and processing of foreign antigens by the immune system. HLA DR2 is implicated in several autoimmune diseases such as insulin-dependent diabetes or multiple sclerosis. Four alleles on HLA class II corresponding to DRB1*1501, DRB5*0101, DQA1*0102, and DQB1*0602 constitute the susceptibility haplotype associated with human narcolepsy in Caucasian populations (Mignot et al. 2001). Shortly after this discovery, it was demonstrated that the DR2 association was weaker (60 to 65%) in African-American narcoleptics (Mignot et al. 2001), while the DQA1*0102, DQB1*0602 haplotype typically in linkage disequilibrium with DRB*1503 and DRB1*1101 in African-Americans showed the strongest association, suggesting that the susceptibility gene should be closer to the DQ loci.

The most specific marker of narcolepsy with cataplexy in the different ethnic groups studied to date is DQB1*0602, an allele found in 85 to 95% of cases (Fig. 20.3). Furthermore, being homozygous for the HLA DQB1*0602 genotype doubles or quadruples the risk for narcolepsy (Pelin, Guilleminault, Risch, Grumet, and Mignot 1998).

In addition, the relative risk for narcolepsy varies in heterozygote subjects according to the alleles associated with DQB1*0602; DQB1*0301 increases susceptibility to narcolepsy whereas DQB1*0501 and DQB1*0601 are protective (Mignot et al. 2001).

This finding is of interest since only minor variations in the peptide binding pockets of DQB1*0602 and *0601 were found (Siebold et al. 2004), with major consequences in term of risk of hypocretin neurons degeneration.

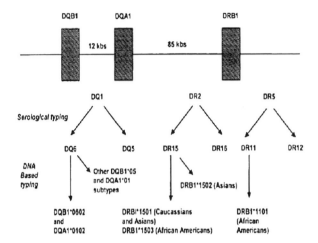

FIGURE 20.3. HLA DR and HLA DQ alleles typically observed in narcoleptic subjects. At the upper part of the figure, the HLA DQB1 and DQA1 genes located 85 kb centromeric to the gene DRB1, on chromosome 6p21. Below, the corresponding DR2 and DQ1 antigens found in most of patients with narcolepsy. HLA antigens have been split into two subtypes, DR15 and DR16 on the one hand, DQ5 and DQ6 on the other hand, with the use of serological typing techniques. Most of narcoleptic patients are positive for DR15 and DQ6. Below, molecular subtypes of DR15 and DQ6 were identified using DNA sequencing or oligotyping. Caucasians and Asians are DRB1*1501, DQA1*0102–DQB1*0602 while African-Americans are more often DRB1*1503, DQA1*0102–DQB1*0602. (Reprinted with permission from Dauvilliers, Y. et al., *Clin Neurophysiol*, Vol. 114, 2010, Copyright 2003.)

Twelve percent of the general population in Asians, 25 in Caucasians and 38 in African-Americans caries HLA DQB1*0602 while only a small fraction suffers from narcolepsy (Chabas et al. 2003; Mignot et al. 2001), indicating that this allele is neither necessary nor sufficient to trigger narcolepsy, especially in narcolepsy without cataplexy and in familial cases. Accordingly, other non-HLA genes may also confer susceptibility to narcolepsy (Chabas et al. 2003; Dauvilliers and Tafti 2006; Taheri and Mignot 2002). Most diseases with a tight HLA association are autoimmune, therefore several studies sought for an association between narcolepsy and immune-related genes including tumor necrosis factor (TNF) alpha and TNF receptor 2 (TNFR2) genes.

An association between TNF, TNFR2 genes and human narcolepsy has been reported (Hohjoh, Terada, Kawashima, Honda, and Tokunaga, 2000; Wieczorek et al. 2003), suggesting the role of the TNF-alpha pathway, involved in a possible inflammatory mechanism. Due to relation to apoptosis/immunity hypothesis targeting hypocretin neurons, the possibility of an association between apolipoprotein E4 and narcolepsy has been tested but without any positive result

(Gencik, Dahmen, Wieczorek, Kasten, Gencikova, and Epplen 2001). In addition, consistent with the imbalance between monoamines and acetylcholine observed in pharmacological and biochemical studies in canine narcolepsy, an association with polymorphisms in candidate genes targeting the monoaminergic pathways was tested (Dauvilliers, Neidhart, Lecendreux, Billiard, and Tafti 2001b; Koch, Craig, Dahlitz, Denney, and Parkes 1999; Wieczorek, Jagiello, Arning, Dahmen, and Epplen 2004), with a sexual dimorphism found in the functional polymorphism of the gene encoding the catechol-*O*-methyltransferase (Dauvilliers et al. 2001b). This polymorphism also modulated the severity of daytime sleepiness (Dauvilliers et al. 2001b) and the response of patients to stimulant treatment with modafinil (Dauvilliers, Neidhart, Billiard, and Tafti 2002).

Familial narcoleptic with or without cataplexy are negative for HLA DQB1*0602 in 25 to 35% of cases, supporting the existence of other genes with high penetrance not associated with HLA (Mignot 1998; Mignot et al. 2001). Only two genetic linkage studies have been performed on familial forms of narcolepsy. A first genome-wide mapping study reported a suggestive linkage (LOD score of 3.09) to chromosome 4p13-q21 in eight small multiplex Japanese families affected with narcolepsy–cataplexy but no mutation search has been conducted to date (Nakayama, Miura, Honda, Miki, Honda, and Arinami, 2000). We performed a genome-wide linkage analysis in a large French family with four members affected by narcolepsy with clear-cut cataplexy and 10 others with isolated recurrent naps with evidence for a linkage identified in a 5 Mb region on chromosome 21q (LOD score of 4.00), without any positive result for mutations or polymorphisms in candidate genes within the region (Dauvilliers et al. 2004a).

In summary, the above genetic association studies in sporadic and familial narcolepsies, although preliminary and needing confirmation and replication in independent samples, support a complex pathogenetic model including disturbances in both the neurotransmission and the immune/autoimmune pathways.

20.2.3 Autoimmunity Hypothesis in Human Narcolepsy

Most of human narcolepsy studies support a multifactorial model for the development of narcolepsy, with a strong influence of environmental factors, acting in combination with genetic factors (at least HLA DQB1*0602), possibly triggering an autoimmune process with irreversible damage to the hypocretin system. A selective destruction of hypocretin neurons is the main etiology of human narcolepsy with cataplexy, however the cause of the hypocretin cell death remains unknown.

The presence of hypothalamus gliosis in human narcolepsy with cataplexy is still a controversial subject (Peyron et al. 2000; Thannickal et al. 2000), but its discovery may correspond to a previous inflammatory event. Although an autoimmune hypothesis is likely, a focal inflammatory process targeting hypocretin neurons has never been confirmed. Studies on cellular and immune CSF markers reported no clear evidence for association with narcolepsy especially the absence of specific IgG oligoclonal bands (Mignot et al. 1995; Carlander et al. 2001; Chabas et al. 2003; Lin, Hungs, and Mignot 2001; Fredrikson, Carlander, Billiard, and Link 1990). However, those studies were performed many years after disease onset.

Patients with typical narcolepsy–cataplexy and HLA DQB1*0602 positivity have no significant pathogenic detectable serum or CSF autoantibodies against prepro-hypocretin (Black et al. 2005b; Overeem et al. 2006; Tanaka, Honda, Inoue, and Honda 2006). Although autoantibodies have been detected against the hypocretin system in few patients with narcolepsy with cataplexy, there is no evidence for any autoantibody-mediated dysfunction in the hypocretin system involved in its pathophysiology (Black et al. 2005a, 2005b; Overeem et al. 2006; Tanaka et al. 2006). One may also hypothesized that other components of hypocretin-producing neurons are involved in the immune process. Actually, a study reported that CSF from narcoleptic subjects showed immunoreactivity to rat hypothalamic protein (Black et al. 2005a). In contrast, another study using immunohistochemistry on human hypothalamic tissue failed to confirm the presence of any specific antibody against lateral hypocretin neurons in the clinical stage of narcolepsy (Overeem et al. 2006). In addition, no antibodies against several neuron-specific and non-neuron-specific antigens were found in serum or CSF in narcoleptic patients when compared to controls (Overeem, Geleijns, Garssen, Jacobs, van Doorn, and Lammers 2003).

Hypocretin deficiency is rare among immune-mediated brain disorders (Mignot et al. 2002; Dauvilliers et al. 2003a; Nishino and Kanbayashi 2005). However, one study found reduced CSF hypocretin-1 levels in patients with Ma2 antibody-associated paraneoplastic encephalopathy, being a rare but interesting model of autoimmune encephalitis associated with hypersomnia and therefore hypocretin dysfunction (Overeem et al. 2004). However, anti-Ma2 autoantibodies were not detected in patients with narcolepsy with cataplexy (Overeem et al. 2004). A significant degree of hypocretin deficiency was also noted in several cases of Guillain–Barré syndrome (presumed autoimmune disorder affecting both peripheral and central nervous system), being another model for studying possible autoimmune neuron damage in narcolepsy.

Recently a very controversial study revealed the presence of a functional autoantibody in human narcolepsy with cataplexy, enhancing cholinergic neurotransmission and detectable upon passive transferring IgG serum from patients to mice (Smith, Jackson, Neufing, McEvoy, and Gordon 2004). However, many methodological problems could be mentioned regarding the inclusion criteria, the absence of details of mice behavior and EEG changes after IgG transferring and the atypical choice of bladder smooth muscle target (Smith et al. 2004). Overall, this study is an interesting report regarding the autoimmune hypothesis in narcolepsy but with several limitations. The absence of any replication findings leads us to consider this discovery as improbable. The presence of yet undetermined pathogenic autoantibodies for other autoantigens remains still possible in narcolepsy with cataplexy. Recent human pharmacological studies using immune-therapies including intravenous immunoglobulins (IVIg) and plasmapheresis have been tested in narcolepsy with substantial clinical success (Dauvilliers, Carlander, Rivier, Touchon, and Tafti 2004a; Chen, Black, Call, and Mignot 2005). If applied at disease onset, those treatments may favorably modify the course of the disease, however those studies lacked of well-design trial.

The absence of direct evidence for humoral immunity in narcolepsy does not exclude the possibility that a transient autoimmune reaction restricted to the brain may have occurred around disease onset and disappeared shortly thereafter. Autoimmune disease may also be mediated by T-cell immune reaction toward autoantigen-expressing tissue. One may hypothesize that T lymphocytes may be involved in initiating, controlling and driving the specific immune responses targeting the hypocretin system. Only one study on cellular autoimmunity has reported in narcoleptic patients with a higher IL-6 secretion by monocytes in comparison to HLA-DR2-positive controls, without any difference in other cytokine levels or abnormalities in T-cell function (Hinze-Selch et al. 1998). However, T-cell mediated autoimmunity restricted to the brain may still be hypothesized in narcolepsy. The detection of such an autoimmune process with or without inflammation remains technically difficult especially if that reaction occurs only at disease onset (during the preclinical phase) and disappears after. In addition, other mechanisms may also be involved in hypocretin neurodegeneration such as a viral or bacterial infection or a spontaneous neurodegenerative process.

In the variant form of narcolepsy named "narcolepsy without cataplexy," the association with HLA DQB1*0602 and the decrease in CSF hypocretin level are less frequently encountered, therefore a common pathophysiology between the two conditions is debated but remains improbable (Dauvilliers et al. 2003a). A continuum of severity progressing from narcolepsy without cataplexy to narcolepsy with cataplexy with a defective hypocretin neurotransmission may be hypothesized in rare cases. However, hypocretin ligand deficiency appears not the major cause for hypersomnias and a pathophysiological continuum between narcolepsy without cataplexy and idiopathic hypersomnia without long sleep time could be suggested. Therefore, if the autoimmune hypothesis (although needed confirmation) may still be proposed in narcolepsy with cataplexy; what is the main etiology of the variant form without cataplexy?

20.3 Treatment

Significant changes have been made in the treatment of daytime sleepiness and cataplexy in the last few years. However, despite recent advances in our understanding of the neurobiological basis of narcolepsy, there is no cure for narcolepsy and current therapies are only symptom-based (Dauvilliers et al. 2003b; Scammell 2003; Dauvilliers and Tafti 2006). However, the most disabling symptoms can be controlled in most patients by stimulants (mostly dopaminergic) for daytime sleepiness and sleep attacks, antidepressants (mostly noradrenergic) for cataplexy and other REM-associated symptoms, and hypnotics for disturbed nighttime sleep (Dauvilliers et al. 2003b; Scammell 2003; Dauvilliers and Tafti 2006). Recently, a single drug named sodium oxybate was shown effective in controlling sleepiness, cataplexy, and disturbed nighttime sleep (Dauvilliers et al. 2003b; Scammell 2003; Dauvilliers and Tafti 2006).

Perspective on therapeutic may concern replacing hypocretin, being an ideal therapy. Such treatments are effective when administered in the brain in hypocretin-deficient mouse models (Mieda, Willie, Hara, Sinton, Sakurai, and Yanagisawa 2004). Attempts at using hypocretin-based therapies after peripheral administration have been disappointing, as the peptides do not cross the blood–brain barrier (Abad and Guilleminault 2004; Mignot and Nishino 2005). However, hypocretin-based therapies such as hypocretin agonists and hypocretin neuron transplantation are part of the research projects for the development of new treatments in narcolepsy (Abad and Guilleminault 2004; Mignot and Nishino 2005).

348   Dauvilliers

Based on the autoimmune hypothesis of narcolepsy, immune-based therapies including IVIg and plasmapheresis have been rested and seem to be effective on the number and severity of cataplexy if applied at early stages of narcolepsy–cataplexy development (Fig. 20.4) (Dauvilliers et al. 2004a; Chen et al. 2005; Lecendreux, Maret, Bassetti, Mouren, and Tafti 2003). The 2-year following of four narcoleptic-cataplectic patients treated with IVIg revealed a safety issue and an impression of long term effectiveness in controlling cataplexy (Dauvilliers 2006). The targets of immunotherapy depend on the pathogenesis of the disease. In early onset of narcolepsy, high doses of IVIg may down-regulate T-cell functions and pathogenic cytokines, and interfere with autoantigen recognition through HLA DQB1*0602 at the induction phase of the immune stimulation.

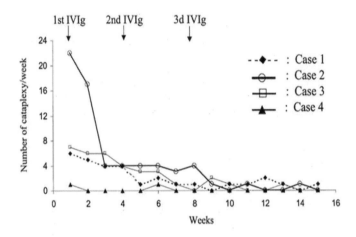

FIGURE 20.4. Cataplexy response to IVIg treatment in 4 patients with narcolepsy. The number of cataplectic attacks per week was scored by each patient during a 15-week follow-up. (Reprinted with permission from *Ann Neurol*, Dauvilliers et al. Successful management of cataplexy with intravenous immunoglobulins at narcolepsy onset. Vol. 56, 907, Copyright 2004.)

These observations point to the importance of early diagnosis of narcolepsy, which once treated quickly, may modify its long-term outcome, i.e., when the "autoimmune" process targeting the cataplexy circuitry is not too advanced and can be at least partially reversed (Dauvilliers et al. 2004a). These recent findings obtained in only few patients in open trials, although needing replication and extension in well-designed trials, support the efficacy of immune-modulating treatments on cataplexy that may also modify the course of the disease.

20.4 Conclusion

Narcolepsy is genetically complex and environmentally influenced. Although the HLA DQB1*0602 remains the unique established genetic risk factor, future studies may focus in identifying new genetic susceptibility factors, environmental factors and the study of gene–environment interactions, involving in hypocretin cell death. The cause of this focal neurodegeneration remains a mystery although an autoimmune hypothesis is likely. A transient and focal inflammatory process targeting hypocretin neurons may have occurred around disease onset and disappeared shortly thereafter.

Further studies that focus on reactivity of serum and/or CSF to other components of the hypocretin neurotransmission system, and on T-cell autoimmune response directed against neuroantigens are warranted. Finally, the efficacy of immune-based therapies in narcolepsy needs to be confirmed in larger double blind placebo controlled trials.

References

Abad, V.C., and Guilleminault, C. (2004) Emerging drugs for narcolepsy. *Expert Opin Emerg Drugs* 9, 281–291.

American Academy of Sleep Medicine (2005) *The International Classification of Sleep Disorders—Revised.* American Academy of Sleep Medicine, Chicago, IL.

Billiard, M., Pasquie-Magnetto, V., Heckman, M. et al. (1994) Family studies in narcolepsy. *Sleep* 17, S54–S59.

Billiard, M., and Seignalet, J. (1985) Extraordinary association between HLA-DR2 and narcolepsy. *Lancet* 1, 226–227.

Black, J.L., Avula, R.K., Walker, D.L. et al. (2005a) HLA DQB1*0602 positive narcoleptic subjects with cataplexy have CSF IgG reactive to rat hypothalamic protein extract. *Sleep* 28, 1191–1192.

Black, J.L., Silber, M.H., Krahn, L.E. et al. (2005b) Analysis of hypocretin (orexin) antibodies in patients with narcolepsy. *Sleep* 28, 427–431.

Blouin, A.M., Thannickal, T.C., Worley, P.F., Baraban, J.M., Reti, I.M., and Siegel, J.M. (2005) Narp immunostaining of human hypocretin (orexin) neurons, loss in narcolepsy. *Neurology* 65, 1189–1192.

Carlander, B., Dauvilliers, Y., and Billiard, M. (2001) Immunological aspects of narcolepsy. *Rev Neurol* 157, S97–S100.

Chabas, D., Taheri, S., Renier, C., and Mignot, E. (2003) The genetics of narcolepsy. *Annu Rev Genom Hum Genet* 4, 459–483.

Chemelli, R.M., Willie, J.T., Sinton, C.M. et al. (1999) Narcolepsy in orexin knockout mice, molecular genetics of sleep regulation. *Cell* 98, 437–451.

Chen, W., Black, J., Call, P., and Mignot, E. (2005) Late-onset narcolepsy presenting as rapidly progressing muscle weakness, response to plasmapheresis. *Ann Neurol* 58, 489–490.

Crocker, A., Espana, R.A., Papadopoulou, M. et al. (2005) Concomitant loss of dynorphin, NARP, and orexin in narcolepsy. *Neurology* 65, 1184–1188.

Dauvilliers, Y. (2006) Follow-up of four narcolepsy patients treated with intravenous immunoglobulins. *Ann Neurol* 26, 153.

Dauvilliers, Y., Baumann, C.R., Carlander, B. et al. (2003a) CSF hypocretin-1 levels in narcolepsy, Kleine-Levin syndrome, and other hypersomnias and neurological conditions. *J Neurol Neurosurg Psychiatry* 74, 1667–1673.

Dauvilliers, Y., Billiard, M., and Montplaisir, J. (2003b) Clinical aspects and pathophysiology of narcolepsy. *Clin Neurophysiol* 114, 2000–2017.

Dauvilliers, Y., Blouin, J.L., Neidhart, E. et al. (2004a) A narcolepsy susceptibility locus maps to a 5 Mb region of chromosome 21q. *Ann Neurol* 56, 382–388.

Dauvilliers, Y., Carlander, B., Molinari, N. et al. (2003c) Month of birth as a risk factor for narcolepsy. *Sleep* 26, 663–665.

Dauvilliers, Y., Carlander, B., Rivier, F., Touchon, J., and Tafti, M. (2004a) Successful management of cataplexy with intravenous immunoglobulins at narcolepsy onset. *Ann Neurol* 56, 905–908.

Dauvilliers, Y., Gosselin, A., Paquet, J., Touchon, J., Billiard, M., and Montplaisir, J. (2004b) Effect of age on MSLT results in patients with narcolepsy-cataplexy. *Neurology* 62, 46–50.

Dauvilliers, Y., Maret, S., Bassetti, C. et al. (2004c) A monozygotic twin pair discordant for narcolepsy and CSF hypocretin-1. *Neurology* 62, 2137–2138.

Dauvilliers, Y., Montplaisir, J., Molinari, N. et al. (2001a) Age at onset of narcolepsy in two large populations of patients in France and Quebec. *Neurology* 57, 2029–2033.

Dauvilliers, Y., Neidhart, E., Billiard, M., and Tafti, M. (2002) Sexual dimorphism of the catechol-O-methyltransferase gene in narcolepsy is associated with response to modafinil. *Pharmacogenom J* 2, 65–68.

Dauvilliers, Y., Neidhart, E., Lecendreux, M., Billiard, M., and Tafti, M. (2001b) MAO-A and COMT polymorphisms and gene effects in narcolepsy. *Mol Psychiatry* 6, 367–372.

Dauvilliers, Y., and Tafti, M. (2006) Molecular genetics and treatment of narcolepsy. *Ann Med* 38, 252–262.

de Lecea, L., Kilduff, T.S., Peyron, C. et al. (1998) The hypocretins, hypothalamus-specific peptides with neuroexcitatory activity. *Proc Natl Acad Sci U S A* 95, 322–327.

Dement, W., Carskadon, M., and Ley, R. (1973) The prevalence of narcolepsy II. *Sleep Res* 2, 147.

Fredrikson, S., Carlander, B., Billiard, M., and Link, H. (1990) CSF immune variables in patients with narcolepsy. *Acta Neurol Scand* 81, 253–254.

Gencik, M., Dahmen, N., Wieczorek, S., Kasten, M., Gencikova, A., and Epplen, J.T. (2001) ApoE polymorphisms in narcolepsy. *BMC Med Genet* 2, 9.

Hara, J., Beuckmann, C.T., Nambu, T. et al. (2001) Genetic ablation of orexin neurons in mice results in narcolepsy, hypophagia, and obesity. *Neuron* 30, 345–354.

Hinze-Selch, D., Wetter, T.C., Zhang, Y. et al. (1998) In vivo and in vitro immune variables in patients with narcolepsy and HLA-DR2 matched controls. *Neurology* 50, 1149–1152.

Hohjoh, H., Terada, N., Kawashima, M., Honda, Y., and Tokunaga, K. (2000) Significant association of the tumor necrosis factor receptor 2 (TNFR2) gene with human narcolepsy. *Tissue Antigens* 56, 446–448.

Honda, Y. (1979) Census of narcolepsy, cataplexy and sleep life among teenagers in Fujisawa city. *Sleep Res* 8, 191.

Honda, Y., Asaka, A., Tanaka, Y., and Juji, T. (1983) Discrimination of narcolepsy by using genetic markers and HLA. *Sleep Res* 12, 254.

Hublin, C., Kaprio, J., Partinen, M. et al. (1994) The prevalence of narcolepsy, an epidemiological study of the Finnish Twin Cohort. *Ann Neurol* 35, 709–716.

Hungs, M., Lin, L., Okun, M., and Mignot, E. (2001) Polymorphisms in the vicinity of the hypocretin/orexin are not associated with human narcolepsy. *Neurology* 57, 1893–1895.

Khatami, R., Maret, S., Werth, E. et al. (2004) Monozygotic twins concordant for narcolepsy-cataplexy without any detectable abnormality in the hypocretin (orexin) pathway. *Lancet* 363, 1199–1200.

Koch, H., Craig, I., Dahlitz, M., Denney, R., and Parkes, D. (1999) Analysis of the monoamine oxidase genes and the Norrie disease gene locus in narcolepsy. *Lancet* 353, 645–646.

Kotagal, S., Krahn, L.E., and Slocumb, N. (2004) A putative link between childhood narcolepsy and obesity. *Sleep Med* 5, 147–50.

Lavie, P., and Peled, R. (1987) Narcolepsy is a rare disease in Israel. *Sleep* 10, 608–609.

Lecendreux, M., Maret, S., Bassetti, C., Mouren, M.C., and Tafti, M. (2003) Clinical efficacy of high-dose intravenous immunoglobulins near the onset of narcolepsy in a 10-year-old boy. *J Sleep Res* 12, 347–348.

Lin, L., Faraco, J., Li, R. et al. (1999) The sleep disorder canine narcolepsy is caused by a mutation in the hypocretin (orexin) receptor 2 gene. *Cell* 98, 365–376.

Lin, L., Hungs, M., and Mignot, E. (2001) Narcolepsy and the HLA region. *J Neuroimmunol* 117, 9–20.

Mieda, M., Willie, J.T., Hara, J., Sinton, C.M., Sakurai, T., and Yanagisawa, M. (2004) Orexin peptides prevent cataplexy and improve wakefulness in an orexin neuron-ablated model of narcolepsy in mice. *Proc Natl Acad Sci U S A* 101, 4649–4654.

Mignot, E. (1998) Genetic and familial aspects of narcolepsy. *Neurology* 50, S16–S22.

Mignot, E., Lammers, G.J., Ripley, B. et al. (2002) The role of cerebrospinal fluid hypocretin measurement in the diagnosis of narcolepsy and other hypersomnias. *Arch Neurology* 59(10), 1553–1562.

Mignot, E., Lin, L., Rogers, W. et al. (2001) Complex HLA-DR and -DQ interactions confer risk of narcolepsy-cataplexy in three ethnic groups. *Am J Hum Genet* 68, 686–699.

Mignot, E., and Nishino, S. (2005) Emerging therapies in narcolepsy-cataplexy. *Sleep* 28, 754–763.

Mignot, E., Tafti, M., Dement, W.C., and Grumet, F.C. (1995) Narcolepsy and immunity. *Adv Neuroimmunol* 5, 23–37.

Mignot, E., Wang, C., Rattazzi, C. et al. (1991) Genetic linkage of autosomal recessive canine narcolepsy with a mu immunoglobulin heavy-chain switch-like segment. *Proc Natl Acad Sci U S A* 88, 3475–3478.

Nakayama, J., Miura, M., Honda, M., Miki, T., Honda, Y., and Arinami, T. (2000) Linkage of human narcolepsy with HLA association to chromosome 4p13-q21. *Genomics* 65, 84–86.

Nishino, S., and Kanbayashi, T. (2005) Symptomatic narcolepsy, cataplexy and hypersomnia, and their implications in the hypothalamic hypocretin/orexin system. *Sleep Med Rev* 9, 269–310.

Ohayon, M.M. (2000) Prevalence of hallucinations and their pathological associations in the general population. *Psychiatry Res* 97, 153–164.

Ohayon, M.M., Priest, R.G., Zulley, J., Smirne, S., and Paiva, T. (2002) Prevalence of narcolepsy symptomatology and diagnosis in the European general population. *Neurology* 58, 1826–1833.

Orellana, C., Villemin, E., Tafti, M., Carlander, B., Besset, A., and Billiard, M. (1994) Life events in the year preceding the onset of narcolepsy. *Sleep* 17, S50–S53.

Overeem, S., Dalmau, J., Bataller, L. et al. (2004) Hypocretin-1 CSF levels in anti-Ma2 associated encephalitis. *Neurology* 62, 138–140.

Overeem, S., Geleijns, K., Garssen, M.P., Jacobs, B.C., van Doorn, P.A., and Lammers, G.J. (2003) Screening for anti-ganglioside antibodies in hypocretin-deficient human narcolepsy. *Neurosci Lett* 341, 13–16.

Overeem, S., Verschuuren, J.J., Fronczek, R. et al. (2006) Immunohistochemical screening for autoantibodies against lateral hypothalamic neurons in human narcolepsy. *J Neuroimmunol* 174, 187–191.

Pelin, Z., Guilleminault, C., Risch, N., Grumet, F.C., and Mignot, E. (1998) HLA-DQB1*0602 homozygosity increases relative risk for narcolepsy but not disease severity in two ethnic groups. US Modafinil in Narcolepsy Multicenter Study Group. *Tissue Antigens* 51, 96–100.

Peyron, C., Faraco, J., Rogers, W. et al. (2000) A mutation in a case of early onset narcolepsy and a generalized absence of hypocretin peptides in human narcoleptic brains. *Nat Med* 6, 991–997.

Peyron, C., Tighe, D.K., van den Pol, A.N. et al. (1998) Neurons containing hypocretin (orexin) project to multiple neuronal systems. *J Neurosci* 18, 9996–10015.

Picchioni, D., Mignot, E.J., and Harsh, J.R. (2004) The month-of-birth pattern in narcolepsy is moderated by cataplexy severity and may be independent of HLA-DQB1*0602. *Sleep* 27, 1471–1475.

Ripley, B., Fujiki, N., Okura, M., Mignot, E., and Nishino, S. (2001) Hypocretin levels in sporadic and familial cases of canine narcolepsy. *Neurobiol Dis* 8, 525–534.

Sakurai, T., Amemiya, A., Ishii, M. et al. (1998) Orexins and orexin receptors, a family of hypothalamic neuropeptides and G protein-coupled receptors that regulate feeding behavior. *Cell* 92, 573–585.

Scammell, T.E. (2003) The neurobiology, diagnosis, and treatment of narcolepsy. *Ann Neurol* 53, 154–166.

Siebold, C., Hansen, B.E., Wyer, J.R. et al. (2004) Crystal structure of HLA-DQ0602 that protects against type 1 diabetes and confers strong susceptibility to narcolepsy. *Proc Natl Acad Sci U S A* 101, 1999–2004.

Silber, M.H., Krahn, L.E., Olson, E.J., and Pankratz, V.S. (2002) The epidemiology of narcolepsy in Olmsted County, Minnesota, a population-based study. *Sleep* 25, 197–202.

Smith, A.J., Jackson, M.W., Neufing, P., McEvoy, R.D., and Gordon, T.P. (2004) A functional autoantibody in narcolepsy. *Lancet* 364, 2122–2124.

Taheri, S., and Mignot, E. (2002) The genetics of sleep disorders. *Lancet Neurol* 1, 242–250.

Tanaka, S., Honda, Y., Inoue, Y., and Honda, M. (2006) Detection of autoantibodies against hypocretin, hcrtrl, and hcrtr2 in narcolepsy, anti-Hcrt system antibody in narcolepsy. *Sleep* 29, 633–638.

Thannickal, T.C., Moore, R.Y., Nienhuis, R. et al. (2000) Reduced number of hypocretin neurons in human narcolepsy. *Neuron* 27, 469–474.

Wieczorek, S., Gencik, M., Rujescu, D. et al. (2003) TNFA promoter polymorphisms and narcolepsy. *Tissue Antigens* 61, 437–442.

Wieczorek, S., Jagiello, P., Arning, L., Dahmen, N., and Epplen, J.T. (2004) Screening for candidate gene regions in narcolepsy using a microsatellite based approach and pooled DNA. *J Mol Med* 82, 696–705.

Willie, J.T., Chemelli, R.M., Sinton, C.M. et al. (2003) Distinct narcolepsy syndromes in Orexin receptor-2 and Orexin null mice, molecular genetic dissection of non-REM and REM sleep regulatory processes. *Neuron* 38, 715–730.

Index

Printed in the United States of America